SUPERNOVA REMNANTS AND THEIR X-RAY EMISSION

INTERNATIONAL ASTRONOMICAL UNION

UNION ASTRONOMIQUE INTERNATIONALE

SYMPOSIUM No. 101

HELD IN VENICE, ITALY, 30 AUGUST – 2 SEPTEMBER 1982

SUPERNOVA REMNANTS AND THEIR X-RAY EMISSION

EDITED BY

JOHN DANZIGER

European Southern Observatory, Garching bei München, F.R.G.

and

PAUL GORENSTEIN

Center for Astrophysics, Cambridge, Massachusetts, U.S.A.

D. REIDEL PUBLISHING COMPANY

A MEMBER OF THE KLUWER  ACADEMIC PUBLISHERS GROUP

DORDRECHT / BOSTON / LANCASTER

Library of Congress Cataloging in Publication Data

Main entry under title:

Supernova remnants and their X-ray emission.

        At head of title: International Astronomical Union, Union
astronomique internationale.
        Includes index.
        1. Supernova remnants–Congresses.   2.   X-ray astronomy–
Congresses.   I.   Danziger, John, 1936–   . II.   Gorenstein, Paul.
III.   International Astronomical Union.
QB843.S95S95   1983   523.8'446   83–17657
ISBN-13: 978-90-277-1667-5   e-ISBN-13: 978-94-009-7231-5
DOI: 10.1007/978-94-009-7231-5

Published on behalf of
the International Astronomical Union
by
D. Reidel Publishing Company, P. O. Box 17, 3300 AA Dordrecht, Holland

Sold and distributed in the U.S.A. and Canada
by Kluwer Academic Publishers,
190 Old Derby Street, Hingham, MA 02043, U.S.A.

In all other countries, sold and distributed
by Kluwer Academic Publishers Group,
P. O. Box 322, 3300 AH Dordrecht, Holland

TABLE OF CONTENTS

III. CENTER FILLED MORPHOLOGIES

SCIENTIFIC ORGANIZING COMMITTEE

P. Gorenstein (Chairman)
R.A. Chevalier
J. Danziger
W.M. Goss
D. Mathewson
S. D'Odorico
F. Pacini
K. Pounds
V. Radhakrishnan
F.D. Seward
I.S. Shklovsky
S. van den Bergh
L. Woltjer

LOCAL ORGANIZING COMMITTEE

S. D'Odorico, Chairman, ESO
E. Talentino, Cini Foundation, Venice
S. Ortolani, University of Padua
P. Rafanelli, University of Padua·

PREFACE

IAU Symposium 101, Supernova Remnants and Their X-ray Emission, was held on the Island of San Giorgio, Venice, 30 August - 2 September 1982. It was co-sponsored by the National Research Council, Italy, the University of Padua, the Observatory of Padua, and the International Astronomical Union, and was hosted by the Cini Foundation.

The contents of this volume show the wide range of disciplines that are involved in supernova remnant research. Many new results were presented, not only from the X-ray observations from the Einstein Observatory but also from observations at optical and radio wavelengths. This has led to the stimulation of theoretical work, much of which attempts to accommodate in a more unified way all of these observations. Research on supernova remnants of all ages was reported. Perhaps the most impressive part of all this work is the way in which observations at all wavelengths have extended well outside the Galaxy to other members of the Local Group and beyond.

The Symposium was attended by scientists from 15 countries. Twenty-five invited papers and sixty-eight shorter contributions were presented during the 4-day meeting. Thirty-three of these shorter contributions were presented in poster sessions. This volume contains almost all (89) of those contributions. They are followed by discussions which took place after each verbal presentation. Since the availability of the discussions was left to the individual contributors, they are not complete, but those contained in this volume convey some idea of the nature of the exchanges.

An evening boat excursion was provided for the participants, followed by a memorable dinner at a typical Venetian restaurant on the Island of Burano. On the day following the conference a trip to Padua of artistic and historic interest was provided by our hosts at the University of Padua.

Special thanks are due to the Local Organizing Committee and especially Sandro D'Odorico for the smooth organization of the meeting and the very special social programme accompanying it. Thanks are also due to Gianna Genovese and Rita Verschuren for their able secretarial help. Adriana Ciani and Antonella Nota, students from the University of Padua and the only true Venetians at the Symposium, provided valuable assistance during the scientific sessions.

J. Danziger
P. Gorenstein

The participants of the symposium on the Island of San Giorgio, Venice.

Acronyms for institutions are given at the end of the listing.

Bandiera, R.	Osservatorio Astrofisico di Arcetri, Florence, Italy
Becker, R.	VPI+SU, Blacksburg, USA.
Begelman, M.	Institute of Astronomy, Cambridge, UK
Beltrametti, M.	MPIfA, Garching, FRG
Benford, G.	University of California, Irvine, USA
Benvenuti, P.	European Space Agency, Vilspa, Madrid, Spain
Biermann, P.	MPIfR, Bonn, FRG
Bisnovatyi-Kogan, G.S.	Institute for Space Research, Moscow, USSR
Blair, W.P.	CFA, Cambridge, USA
Bleeker, Ir. J.A.M.	Huygens Laboratorium, Leiden, NL
Braun, R.	Sterrewacht Hugyens Lab, Leiden, NL
Canizares, R.	MIT, Cambridge, USA
Chechetkin, V.M.	USSR Academy of Sciences, Moscow, USSR
Chevalier, R.	University of Virginia, Charlottesville, USA
Cox, D.P.	University of Wisconsin, Madison, USA
D'Amico N.	Istituto di Fisica, Palermo, Italy
Danziger, I.J.	ESO, Garching, FRG
Davelaar, J.	Estec, Noordwijk, NL
Dennefeld, M.	Institut d'Astrophysique, Paris, France
Dickel, J.R.	Radiosterrenwacht, Dwingeloo, NL
D'Odorico, S.	ESO, Garching, FRG
Dopita, M.A.	MSSSO, ANU, Canberra, Australia
Durouchoux, P.	CEN Saclay, Gif-sur-Yvette, France
Epstein, R.	Nordita, Copenhagen, Denmark
Fabian, A.C.	Institute of Astronomy, Cambridge, UK
Falle, S.A.E.G.	The University of Leeds, Leeds, UK
Fedorenko, V.N.	A.F. Ioffe Physical Technical Institute, Leningrad, USSR
Fesen, R.A.	GSFC, Greenbelt, USA
Fürst, E.	MPIfR, Bonn, FRG
Fusco-Femiano, R.	Istituto Astrofisica Spaziale, Frascati, Italy
Garmire, P.	Pennsylvania State University, University Park, USA
Giovanelli, F.	Istituto Astrofisica Spaziale, Frascati, Italy
Gorenstein, P.	CFA, Cambridge, USA
Goss W.M.	Sterrenkundig Lab Kapteyn, Groningen, NL
Graham J.	Imperial College, London, UK
Green, D.A.	Cavendish Laboratory, Cambridge, UK
Gregory, P.C.	University of British Columbia, Vancouver, Canada
Grindlay, J.E.	Harvard College Observatory, Cambridge, USA
Gull, S.F.	Institute of Astronomy, Cambridge, UK
Gull, Th.R.	GSFC, Greenbelt, USA

Habets, G.M.H.J.   Universiteit van Amsterdam, Amsterdam, NL
Hamilton, A.   Institute of Astronomy, Cambridge, UK
Harnden, F.R. Jr.   CFA, Cambridge, USA
Hawkins, N.R.S.   Royal Observatory, Edinburgh, UK
Heidmann, J.   Observatoire de Meudon, Meudon, France
Heiles, C.   University of California, Berkeley, USA
Helfand, D.J.   Columbia Astrophysics Lab, New York, USA
Higgs, L.A.   DRAO, Penticton, Canada
Holt, S.S.   GSFC, Greenbelt, USA
Hughes, V.A.   Queen's University, Kingston, Canada
Itoh, H.   University of Kyoto, Kyoto, Japan
Jones, E.M.   Los Alamos National Lab., Los Alamos, USA
Jurkevich I.   Annapolis, USA
Kirk, J.G.   MPIfA, Garching, FRG
Kirschner, R.   University of Michigan, Ann Arbor, USA
Koch-Miramond, L.   CEN Saclay, Gif-sur-Yvette, France
Kronberg, P.   University of Toronto, Toronto, Canada
Ku, W.H.-M.   Columbia Astrophysics Lab, New York, USA
Landini, M.   Osservatorio Astrofisico di Arcetri, Florence,
                          Italy
Lattimer, J.   SUNY, Stony Brook, USA
Long, K.S.   John Hopkins University, Baltimore, USA
Lyne, A.G.   NRAL, Jodrell Bank, UK
Ma, E.   Sterrewacht Hugyens Lab, Leiden, NL
Manchester, R.N.   Radiophysics, CSIRO, Sydney, Australia
Mantovani, F.   Istituto di Radioastronomia, Bologna, Italy
Mathewson, D.S.   MSSSO, ANU, Canberra, Australia
McKee, C.F.   University of California, Berkeley, USA
Mills, B.Y.   University of Sydney, NSW 2006, Australia
Montmerle, T.   CEN Saclay, Gif-sur-Yvette, France
Morini, M.   IFCI/CNR, Istituto di Fisica, Palermo, Italy
Murdin, P.G.   Royal Greenwich Observatory, Hailsham, UK
Murray, S.S.   CFA, Cambridge, USA
Nomoto, K.   MPI für Physik und Astrophysik, Garching, FRG
Ortolani, S.   Osservatorio Astrofisico, Asiago, Italy
Pacini, F.   Istituto di Astronomia, Firenze, Italy
Péquignot, D.   Observatoire de Meudon, Meudon, France
Pizzichini, G.   Ist. Tesre/CNR, Bologna, Italy
Pravdo, S.H.   JPL, Pasadena, USA
Preite-Martinez, A.   Istituto Astrofisica Spaziale, Frascati, Italy
Radhakrishnan, V.   Raman Research Institute, Bangalore, India
Rafanelli, P.   Osservatorio Astronomico, Padova, Italy
Raymond, J.C.   CFA, Cambridge, USA
Re, S.   IFCI/CNR, Istituto di Fisica, Palermo, Italy
Reich, W.   MPIfR, Bonn, FRG
Rosado, M.   Instituto di Astronomia, UNAM, Mexico
Rothenflug, R.   CEN Saclay, Gif-sur-Yvette, France
Ruiz, M.T.   Universidad de Chile, Santiago, Chile
Salvati, M.   Istituto Astrofisica Spaziale, Frascati, Italy
Sanders, W.   University of Wisconsin, Madison, USA
Sarazin, C.L.   Institute for Advanced Study, Princeton, USA

Sarkar, S.   Rutherford and Appleton Laboratory, Chilton, UK
Seward, F.   Institute of Astronomy, Cambridge, UK
Shapiro, M. M.   Naval Research Lab., Washington D.C., USA
Shull, M.   University of Colorado - JILA, Boulder, USA
Shull, P.   MPI für Astronomie, Heidelberg-Königstuhl, FRG
Sieber, R.W.   MPIfR, Bonn, FRG
Smith, A.   Cambridge, UK
Smith, M.D.   University of Maryland, College Park, USA
Spencer, R.E.   NRAL, Jodrell Bank, UK
Srinivasan, C.   Raman Research Institute, Bangalore, India
Stewart, G.   Institute of Astronomy, Cambridge, UK
Strom, R.G.   Radiosterrenwacht, Dwingeloo, NL
Szymikowiak, A.   NASA/GSFC, Greenbelt, USA
Tanaka, Y.   Institute of Space and Astronautical Science,
                         Tokyo, Japan
Tenorio-Tagle, G.   MPIfA, Garching, FRG
Tsygan, A.I.   A.F. Ioffe Physical Technical Institute,
                         Leningrad, USSR
Tuffs, R.   Cavendish Lab., Cambridge, UK
Tuohy, I.R.   MSSSO, ANU, Canberra, Australia
Van Riper, K.A.   Los Alamos National Lab., Los Alamos, USA
van den Bergh, S.   DAO, Victoria, Canada
Velusamy, T.   Radio Astronomy Centre, Ootacamund, India
Watson, M.G.   CFA, Cambridge, USA
Weiler, K.W.   NSF, Washington D.C., USA
Westergaard, N.J.   Danish Space Research Institute, Lyngby,
                         Denmark
Winkler, P.F.   Middlebury College, Middlebury, USA
Woltjer, L.   ESO, Garching, FRG
Wood, R.   Royal Greenwich Observatory, Hailsham, UK
Yorke, H.W.   Universitäts-Sternwarte, Göttingen, FRG
Zealey, W.J.   United Kingdom Infrared Telescope, Hilo,
                         Hawaii, USA

ANU   Australian National University
CFA   Harvard-Smithsonian Center for Astrophysics
CSIRO   Commonwealth Scientific & Industrial Research Organization
DAO   Dominion Astrophysical Observatory
DRAO   Dominion Radio Astrophysical Observatory
ESO   European Southern Observatory
GSFC   Goddard Space Flight Center
JPL   Jet Propulsion Laboratory
MIT   Massachusetts Institute of Technology
MPIfA   Max-Planck-Institut für Astrophysik
MPIfR   Max-Planck-Institut für Radioastronomie
MSSSO   Mount Stromlo and Siding Spring Observatory
NRAL   Nuffield Radio Astronomy Laboratory
UNAM   Universidad National Autonome de Mexico

THE MASS AND STRUCTURE OF THE REMNANT OF TYCHO'S SUPERNOVA

Paul Gorenstein, Frederick Seward, and Wallace Tucker
Harvard/Smithsonian Center for Astrophysics,
Cambridge, MA 02138

ABSTRACT

A high resolution X-ray image of Tycho's supernova remnant obtained from the Einstein Observatory reveals three components of X-ray emission that we identify with shocked interstellar material, diffuse ejecta, and clumpy ejecta. This picture is applied to derive the mass of X-ray emitting material. Assuming a distance of 3 kpc, an absorbing column density of 3×10^{21} atoms/cm^2, and using an ion-electron non-equilibrium calculation for the emissivity, we find the average density of the ISM is 0.4 atoms/cm^3, and the energy contained in the remnant is 1.4×10^{51} ergs. The total mass of X-ray emitting material in the remnant is ≈ 4 M$_\odot$, ≈ 2 M$_\odot$ ejecta and ≈ 2 M$_\odot$ swept up, putting the remnant at an intermediate state between a free expansion and the Sedov phase. There is no evidence for neutron star. The upper limit on the surface temperature is in the range 1.1 to 1.8×10^6K.

1.0   INTRODUCTION

The remnant of the supernova explosion of 1572 described by Tycho (Brahe 1573, Clark and Stevenson 1977) occupies a unique and important position in the study of SNRs. The age of the remnant is known, and the light curve was well-measured (Baade 1945) and firmly establishes the supernova as type I. Indeed, when a "typical" example of a type I SN is mentioned in the literature, it is usually Tycho's. Information concerning the mass of ejecta in this SNR are therefore vital to determining the nature of the type I SN explosion.

Much of the recent theoretical work has centered on low mass stars as progenitors of Type I SN. A number of authors (Arnett 1979, Chevalier 1981, Nomoto 1981, and Wheeler 1982) have considered accreting white dwarfs of ~ 1.5M$_\odot$ which are disrupted completely by the explosion, leaving no compact remnant and producing a large amount of Ni 56. In a more general way, Lasher (1975) showed that the narrowness of the peak of the early Type I SN light curve required $\lesssim 2$M$_\odot$ of ejecta. This is

1

consistent with the observed lack of correlation of Type I supernovae with spiral arms which points to an older stellar population of low mass stars (see, e.g., Maza and van den Bergh 1976). On the other hand, Oemler and Tinsley (1979) have argued that Type I supernovae are correlated with star formation rates and are therefore associated with stars having lifetimes less than about a billion years, or masses greater than about two solar masses. Weaver, Axelrod and Woosley (1980) have shown how the evolution of a 9 $M_\odot$ star can lead to a Type I supernova event. So, the question as to the mass of the progenitor remains open.

Another approach to this problem is to study the dynamics of the remnant. Strom, Goss and Shaver (1982) have recently made an important contribution in this connection. Using two radio maps taken eight years apart, they have shown that the expansion velocity of Tycho's SNR is 3600±360 km/s at a radius of 221 arc seconds or 0.99 x 10^{19} cm, assuming a distance of 3 kpc. This is to be compared with the average velocity of expansion which is 7640 km/sec over the 410 year lifetime of Tycho's SNR. The ratio of the present to the average velocity is 0.47±.044, in agreement with the results of Kamper and van den Bergh (1976) which were based on optical observations. These results have generally been taken as proof that Tycho's supernova remnant is in the adiabatic phase, with the swept-up mass much greater than the ejected mass. In this phase the radius $R \propto t^{2/5}$, so the velocity $\dot{R} = V = \frac{2}{5}\frac{R}{t} = 0.4\ \overline{V}$. However, a comparison of the instantaneous and average velocities of expansion may be misleading if a reverse shock (McKee 1974) is present. The reverse shock, moving back into the expanding ejecta, appears to an external observer to be expanding at a lower velocity than the primary shock propagating into the ISM. If the reverse shock is bright, the observed expansion will be closer to that of the reverse shock than that of the shock in the ISM.

Yet another approach is to attempt to infer the mass by means of X-ray observations. All heated material is visible through its X-ray emission. The observed morphology, a model for the three-dimensional structure, and the measured X-ray surface brightness and temperature can be used to calculate the amount of X-ray emitting plasma. Spectra taken with the Einstein SSS (Becker et al. 1980) show surprisingly strong emission lines, particularly from Si and S. Both equilibrium and non-equilibrium models require several times solar abundance of Si group elements (Shull 1982a). The spectrum obtained from HEAO-1 (Pravdo et al. 1980) shows X-ray emission to at least 25 keV, requiring high temperatures in at least part of the remnant. An Einstein IPC observation was reported by Reid, Becker and Long (1982) based upon substantially lower spatial resolution than the HRI measurements. They derive a mass of X-ray emitting material of 15$M_\odot$ and a density of 2.3 atoms/cm^3 for the ISM, typical of past results for this remnant. The better resolution of the HRI, a lower value of interstellar absorption, and a recent non-equilibrium emissivity calculation lead to an ISM density of only 0.4 (this paper) and to a lower calculated mass of the remnant. The X-ray luminosity is consistent with this lower mass because of high emissivity from non-equilibrium effects, enrichment of metal abundances, and the presence of clumps in the ejecta.

In the following sections, we will show that the high resolution Einstein image has features which can be interpreted as a shock heated shell in the ISM and an inner shell containing ejecta which has broken into clumps. The mass of plasma in each region will be calculated from the observed surface brightness using emissivities generated from a model in which ions and electrons are not in equilibrium. The result places the remnant in a phase intermediate between a free expansion and the Sedov phase.

2.0 THE EINSTEIN HRI OBSERVATION

2.1 Overview

The Einstein telescope was pointed at Tycho's SNR for 22 hours starting 1940 UT, 8 February 1979. The resulting image contained 14 hours of good data and is shown in Figure 1 at various levels of exposure. Several features are apparent:

(1) The remnant is almost circular with diameter of 8 arcminutes. There is limb brightening, not uniform around the circumference but indicative of emission from a shell, which varies from a maximum in the NW to a minimum in the SE where there is almost no limb brightening at all. There is a discontinuity in the SE which is exactly that observed in the radio region (Duin and Strom 1975, Dickel et al. 1982).

(2) A thin shelf of emission can be seen at the outer edge of the remnant, outside the region of maximum brightness, around most of the circumference. We assume that this feature is produced by radiation from a shock wave propagating into the interstellar medium. The interstellar shock wave is clearest on the west side and the circles shown in Figure 1D are centered with respect to this outer shock.

(3) There is no emission detected from a central compact object or from any point-like source within the remnant.

(4) Most of the emission is from small, irregular, patchy regions. These must be clumps of material having high X-ray luminosity either because of greater-than-average density or greater-than-average emissivity. Temperature variations will also affect the brightness but not as strongly. We assume that this material is supernova ejecta. These clumps appear to be arranged inside a spherical shell (with the exception of the SE discontinuity), but the distribution within the shell is far from uniform. Individual clumps can be discerned in the center of the remnant (where we are looking approximately normal to the surface of the shell) and in the SE where the density of clumps is low. Maximum surface brightness of the remnant is in the NW, where the density of clumps is highest. The brightest regions are due to limb brightening from many overlapping high emissivity clumps.

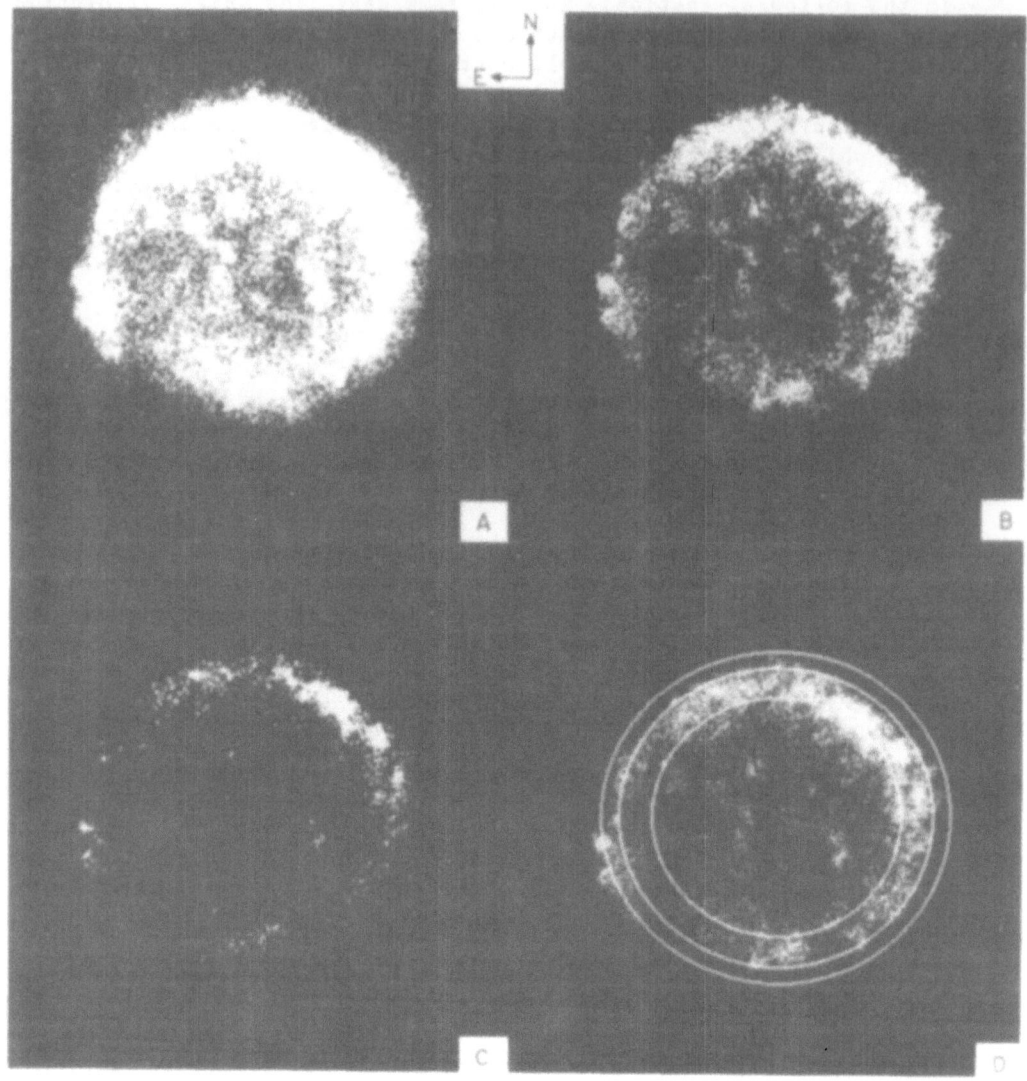

Figure 1:   The Einstein HRI image of Tycho's SNR exposed to show (A) the
faint   shock   heated   material   on   the   outside of the remnant, (B) the
clumpy appearance of the X-ray emitting material, and (C) the   brightest
regions   around   the   limb.   (D) Circles with radii 172″, 216″, and 240″
are centered at RA $0^h22^m30^s9$, Dec 63°51′45″.   These illustrate   the   two
shells described in the text.

2.2 Details

The HRI image of a point source consists of a high resolution core FWHM ~4 arcseconds and a rather broad tail several arcminutes in extent due to scattering from imperfections in the mirrors. At 2 keV, 45% of the focused energy is contained in a circle of diameter 12 arcseconds, and this fraction decreases as photon energy increases. The effect of scattering appears in the image as an increased brightness in the center of the remnant and as faint diffuse emission outside the shell. Figure 2 is a contour plot of an image that has been deconvolved by a maximum entropy technique.

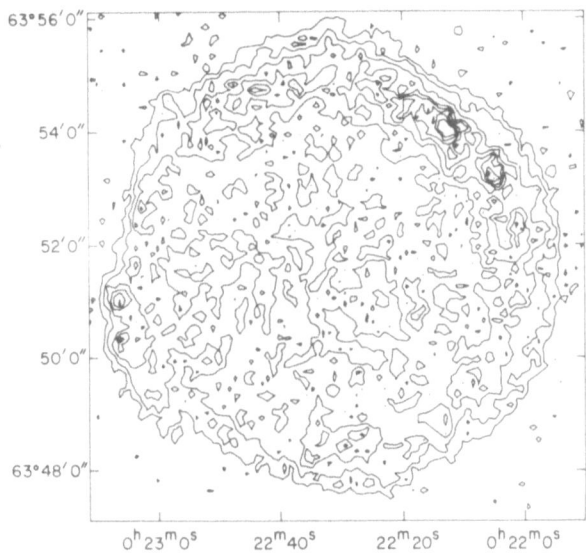

Figure 2: Contour map of Tycho. The spacing between the contours of constant surface brightness is 1×10^{-5} HRI counts/s.arcsec^2 except that the spacing of the first two contours is half this.

Figure 3 shows radial profiles of surface brightness (prior to deconvolution) for 12 equal segments, each spanning 30° in position angle. It illustrates the gross clumpiness in the interior of the remnant and the variation in limb brightening around the circumference. The center of these radial profiles is RA $0^h22^m30\overset{s}{.}9$, Dec 63°51'45" as illustrated in Figure 1D. The maximum extent is exhibited by the SE discontinuity at PA 90°–120°, and the minimum extent by the adjoining region at PA 150°–180°. The shock appears in some of these profiles as a small inflection superimposed on the generally smooth decrease in surface brightness going outward from maximum emission at the limb. The gradually decreasing background starting at a radius of 4 arcminutes is due to the scattering wings of the instrument response. Variations in the radius of the remnant over the 30° segments, certainly over the 360° average, wash out the characteristic structure associated with the shock. Figures 1, 2, and 3 may be compared with the IPC data given by Reid et al. (1981) which have a resolution of ~1.5 arcmin. The IPC is more sensitive than the HRI at higher X-ray energy. The basic features of the remnant are the same in the IPC and in the HRI image implying that there are no gross spectral differences in emission from different

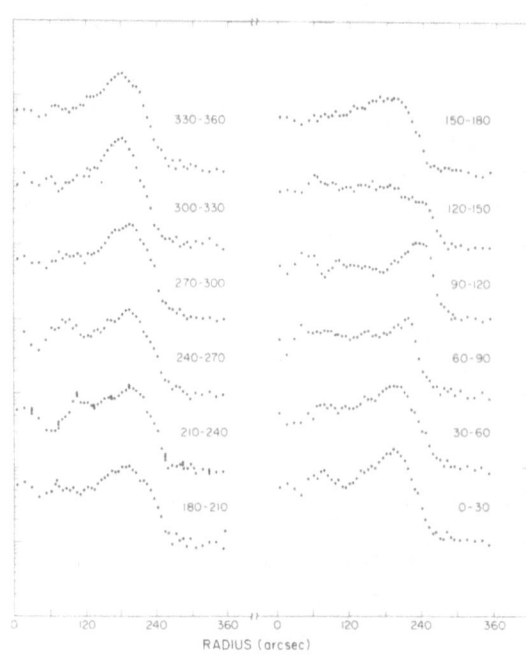

RADIUS (arcsec)

Figure 3: Surface brightness as function of radius at 12 different position angles.

parts of the remnant. The resolution of the IPC was not good enough to distinguish the shock or the clumps of material. Maximum limb brightening measured in the NW with the HRI is ≈4 compared with ≈2.5 with the IPC. Figure 4 shows surface brightness averaged over only 10° in position angle centered at position angles of 235° and 315° (measured from N through E) where the shock is prominent. The HRI response to a uniform ring of inner and outer radii 156" and 204" (an approximation to the projection of a shell) is shown as a solid curve. Note that the broad wings of the HRI response to this ring (containing 90% of the emission) contribute considerable brightness over the rest of the remnant. These data are from the HRI image with no deconvolution and show the strength of the various signals and backgrounds. After subtraction of background due to instrument response to the bright limb, we calculated surface brightness of the shock and from the interior. Since densities to be calculated depend on the square root of the surface brightness, small uncertainties in background subtraction are not important.

3.0 MASS DISTRIBUTION MODEL

The high resolution image allows us to identify three components of the X-ray emission: (1) emission from an outer shell of radius $R_0 = 240"$ and thickness $\Delta R/R \sim .1$; this we identify with shocked interstellar matter; (2) emission from diffuse material in a shell of outer radius $R_1 \approx 216"$ and thickness $\Delta R/R \sim 0.2$; this we identify with supernova ejecta; (3) emission from clumps distributed in the same shell as the diffuse ejecta; this we also identify with supernova ejecta. The contribution of these three components is summarized in Table 1.

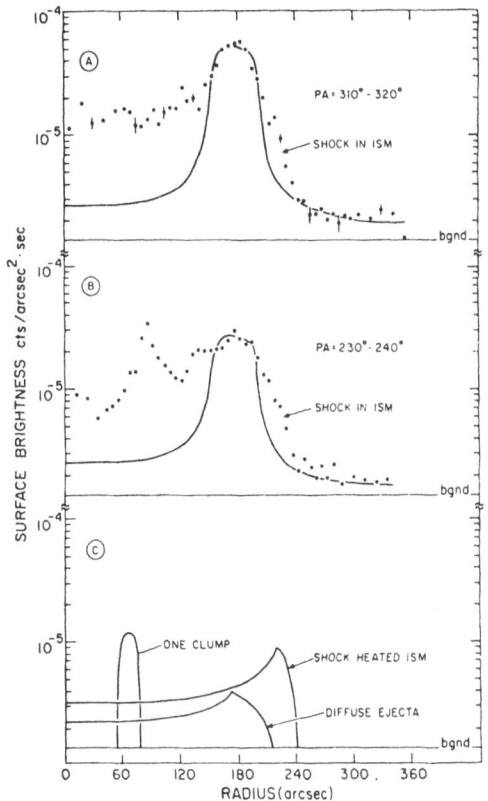

Figure 4: Surface brightness at 2 position angles, one with maximum limb brightening, one with a very bright clump. Solid curve shows instrument response to bright ring of emission. Solid curves in lower plot show surface brightness of shell of shock heated material, inner shell of diffuse emission and of a typical clump. Lower curves refer only to source, instrument response is not included.

Table 1. Data From HRI Image

Region	Surface Brightness (counts/s.arcsec^2)	Geometry	Signal (counts/s)
shock	7.4 (−6) at limb	shell r_o=240" r_i=216"	0.80
diffuse ejecta	0.8 (−6) at center	shell r_o=216" r_i=173"	0.21
one clump	10.3 (−6) at maximum	sphere r=12"	0.0047
all clumps		400 distributed in diffuse ejecta shell	1.86
total remnant			2.87

We use the observed surface brightness of the shock to calculate the swept up mass, M_s, and density of the ISM. We then calculate the ejecta mass, M_e, and compare the ratio of M_e/M_s with that determined from the measured dynamic state of the SNR. The shock in the ISM is assumed to be spherical, and composed of material of normal cosmic abundance that has been snowplowed and heated by the shock to a temperature which is dependent on the shock velocity. At the end of the next section, we will deduce a shock velocity ≈ 6000 (d/3 kpc), implying a temperature of $\approx 4 \times 10^8 K = 36$ keV. The high energy X-rays detected by Pravdo and Smith (1979) imply the existence of some high temperature material but do not support the bulk of shock heated material being at 36 keV. A lower temperature gives a better fit to the high energy spectrum, and we use a temperature of 7 keV. This is consistent with the idea that the electrons and ions have yet to come to equilibrium, so that the electron kinetic temperature is lower than the ion kinetic temperature (Itoh 1977).

As can be seen in Figure 1, 70% of the detected X-rays come from a clumpy shell just behind the shock heated material. We assume this shell contains the stellar debris or ejecta which has been greatly enhanced in silicon group elements as required by the X-ray spectrum (Becker et al. 1980, Shull 1982a). The ejecta (or ejecta-containing regions) are clumpy and the size distribution of clumps can be estimated from the X-ray image.

The observed clumps are not uniform, and the distribution in size can be estimated from the central part of the remnant where isolated clumps are observed. Since no unresolved point sources appear, clump dimensions can be measured. We approximate the actual distribution with two components: uniform density clumps with diameter 24" (0.34 pc), and a uniform distribution of material filling a shell of outer radius 216" and of thickness 0.2 of this radius.

4.0 CALCULATION OF SWEPT UP AND EJECTA MASS

We take the distance to the remnant as 3 kpc. Estimates in the literature range between 2 and 6 kpc, and recent references are briefly reviewed by Strom et al. (1982) and Reid et al. (1982). The dependence of calculated mass on distance, d, is $d^{5/2}$. Thermal energy also goes as $d^{5/2}$, and kinetic energy is proportional to $d^{9/2}$. Using the model described above (i.e. 2 shells of diffuse emission plus clumps), the X-ray surface brightness observations can be used to calculate density of material in the different components of the remnant and hence their mass. The major uncertainties in the calculation concern the emissivity of the plasma and the absorption of X-rays in the ISM between the SNR and Earth. The ISM column density is taken as 3×10^{21} atom/cm^2 with solar composition. This is the neutral H absorption measure of Hughes, Thompson and Colvin (1971). If the column density were doubled to 6×10^{21} atoms/cm^3, our calculated masses would increase by 30-40%. Considerations of ionization, recombination, and expansion time scales indicate that the ions and electrons are not in thermal equilibrium (Itoh 1977). The shock heated electrons probably achieve equilibrium, but the ions are collisionally ionized more slowly and "ionization" temperature lags behind electron temperature which lags behind the ion kinetic temperature. Shull (1982a) has calculated the emissivity of a Sedov model for Tycho's SNR containing an enriched plasma (abundances of elements relative to solar are 0.4 (Ne), 2 (Mg), 8 (Si), 6 (S), 3 (Ar), 3 (Ca), and 2 (Fe)) and shock heated to an (electron) temperature of 7 keV. Parameters were adjusted to fit the SSS spectral data. We have used these emissivities under the assumption that emissivity of the Sedov model is approximately that of the recently shocked material. Shull (1982b) has also kindly provided an identical calculation for material of solar composition.

The electron density, n, in a diffuse, thin source of thermal X-rays can be calculated from the observed surface brightness using the expression

$$n^2 = 5.35 \times 10^{11} \frac{1}{H} \frac{\varepsilon}{T_r P(T)} S\frac{1}{t}.$$

S is observed surface brightness in counts/arcsec^2.sec; t is depth of emitting region in cm; T_r is transmission of ISM and has value $\langle 1$; P(T) is emissivity in the .2–4 keV band of 1 cm^3 of plasma in ergs cm^3/s; ε is detector efficiency in erg/cm^2/count; and H is a factor, varying between 1.0 for a point source and 2.0 for a large diffuse source, which takes into account the spatial response function of the detector.

The mass, M, and X-ray luminosity, L_x, depend on the density and volume, V, of the source

$$M = \mu n V \text{ and } L_x = n^2 P(T) V,$$

where μ is the average ion mass per electron and n is the number of electrons/cm^3.

4.1 Swept-up Mass

We first calculate the parameters of the interstellar shock. Subtraction of HRI background and instrument response to the bright inner ring gives the surface brightness of shocked material at the point of maximum limb brightening within the shocked ISM. Since it is generally accepted that the gas behind the shock is not in equilibrium, we use the non-equilibrium solar calculations of Shull (1982b) for the emissivity. The density in the shock heated ISM is found to be 1.44 cm^{-3}. Assuming a compression factor of 3.7 behind the shock ($\Delta R/R = 0.10$), the ambient interstellar electron density is 0.39 cm^{-3}, corresponding to a baryon density N = 0.50 cm^{-3}. The swept up mass is 2.2 $M_\odot$.

4.2 Diffuse Component of Ejecta Shell

The diffuse component of the ejecta shell is determined by the minimum measured surface brightness at the center of the remnant. There are regions in the center, apparently free of clumps, which have a residual brightness after subtraction of the HRI response to the bright regions in the rim of the remnant and subtraction of the contribution of the shocked ISM shell. Assuming that (a) this material is distributed in a shell of outer radius R_c = 216″ and thickness $\Delta R/R$ = 0.2, (b) the emissivity is given by the non-equilibrium enriched calculation of Shull (1982a), we calculate an electron density of 0.61 cm^{-3} and a mass of 1.2 M_Θ. If we assume that the temperatures of the electron and ion components in the diffuse shell are the same as in the swept up material, then the pressure in the diffuse ejecta shell is less than in the shell of swept up material, consistent with the idea that the ejecta shell is decelerating.

4.3 Clumpy Ejecta

The measured size and surface brightness of a few individual clumps were used to calculate the average characteristics of a clump. The remaining emission of the remnant, after subtracting contributions of the shock and the diffuse ejecta, was entirely assigned to clumps with no restrictions on the arrangement of the clumps within the remnant. The density of the clumpy ejecta, assuming a non-equilibrium plasma enriched in silicon group elements, and pressure equilibrium between the diffuse and clumpy ejecta, is found to be 2.5 cm^{-3}. The mass of the clumpy ejecta is 0.7 M_Θ. Pressure equilibrium fixes the temperature of the material in clumps at 2 keV. Table 2 summarizes the masses derived for the different components. For comparison with previous work, and to illustrate two other possibilities (which at present are not thought to be as likely as the non-equilibrium model), we have included in the table results from an equilibrium calculation with slightly enriched ejecta and super-enriched ejecta.

The ejecta mass resulting from an equilibrium calculation with temperatures derived from the SSS spectrum (Becker et al. 1980) is 3.6 M_Θ. The only way of reducing this mass considerably is to assume that the ejecta are bright in X-rays, not because their density is high, but because the emissivity is high. The emissivity can be increased by increasing the abundance of the medium heavy elements by a large factor. For example, if it is assumed that the medium heavy elements are enhanced by a factor of 1000, then the emissivity will be enhanced by a factor of about 10-100 depending on temperature (cf., Long, Dopita, and Tuohy 1982). The computed ejecta mass would then be reduced to 1.4 M_Θ, and the ratio of swept up to ejected mass would be about 4. There are two difficulties with this approach. First, if we increase the emissivity by too large a factor, then the computed density will be low, and the clumps would not be in pressure equilibrium with the surrounding gas, so it is difficult to see how they could exist. Second, if the mass of the ejecta gets too small, the SNR would be in the Sedov phase,

Table 2. Calculated Physical Parameters

Region	Material	Assumed Temp. (keV)	0.2-4 keV P(T) (ergs cm³/s)	n (elec/cm³)	μ	Mass (M⊙)	0.2-4 keV L_x (ergs/s)
Non-equilibrium							
shock	non-equilibrium solar	7.	7 (-23)	1.44	1.3	2.2	1.9 (35)
diffuse ejecta	non-equilibrium enriched[a]	7.	10 (-23)	0.61	1.3	1.2	0.7 (35)
ejecta clumps	non-equilibrium enriched[a]	2.	20 (-23)[b]	2.5	1.3	0.7	3.1 (35)
Ionization Equilibrium							
shock	solar	5.	1.2 (-23)	3.2	1.3	5.0	1.8 (35)
diffuse ejecta	enriched[a]	5.	2.4 (-23)	1.15	1.3	2.3	0.6 (35)
ejecta clumps	enriched[a]	0.5	5.6 (-23)	4.6	1.3	1.3	3.1 (35)
Ionization Equilibrium and Super-enriched Ejecta							
shock	solar	5.	1.2 (-23)	3.2	1.3	5.0	1.8 (35)
diffuse ejecta	metal abundance	5.	2.4 (-22)	0.36	2.0	1.1	0.6 (35)
ejecta clumps	1000 x solar	0.5	2.5 (-21)	0.68	2.0	0.3	3.1 (35)

[a] Elemental enrichment relative to solar are: 0.4(Ne), 2(Mg), 8(Si), 3(Ar), 3(Ca), and 2(Fe) after Shull, 1982a.

[b] Emissivity assumed to scale ∝ $T^{-1/2}$ between 7 and 2 keV.

the ejecta would be well mixed with swept up ISM, and we would not expect to see the two-shell structure that we see. In summary, it seems that the observations of the spatial structure constrain the mass of ejecta to be approximately equal to that of the swept up mass, and favor a non-equilibrium model in which the medium heavy abundances are enhanced only about an order of magnitude.

The non-equilibrium results given in Table 2 show that the mass of the ejecta is approximately equal to the mass of the swept up matter. This means that Tycho is in a stage intermediate between uniform expansion and the adiabatic phase. Theoretical work on the dynamics of supernova remnants indicates that this intermediate stage is characterized by: (1) a shock wave moving into the interstellar medium with a velocity v_s (Rosenberg and Scheuer 1973); (2) a reverse shock wave moving back into the ejecta with a velocity less than v_s (McKee 1974, Chevalier 1982a,b); and (3) clumps of material created by instabilities in the decelerating ejecta (Gull 1975; Jones, Smith, and Straker 1981).

This theoretical picture is not far from what is observed in the high resolution X-ray image of the Tycho SNR. The observed ratio of radii of the reverse and interstellar shock waves is 0.72, compared with 0.77 derived from Chevalier (1982b). Chevalier (1982a) calculates the density profile of a shock and reverse shock in a young SNR expanding into a uniform radius. Our observation is close to this, but the shocked ejecta shell is about three times as thick as his calculation, extending more both towards the ISM shock and back toward the center of the remnant. We attribute this to the breakup of the ejecta shell into clumps as calculated by Gull (1975) and, as expected, the size of the observed clumps is comparable with the thickness of the reversed-shocked shell. The theoretical work is not detailed enough to satisfactorily calculate the energy and velocities from our data. Rosenberg and Scheuer (1973) calculate the propagation of a shock into a uniform ISM. By scaling Tycho's apportioned remnant to their work at $M_{ej} = M_{sw}$, we derive $E_0 = 1.4 \times 10^{51}$ ergs, 0.2×10^{51} ergs thermal and 1.2×10^{51} ergs kinetic. The velocity of the shock in the ISM is 6000 km/s, and the velocity of the material behind the shock as well as the contact discontinuity (the surface between the shocked ISM and the shocked ejecta) is 4500 km/s. On the other hand, Chevalier assumes that the ISM shock velocity is $\propto t^{4/7}$ or 4700 km/s for Tycho's SNR and, at the present time, ($t = 1.28 \, t_c$ in his notation) the velocity of the reverse shock is 2700 km/s.

The expansion velocities of 3000-3600 km/sec as measured by Strom et al. (1982) in radio and by Kamper and van den Bergh (1978) in the optical (when adjusted to a 3 kpc distance) is consistent with our analysis, if we interpret these measurements as referring to the shocked ejecta or pre-existing material and not to the interstellar shock wave. It appears that most radio emission is from the region we identify with the diffuse ejecta shell (Dickel et al. 1982). Thus, it is likely that the velocity of the ISM shock is considerably greater than the radio expansion velocity and that the remnant is not as close to the Sedov phase as they have concluded. The bright optical filaments are found in regions where the outermost radial contour is at a local minimum in distance from the center of Tycho. They may correspond to regions where the shock has encountered pre-existing interstellar material. We suggest that the expansion velocities that are seen are those of accelerated interstellar material rather than of the primary shock. In fact, we might expect the brightest and most conspicuous optical and radio filaments to occur under these local conditions and not be indicative of the higher global velocity of the interstellar shock.

The ejected mass of 1.9 $M_\odot$ is somewhat larger than the value of 1.4 $M_\odot$ expected from exploding white dwarf models (Chevalier 1981 and references). However, a small reduction in the assumed distance would bring the computed mass below 1.4 $M_\odot$, so this discrepancy is perhaps not serious. On the other hand, an increase in distance or in absorbing ISM column density will increase the mass considerably above this theoretical expectation.

5.0 NO NEUTRON STAR DETECTED

Since no point sources were detected inside the remnant, we can set an upper limit to the surface temperature of a neutron star which may have been formed in the explosion. We consider two positions: at the center where the surface brightness is low, and as a pessimistic upper limit, the bright knot ~1.5' N of the center. The 3σ upper limit to the signal of a point source at the center is 30 counts. If the source were located in the bright knot, the signal could be as high as 90 counts. There are several such knots or clumps inside this remnant. None appear as unresolved point sources, and there are no features which might distinguish one of them as containing a compact object. The upper limit to the surface temperature was calculated by folding black body spectra as observed through the ISM and the detector response over a range of temperatures. The radius of the neutron star was assumed to be 11 km. The limit is also dependent on the distance and column density of ISM. At a distance of 3 kpc, the limit ranges from 1.1×10^6K ($N_H = 3 \times 10^{21}$, center) to 1.8×10^6K ($N_H = 9 \times 10^{21}$, knot). The significance of this has been discussed at length by Nomoto and Tsuruta (1981) and by Van Riper and Lamb (1981).

6.0 SUMMARY

A high resolution X-ray image of the Tycho SNR reveals three emission components:

(1) an outer shell of radius 240", $\Delta R/R \approx 0.1$,
(2) a diffuse inner shell of radius 216", and
(3) ~400 bright clumps of material distributed in a shell of radius 216", $\Delta R/R \approx 0.2$.

We identify these components with

      (1) shocked interstellar matter,
      (2) diffuse supernova ejecta, and
      (3) clumpy supernova ejecta.

The mass of these components is calculated to be 2.2 $M_{\odot}$, 1.2 $M_{\odot}$ and 0.7 $M_{\odot}$, respectively.

The swept up mass is approximately equal to the ejected mass, so Tycho's SNR must be a stage intermediate between the uniform expansion and adiabatic stages. The observed morphology of the remnant is in reasonable agreement with theoretical expectations, but no numerical calculations have been published showing the reverse shock at this stage of evolution. The mass estimates scale as the distance according to $d^{5/2}$, so a reduction from the assumed distance of 3 kpc to 2.5 kpc would reduce the ejected mass to 1.4 $M_{\odot}$.

At a distance of 3 kpc, the upper limit on the surface temperature of a central neutron star is found to be in the range 1.1 to 1.8 million degrees.

REFERENCES

Arnett, W.D. 1979, Ap.J. (Letters), 230, L37.

Baade, W. 1945, Ap.J., 96, 188.

Becker, R.H., Holt, S.S., Smith, B.W., White, N.E., Boldt, E.A., Mushotzky, R.F., and Serlemitsos, P.J. 1980, Ap.J., 235, L5.

Brahe, T. 1573, De Nova Stella, Laurentius Benedictus Copenhagen, 1573. Facsimile Edition, Danish Royal Scientific Society, Copenhagen, 1901.

Chevalier, R. 1981, Ap.J., 246, 267.

Chevalier, R. 1982a, Ap.J., 258, 790.

Chevalier, R. 1982b, Ap.J. (Letters), 259, L85.

Clark, D.H. and Stevenson, F.R. 1977, The Historical Supernovae, (Pergamon Press, Oxford).

Dickel, J.R., Murray, S.S., and Morris, J. 1982, Ap.J., 257, 145.

Duin, R.M. and Strom, R.G. 1975, Ast. and Ap., 39, 33.

Gull, S.F. 1975, MNRAS, 171, 263.

Hill, R.W., Burginyon, G.A., and Seward, F.D. 1975, Ap.J., 200, 158.

Hughes, M., Thompson, A., and Colvin, R. 1971, Ap.J. Suppl., 23, 323.

Itoh, H. 1977, Pub. Ast. Soc. Japan, 29, 813.

Jones, E.M., Smith, B.W., and Straker, W.C. 1981, Ap.J., 249, 185.

Kamper, K.W. and van den Bergh, S. 1978, Ap.J., 224, 851.

Lasher, G. 1975, Ap.J., 201, 194.

Long, K.S., Dopita, M.A., and Tuohy, I.R. 1982, Columbia Astrophysics Laboratory preprint #213.

Maza, J. and van den Bergh, S. 1976, Ap.J., 204, 519.

McKee, C. 1974, Ap.J., 188, 335.

Nomoto, K. 1981, in Fundamental Problems in the Theory of Stellar Evolution, ed. D. Sugimoto, D.Q. Lamb, and D.N. Schramm (Dordrecht:Reidel) p. 295-315.

Nomoto, K. and Tsuruta, S. 1981, Ap.J. (Letters), 250, L19.
Oemler, Jr., A. and Tinsley, B.M. 1979, A.J., 84, 985.
Pravdo, S.H. and Smith, B.W. 1979, Ap.J. (Letters), 234, L195.
Pravdo, S.H., Smith, B.W., Charles, P.A., and Tuohy, I.R. 1980, Ap.J. (Letters), 235, L9.
Reid, P.B., Becker, R.H. and Long, K.S. 1982, Ap.J., 261, 485.
Rosenberg, I. and Scheuer, P.A.G. 1973, MNRAS, 161, 27
Shull, J.M. 1982a, Ap.J. (Letters) (in press, Vol. 262).
Shull, J.M. 1982b, private communication.
Strom, R.G., Goss, W.M., and Shaver, P.A. 1982, MNRAS, 200, 473.
Van den Berg, S., Marscher, A.P., and Terzian, Y. 1973, Ap.J. Suppl., 26, 19.
Van Riper, K.A. and Lamb, D.Q. 1981, Ap.J. (Letters), 244, L13.
Weaver, T.A., Axelrod, T.S., and Woosley, S.E. 1980, in Proc. Texas Workshop on Type I Supernovae, ed. J.C. Wheeler (Austin: University of Texas Press), p. 113.
Wheeler, J.C. 1982, in Proc. NATO Advanced Study Institute on Supernovae (Cambridge: Institute of Astronomy), p. 167.

DISCUSSION

KIRSHNER: Is there a problem with clumps that might be present on a scale smaller than your resolution? Would it be possible to get away with less mass if the material were clumpy on all scales?

GORENSTEIN: If there were clumps smaller than our resolution, we would mistake them for diffuse emission and overestimate the mass. However, there is no reason to believe they exist because the typical clump we do see is several times larger than our resolution.

DICKEL: Could some of the clumps be overtaken interstellar ones which are now evaporating rather than lumps of ejection?

GORENSTEIN: The clumps are probably responsible for the enrichment in Si, S, Ca, and A. This indicates that they are primaily ejecta.

CHEVALIER: Were the non-equilibrium calculations used in the mass estimates consistent with most of the emission coming from the reverse shock region?

GORENSTEIN: We used the non-equilibrium ionization model of Shull which is obtained by fitting the theory to the SSS intensity and spectrum of Tycho. We did not consider a complete non-equilibrium model that provides an independent prediction for the X-ray intensity of the reverse shock.

WINKLER: Why have you taken the temperature of the inner diffuse shell of ejecta to be the same (7 keV) as that of the outer shell of ISM?

GORENSTEIN: This was an assumption. However, the value of the mass is relatively insensitive to temperature.

X-RAY SPECTRA OF YOUNG SUPERNOVA REMNANTS

S.S. Holt
Laboratory for High Energy Astrophysics
NASA/Goddard Space Flight Center
Greenbelt, Maryland 20771

A. HISTORICAL NOTES

The early identification of the strong X-ray source in Taurus with the Crab nebula (Bowyer et al. 1964) was the first milestone in the association of X-ray emission with supernova remnants. Unfortunately, it proved to be "red herring" which clouded the interpretation of X-ray emission from supernova remnants for a decade. Because the Crab was one of the brightest X-ray sources in the sky at a few keV, the interrogation energy of the early surveys, and because it was the first (and for several years the only) X-ray source conclusively identified, the potential association of a supernova origin with the large body of unidentified X-ray sources was not an unreasonable hypothesis.

By 1970 proportional counter spectra from rocket-borne observations, with resolving power $E/\Delta E < 2$, were available for Cas A and Tycho as well as the Crab (Gorenstein, Kellogg and Gursky 1970). Since the statistical significance of the data points were exposure-limited, they were fittable with either bremsstrahlung continua with temperatures of $\sim 10^7$ K or power-law spectra; the analogy with the Crab suggested to many interpreters of the data that the latter representations of the spectra were more appropriate. By that time the Crab spectrum was well-measured to be a power-law with $\alpha \sim 1$ between 1-100 keV, and relatively well-understood (from the point of view of available energy, if not energy transfer) as arising from the non-thermal conversion of rotational energy from its pulsar. Similar pulsars which were unobservable because of their beaming characteristics were suggested for other young supernova remnants, since only young pulsars could provide the required rotational energy loss (e.g. Pacini 1971). Similarly, models featuring hidden pulsars in sources not associated with traditional supernova remnants (e.g. Davidson, Pacini and Salpeter 1971) were relatively popular.

The discovery of "X-ray pulsars" (Giacconi et al. 1971), distinguishable from radio pulsars like the Crab by their inability to account for the source luminosity from rotational energy loss, provided the key to understanding the energy generation mechanism of the non-SNR

17

J. Danziger and P. Gorenstein (eds.), Supernova Remnants and their X-Ray Emission, 17–27.

sources in terms of mass accretion onto neutron star components of
binary systems, as had been suggested by Shklovsky (1967). But what
about the young SNRs Cas A and Tycho? An early suggestion by Heiles
(1964), that the SN blast wave sweeping up the interstellar medium into
a shell of shocked material would provide an observable X-ray source,
seemed more appropriate to older remnants in which enough mass had been
swept up to provide an adequate emission measure, and enough time had
elapsed to allow the radiating shell to equilibrate at a temperature low
enough to cool efficiently (e.g. Woltjer 1972). Nevertheless, a blast
wave interpretation for the young SNRs increased in popularity as the
newer data indicated better agreement with a thermal spectrum. The key
experimental result was the discovery of Fe line emission from Cas A
(Serlemitsos et al. 1973) which, although originally interpreted by its
discoverers as arising from charge exchange with ~ 10 MeV/nucleon cosmic
rays, was quickly recognized as the signature of thermal processes in
the remnant. As data from satellite experiments became available
(Charles et al. 1975; Pravdo et al. 1976; Davison, Culhane and Mitchell
1976) this picture was refined to require two separate thermal
components for consistency with these higher statistical precision
data. Itoh (1977) suggested that the harder of these two components was
associated with the blast wave, while the softer was associated with a
thin, dense shell of ejecta accumulated from the "reverse shock"
postulated by McKee (1974). The history of Cas A and Tycho measurements
in the decade preceding the first spectral data from Einstein is given
in Table 1.

Table 1

EARLY THERMAL FITS TO CAS A AND TYCHO

Ref	Cas A		Tycho	
	T_{low}	T_{high}	T_{low}	T_{high}
Gorenstein et al. 1970	----	1.25± .34	----	2.2±1.0
*Brisken 1973	.83±.06	----	.45±.05	----
Hill et al. 1975	----	2.3	----	1.9
Charles et al. 1975	.73±.20	2.6 ±1.0	----	----
Pravdo et al. 1976	< .7	3.9 ± .6	----	----
Davison et al. 1976	1.08±.03	5.2 ± .6	.55±.04	3.5± .3

*refined analysis of data of Serlemitsos et al. (1973)

B.   THE EINSTEIN SSS DATA

The newer Einstein data summarized here are from the Solid State
Spectrometer (SSS), a cryogenically-cooled Si(Li) device with a FWHM
energy resolution of ~ 160 eV over its effective 0.5-4.5 keV energy
range (see Holt et al. 1979 for experiment details). The spectra I
shall discuss have already been preliminarily reported in the
literature. The new information presented here will be of two kinds: a

refined set of experimental spectra with all sources of background over the full 0.5-4.5 keV band now properly accommodated, and some remarks about the progress which has been made in the physical modelling of SNR to reproduce these spectra.

A Crab spectrum is presented in Figure 1; it represents one of $\sim$ 100 such spectra obtained over the lifetime of the experiment, all of which were used for calibration purposes to ensure uniformity in the interpretation of detector response with time. The input spectrum is presumed to be a featureless power law of index α = 1.1 observed through an interstellar medium of cosmic abundance with column density N_H = 3 x 10^{21} atoms cm^{-2}. Note that the experiment introduces an apparent kink in the spectrum at $\sim$ 0.8 keV (arising from the nickel with which the telescope is coated), but that any other features (particularly ones assocaited with silicon, of which the detector is made, and which might be expected near 1.8 keV) are much smaller than the statistical precision of one of the displayed pulse-height channels in a typical measurement. The Crab spectrum is illustrative, therefore, of the fact that any potential line features in the spectra are not systematically induced in the detector.

The line features which are observed in young SNR spectra are indicated in Figure 2, where are plotted the pulse height data from an exposure to the Tycho (SN 1572) remnant, together with a comparison spectrum for illustrative purposes. The comparison spectrum is that expected from a T = 0.5 keV solar-abundance plasma in thermal equilibrium, viewed through an interstellar column density of 2 x 10^{21} atoms cm^{-2}; this comparison histogram is not convolved through the detector response, but is smeared to a FWHM resolution of $\sim$ 50 eV with exactly the same bin widths as the experimental data. The shaded portion of the comparison spectra is the contribution from Fe-L emission from the isothermal plasma, and the darkened portion the contribution from Si-K emission; the four blackened bumps represent Si^{+12} Kα, Si^{+13} Kα, Si^{+12} Kβ and Si^{+12}Kγ in order of increasing energy. From the obvious line identifications, therefore, it is clear that the observed lines are characteristic of a relatively low temperature (e.g. T = 0.5 keV) plasma, in the sense that the helium-like lines of the Z > 10 constituents are evident while the hydrogen-like lines are not, but the equivalent continuum widths of the observed lines are considerably in excess of the equilibrium values for solar abundances. Most of the remainder of this paper will be devoted to a discussion of the modelling of such spectra.

C.   SPECTRAL MODELLING

Figure 1 displays the SSS spectrum from Cas A. As originally reported by Becker et al. (1979), the data cannot be fitted with a single equilibrium temperature with any abundance distribution; instead, it can be reasonably well-fitted with two equilibrium components. The lines are consistent with an equilibrium plasma at T $\simeq$ 0.65 keV (albeit

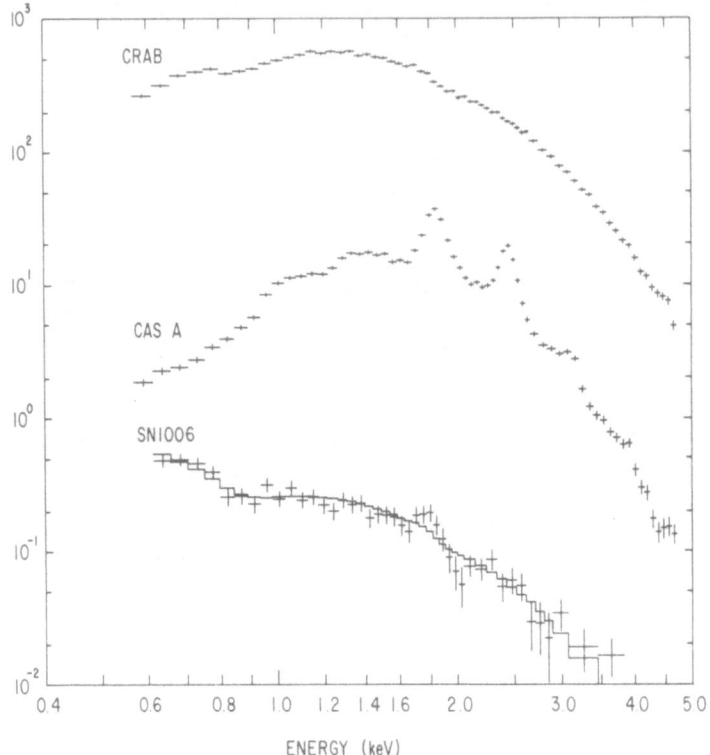

Figure 1. Experimental SSS spectra for the Crab nebula, CasA and SN1006. The solid histogram through the SN1006 data points represents a power law of index $\alpha = 1.2$ observed through a column density of 9×10^{20} atoms cm^{-2}.

with overabundances in the line-emitting components), but a higher temperature component is required to replicate the harder continuum observed and to smoothly connect to the higher energy measurements; because the bremsstrahlung continua from any temperatures ~ 3 keV are indistinguishable in the SSS 0.5-4.5 keV bandpass, a fixed 4 keV was chosen to approximately connect to the higher energy measurements.

The data presented in Figure 1 are somewhat more refined than originally presented in Becker et al. (1979), but the important source-fitting parameters are not substantially changed. In particular, the best-fit low temperature component (for a 2-temperature equilibrium fit) is still $T \simeq .65$ keV, and the Si abundance is ~ 1.5 times solar, as determined from the simultaneous fitting of the lines and the continuum. Relative to this inferred Si abundance, the inferred abundances of other important constituents relative to solar proportions are S/Si($\sim$2), Ar/Si($\sim$4), Ca/Si($\sim$2) and Mg/Si($\sim$.1). The only significant difference between the results of the present modelling and that originally reported is in the Fe abundance, as the model extension to

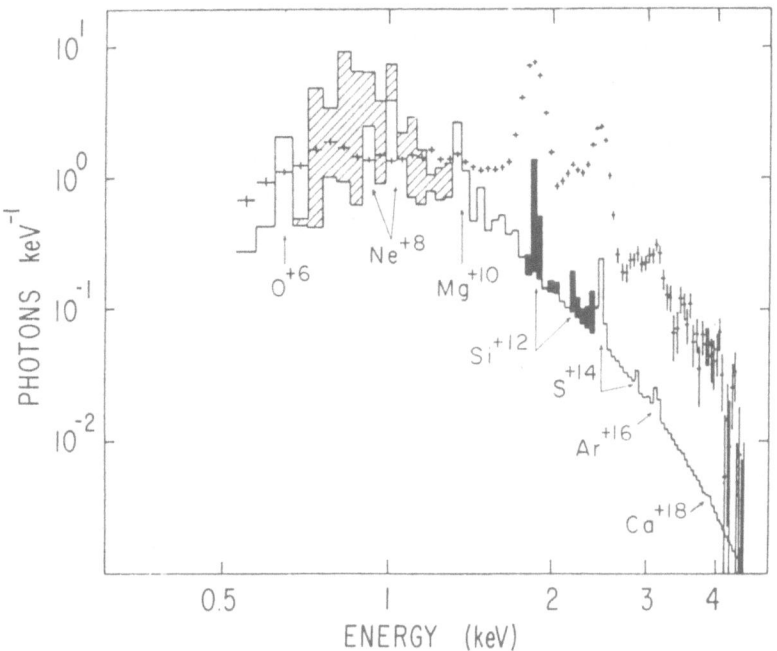

Figure 2. Experimental SSS spectrum of the Tycho SNR, compared with the
expected spectrum from a solar-abundance equilibrium plasma at T = 0.5
keV as it would be observed with a detection system which has an
energy-independent response (see text for additional details).

lower energies changes what was previously a measured Fe/Si ~ .5,
relative to solar, to an upper limit. The lack of a pronounced increase
with Z in the relative overabundances of Si, S, Ar and Ca suggests that
the standard qualifications regarding non-equilibrium ionization may
obscure the important qualitative conclusions that may be drawn from the
spectrum. If, as suggested by Fabian et al. (1980) from an
interpretation of the X-ray morphology, the remnant is in the free
expansion phase and the low energy X-ray emission is dominated by the
ejecta, the clumpiness of the emission suggests a situation closer to
equilibrium (at least within the clumps) than would a smoother
distribution, as the Rayleigh-Taylor-induced density enhancements should
equilibrate more rapidly than would the unclumped ejecta (see Chevalier
1982 for a discussion of the unclumped density distribution which might
be expected from a system like Cas A with pre-SN mass loss). The modest
abundance enhancement of Si-group material implied by the equilibrium
simplification may, therefore, better characterize the spectrum than
would a detailed (non-unique) calculation, at least until such time as
measurements with better spectral and spatial resolution are available.

        In contrast to Cas A, the three Type I remnants Tycho, Kepler and
SN1006 display a clear necessity for non-equilibrium effects. The
spectra of Tycho and Kepler bear marked similarities to each other (see

Figure 3); utilizing the same two-temperature equilibrium model as a
starting point, both have low temperatures of T ≃ .5 keV. Tycho
requires a Si abundance which is approximately an order of magnitude in
excess of solar (Kepler requires about half the Tycho Si abundance), and
for both remnants the allowable relative abundances of S, Ar and Ca
increase with atomic number (in the newer equilibrium fits, these
implied abundances are even higher than those in Becker et al. 1980a and
Becker et al. 1980b), in accordance with the expectation for a
non-equiiibrium plasma which is still ionizing. The application of a
Sedov approximation (Shull 1983; Hamilton, Sarazin and Chevalier 1983)

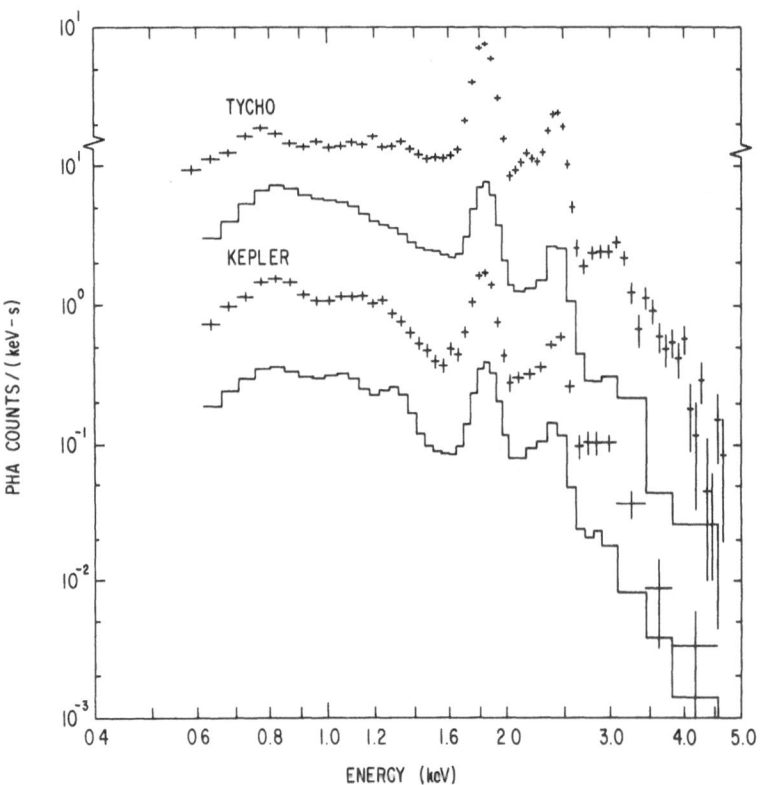

Figure 3 - Experimental SSS spectra of the Tycho and Kepler SNRs. The
solid histograms, arbitrarily placed above and below the Kepler
spectrum, represent different fits to Kepler of comparable statistical
adequacy. The upper fit is a two-temperature equilibrium model for
which the abundances are free parameters (see text), while the lower fit
is a Sedov model with similarly unconstrained abundances; many such
models (with differing n, electron fitted with comparable statistical
adequacy, but all require S_i overabundances relative to solar.

relieves this latter problem, but its applicability may be
problematic. A variety of Sedov models with and without clumping (and
with and without the assumption of rapid equilibration of the electrons
with the shock) all yield pronounced overabundances of Si, comparable to
those found in the equilibrium fits. The relative abundance ratios for
the higher-Z constituents (i.e. those which are identifyable by their
K-emission) are no longer unreasonably high, however; S/Si is still ~ 2
(as it appears to be for all SNR fits), but Ar/Si and Ca/Si need not be
much higher than unity (i.e. solar). The utilization of a Sedov model
(properly constrained to the observable luminosity at the best-estimated
distance, with reasonable values of $\eta = E_n n_0^2$) is questionable a priori
if the interpretation of the X-ray morphology given by Seward,
Gorenstein and Tucker (1983) is correct, i.e. if the large fraction of
the X-ray luminosity arises from clumped reverse-shocked ejecta rather
than from the blast wave. Interestingly, Hamilton (this volume) has
demonstrated that the emission from the ejecta can be modelled with a
Sedov approximation (where the parameter η is a bit more obscure),
allowing fits which better match the data in detail (Sarazin, Hamilton
and Chevalier, this volume).

A continuing feature of all the fits is the lack of marked
overabundances of Fe, as might be inferred from its L-emission (see
Figure 2) in accordance with current models of Type I SN which release a
large fraction of a solar mass of Fe-group material. The primary
difference between the Tycho and Kepler spectra of Figure 3 is in the Fe
L-emission; Fe is virtually undetectable from Tycho, but is required at
about the level Fe/Si ~ 1 (relative to solar) in the various equilibrium
and non-equilibrium Kepler fits. Even in Kepler, this abundance is at
least an order of magnitude below that expected if the Fe-group material
is well-mixed into the X-ray-emitting plasma. Clearly, the retention of
the Fe-rich model requires the not-unreasonable assumption that the
heavier ejecta is interior to the Si-group ejecta, and has not yet been
significantly reverse-shocked.

The X-ray spectrum of SN1006 is even more puzzling. Unlike the
younger Type I remnants discussed above, there is no statistically
significant evidence for K-line emission. In fact, the X-ray spectrum
can be fit with a Crab-like power law (α ~ 1) from 0.5 keV out to at
least 20 keV (Becker et al. 1980c). It is worth noting that the higher
energy X-ray spectra of other SNR can be relatively well-fit by power
law approximations, but with spectral indices which are much larger
(i.e. the younger and more initially energetic SNR Cas A can be fit
with α ~ 3, while Tycho can be fit with α ~ 4; see Figures 1 and 2 of
Pravdo and Smith 1979). Fabian, Stewart and Brinkmann (1982) have
suggested that the appearance of a featureless power law may arise from
the superposition of emission from differing thermal shells. If the
intermediate-Z metals are substantially locked in grains, and if the
reverse-shock has long-since passed through the Fe-rich ejecta (so that
the remnant interior has cooled to ~ 10^6 K), the absence of the
prominent lines observed in the younger remnants may be reconciled. It
should be noted, however, that the spectral hardness of the high energy
continuum may be more easily explained with non-thermal processes, e.g.

synchrotron radiation from electrons accelerated near the blast wave (Reynolds and Chevalier 1981) or possibly even from the naive picture of a hidden pulsar which was so popular ten years ago.

D.   FINAL REMARKS

The combination of higher-energy (> 5 keV) spectral data from pre-Einstein investigations (see Pravdo, this volume) with morphological and spectral data from all the Einstein instrumentation provides a powerful tool for studying SNRs. This evidence is not totally conclusive in all particulars, however. The modelling of the SSS data requires a complete time-dependent coupled solution of the hydrodynamics and ionization, including whatever density inhomogeneities may exist in each specific remnant, since the experiment aperture is 6 min in diameter. Better spectral resolution than its 160 eV FWHM would be useful (insofar as K-emission contributions from differing multiplet components would then be resolvable), but the present resolution is sufficient to distinguish H-like and He-like ionization states. Much more important would be the application of the available spectral resolution to much more localized regions of the SNR, so that differential information would not have to be extracted from the attempted separation into components of the whole-remnant spectrum.

The available procedure is necessarily non-unique, and cannot unambiguously distinguish whether the electrons in the recently shocked plasma can quickly equilibrate with the heavier constituents (as it least some fraction must, from > 10 keV measurements; Pravdo and Smith 1979) or whether they require a Coulomb equilibration timescale (typically longer than the remnant age) to do so; the two cases yield very different bremsstrahlung continua and, hence, can result in very different blast/ejecta proportions and abundances. Unfortunately, fits to the whole remnant data with either assumption for the electron equilibration timescale can yield results of comparable statistical significance. With no independent reason for fixing such otherwise indeterminate parameters, therefore, model uniqueness is not assured. Until such time as more independent information exists (e.g. from the application of like quality spectral data with much finer spatial resolution, so that some separability of otherwise degenerate fitting parameters can be obtained), a satisfactory fit to the whole-remnant spectral data cannot guarantee a unique detailed explanation.

My thanks to the many members and associates of the GSFC X-ray group, past and present, who have contributed to our ongoing program of research. Andrew Szymkowiak, in particular, is responsible for the large fraction of the previously unpublished work on which much of this discussion is based.

REFERENCES

Becker, R.H., Holt, S.S., Smith, B.W., White, N.E., Boldt, E.A., Mushotzky, R.F., and Serlemitsos, P.J. 1979, Ap. J. (Letters) 234, L73.

Becker, R.H., Holt, S.S., Smith, B.W., White, N.E., Boldt, E.A.,
    Mushotzky, R.F., and Serlemitsos, P.J. 1980a, Ap. J. (Letters) 235,
    L5.
Becker, R.H., Boldt, E.A., Holt, S.S., Serlemitsos, P.J., and White,
    N.E. 1980b, Ap. J. (Letters) 237, L77.
Becker, R.H, Szymkowiak, A.E., Boldt, E.A., Holt, S.S., and Serlemitsos,
    P.J. 1980c, Ap. J. (Letters) 240, L33.
Bowyer, C.S., Byram, E.T., Chubb, T.A., and Friedman, H. 1964, Science
    146, 912.
Brisken, A.F. 1973, PH.D. Thesis (University of Maryland).
Charles, P.A., Culhane, J.L., Zarnecki, J.C., and Fabian, A.C. 1975, Ap.
    J. (Letters) 197, L61.
Chevalier, R.A. 1982, Ap. J. 258, 790.
Davidson, K., Pacini, F. and Salpeter, E.E. 1971, Ap. J. 168, 45.
Davison, P.J.N., Culhane, J.L., and Mitchell, R.J. 1976, Ap. J.
    (Letters) 206, L37.
Fabian, A.C., Stewart, G.C., and Brinkmann, W. 1982, Nature 295, 508.
Fabian, A.C., Willingale, R., Pye, J.P., Murray, S.S., and Fabbiano, G.
    1980, Mon. Not. R. Astron. Soc. 193, 175.
Gorenstein, P., Kellogg, E.M., and Gursky, H. 1970, Ap. J. 160, 199.
Hamilton, A.J.S. 1983, this volume, p. 113.
Hamilton, A.J.S., Sarazin, C.L., and Chevalier, R.A. 1983, Ap. J.
    Suppl., in the press.
Heiles, C. 1964, Ap. J. 140, 470.
Hill, R.W., Burginyon, G.A., and Seward, F.D. 1975, Ap. J. 200, 158.
Holt, S.S., White, N.E., Becker, R.H., Boldt, E.A., Mushotzky, R.F.,
    Serlemitsos, P.J., and Smith, B.W. 1979, Ap. J. (Letters) 234, L65.
Itoh, H. 1977, Publ. Astron. Soc. Japan 29, 813.
Pacini, F. 1971, Ap. J. (Letters) 163, L17.
Pravdo, S.H., Becker, R.H., Boldt, E.A., Holt, S.S., Rothschild, R.E.,
    Serlemitsos, P.J., and Swank, J.H. 1976, Ap. J. (Letters) 206, L41.
Pravdo, S.H. and Smith, B.W. 1979, Ap. J. (Letters) 234, L195.
Pravdo, S.H. 1983, this volume, p. 29.
Reynolds, S.P. and Chavalier, R.A. 1981, Ap. J. 245, 912.
Sarazin, C.L., Hamilton, A.J.S., and Chevalier, R.A. 1983,
    this volume, p. 102.
Serlemitsos, P.J., Boldt, E.A., Holt, S.S., Ramaty, R., and Brisken,
    A.F. 1973, Ap. J. (Letters) 184, 1.
Seward, F., Gorenstein, P., and Tucker, W. 1983
Shklovsky, I.S. 1967, Ap. J. (Letters) 148, L1.
Shull, J.M. 1982, Ap. J. (Letters), in the press.
Woltjer, L. 1972, Ann. Rev. Astron. Ap. 10, 129.

DISCUSSION

KIRSHNER: The principal optical result on Cas A is that oxygen is
greatly enhanced. The SSS spectra only give the ratio of Si, S and the
rest to oxygen. Are the X-ray data consistent with the extraordinary
abundances seen in the optical filaments?

HOLT: Because the energy resolution of the SSS does not allow us to
completely separate line and continuum contributions to the experimental
spectrum, and because the bremsstrahlung continuum is dominated by

oxygen (if present in at least solar proportions to hydrogen), it is
not possible to define a totally unambiguous reference for the
"abundances"--even before addressing the problem of how appropriate an
equilibrium model should be for the whole-remnant spectrum of Cas A.
If we assume that the portion of the spectrum below ~ 3 keV is dominated
by the ejecta in the fast-moving knots, the implied Si/O ratio is well
within a factor of two of the values given in Table 5 of Chevalier and
Kirschner (1978).

TUFFS: Is a shock velocity of, say, 2800 km s^{-1} for Cas A consistent
with the observed line ratios?

HOLT: The shock temperature defined by the shock velocity (for a Sedov
solution) is > 10^8 K, or much too high to match the observed line
ratios. The experimental preponderance of helium-like to hydrogen-like
lines forces the equilibrium fits to the spectra of Tycho and Kepler,
in addition to that of Cas A, to contain a component at < 10^7 K which
dominates below ~ 3 keV. At higher energies, however, there is continuum evidence for plasma approaching equilibrium with the blast
wave (Pravdo and Smith 1979).

CHEVALIER: With regard to the non-uniqueness of the X-ray spectral
fits, do all the models have the proper ages, radii and X-ray
luminosities?

HOLT: The Sedov models we have attempted to fit to Tycho and Kepler,
such as that displayed in Figure 3 for Kepler, are those for which the
ages are correct and the radii/luminosities are consistent with
reasonable distance estimates. We don't take seriously any fortuitous
spectral fits with unreasonable model implications.

FEDORENKO: Are the power-law X-ray spectra consistent with the
equilibrium particle distribution functions? What is the mechanism of
radiation?

HOLT: I presume that you are referring to the spectrum of SN1006, for
which we find that a power-law is an adequate fit to our data. It is
certainly possible to synthesize power-law fits from combinations of
equilibrium particle distribution functions (see the early attempt by
Sartori and Morrison 1967 to so reproduce the spectrum of the Crab),
but the consistency of the same power law over two orders of magnitude
in energy is suggestive of a synchrotron origin for the X-ray emission.

FABIAN: Since the electron-ion equilibration time is often greater
than the age of the remnant, how do you know what high energy spectral
form should be fit to the continuum data (e.g. in SN1006)?

HOLT: In the case of Sedov models, we utilize computer-generated
differential spectra instead of limiting analytic forms, which are
dependent upon the usual model parameters (including time) as well as
upon physical assumptions such as whether electron-ion equilibration

can be more rapid than specified by Coulomb interactions alone.
Although a power-law appearance is certainly possible for the superposition of non-equilibrium as well as equilibrium thermal spectra,
the extension of the form to at least 20 keV for SN1006 argues against
it because such a hard spectrum ($\alpha \sim 1$) suggests a blast wave with
much higher velocity than that of the younger and more energetic Cas A,
for example (with $\alpha \sim 3$); the appearance of any photons at all at
20 keV requires a current blast wave velocity of 4000 km s^{-1}, but the
maintenance of the hard spectral form requires that the blast wave
velocity is considerably larger than that. The assumption of a slow
equilibration time only exacerbates the problem, as it is then even
more difficult to produce X-rays with energies comparable to the value
for equilibrium with the blast wave and, hence, more difficult to
produce both a hard spectrum and its extension out to 20 keV.

THE HIGH ENERGY X-RAY SPECTRA OF SUPERNOVA REMNANTS[1]

Steven H. Pravdo
Jet Propulsion Laboratory
California Institute of Technology
    and
John J. Nugent
Downs Laboratory of Physics
California Institute of Technology

ABSTRACT

We present the results of fitting an ionization nonequilibrium (NIE) model to the high energy (> 5 keV) X-ray spectra of the young supernova remnants Cas A and Tycho. As an additional constraint, we demand that the models simultaneously fit lower energy, higher resolution data. For Cas A, a single NIE component can not adequately reproduce the features for the entire X-ray spectrum because 1) the ionization structure of iron ions responsible for the K emission is inconsistent with that of the ions responsible for the lower energy lines, and 2) the flux of the highest energy X-rays is underestimated. The iron K line and the high energy continuum could arise from the same NIE component but the identification of this component with either the blast wave or the ejecta in the "standard" model is difficult. In Tycho, the high energy data rule out a class of models for the lower energy data which have too large a continuum contribution.

1. INTRODUCTION

The high energy X-ray spectrum of a young supernova remnant provides a measure of the X-ray continuum temperature and intensity, the most direct determinations of the electron temperature and emissivity. Also present is the iron K line, the only prominent line with so high an ionization energy. The most sensitive measurements to date of several high energy spectra were performed with the HEAO 1 A2 spectrometers[2] (HED). Previous analyses of the continua have indicated the presence of electron-ion temperature equilibirium (Pravdo and Smith 1979, Nugent 1982), but quantitative analyses of the line emission (Mason et al. 1979, Pravdo et al. 1980, and Winkler 1979) have been hampered by the absence of general theoretical models which go beyond "simple" collisional ionization equilibrium (CIE; e.g. Raymond and Smith 1977). Indeed, the conclusion from many of the above references was that a CIE model was not adequate and that

29

J. Danziger and P. Gorenstein (eds.), Supernova Remnants and their X-Ray Emission, 29–36.
© 1983 by the IAU.

an ionization nonequilibirum (NIE) model such as described by Itoh (1977) was needed.

A number of workers have now rushed to fill this need. We will discuss herein a reanalysis of the Cas A and Tycho spectral data using the NIE model developed by Nugent (1982). A recurring problem with complex models such as these, is that there are too many free parameters to uniquely fit the data. In order to further constrain the class of possible good fits, we have simultaneously fit both the A2 data and lower energy, higher resolution data obtained with the Solid State Spectrometer (SSS) onboard the Einstein Observatory. The SSS data were also previously analyzed in the context of the CIE model (Becker et al. 1979, 1980).

The NIE models described below are in the Sedov limit; i.e., the X-ray emission generated by interstellar material swept up by the blast wave dominates that from the ejecta. However, little generality is lost in this approach since a blast wave component can mimic an ejecta component if the parameters have a free rein (Hamilton and Sarazin, 1982, this Symposium). In addition, electron-ion temperature equilibrium is assumed.

2. PARAMETERS OF THE MODELS

This section will not attempt to describe the NIE model. For this the interested reader must see Nugent (1982). Instead we will list and give mnemonics for the parameters which are discussed in later sections.

Three "reduced" parameters describe the continuum spectrum in the Sedov limit. The first is C_{ev}, which is given in units of 10^{13} cm^{-5}, and can be thought of as a measure of the X-ray surface brightness. The second is η, a density-like parameter, and the last is τ, a time-like or inverse temperature-like parameter. Their product, $\eta\tau$, is proportional to the evolutionary progress of the ionization structure. Higher values of this quantity result in heavy element ionization structures which are closer to CIE, because the density is higher, the time is longer, or both. Similar parameters in the recent NIE model of Hamilton, Sarazin, and Chevalier (1982; HSC) can be obtained through the relations $\eta_{HSC} = 10^{51} \eta^3$ ergs cm^{-6} and T_s (the "post-shock" temperature) = $6 \times 10^7 \tau^{-6/5}$ K. The fitted quantities C_{ev}, η, and τ can be inverted to yield the following familiar quantities (plus an undetermined filling factor f): the initial ambient interstellar density (divided by f); the remnant age (times f); the distance (times f); and the total blast wave energy (times f^2).

Another set of free parameters is the elemental abundances. These are represented as A_Z, the abundance relative to hydrogen of the element with atomic number Z, normalized to the solar values

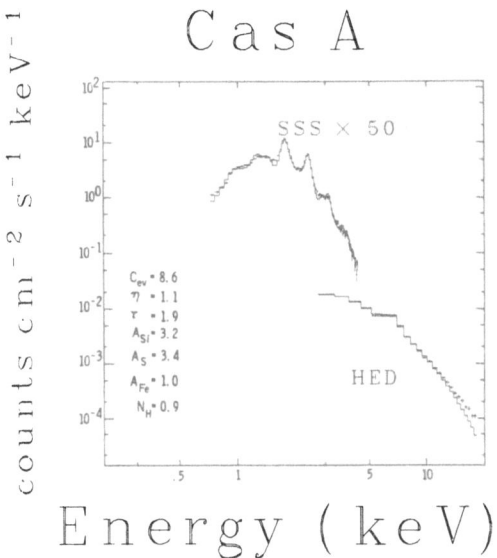

Fig. 1. NIE$_{ALL}$ model histogram.

(Allen 1973). The elements included are Ar+Ca, Mg, Si, S, and Fe. As we are not primarily concerned in this work with determinations of elemental abundances except, perhaps, of iron, we only present the Si and S value as representative of the fits, in addition to the Fe.

A final, somewhat diverse set of free parameters is unrelated to the NIE model. First, is the interstellar, neutral hydrogen column density (Fireman 1974), N_H, in units of 10^{22} cm^{-2}. Next, in some cases, an additional thermal bremmstrahlung continuum component is added. Lastly, the relative normalization between the A2 and SSS data sets is left as a free parameter and afterwards compared to an initial estimate.

Acceptable values of chi squared (χ^2) are never achieved for these fits. This is in part because the systematic uncertainties in the detector response functions can exceed the statistical uncertainties in many energy channels. Moreover the atomic physics used to calculate the relative line strengths is significantly uncertain in some cases. Nevertheless, comparisons between χ^2 values are meaningful.

3. CAS A

Figure 1 shows a histogram of the best-fit NIE model superposed on the spectral data from Cas A. The relative normalization between these A2 and SSS data is expected to be unity since both detectors contain the entire remnant within their respective fields of view. The best-fit value of 1.1 indicates that the detector normalizatins have been adequately evaluated. Three comments can be made about this model fit. First, the 0.7-20. keV spectrum is grossly approximated by a single component model. This is in contrast with CIE models which require two components at different temperatures (e.g., Davison, Culhane, and Mitchell, 1976). Much of this difference is due to the enhanced low energy line emission in the NIE compared to the CIE model.

Second, the high energy continuum emission is underestimated. This reflects the fact that the low energy line emission drives the spectral fitting, and is inconsistent with a high enough continuum "temperature" to match the data.

Last, a major contribution to the χ^2 comes from the Fe K line fit. From a total χ^2 of 1581 for 103 energy channels, the 7 Fe K channels contribute 481. This is illustrated in Figure 2a where this model and the data in the region near 6.8 keV are magnified. The observed line energy is clearly too high to be consistent with the rest of the data and indicates that the iron is more highly ionized than this model would predict.

The fact that both aspects of the high energy data, the continuum and the Fe K line, are inconsistent with the overall best fit suggests an attempt to fit a comparison NIE model to the A2 data alone. In this case the only free parameters are the three NIE parameters, the iron abundance, and a normalization and temperature for a low temperature continuum component. This last component is designed to account for the anticipated excess at the low energy end of the A2 data (near 3 keV). The resultant temperature is 2×10^7 K, an a posteriori confirmation that this component performs the desired function and no more.

The NIE model fit to the A2 data alone (NIE_{Fe}) succeeds where the overall best-fit model (NIE_{ALL}) fails. Figure 2b illustrates the fit to the Fe K line. The contribution to χ^2 from the 7 Fe K channels is now 48, a reduction by a factor of 10. In addition, the high energy continuum is well represented by the model. For this model the NIE parameters are $C_{ev} = 1.5$, $\eta = 7.5$, and $\tau = 0.8$, where the units are as described in Section 2.

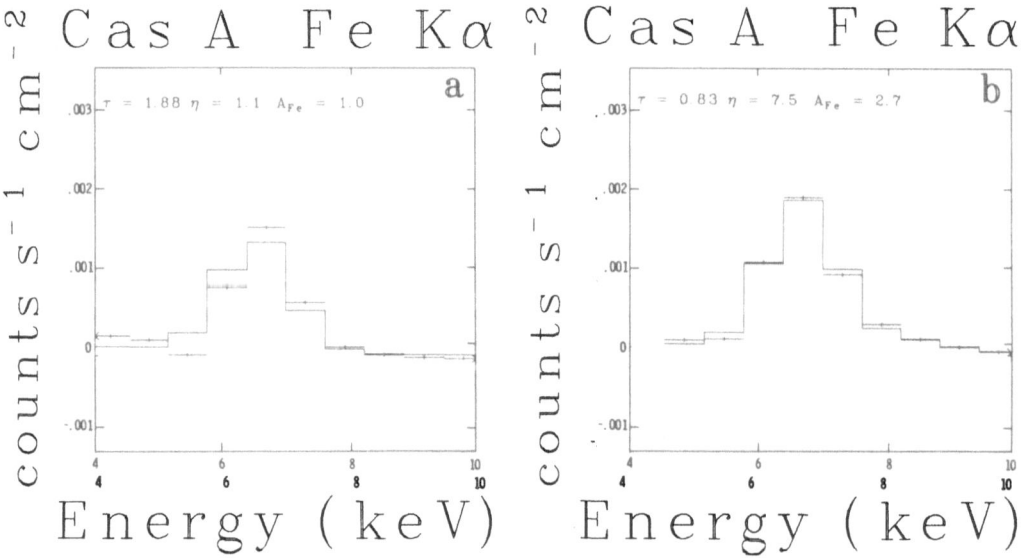

Fig. 2.a. NIE_{ALL} model histogram.   b. NIE_{Fe} model histogram.

A comparison between the NIE_{Fe} and the NIE_{ALL} models is instructive. The product, $\eta\tau$, is three times larger in the former. This indicates that the ionization structure of the iron K line has effectively evolved three times longer than the lower energy lines. The age derived from the NIE_{ALL} model is 190 f years. Since f cannot be larger than unity and the age of Cas A is about 300 years, this model is already in trouble. However, it is evident that the NIE_{Fe} model requies a high filling factor. In contrast, the age derived from the NIE_{Fe} model is 17000 f, implying a filling factor closer to 0.02. Similarly, in order for the derived distances and blast wave energies to approach typical estimates, the NIE_{Fe} model requires a high f and the NIE_{ALL} model requires a low f. The densities derived from these values of f are 11 cm^{-3} (NIE_{ALL}) and 21 cm^{-3} (NIE_{Fe}). The iron abundance is similar in both models and close to the solar value.

The NIE_{Fe} component can produce the Fe K line and the high energy X-ray continuum, and has a low filling factor. Its longer effective ionization time makes it closer to CIE than the NIE_{ALL} model. The latter is more characteristic of the low energy continuum. In the standard interpretation for the Cas A X-ray spectrum the high energy emission is associated with the blast wave and the low energy emission with ejecta. Unfortunately for this scenario the NIE_{Fe} component does not look like a blast wave, which is instead expected to have a large f and a low density. Ejecta is present in high density clumps with a low filling factor but cannot be heated to a high enough temperature to create the highly ionized iron without, for example, significant heat conduction between the blast wave and the ejecta (Chevalier, 1975). A mechanism of this sort appears necessary to create an X-ray component with these properties.

The X-ray spectrum of the supernova remnant RCW 86 (MSH 14-63) exhibits many of the same properties as Cas A, although it is not as extreme a case. Nugent (1982) discusses the NIE analysis of RCW 86.

4. TYCHO

We know from previous results that the high energy X-ray spectrum of Tycho differs from that of Cas A in one important respect: the Fe K line energy is significantly lower (Pravdo et al. 1980).[3] Our NIE modelling for Tycho reflects this fact with the result that there is no significant difference between the effective ionization times of the corresponding NIE_{Fe} and NIE_{ALL} models.

The Tycho data and best-fit NIE_{ALL} model are illustrated in Figure 3. The high energy continuum intensity is adequately reproduced. For Tycho the SSS data must be scaled by a relative normalization factor larger than unity because the remnant is larger than the SSS field of view, but still smaller than the A2 field of view. A maximum of 50% of remnant is in the SSS field of view at

any one time, but the average value from different pointing positions
must be less. The averaged SSS data from two positions is used here.
A simple scaling of the SSS data relative to the A2 data is
approximately correct since there is no significant spectral
variability in SSS spectra from a number of positions in Tycho
(Becker et al. 1980). The best-fit value for the relative
normalization (the inverse of the fraction of the remnant viewed) is
3.5 in good agreement with the estimate.

Using the same procedure as above, this model implies a filling
factor of 0.1–0.3 and a density of 1–3 cm^{-3}. In contrast with Cas A,
the abundances of Si and S are significantly enhanced over the solar
values.

In a recent model of Shull (1982) a fit to the Tycho SSS data
gives Si and S abundances a factor of three lower than these. The
ambiguity of the continuum level in the SSS energy range allows a
tradeoff between the abundances and the continuum. The Shull model
was accurately reproduced by fixing η, τ, and A_z at the published
values and allowing only the scale factor, C_{ev}, to vary. However, an
extrapolation of this model into the A2 energy range reveals that the
model continuum level is, in fact, too high. This is shown in Figure
4. A fixed relative normalization factor of 2.8 is chosen for this
model so that consistency is obtained in the data overlap region of
the two detectors, ~2.5 to 3 keV. If the factor is treated as a free
parameter, an unreasonably low value of 1.6 is obtained and the fit
underestimates the low energy A2 data while still overestimating the
high energy data. This does not constitute a proof that the
elemental abundances must be as high as those derived herein, multi-
component models have more flexibility, but it is a caveat against
seemingly good models which cover a limited energy range.

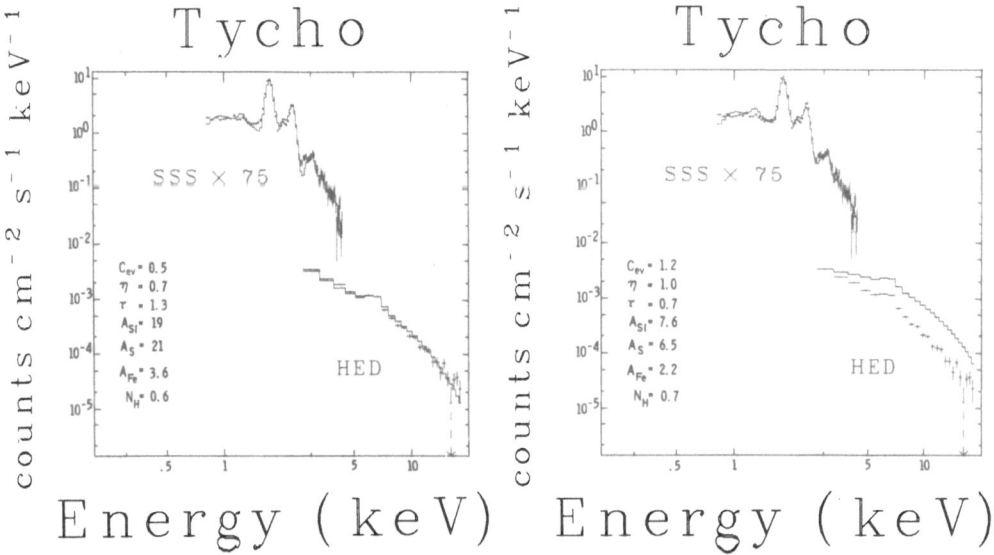

Fig. 3. NIE$_{ALL}$ model histogram.   Fig. 4. Shull (1982) model.

As with Cas A, although the single-component NIE$_{ALL}$ model apporoximately fits the Tycho data, it does not do so in detail. Again, the Fe K line region is poorly fit, although not for the same reason as in Cas A. For Tycho, the model is not inconsistent with the line energy, but rather with the continuum shape near the line. The reason that the iron line in Tycho is at a lower energy than in Cas A is still not clear, but may be related to the lower densitites derived for Tycho.

5. SUMMARY AND CONCLUSIONS

The high energy X-ray spectra of Cas A and Tycho have been reanalyzed in the context of an NIE model for supernova remnant emission. Single-component NIE models cannot reproduce all the features of either spectra. The component responsible for the Fe K line (and perhaps the high energy continuum) in Cas A has some properties which are not consistent with either the standard blast wave or ejecta model. In general, the high energy continuum data is a crucial constraint for a determination of the continuum contribution over the entire X-ray spectrum.

One could proceed by fitting multi-component NIE models to the spectral data and, without doubt, improved fits would be obtained. However, the number of free parameters is then so large that equally good fits with widely varying parameter sets are possible. Unique, and therefore, meaningful, values for the parameters could not be specified. The problem may be that because the detector fields of view are large compared to the remnant size scales, the contemporary data contain contributions from several spectral components. Spectral data from a next-generation, high spatial resolution, broadband detector will be a sufficient challenge for future analysts.

We thank several colleagues at the Symposium for interesting discussions.

[1]The research described in this paper was carried out by the Jet Propulsion Laboratory, under contract with the National Aeronautics and Space Administration.

[2]The A2 experiment on HEAO-1 is a collaborative effort led by E. Boldt of GSFC and G. Garmire of CIT (now PSU), with collaborators at GSFC, CIT, JPL, and UCB.

[3]Please note that in this reference the intensity units on page L10, second paragraph, should be 10^{-10} ergs cm^{-2} s^{-1}.

6. REFERENCES

Allen, C.W., 1973, Astrophysical Quantities (3d ed.; London: Athlone Press).

Becker, R.H., Holt, S.S., Smith, B.W., White, N.E., Boldt, E.A., Mushotzky, R.F., and Seflemitsos, P.J., 1979, Ap. J. (Letters), 234, p. L73.

------. 1980, Ap. J. (Letters), 235, p. L5.

Chevalier, R.A., 1975, Ap. J., 200, p. 698.

Davison, P.J.N., Culhane, J.L., and Mitchell, R.J., 1976, Ap. J. (Letters), 206, L37.

Fireman, E.L., 1974, Ap. J., 187, p. 57.

Hamilton, A.J.S., Sarazin, C.L., and Chevalier, R.A., 1982, preprint.

Mason, K.O., Pravdo, S.H., Charles, P.A., Smith, B.W., and Raymond, J.C., 1979, unpublished spectral analysis of Cas A.

Nugent, J.J., 1982, PhD. Thesis, California Institute of Technology.

Pravdo, S.H., and Smith, B.W., 1979, Ap. J. (Letters), 234, p. L195.

Pravdo, S.H., Smith, B.W., Charles, P.A., and Tuohy, I.R., 1980, Ap. J. (Letters), 235, p. L9.

Raymond J.C., and Smith, B.W., 1977, Ap. J. (Suppl.), 35, p. 419.

Shull, M., 1982, preprint.

Winkler, P.F., 1979, in Proc. HEAO Science Symp., eds, C. Dailey and W. Johnson (NASA CP-2113), p. 244.

STRUCTURAL CHANGES IN CAS A, TYCHO AND OTHER YOUNG SUPERNOVA REMNANTS.

Richard G. Strom
Netherlands Foundation for Radio Astronomy,
Dwingeloo, Netherlands

INTRODUCTION

Structural changes have been found and investigated in a handful of young supernova remnants (SNR). The variations in the Crab Nebula (Trimble, 1968) have been known for some time. Here I am concerned with young shell remnants of which the best studied is Cas A. It is, like the Crab nebula, undergoing an overall radial expansion, although other forms of motion have also been observed. Simple expansion has been found in Tycho (3C10) and the remnant of SN 1006. Optical measurements of the remnant of Kepler's supernova (SN 1604), however, have failed to show motion of the magnitude expected from this relatively young SNR.

PRELIMINARY CONSIDERATIONS

From the following line of reasoning one can infer that motion might be observable in young SNR. A typical member of the class is at least 3´ arc in diameter and has an age of, say, 500 yr. The average expansion speed at the rim is, then, about 100" arc/500 yr = 0."2 arc/yr. If the motion has been undecelerated, this will also be the present speed, a value which can be measured on a time scale of 5-10 yr using modern techniques.

A model applicable to SNR which are strongly interacting with an ambient medium was considered and discussed by Oort (1946). He suggested that nearly circular nebulae like the Cygnus Loop might indeed be remnants of supernovae which are now sweeping up interstellar material like a snow-plow. In this case momentum is conserved, although energy is not (the gas heated at the interaction front is not so hot that radiation losses can be neglected), and it is easy to show that the expansion speed should be one-quarter of the average value.

A third possibility, intermediate to the other two, describes the stage where the kinetic and thermal energies are independently conserved. A similarity solution due to Sedov (1959) treats the case of a spherical shock wave expanding into a uniform medium. The amount of interstellar material encountered much exceeds the ejecta mass, and the shock front moves at two-fifths of the average expansion speed.

J. Danziger and P. Gorenstein (eds.), Supernova Remnants and their X-Ray Emission, 37–48.
© 1983 by the IAU.

Woltjer (1972) considers these phases to represent the first three stages of SNR evolution (in the final stage the remnant loses its identity and disperses). Of particular interest to SNR motion studies is the proportionality constant, η , in the relationship between the expansion speed, v, the remnant radius, r, and its age, t:

$$v = \eta \frac{r}{t}$$

For the phases of SNR evolution outlined above, we expect (Woltjer, 1972):

Phase I (undecelerated motion): $v = \frac{r}{t}$

Phase II (energy conserving): $v = \frac{2}{5}\frac{r}{t}$

Phase III (momentum conserving): $v = \frac{1}{4}\frac{r}{t}$

A primary goal of motion studies is to test these relationships in the light of other information about a given remnant, and in particular determine the value of η.

INDICATORS OF MOTION IN SNR

The evidence for motion in SNR falls into two categories: direct and indirect. In the former a property is measured which directly gives a speed, while the latter involves measurements from which one can infer that different parts of the remnant are in relative motion.

Direct evidence

(a) Determine the displacement of identifiable features. A major advantage of this approach is that a single measurement near the limb of an SNR is sufficient to determine the expansion rate. Furthermore, the fact that shell remnants have a sharp, prominent outer boundary makes them particularly well-suited for such measurements. However, the time required for the SNR boundary to advance by a detectable amount usually dictates considerable patience in the observer. It may, in turn, lead to difficulties if identifiable features change on a similar time scale, which is a major pitfall.
(b) Spectroscopic determination of velocity. The advantage of this technique over a proper motion measurement is that one obtains the velocity instantaneously. However, conclusive evidence for motion requires that different velocities be found in a remnant or that we have at least an indication of its systemic velocity. An unambiguous determination of the expansion speed requires that spectra be obtained from material near the remnant's center in both the approaching and receding hemispheres. For many shell remnants such material is difficult to observe.
It may be worth noting that techniques (a) and (b), if reliably applied to a single SNR, can be used to determine its distance. This has been done by Van den Bergh (1971) for Cas A. Its inherent uncertainty

lies in the tacit assumption of a uniformly expanding spherical shell.

Indirect evidence

(a) Changes in flux density or surface brightness. Such variations can imply compression or expansion of a gaseous component, and hence motion. The best example is probably the adiabatic expansion of the relativistic plasma responsible for radio emission from young SNR, which should, as Shklovsky (1960) correctly predicted, produce a measurable decrease in the flux density of Cas A. The observational confirmation (Högbom and Shakeshaft, 1961) provided early evidence that Cas A probably was expanding. The Shklovsky model assumes that the adiabatic losses dominate and that the relativistic particles have no source of additional energy; even in the case of Cas A it was evident from the start that this is probably not true.
(b) Frequency changes in spectral line features. If it can be shown that such a change occurs in one spatial component (and is not due to, for example, brightness changes in a blend or superposition of features) then it implies acceleration of the gas and thereby motion.
(c) Changes in the radio polarization. Polarized radiation at long wavelengths is, through the Faraday effect, quite sensitive to changes in the magnetoionic medium through which it propagates. The observation of variations in the linearly polarized component may indicate changes in the density of material and hence motion within the remnant.
(d) In general, any temporal change observed in an SNR makes it a candidate for motion studies.

MOTION IN CAS A

The wealth of observational information which has been collected on Cas A is largely due to the efforts of Sidney van den Bergh and his coworkers over many years. Starting with plates taken by Baade they have obtained data of high uniform quality spanning well over a quarter of a century. The existence of this optical treasure-trove has no doubt helped spur researchers in other wavelength bands to peruse their data for corresponding changes.

Optical studies

The situation through the mid-1970s has been summarized by Kamper and Van den Bergh (1976). From the standpoint of motion in Cas A, there are two populations of filaments: the fast-moving knots, and the quasi-stationary flocculi. The former indicate an expansion lifetime for the remnant of some 300 yr, while the latter suggest a dynamical age some 35 times larger. In their analysis of the fast-moving knots, Kamper and Van den Bergh find a close correlation between position and velocity. This smooth linear relationship means, they argue, that there has been essentially no deceleration of this component, although there are small systematic differences between groups of filaments such as the "flare" to the northeast, which is moving faster than the rest.
    In their latest analysis, Kamper and Van den Bergh (reported

elsewhere in this volume) find that the longest lived knots, if their motions are linearly extrapolated back in time, yield a date for the supernova explosion of 1658+3 yr. The finite lifetimes of the knots - both increases and decreases in brightness have been observed, and the changes are consistent with an e-folding time of 25 yr - provide a not insignificant obstacle to analyses of the dynamics. Brightness and shape changes in individual filaments certainly can be a source of error.

Radio Studies

As with the optical filaments, radio knots in Cas A also exhibit substantial brightness changes over similar time scales. Furthermore, radio velocity determinations which different investigators using a variety of instruments have made disagree as to whether overall expansion is present, although they all find evidence for motion.

Before considering the attempts to observe expansion in Cas A it may be useful to remember that the steady decline observed in flux density (Högbom and Shakeshaft, 1961), which predates the synthesis maps, provides strong circumstantial evidence that the remnant is indeed expanding in accordance with Shklovsky's (1960) model calculation. In addition, reports of a low frequency "flare" (Erickson and Perley, 1975; Read, 1977a, 1977b) with a time scale of 2-6 yr suggest activity, if not motion, in Cas A.

Bell (1977) used maps made five years apart with the Cambridge One-Mile and 5-km telescopes to observe changes in some thirty compact knots at a frequency of 5 GHz. He found evidence for expansion although there was large scatter and a substantial degree of tangential motion.

Shortly thereafter, Dickel and Greisen (1979) using 2.7 GHz observations made with the NRAO three element interferometer in 1967 and 1976, were unable to find evidence for systematic expansion in the pattern of essentially random motions they observed. The result was, however, not fully inconsistent with Bell's determination considering the formal errors. Dickel and Greisen suggest that the discrepancy may be (partially) attributed to brightness changes in blends of multiple knots which give rise to apparent motion if the angular and temporal sampling are not propitious for resolving this effect.

Various criticisms can be levelled at both sets of measurements. The most significant are that Bell compared data obtained from two different instruments, while Dickel and Greisen used measurements which did not sample the visibility plane well. Of these, the latter may be the most telling. Nevertheless, it is worth noting that both studies agree that the fine scale structure in Cas A is in motion, and that it has a substantial, if not dominant, random component.

The most extensive and determined effort to study motion in Cas A, which avoids both of these deficiencies, has now been carried out by Tuffs (reported elsewhere in this volume). Using 5 GHz observations made with the Cambridge 5-km telescope in 1974 and 1978 he has determined the motion of nearly 350 compact features as well as several extended ridges. The main result is that an overall pattern of expansion is clearly present despite a very considerable random component, confirming Bell's main conclusion. The mean radial expansion speed for all measured compact features corresponds to an expansion "age" (ignoring deceleration) of 950 yr. The upper envelope in a radial velocity vs. distance from center diagram – the expansion of the fastest radio features – is in reasonable agreement with the motion observed in the system of fast moving optical filaments. This suggests that if the observed motion of radio features corresponds to material transport, we are witnessing acceleration and/or deceleration in Cas A on a large scale. The implications are discussed in greater detail in Tuffs' contribution.

The episode would thus appear to be neatly concluded and the dispute settled, were it not for observations recently carried out with the VLA. Angerhofer and Perley (1982) have begun a program to monitor changes in Cas A. In their first results, based on a comparison of observations separated by 21 months, they have measured the displacement of some 120 radio knots. They find radial and tangential motions of about the same magnitude, but observe no pattern of overall expansion, a conclusion which appears to be in qualitative agreement with Dickel and Greisen, but not with Bell or Tuffs.

A definitive resolution of this discrepancy will probably be forthcoming only if the different groups are able to compare their measurements of the same radio features. Such a comparison cannot be done in more than a limited way with the existing data which were obtained at different (and not necessarily overlapping) times over intervals ranging from less than 2 to more than 9 yr. The pair of American determinations suffer from an undersampling of the visibility plane, and as a result they observe only a fraction of the emission from Cas A. Considering both this effect and the difference in epoch, it is not too much of an exaggeration to maintain that the Cambridge and American groups have not actually been studying the same thing. This is surely the best place to start searching for the cause of the discrepancy.

Though the problem may not be resolved, it is clear from all the radio studies that the radio knots do not generally share the motion of the fast moving optical filaments. A detailed radio-optical comparison would nevertheless be fruitful, even if it were to do no more than demonstrate that none of the features is spatially coincident. With the resolution which synthesis telescopes now achieve such a comparison is feasible.

Ultimately, we will want to know the relationship of the radio and optical emission to the shock front(s) which bound Cas A, for it is only the velocity of the latter which we can simply apply to the equations of motion. It seems a safe bet to assume that the shock is moving at least as fast as the fastest filaments, but an unambiguous determination of shock velocity must await an X-ray proper motion study. Thus far, the only changes which have apparently been observed in the X-ray are possible brightness variations (Murray et al., elsewhere in this volume). An improvement in this situation will probably have to await an X-ray observatory with high resolution imaging capability which remains in orbit for at least five years.

MOTION IN THE REMNANT OF SN 1572 (TYCHO)

The situation in Tycho's remnant (3C10) is refreshing in its simplicity
when contrasted with the complexity observed in Cas A. This difference
may not be too surprising for although both objects are young shell SNR,
they differ significantly in many radio and optical properties. In fact,
both optical (Kamper and Van den Bergh, 1978) and radio (Strom et al.,
1982) measurements show clear expansion in 3C10 and are in good
agreement.

Optical expansion

The optical determination is based upon a series of Hale 5 m telescope
plates beginning with one obtained by Baade in 1949. Kamper and Van den
Bergh (1978) have analyzed observations spanning 28 years. Although the
filaments are quite faint, they do lie near the limb so their speed
should directly give the rate of expansion. Motion can, in fact, be
readily seen in the brightest filaments along the eastern rim. The
analysis involved determining the radial position (with respect to the
center of 3C10) of the nebulosity at locations where it is sufficiently
sharp, from which the weighted mean was taken. The solution is thus
strongly influenced by the brightest filaments, a point which I will
return to shortly.
     The average expansion speed found is 0."21 arc/yr. This agrees well
with the value predicted by the Sedov solution for a 400 yr old shell
provided the shock radius can be ascertained from the size of the radio
remnant. Such an assumption is necessary because the nebulosity is so
irregular, and is distributed along less than half of the limb, that an
optical determination of the radius would be quite uncertain. In the
same vein, it is interesting to note that the radio map (Duin and Strom,
1975) was also used to determine the geometrical center which was an
essential reference point for intercomparing the optical plate material.
In several respects, then, radio data played a significant role in
interpreting the optical measurements.

Radio expansion.

A prime reason for carrying out a radio study of motion in 3C10 is the
symmetry and hence apparent completeness of the shell, and the strength
and sharpness of the outer rim. While this contrasts favorably with the
optical situation, a disadvantage is the lower radio angular resolution.
Nevertheless, Strom et al. (1982) have determined the expansion rate at
regular intervals around the limb of 3C10 from 1.4 GHz observations made
eight years apart with the Westerbork Synthesis Radio Telescope (WSRT).
The result of these measurements is summarised in Fig. 1, where the
optical determinations are also shown for comparison.
     It is clear from Fig. 1 that the optical and radio measurements are
in good agreement, and that several values in the east are significantly
low. Ignoring the lowest radio point, one obtains an average expansion
speed of 0."26 arc/yr. This is higher than the optical value of 0."21
arc/yr because of the low points in the east, suggesting that the

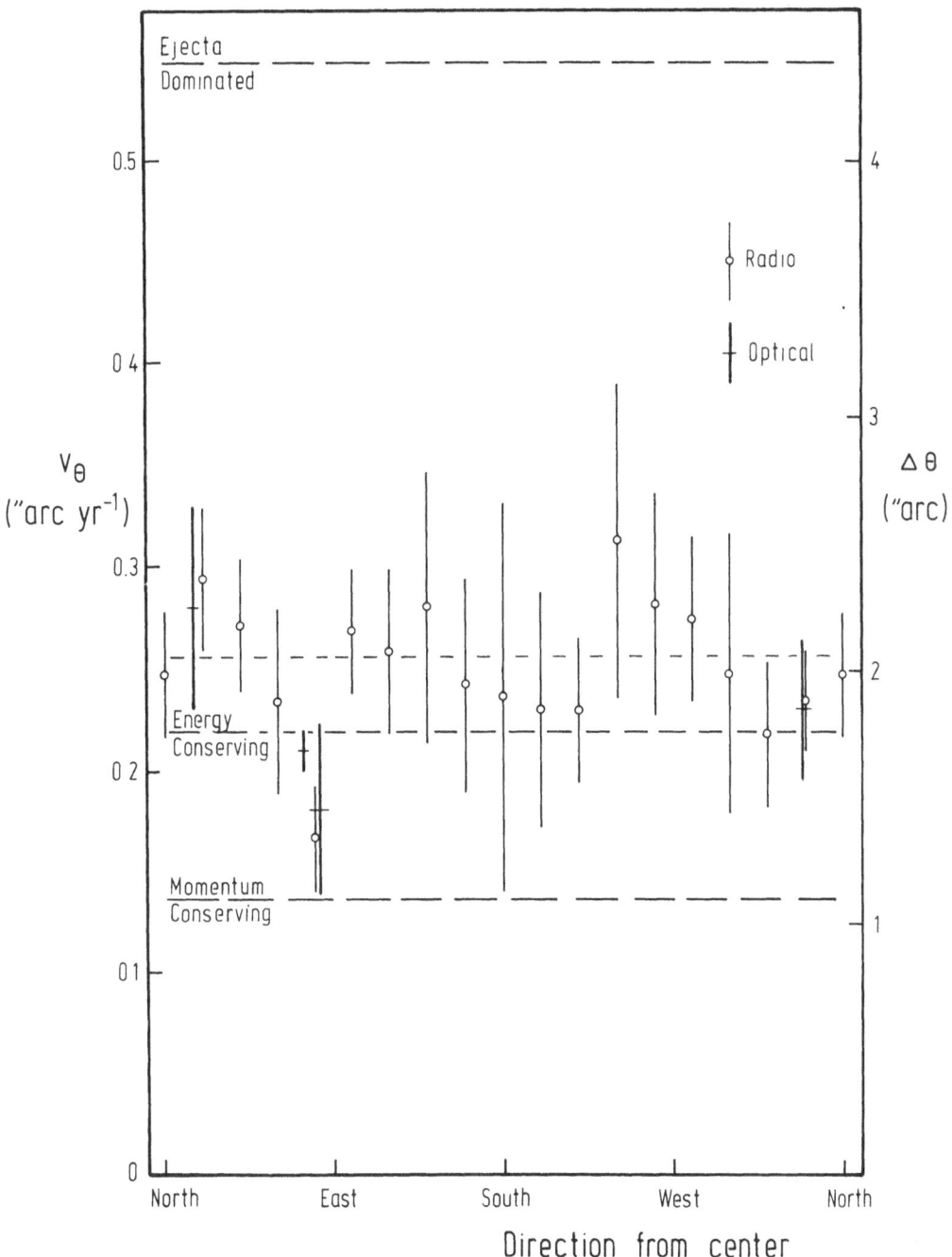

Figure 1. The expansion speed measured around the periphery of Tycho. Short dashes show the average radio value, excluding the anomalous point in the east.

expansion speed of the brighter nebulosity may be anomalously low. This, in turn, has implications both for optical determinations of SNR expansion speeds and for interpreting the dynamics in Tycho.

Considering this last point briefly, it is probably significant that this eastern region of brighter filaments and low expansion speed coincides with a prominent dent in the otherwise nearly circular radio shell (Strom et al., 1982). The most likely explanation is that we are witnessing an encounter between the shock front and a region of higher than average density, perhaps a discrete interstellar cloud. It will be interesting to see whether other changes can be observed here in coming years.

Further analysis of the existing 1.4 GHz data, although somewhat preliminary, has turned up additional information. New maps have been made in which the longest baselines are not weighted down, so as to obtain higher angular resolution (at the cost of an increased sidelobe level). They show that discrete emission peaks in the interior also participate in the overall expansion, with a speed which appears to increase with distance from the center. Furthermore the polarization maps, while generally in good agreement, show several areas where the position angles appear to have changed substantially over eight years. This work, including a comparison of two 5 GHz maps made with the WSRT, is continuing.

We must now consider how the radio/optical expansion relates to the shock dynamics. In view of the compatibility between the two sets of measurements the conclusions we draw should be less ambiguous than was the case for Cas. Nevertheless, we need to know something about the location of the shock front. In a recent preprint, Seward, Gorenstein and Tucker (see also Gorenstein et al., elsewhere in this volume) present and discuss a high resolution X-ray map of Tycho. The shape of the outer boundary bears a remarkable similarity to the radio map, and a detailed comparison shows that the two images are nearly congruent. This degree of coincidence suggests that the relativistic particles are largely accelerated in, or directly behind, the shock front. If this is correct, the radio expansion measurements of the outer edge should faithfully reflect the shock speed.

The technique used for determining the radio expansion (Strom et al., 1982) is most sensitive to the steepest gradient. As this occurs along the outer boundary, it should ensure that the measurement essentially refers to the motion of that boundary, and hence of the shock. (It should be pointed out that the radio measurements compare the positions of the most recently accelerated particles, and the "motion" observed does not necessarily reflect flow of the relativistic plasma, although it does follow the progression of the shock.) The sharpest optical filaments are found at the very edge of the radio remnant, so it is not surprising that they share the radio expansion observed.

The expected expansion rate for the three phases of SNR evolution – the undecelerated, adiabatic and momentum conserving stages – are indicated on Fig. 1. The observations are clearly only consistent with the adiabatic or energy conserving phase, and Strom et al. (1982), using the calculations of Rosenberg and Scheuer (1973), estimate that the swept-up material exceeds the ejecta mass by a factor of 8. This may

conflict with the mass ratio of 2 which Seward et al. obtain from their
X-ray data, although the uncertainties in that estimate could be as much
as a factor 2 in each component.

The ratio of instantaneous to average speed found by Strom et al.,
$\eta = 0.47$, is larger than the value of 0.4 predicted from the Sedov
solution by a marginally significant amount. If the effect is real, it
may mean, as suggested by the mass ratio estimate mentioned above, that
the amount of swept-up material is not quite sufficient to bring Tycho
fully into the energy conserving phase. There are, however, other
possibilities.

Chevalier (1982) has shown how a power law density profile for the
outer atmosphere of the supernova precursor will modify the value of η
in a Sedov-like solution. For a Type I supernova like Tycho he suggests
a value somewhat in excess of 0.4. Another possible modification to the
dynamics would be caused by the evaporation of dense cloudlets behind
the SNR shock front (McKee and Ostriker, 1977). This gives a value of
between 0.4 and 0.6, although the number of evaporating cloudlets
encountered by a young remnant may be too small to have much effect.

MOTION IN OTHER YOUNG SHELL REMNANTS

For two other shell remnants - those of SN 1006 and SN 1604 (Kepler) -
the association with a historical supernova is convincing. Both have
been the subject of optical motion studies with markedly different
results. Van den Bergh and Kamper (1977) find very little evidence for
expansion in Kepler, the velocities being much smaller than expected and
not conforming to an overall pattern. They suggest a dynamic age which,
as in the case of the quasi-stationary flocculi in Cas A, is a great
deal larger than the probable age of the remnant.

In the remnant of SN 1006, Hesser and Van den Bergh (1981) have
measured the speed of sharp filaments along the limb. Their result gives
$\eta = 0.47$, suggesting that this remnant is also in the Sedov phase. In
neither SN 1006 or Kepler has it been possible to carry out radio motion
studies. Their southern declinations have so far made them inaccessible
to high resolution synthesis telescopes, all of which are sited in the
northern hemisphere.

CONCLUSIONS

It should be clear from this brief review that attempts to measure and
interpret motion in young SNR have achieved a considerable degree of
success. An optimist might be tempted to emphasize that in two of the
objects, the remnants of SN 1006 and SN 1572, effective agreement is
found between the measured expansion rate and the Sedov prediction. He
might also point out that nothing in the X-ray data suggests that either
remnant is in anything other than the adiabatic phase, and in Tycho the
optical and radio determinations of expansion agree well.

Regrettably, I must warn against such a sanguine attitude. Even
attributing the deviations in average expansion speed from the expected
Sedov behavior to measuring uncertainties, we are still faced with real
velocity differences within Tycho. The mechanisms for producing the

radio and optical emission are so different that we really do not know
whether they should display the same (apparent) motion, and if they do
we can only guess why. Moreover, the expansion observed in SN 1006 is
solely based on optical observations of nebulosity in one limited area;
we have no way of knowing to what extent this is representative of the
global motion.

In point of fact of the four young shell remnants in which motion
studies have been attempted, one produces a confusing picture of both
moving and nearly stationary features, one shows practically no motion
at all, while the other two are roughly doing what we expect. A
pessimist could be forgiven for saying that we have almost as many types
of motion as objects studied, a dismaying state of affairs. How might
this be rectified?

On the observational side I think that both the Cas A and
especially Tycho studies demonstrate the value of radio expansion
determinations. Kepler and SN 1006 should be the subject of similar
measurements - both can be observed with the VLA. Work on Cas A and
Tycho should continue, in coordination where possible with new optical
measurements; for Cas A this is particularly important. Where feasible
more velocities via spectroscopy ought to be obtained to enable
exploration of the velocity field. Improved instrumentation and the
Space Telescope should create new opportunities here. It goes almost
without saying that X-ray motion studies are sorely needed, but they are
probably some years in the future.

On the theoretical side there are at least as many problems to be
tackled. Where (not to say how) does particle acceleration occur, and
how should the synchrotron volume emissivity vary with distance from the
shock? Where does the optically emitting material lie with respect to
the shock? Young shell remnants appear to be in either the undecelerated
or energy conserving phase. We need more work on the transition between
these two evolutionary stages. Finally, observers in particular are keen
to know which measurable quantities provide a crucial test for
theoretical models.

REFERENCES

Angerhofer, P.E., Perley, R.A., 1982. Paper presented at the AAS Boulder
        meeting; see also Bull. AAS 13, 841, 1981.
Bell, A.R., 1977. Mon. Not. Roy. Astr. Soc. 179, 573.
Bergh, S. van den, 1971. Astrophys. J. 165, 457.
Bergh, S. van den, Kamper, K., 1977. Astrophys. J. 218, 617.
Chevalier, R.A., 1982. Astrophys. J. 258, 790.
Dickel, J.R., Greisen, E.W., 1979. Astron. Astrophys. 75, 44.
Duin, R.M., Strom, R.G., 1975. Astron. Astrophys. 39, 33.
Erickson, W.C., Perley, R.A., 1975. Astron. J. 200, L83.
Hesser, J.E., Bergh, S. van den, 1981. Astrophys. J. 251, 549.
Högbom, J.A., Shakeshaft, J.R., 1961. Nature 189, 561.
Kamper, K., Bergh, S. van den, 1976. Astrophys. J. Suppl. 32, 351.
Kamper, K.W., Bergh, S. van den, 1978. Astrophys. J. 224, 851.
McKee, C.F., Ostriker, J.P., 1977. Astrophys. J. 218, 148.
Oort, J.H., 1946. Mon. Not. Roy. Astr. Soc. 106, 159.

Read, P.L., 1977a. Mon. Not. Roy. Astr. Soc. 178, 259.
Read, P.L., 1977b. Mon. Not. Roy. Astr. Soc. 181, 63P.
Rosenberg, I., Scheuer, P.A.G., 1973. Mon. Not. Roy. Astr. Soc. 161, 27.
Sedov, L.I., 1959. Similarity and Dimensional Methods in Mechanics,
    Academic Press, New York.
Shklovsky, I.S., 1960. Astr. Zh. 37, 256.
Strom, R.G., Goss, W.M., Shaver, P.A., 1982. Mon. Not. Roy. Astr. Soc.
    200, 473.
Trimble, V.L., 1968. Astron. J. 73, 535.
Woltjer, L., 1972. Ann. Rev. Astron. Astrophys. 10, 129.

DISCUSSION

WOLTJER: What can you say about changes in the total flux and in the
polarization of Tycho?

STROM: As reported in our paper on the 1.4 GHz expansion (Strom et
al., 1982), we find an insignificant decrease of $-0.23 + 0.19$ per cent/yr
in the flux density of Tycho. The Shklovsky model based on adiabatic
expansion and magnetic flux conservation predicts a decrease of -0.49
per cent/yr. Our results are thus marginally inconsistent with a
situation in which expansion losses dominate.
        Turning to the polarization changes, our preliminary results
show significant position angle differences in a few regions although
the two maps are generally in good agreement. The most substantial
changes are found in the northeast quadrant, approximately where the
brightest nebulosity occurs. A rough calculation suggests a change in
the product of electron density, magnetic field strength and path length
of at most about 20 cm^{-3} G pc. This is comparable to the changes
between adjacent regions found by Duin and Strom (1975) in their
analysis of cellular structure observed in the polarization
distribution.

McKEE: Theories of relativistic particle acceleration by a first
order Fermi process in shocks suggest a precursor of relativistic
particles extending ahead of the shock. Electrons in such a precursor
could radiate in the interstellar magnetic field. What limits can be set
on the intensity of such emission around bright radio-emitting SNRs?

STROM: The best limits to halo emission around the brighter remnants
are, as far as I am aware, about 1% of the peak brightness, although
Wilson and Weiler have recently set better limits to extended emission
around the Crab Nebula. Using the WSRT in redundancy mode and applying
closure phase we can now improve upon this by a factor of between 10 and
100. For a bright remnant such as Cas A we should be able to detect
emission with a surface brightness of 2-10 mJy/beam (for a 12" arc
synthesized beam at 21 cm). For a weaker remnant such as Tycho the
measurement would be noise limited to a few tenths of a mJy/beam
(depending on the amount of observing time invested).

TUFFS: There have now been four measurements of the radio expansion of Cas A. The Cambridge results, by Bell and myself show that the remnant is expanding with a time scale of 1000 yrs, whereas the results of Dickel, with the NRAO interferometer, and more recently Angerhofer & Perley, with the VLA show only random motions for the radio morphology. I should like to point out that there are significant differences between the sampling in the aperture plane of the various surveys which points to the Cambridge results being more reliable. Both my observations, and those of Bell, are well oversampled in the aperture plane, and both Cambridge 5 km observations had exactly the same telescope configuration. The NRAO observations also had the same telescope configuration for both epochs, but were severely undersampled in the aperture plane, thus necessitating the use of clean. Clean is a somewhat risky technique to use on Cas A, as the broad scale structure dominates the visibility function. In fact the compact features on 5 km maps contribute only 1/40 of the total flux of Cas, and extreme care is obviously necessary when cleaning the map. This is particularly relevant to the measurement of proper motions when the grid point separation on the map is much greater than the proper motions, as is the case for Cas A.

I have also measured the proper motion of extended shell features in Cas A (up to 1 arc minute in size) and these show the same overall expansion as for the compact features, with a true time scale of ~1000 yr. There is no question, therefore that the proper motion is dependent on the morphology of the radio features; it is not.

DICKEL: I think the major difference between the NRAO and Cambridge results could be, as suggested by Strom, that the ones taken at NRAO (by Greisen and myself with the 3 element interferometer and also with the VLA by Angerhofer and Perley) do miss the short spacings responsible for the diffuse component and so we see only the small scale features. The Cambridge ones include the diffuse component which is probably 2/3 of the total flux density.

The VLA data are well sampled and so can be further processed by "clean" plus other algorithms and the 3 element interferometer data show the same results for the original or "dirty" results.

GULL: I would be intrigued to know the 5 GHz flux of the point source at the centre of 3C10 (it is 4.0 mJy at 2.7 GHz).

STROM: In our recent 5 GHz map we find a flux density for this object of about 2.7 mJy, so there seems to be nothing remarkable about its spectrum.

THE EXPANSION OF CASSIOPEIA A

R.J.Tuffs
Mullard Radio Astronomy Observatory, Cavendish Laboratory,
Cambridge, United Kingdom.

INTRODUCTION

This paper describes a detailed comparison of radio maps of Cas A
obtained with the Cambridge 5 km Radio Telescope in 1974 and 1978.
Accurate proper motions and brightness changes are presented for 342
distinct radio peaks, ranging in angular size from 1 arc minute for
fragments of the intense ring to less than the resolution limit of
2.0×2.3 arc seconds for the compact radio knots.

OBSERVATIONS

Cas A was observed at 5 GHz with the Cambridge 5 km Telescope (Ryle 1972)
during September – December 1974 and October – December 1978. Identical
interferometer baselines were used for both epochs. The spacing increment
is 35.74 m, providing a grating ring radius (5.8 arcmin in RA) which
exceeds the maximum angular size of Cas A (5.4 arcmin). No polarisation
data was taken in 1974 so that the present work is restricted to a
comparison of maps of I–Q. This is not a serious limitation since the
percentage polarisation at 5 GHz mapped across Cas A with the full
resolution of the instrument never rises above ~ 5 percent.

Amplitude and phase corrections were applied to each spacing
according to the procedure outlined by Ryle & Elsmore (1973). Astrometric
uncertainties in proper motions (due principally to tropospheric phase
fluctuations and short term temperature dependent variations in telescope
geometry) are smaller than 0.02 arcsec yr^{-1}. A greyscale representation of
the 1978 observations is presented in fig. 3.

PROPER MOTIONS

The change in position of each peak was measured using a computer program
which shifted one map with respect to the other until the square of the
difference in each pixel summed over a region containing the peak had
been minimised. Proper motions of all 342 radio features are displayed
in fig.1 as arrows with length proportional to the protected velocity.

49

J. Danziger and P. Gorenstein (eds.), Supernova Remnants and their X-Ray Emission, 49–54.
© *1983 by the IAU.*

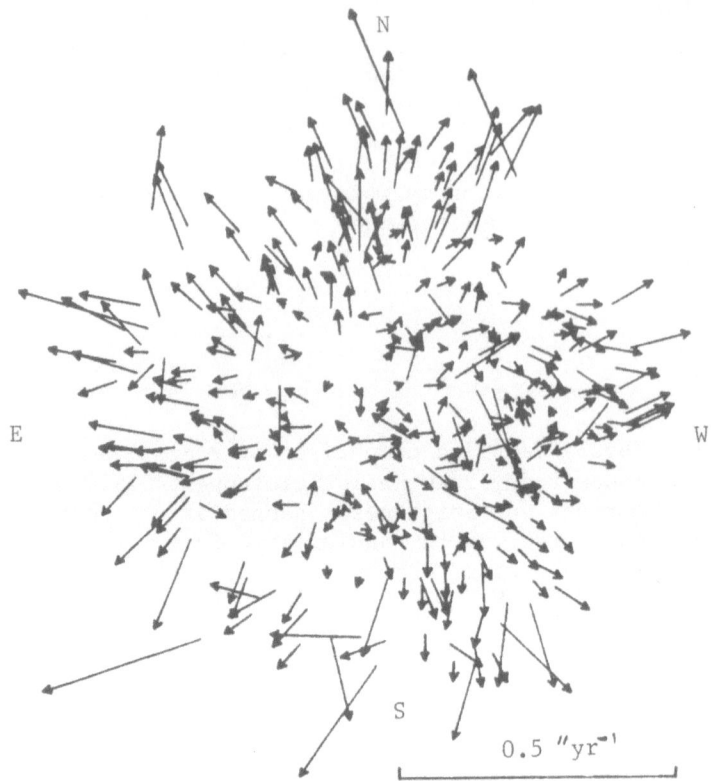

Figure 1. Proper motions of 342 radio features in Cas A.

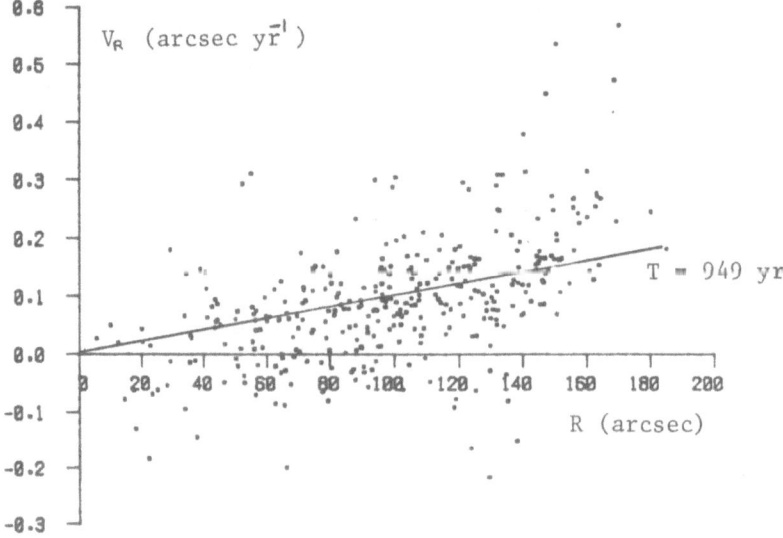

Figure 2.   Radial component of projected velocity vs
            distance from the optical expansion centre.

There is an unmistakable expansion of the remnant. Outwards motion is
shown both by broad scale structure such as the intense ring and by
compact structure such as the bright knots. The expansion timescale is
independent of radio morphology. The centre of expansion of the radio
features is at $\alpha = 23^h 21^m 10\overset{s}{.}2$ ($\pm 0\overset{s}{.}7$), $\delta = 58° 32' 19.7''(\pm 4.4'')$, offset by
1.8" N and 13.8" W from the optical expansion centre (OEC) for the system
of fast filaments (Kamper & van den Bergh 1976).

The variation in projected velocity outwards from the OEC with
projected radius from the OEC is shown in fig. 2. A best straight line
fit to fig. 2 constrained to pass through the OEC has an expansion
timescale of 949 yr with upper and lower one sigma limits (determined by
the intrinsic spread in the data) of 993 and 908 yr. This radio expansion
age is some three times that of the undecelerated optical filaments
(Kamper & van den Bergh 1976) and is in good agreement with Bell's
earlier comparison of observations made in 1969 and 1974 (Bell 1977). It
differs however from the measurements of Dickel & Greisen (1979) who,
using the NRAO interferometer at 2.7 GHz, failed to detect any systematic
outwards motion of radio features over a 9 yr interval. It is not clear
whether the results obtained with the NRAO interferometer are inconsistent
with the present work since this instrument did not have sufficient
coverage of the aperture plane to oversample the visibility function of
Cas A and the effect of such an undersampling on the proper motions is
difficult to predict. In the present work the proximity of the radio
expansion centre to the centre of the intense radio ring and its 25 arcsec
offset from the map centre suggests that the expansion is real and not
due to some unforseen scaling error in one or other of the maps about
the map centre.

Seemingly random motions are superimposed on the general expansion.
Most peaks have substantial tangential components of projected velocity
and several peaks, including the brightest, have components of proper
motion directed towards the OEC. Furthermore there are regions of the
remnant (particularly in the Eastern plateau) in which proper motions
appear to be mutually aligned. These random motions are not accounted
for by measurement errors (due mainly to random brightness variations in
fore and background sources in the remnant) but it is not yet clear
whether shape changes within the peaks are wholly responsible since the
changes in position are still smaller than the beam. Random motions are
generally independent of the size, shape, flux and change in intensity
of the radio features. The rms values for random motions in radial and
tangential directions to the OEC (with measurement errors subtracted) are
0.085 arcsec yr^{-1} and 0.053 arcsec yr^{-1} respectively.

Bell's peak 38, which is almost certainly associated with a quasi-
stationary flocculus (Chevalier & Kirshner 1979) has proper motions in
right ascension and declination of 0.01 ± 0.01 arcsec yr^{-1} and -0.11 ± 0.01
arcsec yr^{-1} respectively. The intensity of this unresolved radio peak has
increased from 174 mJy in 1974 to 218 mJy in 1978.

It should be emphasised that it is by no means clear how the radio

expansion of Cas A may be related to a physically meaningful parameter
such as an outer shock velocity. In contrast to the highly limb-brightened
remnant of Tycho's supernova, whose radio luminosity must have a
substantial contribution from emission associated with the outer shock,
the plateau and compact features in Cas A are distributed over a large
range of deprojected radii so that only a small proportion of the radio
emission may be directly associated with an outer shock. The overall
expansion of the remnant nevertheless suggests that the velocity of the
radio features deep inside Cas A is related in some way to the velocity
of the outflowing matter between back and front shocks. One dimensional
hydrodynamical simulations of young supernova remnants evolving away from
the free expansion phase (eg Gull 1973) predict a steep internal velocity
profile behind the outer shock front. If such simple models may
legitimately be applied to a source as complex as Cas A the average of
949 yr will be an <u>overestimate</u> of the deprojected expansion age of the
radio emission at the periphery of the remnant. Some evidence for this is
provided by the fact that features in the plateau have an average
expansion age of only 750 ± 50 yr.

BRIGHTNESS CHANGES

A subtraction of the 1974 map from the 1978 map (fig. 4 - hereafter called
the difference map) shows the supernova remnant to be divided into regions
of either predominantly increasing or decreasing brightness. A large area
of increasing brightness extends from the NW into the plateau in the SW
and contains the majority of the brightest compact peaks. The main broken
ring of radio emission is decreasing in brightness at 1.32 percent yr^{-1}
and accounts for half the secular decrease in flux of Cas A at 5 GHz
(Baars et al 1977). The remainder of the secular decay is due to the
plateau and its extensions across the face of the remnant.

The difference map shows that broad regions of increasing flux
include most of the brightening compact radio peaks and similarly
decreasing broad scale radio brightness is associated with mainly fading
compact features. This gives credibility to the extended emission on the
difference map and suggests that the broad scale structure in Cas A may
in fact consist of an aggregation of compact features of similar nature
to the conspicuous bright compact knots but of much lower power.

Proper motions of many of the compact features and of parts of the
intense ring are prominent on the difference map as adjacent patches of
positive and negative, separated by the angular size of the feature in
the direction of its motion. Twice as many compact peaks are decaying as
are brightening but although most are resolved there is no evidence that
the decaying features are increasing in size- The average decay rate for
all compact features is 1.57 percent yr but the total integrated power
of a complete sample of compact features with peak fluxes greater than
the average surface brightness of the remnant is insufficient for these
features to make a significant contribution to the overall secular
decay of Cas A.

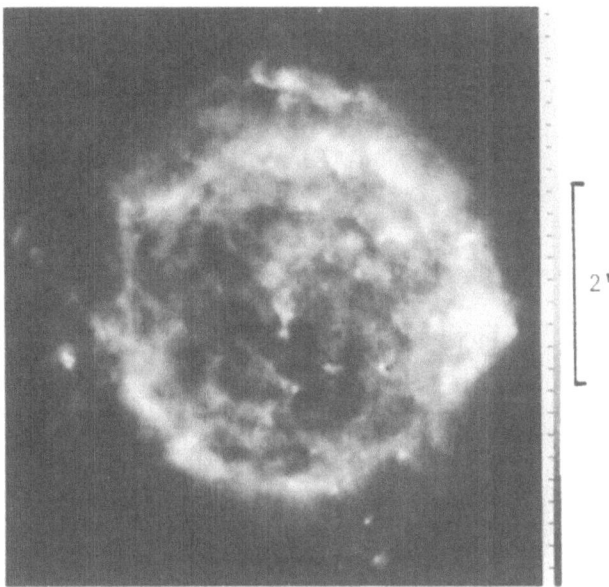

Figure 3. Cas A at 5 GHz in 1978.
Range (black to white) is 0 to 150 mJy beam^{-1}.

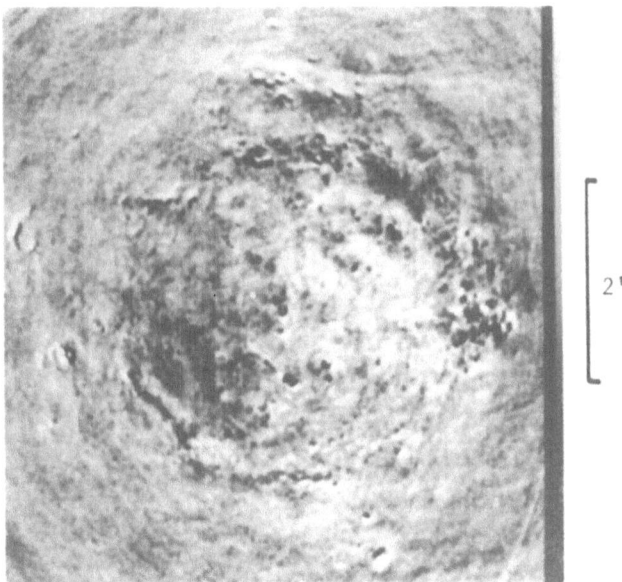

Figure 4. Secular changes within Cas A at 5 GHz.
Range (black to white) is -6 to +6 mJy yr beam^{-1}.

ACKNOWLEDGEMENTS

I am grateful to Drs Kenderdine and Pooley and the staff of the 5 km
Telescope for making the observations of Cas A, to Dr. Kenderdine for
assistance in calibrating the data, to Drs Scheuer and Gull for helpful
discussions, and to Sidney Sussex College, Cambridge and the IAU for
financial support which enabled me to attend the Symposium.   I thank
the SERC for a Research Studentship.

REFERENCES

Baars, J.W.M., Genzel, R., Pauliny-Toth, I.I.K., Witzel, A., 1977.
     Astron. Astrophys., 61, 99.
Bell, A.R., 1977.   Mon. Not. R. astr. Soc., 179, 573.
Chevalier, R.A. & Kirshner, R.P., 1979.   Astrophys. J., 233, 154.
Dickel, J.R. & Greisen, E.W., 1979.   Astron. Astrophys., 75, 44.
Gull, S.F., 1973.   Mon. Not. R. astr. Soc., 161, 47.
Kamper, K. & van den Bergh, S., 1976.   Astrophys. J. Suppl., 32, 351.
Ryle, M., 1972.   Nature, 239, 435.
Ryle, M. & Elsmore, B., 1973.   Mon. Not. R. astr. Soc., 164, 223.

RECENT OBSERVATIONS OF CASSIOPEIA A

K. W. Kamper
David Dunlap Observatory
S. van den Bergh[1]
Dominion Astrophysical Observatory

During the last few years radio and x-ray astronomers have produced high-resolution imagery of the remnant of Cas A. Since the most recent published optical photographs of Cas A date back to 1975 (Kamper and van den Bergh 1976 a, b) it seemed worthwhile to present new optical results based on plates obtained with the 5-m Hale telescope in 1976, 1977 and 1980. A photograph taken in 1980 is shown in Fig. 1.

The major changes that have taken place in Cas A during the last decade are: (1) A broken shell of fast-moving knots has formed along the southern and SW rim of the remnant and (2) A number of blue (oxygen-rich) filaments have developed to the north of the centre of Cas A. Figure 2 shows the locations of a number of knots of blue nebulosity that have become visible between 1977 and 1980. Knots F, G and K are visible in the blue-green [OIII] but not in the red [SII]; the other knots are fainter in the red than they are in the blue.

Measurements of 47 long-lived knots that were observed in 1976, 1977 and 1980 and for which observations had previously been published by Kamper and van den Bergh (1976) yield an explosion date $T_0 = 1658 \pm 3$ for Cas A. All of the knots used in this solution have been observed for a minimum of 10 years; 11 of them have remained visible over the entire 29 year period during which Cas A has been observed at Palomar. As was pointed out by Kamper and van den Bergh (1976) the recently formed blue filaments and knots in the SW part of the supernova shell have motions that deviate systematically from those of the other fast-moving knots. These deviant objects were excluded from the expansion solution given above. The system of blue knots yields $T_0 = 1598 \pm 43$. This indicates that the blue knots are expanding more slowly (at a marginal level of statistical significance) than is the rest of the remnant of Cas A. The observation that some knots are oxygen-rich, whereas others are sulphur-rich (Chevalier and Kirshner 1978) shows that the shells of differing composition in the

---

[1]   Guest observer, Hale Observatories.

55

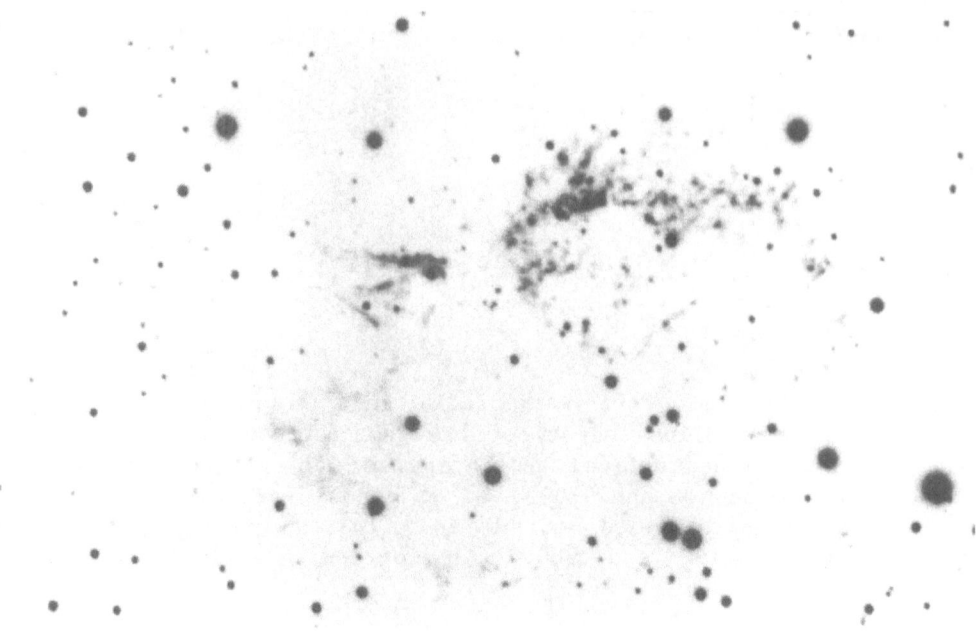

Fig. 1.   Appearance of the optical remnant of Cas A in 1980 on a 100
           min red (098 + RG 645) exposure.

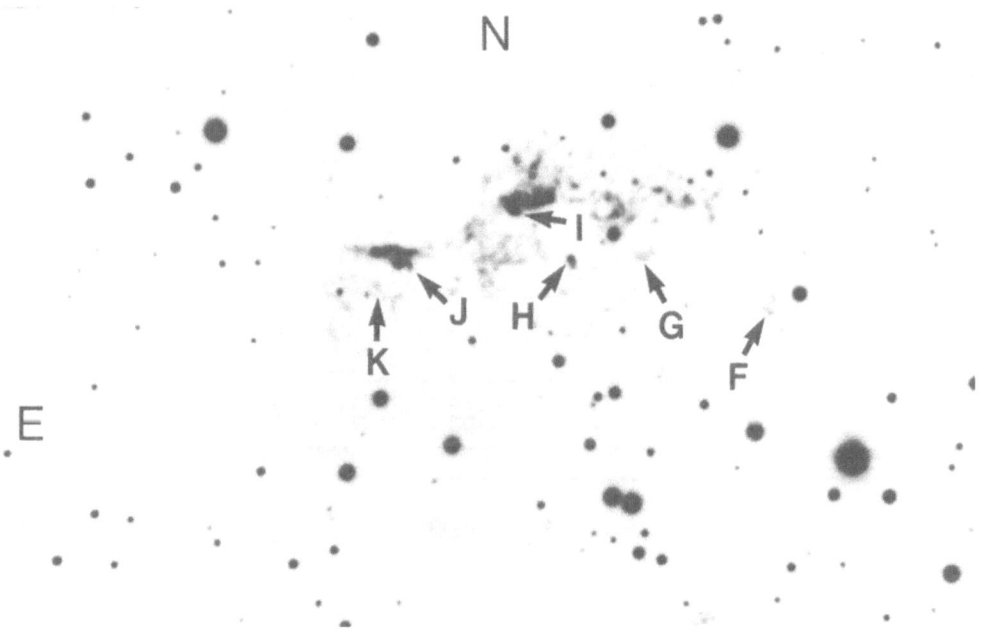

Fig. 2.   Two hour blue (IIIaJ + GG7) exposure of Cas A obtained in
           July of 1980. Blue knots that have appeared since 1977 are
           marked.

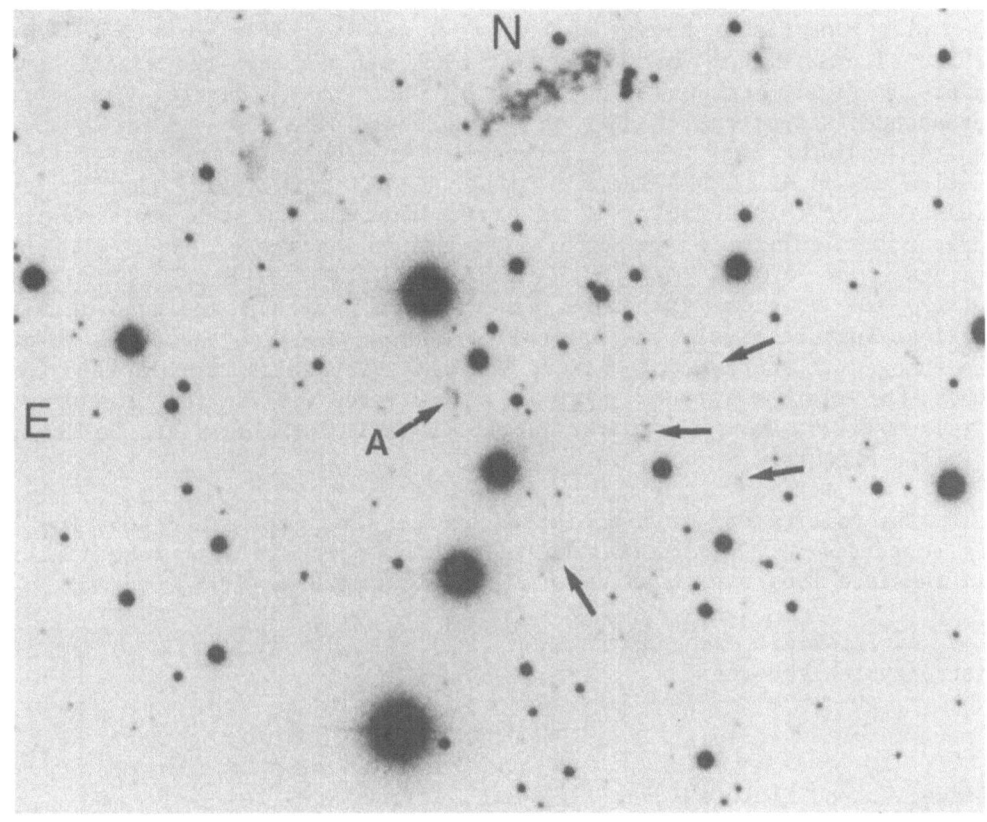

Fig. 3.   Two hour red (098 + RG645) exposure obtained in July of 1976
           of southern part of Cas A showing the faint quasi-stationary
           flocculi outside the main SNR shell.

supernova precursor were not well mixed during its detonation.

Ever since the pioneering investigation of Baade and Minkowski
(1954) it has been known that Cas A contains two distinct types of
objects; fast-moving knots that are oxygen-rich and contain no hydrogen
and quasi-stationary flocculi (QSFs) that emit Balmer lines and [N II].
All previously known QSFs (Kamper and van den Bergh 1976) had R < 130"
and were located within the radio shell of Cas A.   Recent work has,
however, revealed the existence of a small cluster of very faint QSF's
located ~ 150" SSW of the centre of expansion i.e. well outside the
Cas A radio shell.   These objects are indentified in Fig. 3.   The
nature of these objects is established by the fact that they are
visible on broad-band red plates (H + [NII]) but are invisible on an
[SII] interference filter plate and on plates sensitive to the
blue-green [OIII] part of the spectrum.   Confirmation of this
conclusion has been obtained by Chevalier and Kirshner (1979) who find
that a spectrum of feature A (which is the brightest of these objects)
"shows strong emission lines of [NII] and Hα along with weak [OI], just

as is seen in QSF's". Inspection of all available deep red-sensitive 5-m plates of Cas A shows that feature A was invisible between 1951 and 1958. It was first clearly seen in 1967 and has remained visible ever since. Plates taken in good seeing show that feature A exhibits pronounced structural changes on a time-scale of a few years. It seems quite probable that these changes are responsible for the "proper motion" μ_α = $-0\overset{''}{.}22 \pm 0\overset{''}{.}05$ yr^{-1} and μ_δ = $-0\overset{''}{.}20 \pm 0\overset{''}{.}12$ yr^{-1} that we have measured for this object. It is of interest to note that this optical proper motion agrees, to within the errors of measurement, with the values μ_α = $-0\overset{''}{.}09 \pm 0\overset{''}{.}08$ yr^{-1} and μ_δ = $-0\overset{''}{.}13 \pm 0\overset{''}{.}08$ yr^{-1} which Bell (1977) has measured for radio knot No. 38, which is coincident with optical feature A. If the optical "proper motion" of feature A is due to changes in shape and localized intensity variations within this knot then the chaotic proper motions which Bell (1977) and Dickel and Greisen (1979) find for the radio knots in Cas A also might be due to similar effects.

The reality of the association between freature A and knot No. 38 is supported by the observation that this object first became visible in the late 1960's at both radio and optical wavelengths.

A detailed account of this work has been submitted to the Astrophysical Journal.

REFERENCES

Baade, W. and Minkowski, R. 1954 Ap.J. **119, 206.**
Bell, A.R. 1979 M.N.R.A.S. **179,** 573.
Chevalier, R.A. and Kirshner, R.P. 1978 Ap.J. **219,** 931.
Chevalier, R.A. and Kirshner, R.P. 1979 Ap.J. **233,** 154.
Dickel, J.R. and Greisen, E.W. 1979 Astr.Ap. **75,** 44.
Kamper, K. and van den Bergh, S. 1976a ApJ.Suppl. **32,** 351.
Kamper, K. and van den Bergh, S. 1976b Sky and Telescope **50,** 236.

AN X-RAY HALO AROUND CASSIOPEIA A

G.C. Stewart[*], A.C. Fabian[*], F.D. Seward[+]
* Institute of Astronomy, Madingley Road, Cambridge, England
+ Harvard-Smithsonian Center for Astrophysics,
    Cambridge, Mass. U.S.A.

ABSTRACT

We report on evidence for weak large scale ($r \gtrsim 6$ arcmin) X-ray emission from Cassiopeia A. We investigate several mechanisms for producing such an X-ray halo. Further observations will be required to determine which mechanism is operative.

INTRODUCTION

Most previous analyses of the X-ray image of the Cassiopeia A supernova remnant have concentrated on the double shell structure associated with the blast wave and reverse shock. This has a radius of ~ 2 arcmin, similar to the size of the radio source and the optical system of fast moving knots.

We report here on evidence for weak X-ray emission from Cas A which is detectable to a radius of ~ 6 arcmin, comparable in size to the optical HII region known to exist around the remnant (Minkowski 1968, Van den Bergh 1971). We discuss several mechanisms for producing such an X-ray halo.

OBSERVATIONAL EVIDENCE

Plate 1 shows the Einstein HRI (High Resolution Imager) X-ray image of Cas A (Murray *et al.* 1979) binned into 16 arcsec pixels. This was displayed on a television monitor using a non-linear conversion of intensity-to-brightness. An extensive halo is observed around the easily discernible inner shell structure.

Despite the nominal high resolution of the HRI (the full width half maximum (FWHM) of the point spread function (PSF) is ~ 4 arcsec), the telescope mirror contributes large scattering wings to the PSF. Clearly, an accurate estimate of the instrumental contribution to the observed halo is required before any astrophysical explanation is sought.

J. Danziger and P. Gorenstein (eds.), Supernova Remnants and their X-Ray Emission, 59–64.
© *1983 by the IAU.*

Deconvolution of the effects of the PSF from the data by Fourier on Maximum Entropy techniques is not practical as the data have insufficient signal to noise ratio. We have therefore attempted to model the intrinsic emission of Cas A folded through the PSF and compared it with the data.

As the PSF of the telescope-instrument combination varies with energy we had to construct a PSF appropriate to Cas A. We used the pre-flight calibration PSFs made at various energies and weighted these according to the overall spectrum of Cas A modified by the HRI efficiency at the appropriate energy. The resultant PSF proved to be similar to that measured using the Zirconium monochromatic source with an energy of 2.04 keV. Consequently, we used that measured PSF through-out the analysis.

Two different models were used:-

a)   a spherical shell of outer radius 110" and thickness 15"
b)   a series of circularly symmetric rings of emission.

Both models gave essentially equivalent results, so here we shall discuss only the results given by method b.

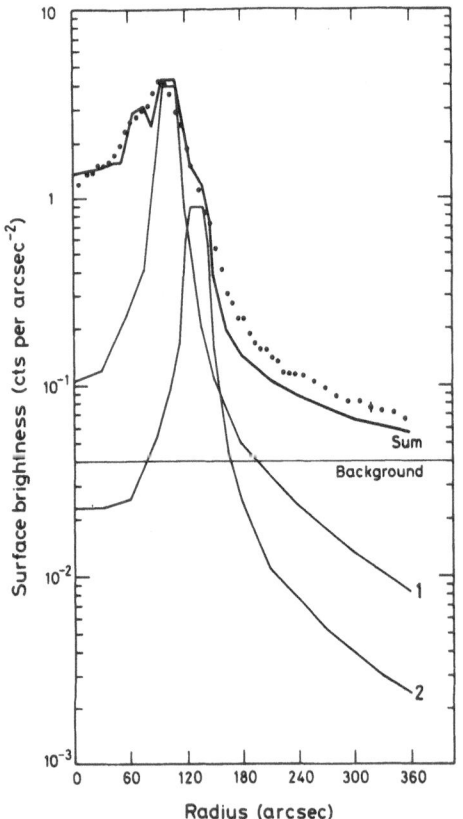

Figure 1 shows the azimuthally binned surface brightness distri-bution of the Cas A image, together with the predicted surface bright-ness profile. Two of the indivi-dual ring contributions are also shown.

Fig. 1. Surface brightness distri-bution of Cas A. Dots are data points. The thick solid line represents the fit to data using method b. Curves for two of the emission rings are labelled 1 and 2 and the background level is shown.

A statistically significant excess is apparent out to a radius of
$\sim$ 6 arcmin. We stress that while the excess is statistically significant, possible systematic effects are largely unknown. However, a
number of tests have been made and in none of these was the instrumental contribution sufficient to completely explain the halo. This
could only be done by assuming a mean photon energy of $\sim$ 3 keV in the
image, clearly unreasonable in view of the HRI efficiency at these
energies, or by postulating a degradation of the mirror since the calibration tests. Post launch trials appear to rule out this latter
effect.

DISCUSSION

The difference between the observed and predicted surface brightness profiles is shown in figure 2. The total luminosity (0.5 - 3 keV)
of the halo is $\sim$ 5 x 10^{34} erg s^{-1} ($\sim$ 2% of the total Cas A luminosity)
assuming a distance of 3 kpc for Cas A. Three mechanisms which might
give rise to the halo emission are thermal, non-thermal emission and
scattering of the intrinsic shell emission by dust along the line of
sight.

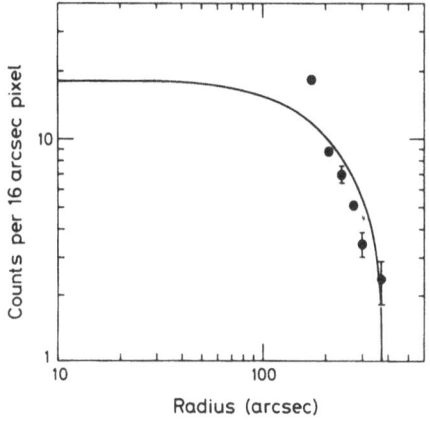

Fig. 2. The excess surface brightness of the Cas A data compared to
model b is shown by the dots. The
solid line represents the fit to
these points of the dust halo model
discussed in the text.

The density of the ISM currently being encountered by the outer
shock is $\sim$ 1 - 2 cm^{-3} (Pravdo & Smith 1980, Fabian et al. 1980). A
sphere of radius 6 pc has a thermal bremsstrahlung luminosity of
$\sim$ 10^{35} nT_7^2 erg s^{-1} which with line emission is sufficient to explain
that observed. It is not clear how this gas may have been heated as
the total thermal energy is $\sim$ 5 x 10^{49} ergs.

Supernova events may produce more than 10^{49} ergs in soft X-rays
(Klein & Chevalier 1978, Canizares et al. 1982). However, the column
density through the halo region of $\sim$ 10^{19} cm^{-2} is insufficient to
absorb much of this energy.

Kinetic heating of the material is also possible. High velocity material ($v \gtrsim 10^4$ km s^{-1}) is seen in the optical jet extending 2 arcmin to the NE of the main shell. A distribution of fragments with high densities and velocities sprayed out through action of the Rayleigh-Taylor instability in the explosion (Bychkov 1974, Falk & Arnett 1973, Chevalier & Klein 1978) could shock heat the matter ahead of the bulk ejecta to $\sim 10^9$ K. This gas may adiabatically expand and cool. This requires $\gtrsim 1\%$ of the energy of the supernova event to be processed efficiently from the kinetic energy of the high velocity fragments to heat. Cooler, higher density gas could explain the optical HII region.

Fast moving thermal electrons could propagate ahead of the shock (Chevalier 1975) and produce bremsstrahlung X-rays. Consideration of the relative volume emissitivities shows that approximately equal numbers of electrons are ahead of and behind the shock front. It remains to be shown that the plasma processes operative in a collisionless shock (see e.g. McKee & Hollenbach 1980) are sufficiently "leaky".

A more efficient process than thermal bremsstrahlung, where the cooling time of the gas is $\sim 10^7$ yrs, is synchroton radiation. The synchroton cooling time to produce 1 keV X-rays is $\approx 10^4 \, B_5^{-3/2}$ yr, where B_5 is the magnetic held strength in units of 2.10^{-6} G. To explain the X-ray observations 10^{46} ergs is the total energy in electrons with $\gamma \sim 10^8$. Hard X-ray observations by Pravdo and Smith (1980) require that the energy spectral index of the halo, α, be greater than ~ 0.6 (unless the emission they observe is from the halo). Assuming that cosmic ray electrons produce synchroton X-rays with $\alpha = 0.7$ and radio waves from the same power law, the 408 MHz radio flux from the source is 15 Jy. The total energy in relativistic electrons above $10^2 \gamma_2$ is then $4.10^{49} \, \gamma_2^{-0.4} \, B_5^{1.7}$ ergs which may not be unreasonable. Detection of the halo at other wavelengths would help determine whether synchroton is the relevant mechanism.

This interpretation requires that cosmic-ray electrons are accelerated to 's $\sim 10^8$ in supernova remnants and leak ahead 3 times faster than the shock velocity, a velocity much greater than the local Alfven speed. A more uniform (radial) field structure allowing rapid streaming to take place may have been created by a stellar wind. An accompanying flux of cosmic ray protons and nuclei would imply a local cosmic ray energy density enhancement of $\sim 10^3$. This would lead to heating of the surrounding gas to $\gtrsim 10^4$ K (see Dalgarno & McCray 1972) and explain the HII region.

Scattering of the main shell emission by dust may also produce an X-ray halo. For a monochromatic point source it has been shown (e.g. Hayakawa 1970) that simple single scattering from spherical grains produces a halo with radius, $\theta \approx 8 \, E_{keV}^{-1} \, a^{-1}$ arcmin, where a is the grain radius in units of $0.1 \, \mu$. The form of the halo is given by $(j_1(x)/x)^2$ where $x = 4\pi a/\lambda \, \sin(\theta/2)$ and j_1 is the first order Bessel function. The total intensity of the halo, I , is given by $I = \sigma_{sc} n_g L$ where σ_{sc} is the scattering cross-section and $n_g L$ is the grain column density to the source.

Convolving such a profile with the emission within 2 arcmin detected from CasA results in the halo profile shown in figure 2. From the normalisations to the observed halo we derive values for a $\approx$ .03 μ and $\sigma_{sc} n_g$ = .006 kpc^{-1} which are not unreasonable for graphite or silicon grains.

CONCLUSION

We believe the extended emission detected to be real and not to be purely an instrumental effect. While we have insufficient information to determine which of the physical processes we discuss is at work the data do already provide useful upper limits to all the processes. Further observations of the halo at different energies, both in X-rays and in other wavebands may provide the solution.

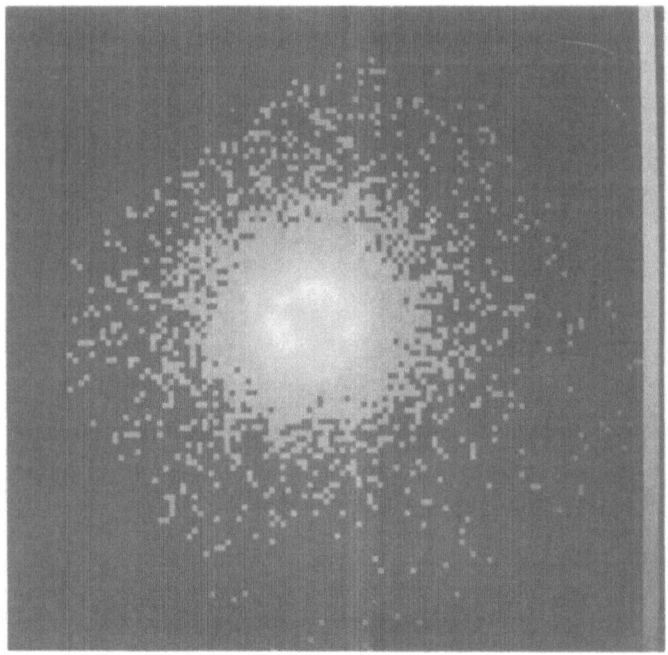

Plate 1 : HRI X-ray image of CasA binned in 16 arcsec pixels and displayed as discussed in the text.

REFERENCES

Bychkov, K.V., 1974. Sov. Astr., 18, 11.
Canizares, C.R., Kriss, G.A., E. Feigelson, E.D., 1982. Astrophys.J., 253, L17.

Chevalier, R.A., 1975. Astrophys.J., 198, 355.
Chevalier, R.A., & Klein, R.I., 1978. Astrophys.J., 219, 994.
Dalgarno, A. & McCray, R.A., 1972. Ann.Rev.Astr.Astrophys., 10, 375.
Fabian, A.C., Willingale, R., Pye, J.P., Murray, S.S. & Fabbiano, G.,
        1980. Mon.Not.R.astr.Soc., 193, 175.
Falk, S.W., Arnett, W.D., 1973. Astrophys.J., 180, L65.
Hayakawa, S., 1970. Prog.Theor.Phys., 43, 1224.
Klein, R.I. & Chevalier, R.A., 1978. Astrophys.J., 223, L109.
McKee, C.F. & Hollenbach, D.J., 1980. Ann.Rev.Astr.Astrophys., 18, 219.
Minkowski, R., 1968. In Stars & Stellar Systems, 7, 623, eds.
        Middlehurst, B.M. & Aller, L.H., Univ. of Chicago Press.
Murray, S.S., Fabbiano, G., Fabian, A.C., Epstein, W. & Giacconi, R.,
        1979. Astrophys.J., 234, L69.
Pravdo, S.H. & Smith, B.W., 1979. Astrophys.J., 234, L195.
van der Bergh, S., 1971. Astrophys.J., 165, 259.

DISCUSSION

GORENSTEIN: As the effect has a scale of 6 arcmin, why did you not
use the IPC data? You would have obtained information on the energy
dependence of the effect and reduced the number of possibilities.

STEWART: Analysis of the IPC data is in progress but is as yet
unfinished. Preliminary results show the existence of the halo in
two energy bands, but we have no information on the energy dependence
of the halo so far.

GRINDLAY: Could Sidney van den Bergh comment on the morphology of the
HII region and how it compares with the X-ray halo.

VAN DEN BERGH: Observations by M. Peimbert and myself show evidence
for some very faint patches of Hα emission nebulosity in the vicinity
of Cas A. In our 1971 paper we speculated that the HII region might
have been excited by a burst of UV radiation produced by the supernova.

VLA OBSERVATIONS OF SGR A

W.M. Goss and U.J. Schwarz
Kapteyn Astronomical Institute
Groningen, the Netherlands
R.D. Ekers and J.H. van Gorkom
NRAO*, Very Large Array
Socorro, New Mexico, U.S.A.

Summary:
   The radio source Sgr A has been mapped with the Very Large Array
(VLA) at 6 and 20 cm with an angular resolution of 5" × 8" arc. In
agreement with the earlier "WORST" map, the non-thermal source Sgr A
East shows a shell structure, while the thermal source Sgr A West
shows a spiral-like morphology. We suggest that Sgr A East is a supernova remnant (SNR) near the galactic centre. Its surface brightness is
the third largest in our galaxy after Cas A and the Crab Nebula. The
diameter is 9 pc and the source fits the surface-brightness diameter
relationship of Clark and Caswell (1976) if a distance of 10 kpc is
assumed.

I.   INTRODUCTION

   Since the discovery of Sgr A in the early 1950's, the nature of
this source has been puzzling. The identification with the centre of
the galaxy now seems certain, based on the associations established by
IR observations. Detailed, high resolution observations of the radio
structure have been difficult from the northern hemisphere because of
the declination of -29°. The multiple nature of the radio source
became evident in the early 1970's (e.g. Downes and Martin, 1972;
Ekers and Lynden-Bell, 1972). The eastern part of Sgr A (Sgr A East)
has a non-thermal spectrum (Dulk and Slee, 1974), while Sgr A West is
predominantly a thermal source which is associated with the IR sources
and the radio recombination lines (e.g. Bregman and Schwarz, 1982).
The compact VLBI radio source discovered by Balick and Brown (1974) is
located within Sgr A West. It has usually been assumed that this
source is the galactic nucleus.
   Jones (1974), Gopal-Krishna and Swarup (1976) and Ekers, Goss,
Schwarz, Downes and Rogstad (1975) have all discussed the possibility
that Sgr A East is a SNR. Given the relative low luminosity of Sgr A

---

* The National Radio Astronomy Observatory is operated by Associated
  Universities Inc., under contract to the National Science
  Foundation

J. Danziger and P. Gorenstein (eds.), Supernova Remnants and their X-Ray Emission, 65-69.
© 1983 by the IAU.

East, it was recognized that this source was not related to the active
sources in the nucleii of galaxies. As an example, Sgr A East would
have a 6 cm flux density of only about 1 mJy at a distance of 3 Mpc.
In addition, the measured brightness of Sgr A East is similiar to
galactic SNR.

Until the present, the best available map of the ~ 5' arc near
Sgr A was the "WORST" (Westerbork, Owens Valley Radio Synthesis Telescope) map at 6 cm published by Ekers et al.(1975). The synthesized
beam was 6 × 34" (α × δ). There was some evidence for shell structure
in Sgr A East. Gopal-Krishna and Swarup (1976) have suggested the
existence of steeper spectrum portions of Sgr A East which form part
of a SNR shell.

In order to elucidate the structure of the possible SNR, the VLA
was used in 1981-1982 to map the Sgr A region at 2, 6 and 20 cm. The
6 and 20 cm observations used scaled arrays (C and B, respectively) so
the angular resolutions are nearly identical at the two wavelengths.
The details of these observations are presented by Ekers, van Gorkom,
Schwarz and Goss (1983); in particular the thermal, spiral component
of Sgr A West is discussed. In this paper, we will discuss the parameters of Sgr A East based on the 6 and 20 cm observations. The H76α
observations (14.7 GHz) with the VLA are discussed by Van Gorkom,
Schwarz and Bregman (in preparation).

II. RESULTS

In the figure we show the 20 cm map with a resolution of 5" × 8"
(α × δ); the galactic plane is indicated. By comparing the 6 and 20 cm
maps the spectral index distribution can be derived and this is shown
as grey scale shading. The total flux densities of Sgr A East in the
6 and 20 cm maps are 31 and 77 Jy, respectively. The expected values
are 90 and 170 Jy (Dulk and Slee, 1974). This difference is due to the
fact that our observations are insensitive to angular scales $\gtrsim$ 3'arc;
thus the larger scale structure underlying the Sgr A complex is not
included in the spectral index estimate. This can cause an offset in
the value of the spectral index, especially at the lower brightness
levels. This could explain the differences between our value of -0.76
for the average spectral index across Sgr A East and the value of
-0.47 deduced from lower resolution observations (Dulk and Slee,
1974). Gopal-Krishna and Swarup (1976) have, however, suggested that
the peaks in Sgr A East have a spectral index of -0.7, in better
agreement with our data.

The shell structure in the non-thermal source Sgr A East is now
obvious for the first time. The projected distance of the centroid
of Sgr A East from Sgr A West is ~ 50"arc. The major and minor axes
are 3.6 and 2.8 arc with an orientation parallel to the galactic
plane. These correspond to 10.5 and 8 pc if the distance is 10 kpc.
At this distance, Sgr A East is in reasonable agreement with the Σ-D
(surface brightness - diameter relationship) for SNR proposal by Clark
and Caswell (1976) and by Milne (1979). As an example the Clark and
Caswell relation would imply a distance of ~ 7 kpc. If the Σ-D rela-

tion proposed by Mills (this conference) is used, a much smaller distance is derived. Based on these considerations, we conclude that Sgr A East is a bright galactic SNR. This SNR is the third brightest SNR in our galaxy after Cas A and the Crab.

III DISCUSSION

The relative locations of Sgr A West and Sgr A East with respect to the strong 40 km s^{-1} molecular cloud is a well-known problem (see Güsten and Downes, 1980). From the figure it is clear that the non-thermal shell of Sgr A East extends to the position of Sgr A West. This structure has important consequences for the interpretation of the 40 km s^{-1} absorption. Interpretations which place this cloud between Sgr A East and West are unlikely. Although the apparent optical depth in the direction of Sgr A West would decrease, there would still be some absorption occurring in the continuum radiation from the portion of the superposed shell structure. Our observations make it clear that the 40 km s^{-1} cloud must have a boundary which causes the large increase in optical depth from Sgr A West to East. Thus, the structure of this absorption line provides no information on the relative locations along the line of sight.

One of the major questions concerning Sgr A East is its location. As discussed in the previous paragraph, there is now no compelling evidence to suggest that the SNR is several hundred pc <u>behind</u> Sgr A West (Güsten and Downes, 1980). Three obvious possibilities remain for the relative location of the SNR and Sgr A West (the galactic centre):
(1) a chance coincidence along the line of sight,
(2) a physical association of the two, and
(3) a location for the SNR within 200-300 pc of the centre inside the nuclear bulge.
The coincidence on the sky of such an unusual SNR is a strong qualitative argument against possibility (1). The choice between (2) and (3) is difficult without further knowledge. The present observations make it clear that the western side of the shell of Sgr A East is coincident with the Sgr A West source and that the southern thermal spiral arm merges smoothly into the shell. Is this evidence for a physical interaction or is it just projection along the line of sight? If a physical interaction with the galactic centre is occurring, the Sgr A West diffuse non-thermal source (see Ekers et al., 1983) can be considered as an enhancement in the SNR shell.

If the Sgr A East SNR is at the galactic centre, then the probability of a SNR in a small volume of the galactic disk becomes a relevant consideration. If the varions Σ-t or Σ-D relations are used (e.g. Caswell and Lerche, 1979; Srinivasan, private communication; Mills, this conference), the derived ages are in the range 140 to 440 years. These ages are only meaningful if the energy in the explosion and the surrounding density in the interstellar medium near Sgr A East are comparable with the conditions for the "average" galactic SNR. The implications of these young ages are discussed by S. van den Bergh in the remarks.

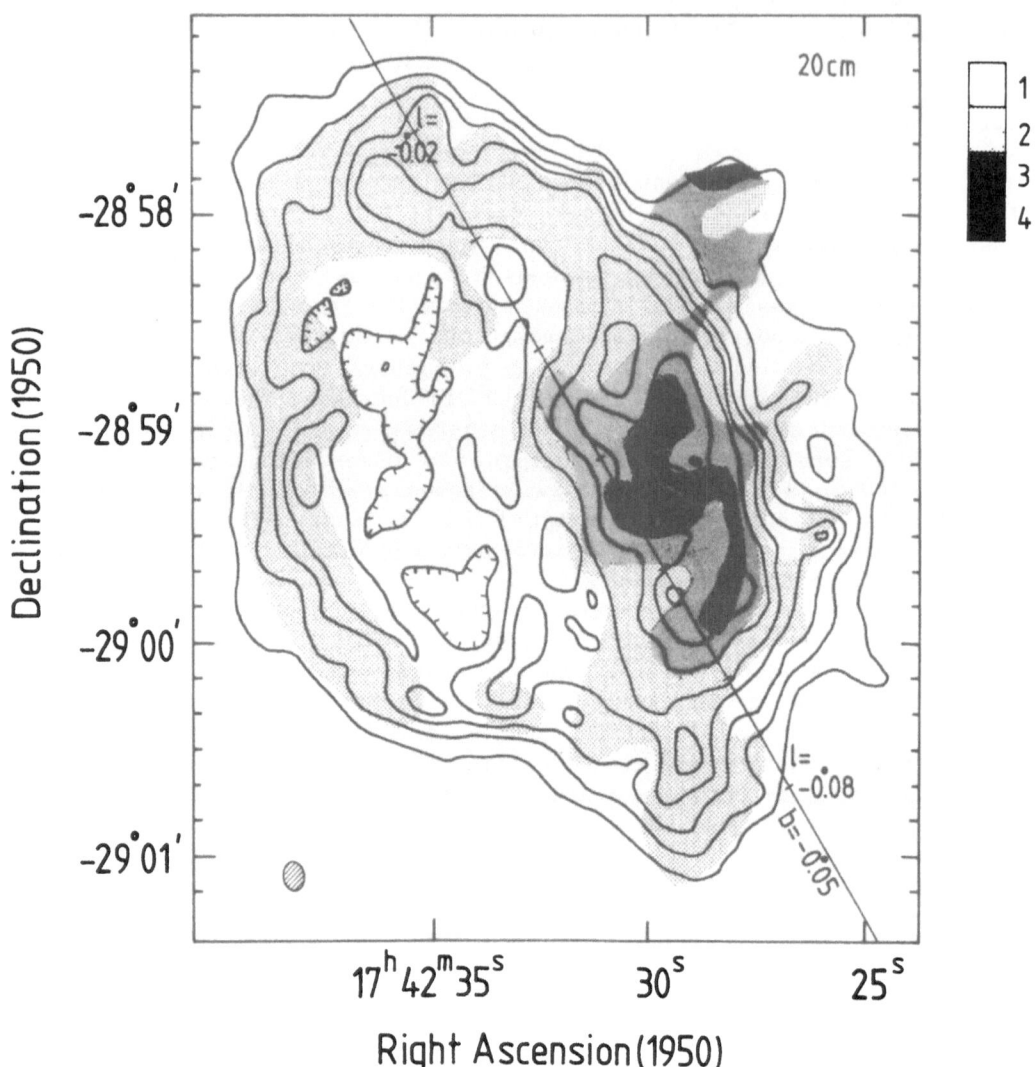

20 cm map of Sgr A as observed with the VLA. The half-power beam width of 5" × 8" is indicated, as well as a line parallel to the galactic plane. The contours are 40, 80, 120, 200, 280, 340, ... mJy/beam. The spectral index between 6 and 20 cm is shown by shading. (1) is $\alpha \leqslant -1$, (2) $-1 < \alpha \leqslant -0.5$, (3) $-0.5 < \alpha \leqslant 0.0$, and (4) $0.0 < \alpha$.

An alternative possibility is that Sgr A East has gone off in a very high density medium (molecular cloud?) in the galactic centre region. In this case, the age can be much larger (see Ekers et al., 1983) since the higher density medium confines the SNR for a longer time.

REFERENCES:

Balick, B., Brown, R.L. 1974, Astrophys. J. 194, 265
Bregman, J.D., Schwarz, U.J. 1982, Astron.: Astrophys. 112, L6
Caswell, J.L., Lerche, I. 1979, Mon. Nat. R. Astrl. Soc. 187, 201
Clark, D.H., Caswell, J.L. 1976, Mon. Nat. R. Astr. Soc. 174, 267
Downes, D., Martin, A.H.M. 1971, Nature 233, 112
Dulk, G.A., Slee, O.B. 1974, Nature 248, 33
Ekers, R.D., Lynden-Bell, D. 1971, Astrophys. Letters 9, 189
Ekers. R.D., Goss, W.M., Schwarz, U.J., Downes, D., Rogstad, D.H.
    1975, Astron. Astrophys. 43, 159
Ekers, R.D., van Gorkom, J.H., Schwarz, U.J., Goss, W.M. 1983,
    Astron. Astrophys., in press
Gopal - Krishna, Swarup, G. 1976, Astrophys. Letters 17, 45
Güsten, R., Downes, D. 1980, Astron. Astrophys. 87, 6
Jones, T.W. 1974, Astron. Astrophys. 30, 37
Milne, D.K. 1979, Ast. J. Phys. 32, 83

DISCUSSION

VAN DEN BERGH: Your conclusion that Sgr A East might be only a few hundred years old is a very exciting one! This object is situated within a region that contains only about 10^{-4} of the total mass of the galaxy. This corresponds to a SNR rate near the galactic nucleus that is ~ 10^3 times the galactic average. In this connection one is reminded of the fact that the only supernova ever seen in M31 occured in a region near the nucleus of the galaxy which only contains ~ 1% of the light of M31. These results suggest the possibility that the nuclear regions of galaxies are particularly supernova prone. Because the central regions of most galaxies are heavily exposed on Schmidt survey plates our sample of more distant supernovae may be heavily biased against the discovery of such a population of "nuclear supernovae".

THE INTERACTION OF SUPERNOVA EJECTA WITH AN AMBIENT MEDIUM

Roger A. Chevalier
Department of Astronomy
University of Virginia

ABSTRACT

Plausible environments for supernovae are the interstellar medium with constant density or a circumstellar medium built up by mass loss with $\rho \propto r^{-2}$. Self-similar solutions for the interaction region between the expanding supernova gas and the ambient gas exist provided that the expanding gas has $\rho \propto r^{-n}$ with $n > 5$. The circumstellar medium case is likely to be important for the early evolution of Type II supernovae because their progenitor stars are probably red supergiants. The radio and X-ray emission observed from extragalactic supernovae may be from this interaction region. The early self-similar solutions can also be applied to the young galactic remnants.

1. SUPERNOVAE AND THEIR SURROUNDINGS

Models for the explosions of Type II supernovae and for their light curves have shown that these events are likely to be the explosions of massive stars (Chevalier 1976a; Falk and Arnett 1977; Weaver and Woosley 1980). The mass range of the progenitor stars is not well known but is likely to be in some range above about 7 $M_\odot$. These massive stars are expected to undergo different phases of mass loss. While they are on and near the main sequence, they are observed to have stellar winds with velocities of about 2000 km s^{-1} and mass loss rates of about 10^{-6} $M_\odot$ yr^{-1} Over the lifetime of the star, this wind can create a low density bubble around the star with a radius of about 20 pc (Weaver et al. 1977). When the star becomes a red supergiant (for the last 10% of its life), it continues to lose mass but the wind properties change. The wind has a velocity of about 10 km s^{-1} and a mass loss rate of 10^{-6} to 10^{-4} $M_\odot$ yr^{-1} (e.g. Zuckerman 1980). This slow wind can create a relatively high density region around the star with a radius of $1(v_w/10$ km s$^{-1})(\tau_{rg}/10^5$ yr) pc, where v_w is the wind velocity and τ_{rg} is the lifetime of the red supergiant phase. It is at the end of this phase that the star explodes as a supernova.

J. Danziger and P. Gorenstein (eds.), Supernova Remnants and their X-Ray Emission, 71–81.
© 1983 by the IAU.

The density structure of the expanding gas in a Type II supernova
depends on the structure of the progenitor star. If the stellar envelope
has a flat density distribution, a shell is ejected (Chevalier 1976a). On
the other hand, Weaver and Woosley (1980) calculated an envelope structure
with decreasing density ($\rho \propto r^{-1.5}$, see Jones, Smith, and Straka 1981)
and most of the ejected envelope had constant density. Outside of the
envelope is a region with a steep density gradient. This region has
generally been calculated with poor resolution, although Jones, Smith,
and Straka (1981) have begun to remedy this situation. They calculated
the explosion of the stellar model of Weaver, Zimmerman, and Woosley
(1978) and included moderately fine zoning in the outer parts of the star.
After the explosion, this region had a steep density gradient ($\rho \propto r^{-12}$).
However, computations with better resolution gave a flatter density profile ($\rho \propto r^{-9}$) (Jones 1981). The calculations do not take into account
the effect of mass loss on the outer parts of the star; this effect could
be substantial.

The question of the progenitors of Type I supernovae is still controversial, although the explosion of a white dwarf in a binary system appears
to be favored. In this model, radioactive energy input is responsible
for the supernova radiation and several 0.1's - 1 $M_\odot$ of Fe are ejected
(see papers in Wheeler 1980). During the evolution of the binary system,
there may be mass loss but it is not known what form this mass loss takes.
Another suggestion for the progenitors of Type I supernovae is that they
are single stars with masses in the range 4-6.5 $M_\odot$ (Tinsley 1979). In
this case, the nature of the star at the end of its life and the nature
of the explosion are similar to the properties of Type II supernovae.

The density distribution of expanding gas in the exploding white
dwarf model for Type I supernovae is different from that expected for
Type II supernovae. In the Type II supernovae, all the energy is released
by the core collapse at the center of the star. In an exploding white
dwarf, the energy is released as a detonation or deflagration wave propagates through the star. Thus the central matter is accelerated by a relatively weak wave and does not achieve a high final velocity. The result
is that a shell is not ejected, but a density profile with the highest
density towards the center. Colgate and McKee (1969) found that the
density profile could be approximated by a constant density for the inner
4/7 of the mass and by $\rho \propto r^{-7}$ for the outer 3/7 of the mass. This
density profile is in accord with Type I supernova light curves near
maximum light (Chevalier 1981).

2.   THEORY OF THE INTERACTION

From the preceding discussion, the expansion of the supernova
ejecta into two types of media is of interest: the interstellar medium
and a circumstellar medium built up by mass loss. The interstellar medium
is assumed to have constant density because the sizes of young supernova
remnants are smaller than the typical distances between clouds. If the
circumstellar medium is created by a stellar wind, $\rho \propto r^{-2}$ is expected.

Circumstellar gas may also be in the form of ejected shells.

Initial work on the interaction of supernova ejecta with a constant density interstellar medium used numerical, finite-difference hydrodynamic methods. Gull (1975), Itoh (1977), and White and Long (1982) presented detailed results for the interaction of constant density, uniformly expanding ejecta with a uniform medium. Their calculations assumed that the ejecta had expanded to a certain radius and had density ρ_i when the interaction with the ambient medium of density ρ_a began. The interaction created a reverse shock wave in the ejecta (e.g. McKee 1974) which initially gave a density $4\rho_i$ at the contact discontinuity. The density at the reverse shock front decreased with time because of the uniform expansion of the ejecta and the density at the contact discontinuity decreased because of adiabatic expansion. A dense region of shocked ejecta formed at the contact discontinuity. The high density was accompanied by a low temperature so that this region was very important for the emission of soft X-rays (Itoh 1977). However, the properties of this region were dependent on the choice of initial conditions for the calculations through the value of ρ_i/ρ_a.

A more appropriate choice of initial conditions would be to include the steep density profile expected at the outer part of the expanding supernova. Jones, Smith, and Straka (1981) have carried out such a numerical calculation by computing the explosion of the star as well as the interaction with the uniform ambient medium. This calculation did not show a region of very high ejecta density as had been found in the previous calculations.

A more complete understanding of the interaction of the steep outer density profile with the ambient medium was provided by the realization that this phase could be described by a self-similar solution if the density profile of the expanding gas is a power law in radius (Chevalier 1982a; Nadyozhin 1982). If the ambient density is given by $\rho = A_1 r^{-s}$ and the ejecta density is given by $\rho = A_2 t^{-3} (r/t)^{-n}$, then the motion of the contact discontinuity between the two shock waves can be expressed as

$$R_c = \left(B \frac{A_2}{A_1} \right)^{\frac{1}{n-s}} t^{\frac{n-3}{n-s}}$$

where B is a constant which depends only on n and s. The self-similar solutions exist for $s < 3$ and $n > 5$. For $n = 5$, the outer shock wave expands as $t^{2/(5-s)}$, the expansion law for a point explosion (Sedov 1959). The most notable difference between the solutions for $s = 0$ and $s = 2$ is that for $s = 0$, $\rho \to 0$ and $T \to \infty$ at R_c while for $s = 2$, $\rho \to \infty$ and $T \to 0$ at R_c.

All of the self-similar solutions contain density gradients that

are subject to the Rayleigh-Taylor. The ultimate outcome of the
instability is not known, but some mixing between the ejecta and the
ambient medium may occur. The only attempted calculation of the Rayleigh-
Taylor instability in supernova remnant evolution is that of Gull (1973).
His treatment was one-dimensional and was analogous to turbulent convection. However, it is not clear whether the motion is fully turbulent or
whether a particular mode dominates the gas motions. The effects of the
instability need further investigation.

The self-similar solutions show that while the reverse shock wave
is in the part of the density profile with n substantially greater than 5,
the thickness of the interaction region is small compared to its radius.
Once it is in the part with n < 5, the reverse shock wave propagates toward the center and the outer shock wave tends toward the blast wave
expansion law. Chevalier (1982c) examined the interaction of a Type I
supernova with an ambient medium with s = 0 or 2 on the assumption that
the expanding gas has a region of constant density inside of a region
with $\rho \propto r^{-7}$ (see section 1). If s = 0, the transition time between
n > 5 evolution and n < 5 evolution is $t_c = 0.362 \ (M^5/E^3 \rho_a^2)^{1/6}$, where M
is the total ejected mass and E is the total energy. The approximate
transition time between the early self-similar expansion law and the
blast wave expansion law for the outer shock wave is $t_s = 4.58 \ t_c$. For
$t < t_s$, $R_1 \propto t^{4/7}$ and for $t > t_s$, $R_1 \propto t^{0.4}$. The time at which the
reverse shock wave reaches r = 0 is $t_r = 3.4 \ t_c$.

The evolution for s = 2 can also be examined. Now $R_1 \propto t^{0.8}$ for
$t < t_s$ and $R_1 \propto t^{2/3}$ for $t > t_s$. The reverse shock wave proceeds slowly
toward r = 0 because the pressure drops to 0 at the center of the blast
wave solution. A general difference between s = 0 and the s = 2 density
profiles is that the density is more strongly peaked at the outer shock
wave for s = 0.

3.   TYPE II SUPERNOVAE

The discussion of section 1 indicates that the early evolution of
Type II supernovae should involve the interaction of the supernova ejecta
with circumstellar gas built up by the slow wind from a red supergiant
star. Chevalier (1982b) suggested that the radio emission observed from
extragalactic Type II supernovae (Weiler et al. 1982,1983) is a result of
this interaction. The observed radio flux can be produced if the ratio
of relativistic electron energy density and magnetic energy density to
thermal density is comparable to that required to produce the radio
emission from the Cas A supernova remnant. The Rayleigh-Taylor instability may play a role in the production of these energy densities (e.g.
Gull 1973). In SN 1979c and SN 1980k, the radio emission was observed
to have a delayed turn-on and the delay was longer at low frequencies
(Weiler et al. 1982,1983). This delay can be attributed to free-free absorption by the unshocked circumstellar gas. In that case, the mass loss rates
from the progenitors of SN 1980k and SN 1979c were about 10^{-5} $M_\odot$ yr^{-1}
and 5 x 10^{-5} $M_\odot$ yr^{-1} respectively, if the velocity of the winds was 10
km s^{-1}.

The interaction of the expanding supernova gas with the circumstellar medium can create hot gas that radiates at X-ray wavelengths. In fact, SN 1980k was observed as an X-ray source (Canizares, Kriss, and Feigelson 1982). The solutions described in section 2 with s = 2 can be applied to this situation and the observed X-ray flux is close to that expected (Chevalier 1982b). The thermal emission is dominated by that from the region of shocked ejecta which has a higher density and lower temperature than does the region of shocked circumstellar gas. Another possible source for the X-ray emission is the inverse Compton mechanism. X-ray spectroscopy is needed to distinguish between these two mechanisms.

The circumstellar gas responsible for the radio and X-ray emission may also be responsible for infrared emission at late times (about one year after maximum light) through radiation of the supernova light by dust (e.g. Bode and Evans 1980). In fact, the late infrared emission from SN 1979c and SN 1980k may be consistent with the presupernova mass loss rates deduced from the radio observations (Dwek 1982). If this interpretation is correct, the high rates of presupernova mass loss apply out to about 10^{18} cm from the star. Under these circumstances, the radio emission should decline slowly over a period of about 30 years and should not show a sudden decline in the near future. The ratio of the timescale for the radio emission to that for the infrared emission is approximately equal to the ratio of the speed of light to the supernova shock velocity. By combined observations of Type II supernovae at various wavelengths (including the ultraviolet), it should be possible to deduce a detailed model for the interaction with circumstellar matter and to determine whether the other emission mechanisms, like a pulsar (Pacini and Salvati 1981; Shklovsky 1981), are important.

The observed radio supernovae have a range of luminosities, which can be attributed to a range of presupernova mass loss rates. Stellar observations do indicate a large range of mass loss rates and the occasional star with a very high mass loss rate might be expected to yield a very luminous supernova. The radio source 41.9 + 58 in M82 may be such an object; its properties have been reviewed by Kronberg, Biermann, and Schwab (1981). The source was observed for over a decade and was found to be decreasing in flux with an e-folding time of 12 years. Its spectrum had a low-frequency turnover at about 1 GHz. These properties can be approximately reproduced by the circumstellar interaction model if the supernova was about 10 years old when it was first observed and the pre-supernova mass loss rate was 4×10^{-4} $M_\odot$ yr^{-1}. Scaling from the observed radio luminosities of SN 1979c and SN 1980k gives about the correct radio luminosity for 41.9 + 58. One possible problem is with VLBI observations of the source. Geldzahler et al. (1977) found a source size of $\sim 0\rlap{.}''0015$, which is too small to be compatible with the supernova model. However, Shaffer and Marscher (1979) found a size ten times larger at a lower frequency. If this size applies to the source and is not the result of scattering, it is compatible with the supernova model. Further VLBI observations should provide the best test of the model. It would be particularly valuable to measure the expansion of the source.

Griffiths (1979) has found that 41.9 + 58 is an X-ray source with a luminosity of $\sim 10^{39}$ ergs s^{-1}. The model of section 2 would give a substantially greater X-ray luminosity from the shocked ejecta. However, in this case the supernova would have interacted with several $M_\odot$ of circumstellar gas and the reverse shock may no longer be in the steep part of the supernova density profile. This could reduce the thermal X-ray luminosity to an acceptable level.

After an expanding Type II supernova has swept up the circumstellar matter from the red supergiant wind, it may expand into a low density region created by the earlier, fast stellar wind. If SN 1054 was a Type II supernova, its remnant may be in this phase. This would explain te absence of thermal X-ray emission (Schattenburg et al. 1980) and radio emission (Wilson and Weiler 1982) in a shell and the presence of a fast optical shell (Murdin and Clark 1981; Henry, MacAlpine, and Kirshner 1982). The X-ray and radio emission require interaction with an ambient medium, while the optical emission can be the result of photoionizing radiation from the central Crab Nebula.

4.   TYPE I SUPERNOVAE

The three historical galactic supernovae SN 1006, SN 1572, and SN 1604 can plausibly be classified as Type I events, although the evidence is weak for SN 1006 (e.g. Clark and Stephenson 1977). All of these super-novae created remnants which are fairly strong X-ray emitters (Pye et al. 1981; Reid, Becker, and Long 1982; White and Long 1982), which implies that they are interacting with moderately dense gas. Although there are uncertainties due to non-equilibrium ionization and heavy element over-abundances, White and Long deduce an ambient hydrogen density $n_o > 0.1$ cm^{-3} for the SN 1604 remnant. This density is higher than might be ex-pected for the ambient interstellar medium, especially because the remnants are some distance from the galactic plane. This has led to the suggestion that the remnants are interacting with circumstellar mass loss (White and Long 1982; see also Fabian, Stewart, and Brinkmann 1982). The required amount of mass loss would be similar to that believed to occur in the progenitors of Type II supernovae (although somewhat more extended) and would be consistent with the moderately massive single star hypothesis for the progenitors of Type I supernovae.

Chevalier (1982c) checked on the circumstellar versus the interstellar model for the Type I remnants using the theory described in section 2. A powerful discriminant between the models is the exponent m in the expan-sion law $R_1 \propto t^m$. For s = 0, m is expected to be between 0.4 and 0.57, and for s = 2, m is expected to be between 0.67 and 0.8. For the incom-plete optical shell of the SN 1572 remnant, m is observed to be 0.38 ± 0.01 (Kamper and van den Bergh 1978) and for the complete radio shell, m is 0.47 ± 0.05 (Strom, Goss, and Shaver 1982). The optical emission is likely to be from the outer shock front and the radio shell expands in the same way as the optical shell where they overlap. The optical fila-ment in the remnant of SN 1006 expands with m = 0.47 ± 0.07 (Hesser and van den Bergh 1981). In both cases, evolution in a medium with s = 0 is implied. Further evidence for this type of evolution comes from the fact that some of the radio and X-ray emission is concentrated toward the outer shock wave in all three elements.

If the progenitor of Type I supernovae do have dense winds, extra-galactic Type I events may be observable as radio sources. Radio observations of SN 1981b have been attempted for a year now and it has not been detected (Weiler et al. 1983). Current observations are consistent with interaction with the interstellar medium and imply that an interstellar density $n_o \geqslant 0.1$ cm^{-3} is fairly pervasive. If $n_o = 0.1$, then the remnants of SN 1572 and SN 1604 are in an early evolutionary stage. The reverse shock wave should be close to the dividing point between the steep and the flat supernova density gradient and much of the ejecta may not yet have been shocked. This situation provides a possible explanation for the lack of strong X-ray Fe line emission from the Type I remnants (Becker et al. 1980; Arnett 1980).

5.   DISCUSSION

The properties of young supernova remnants are generally consistent with the properties of the initial supernova events. Type II supernovae probably initially interact with circumstellar gas from presupernova mass loss and Type I supernovae probably interact directly with the interstellar medium. There may be other stellar explosions in which the entire envelope is lost in presupernova mass loss and the result is neither a Type I nor a Type II supernova. The Cassiopeia A explosion may have been such an event (Chevalier 1976b). There are now several remnants with similar properties to Cas A; a particularly remarkable one is the source in NGC 4449 (Blair et al. 1983).

The interaction with circumstellar material implies that the young remnants of massive star explosions can be quite luminous at radio and X-ray wavelengths. The VLA (Very Large Array) is well suited to detect these objects. Its high spatial resolution allows it to pick out compact, high brightness temperature sources. Studies of regions of star formation, such as that of Sramek (1982), are expected to reveal young remnants. Once radio positions are known, further studies at X-ray and optical wave-lengths will be valuable.

High resolution radio and X-ray studies of the galactic remnants make possible not only analyses of the current structure of the remnants, but also allow the measurement of structural changes in time. As discussed in section 4, these studies are crucial for determining the type of evolu-tion that the remnant is undergoing. Finally, it is clear that X-ray spectroscopy with high spatial resolution will play an important role in distinguishing the shocked ejecta from the shocked interstellar medium.

With regard to the theory of the interaction of ejecta with an ambient, most features of the one-dimensional evolution are now clear. The largest uncertainty in the dynamical evolution is the role of the Rayleigh-Taylor instability, which may be responsible for a widening of the shell of ejecta, for the creation of clumps, and for mixing of the ejecta with the ambient medium. At present, the theory of the Rayleigh-Taylor instability in the nonlinear regime is in a rudimentary state. Further progress will probably require the use of two-dimensional numerical hydrodynamic calculations with high resolution.

The author's research on supernovae is supported by NSF grant AST 80-19569.

REFERENCES

Arnett, W.D.: 1980, Astrophys. J., 240, p. 105.

Becker, R.H., Holt, S.S., Smith, B.W., White, N.E., Boldt, E.A., Mushotsky, R.F., and Serlemitsos, P.J.: 1980, Astrophys. J. Letters, 235, p. L5

Blair, W.P., Kirshner, R.P., Winkler, P.F., Raymond, J.C., Fesen, R.A., and Gull, T.R.: 1983, this volume, p. 591.

Bode, M.F. and Evans, A.: 1980, M.N.R.A.S., 193, p. 21^{p}.

Canizares, C.R., Kriss, G.A., and Feigelson, E.D.: 1982, Astrophys. J. Letters, 253, p. L17.

Chevalier, R.A.: 1976a, Astrophys. J., 207, p. 872.

Chevalier, R.A.: 1976b, Astrophys. J., 208, p. 826.

Chevalier, R.A.: 1981, Astrophys. J., 246, p. 267.

Chevalier, R.A.: 1982a, Astrophys. J., 258, p. 790.

Chevalier, R.A.: 1982b, Astrophys. J., 259, p. 302.

Chevalier, R.A.: 1982c, Astrophys. J. Letters, 259, p. L85.

Clark, D.H. and Stephenson, F.R.: 1977, The Historical Supernovae (Oxford: Pergamon).

Colgate, S.A. and McKee, C.: 1969, Astrophys. J., 157, p. 623.

Dwek, E.: 1982, private communication.

Fabian, A.C., Stewart, G.C., and Brinkmann, W.: 1982, Nature, 295, p. 508.

Falk, S.W. and Arnett, W.D.: 1977, Astrophys. J. Suppl., 33, p. 515.

Geldzahler, B.J., Kellermann, K.I., Shaffer, D.B., and Clark, B.G.: 1977, Astrophys. J. Letters, 215, p. L5.

Griffiths, R.E.: 1979, IAU Joint Discussion, Extragalactic High Energy Astrophysics, p. 20.

Gull, S.F.: 1973, M.N.R.A.S., 161, p. 47.

Gull, S.F.: 1975, M.N.R.A.S., 171, p. 263.

Henry, R.B.C., MacAlpine, G.M., and Kirshner, R.P.: 1982, preprint.

Hesser, J.E. and van den Bergh, S.: 1981, Astrophys, J., 251, p. 549.

Itoh, H.: 1977, Pub. Astr. Soc. Japan, 29, p. 813.

Jones, E.M.: 1981, private communication.

Jones, E.M., Smith, B.W., and Straka, W.C.: 1981, Astrophys. J., 249, p. 185.

Kamper, K.W. and van den Bergh, S.: 1978, Astrophys. J., 224, p. 851.

Kronberg, P.P., Biermann, P., and Schwab, F.R.: 1981, Astrophys. J., 246, p. 751.

McKee, C.F.: 1974, Astrophys. J., 188, p. 335.

Murdin, P. and Clark, D.H.: 1981, Nature, 294, p. 543.

Nadyozhin, D.K.: 1982, preprint.

Pacini, F. and Salvati, M.: 1981, Astrophys. J. Letters, 245, p. L107.

Pye, J.P., Pounds, K.A., Rolf, D.P., Seward, F.D., Smith, A. and
    Willingale, R.: 1981, M.N.R.A.S., 194, p. 569.

Reid, P.B., Becker, R.H., and Long, K.S.: 1982, Astrophys. J., in press.

Schattenburg, M.L., Canizares, C.R., Berg, C.J., Clark, G.W., Markert,
    T.H., and Winkler, P.F.: 1980, Astrophys. J. Letters, 241, p. L151.

Sedov, L.I.: 1959, Similarity and Dimensional Methods in Mechanics
    (New York: Academic Press).

Shaffer, D.B. and Marscher, A.P.: 1979, Astrophys. J. Letters, 233,
    p. L105.

Shklovsky, I.S.: 1981, Sov. Astron. Letters, 7, p. 263.

Sramek, R.A.: 1982, private communication.

Strom, R.G., Goss, W.M., and Shaver, P.A.: 1982, M.N.R.A.S., in press.

Tinsley, B.M.: 1979, Astrophys. J., 229, p. 1046.

Weaver, R., McCray, R., Castor, J., Shapiro, P., and Moore, R.: 1977,
    Astrophys. J., 218, p. 377.

Weaver, T.A., and Woosley, S.E.: 1980, Ann. NY Acad. Sci., 336, p. 335.

Weaver, T.A., Zimmerman, G.B., and Woosley, S.E.: 1978, Astrophys. J.,
    225, p. 1021.

Weiler, K.W., Sramek, R.A., van der Hulst, J.M., and Panagia, N.: 1982,
    in Supernovae, ed. M.J. Rees and R.J. Stoneham (Dordrecht: Reidel),
    p. 281.

Weiler, K.W., Sramek, R.A., van der Hulst, J.M., and Panagia, N.:1983,
    this volume, p. 171.

Wheeler, J.C.: 1980, Type I Supernovae (Austin: Univ. of Texas).

White, R.L. and Long, K.S.: 1982, Astrophys. J., in press.

Wilson, A.S. and Weiler, K.W.: 1982, Nature (submitted).

Zuckerman, B.: 1980, Ann. Rev. Astrophys., 18, p. 263.

DISCUSSION

MATHEWSON: If the circumstellar gas lies in a plane, how much does this affect your model?

CHEVALIER: My model does assume spherical symmetry. If the gas is concentrated toward a plane, the shock wave would be retarded in the plane. Two-dimensional hydrodynamic models would be required for the dynamics. The mass loss rate deduced from the radio absorption would not be correct and the observed properties of a supernova would depend on the viewing angle.

WEILER: Two comments: (1) From IR results you predicted that Type II radio supernovae may be visible for up to 30 years. However, from observations the Type II SN 1980k is already decreasing strongly in the radio at age 2 years and there is very strong evidence from observations that no extragalactic SN remains as strong as Cas A for more than 5-10 years. (2) You suggested that Type I supernovae may also be radio sources. Our present limit on Type I SN 1981b is ~ 60 μJy (1σ) and no detection.

CHEVALIER: (1) My model does predict a slow decrease in radio flux. The recent radio data should be compared with the model to see if there is a contradiction. If so, the interpretation of either the radio or the infrared data will need to be modified. The source 41.9 + 58 in M82 may be in the intermediate age range. (2) The lack of radio emission is consistent with the lack of a dense circumstellar medium around Type I supernovae.

HEIDMANN: With respect to possible future observations of compact sources with the VLA, Dave Heeschen and I just observed at 6 cm a non-nuclear variable source in the clumpy galaxy Mkn 297. It increased by a factor of 3 in 27 months and is now 20 times stronger than SN 1979c in M100. In addition to being possibly related to very powerful SN events, it is the first example of a non-nuclear compact variable very strong radio source.

CHEVALIER: I believe that Type II supernovae can have a wide range of radio luminosities. However, a slow increase in flux is not characteristic of supernovae.

MCKEE: In the case of SN 1006, you suggested that the reverse shock has propagated farther than in the case of SN 1572 and SN 1604. Where is the X-ray emission from the reverse shocked gas in SN 1006?

CHEVALIER: The greater evolution of SN 1006 is based on the assumption that all the young Type I remnants are expanding into a medium of the same density. This is not necessarily the case. Recent ultraviolet observations of a star that may be behind the SN 1006 remnant show evidence for cool rapidly expanding Fe.

TUFFS: Just a short point of information: The ratio of the shock velocity deduced from radio proper motion observations to the average velocity of expansion of the remnant is 0.44. This would place Cas A nearer the s = 0 than the s = 2 density law for its circumstellar medium.

DENNEFELD: You conclude that Type I SN are essentially interacting with the interstellar medium rather than circumstellar one, based on SN 1006, 1572 and 1604. I thought that there was evidence in Kepler of the contrary (high density and nitrogen overabundance in the shocked material)?

CHEVALIER: Kepler's supernova is evidently interacting with small clouds but I believe that they have a small volume filling factor and that the overall dynamics is probably determined by the interaction with the interstellar medium.

FEDORENKO: All the observed SNR are produced by the SN explosion into ISM or into the region of stellar wind. The density is $n \gtrsim 0.1$ cm^{-1}. But it is well known that most of the volume of our Galaxy consists of a hot (T $\sim 10^6$ $^\circ$K), dilute ($n \sim 3 \times 10^{-3}$ cm^{-3}) plasma. What will be the result of explosion of SN into such a coronal ISM?

CHEVALIER: The expansion of a remnant into the hot medium should be faint at radio and X-ray wavelengths until the remnant begins to interact with clouds. However, the question of whether the hot medium occupies most of the volume of the Galaxy is still controversial.

COX: Why do we not see the several young remnants which "should have" been generated in the last few hundred years, using either radio or X-ray techniques?

CHEVALIER: If remnants are interacting with a low density medium, as could be created by a fast presupernova wind, they may be faint at radio and X-ray wavelengths.

SHAPIRO: Comment concerning the chances of "seeing" in radio or X-rays of a new supernova remnant in our Galaxy: This is a reasonable question, especially in the light of G. Tammann's extimate of the supernova frequency in our Galaxy. To within a factor of two, he puts this frequency at $\approx$ 8 per century. Given the obscuration by dust of most supernova events in the Galaxy in the optical channel, should we nevertheless have detected such an event in radio or X-rays? It seems to me not surprising that we haven't if we consider that the effective history of radio astronomy is only 3 or 4 decades, and that of X-ray astronomy is hardly 2 decades.

MURDIN: If you just look at the Poisson statistics of events occurring a few times per century and then put in light travel times from the events across our Galaxy, then gaps of three hundred years are not uncommon. Supernova bunch together and long gaps occur between bunches. Unseen young SNR may be unseen simply because none have happened recently.

VERY HIGH RESOLUTION CALCULATIONS OF VERY YOUNG SUPERNOVA REMNANTS

Eric M. Jones and Barham W. Smith
Los Alamos National Laboratory
University of California
Los Alamos, NM USA

1. Early Supernova Remnant Structure

After the supernova shock wave has swepted up about 8-10 stellar masses of interstellar material, the SNR structure is well described by blast wave theory (eg. Sedov 1959, Chevalier 1977). In fact, both numerical calculations of the early phases (Jones, Smith, and Straka 1981) and small scale, laboratory simulations (Wilke 1982) show transition to blast wave at 8-10 masses. While the late stages have been well understood for some time, the early stages have only been crudely modeled until very recently.

In hindsight, we now know that the transition region between the photosphere (roughly 10^{-9} g/cm^3) and the circumstellar medium (10^{-24} g/cm^3) plays a crucial role. The shock wave is strongly accelerated down the density gradient, putting the shocked material behind into free expansion. When the shock encounters circumstellar material, it begins to decelerate. A second, reverse shock propagates into the stellar material that plows into the shocked circumstellar gas. All this happens on a timescale of days.

The first attempts to include a description of the outer stellar envelope (Chevalier 1976, Falk and Arnett 1973, 1977) were aimed at analysis of the UV and X-Ray bursts produced when the shock wave reaches the photosphere. Falk and Arnett (1977) terminated their calculations before the shock reached the circumstellar gas. Chevalier (1976) mentions a reverse shock forming early but did not go into any details. Apparently, his model was not well enough resolved in the outer regions to detail much of the double-shock behaviour.

For aesthetic reasons, we included the complete transition region in our first calculations (Jones, Smith, and Straka 1981). We noticed and described the double-shock structure but, in hindsight, lacked sufficient resolution to produce the detailed structure between the shocks. Chevalier (1982) derived a similarity solution for the inter-shock region. In this paper we describe very high resolution calculations which reproduce and confirm the Chevalier similarity solution. There are, of course, differences and caveats which must be kept in mind. Nonethless, we have all come a long way in a short time.

J. Danziger and P. Gorenstein (eds.), Supernova Remnants and their X-Ray Emission, 83–86.
© *1983 by the IAU.*

2. Comparison with the Chevalier Solution

Rather than attempt a comparison with a model using a detailed stellar model, we first constructed a set of initial conditions which model the assumptions of the similarity solution. Specifically, we use a perfect gas (=5/3) in free expansion with a density profile proportional to r^{-7}. At its outer edge the gas is in contact with a stationary, homogeneous gas (s=0) of density 1.67e-24 g/cm^3. The initial time is 1.0e+06 seconds. These parameters completely specify the problem. Runs were made with different resolution and with a variety of viscosity prescriptions. Figure 1. is a well-resolved run (100 zones in the piston) with relatively high viscosity. We show here only the structure between shocks at the output time, 1.0e+07 seconds. There are differences, probably due to the finite starting time and to transients. Table I is a brief comparison.

Table I

	Chevalier Solution	Calculation
R_2/R_1	0.722	0.829
ρ_2/ρ_1	1.3	1.65
P_2/P_1	0.47	0.62 (poorly determined)
u_2/u_1	1.253	1.33
R_c	3.83e+16 cm	3.75e+16 cm

For most purposes the Chevalier solution represents a great improvement over previous models, especially when detailed numerical calculations are likely to remain quite expensive for some time.

3. Physical Instabilities in the Similarity Solution

A good rule-of-thumb determining the stability of a hydrodynamic flow (Chevalier 1976) is that Rayleigh-Taylor modes are unstable if the local gradients of pressure and density are of opposite sign. The growth time is given by

$$\frac{1}{\tau^2} = \frac{1}{\rho}\frac{dP}{dr}\frac{d}{dr}\left(\frac{p^{1/\gamma}}{\rho}\right) .$$

Examination of the similarity solution shows that for s=0, the region interior to the contact surface is Rayleigh-Taylor unstable while for s=2 the region outside is unstable. For the n=7/s=0 case discussed above the growth times just inside the contact surface are of order 2.0e+6 seconds. They will be longer for the corresponding s=2 case since the gradients are shallower. Nonetheless, the likely prospect that large amplitude instabilities will grow on timescales of weeks must be taken into account when using either the Chevalier solution or numerical calculations to compare with observations.

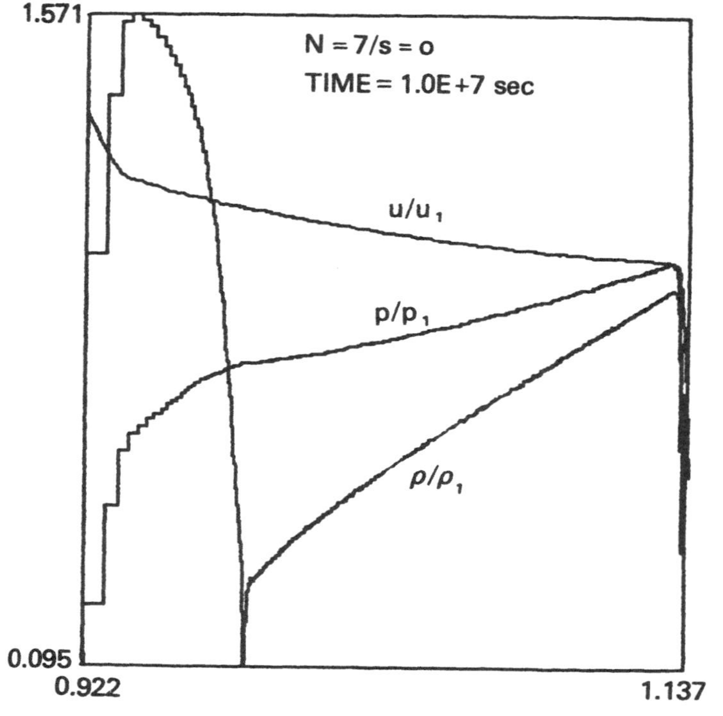

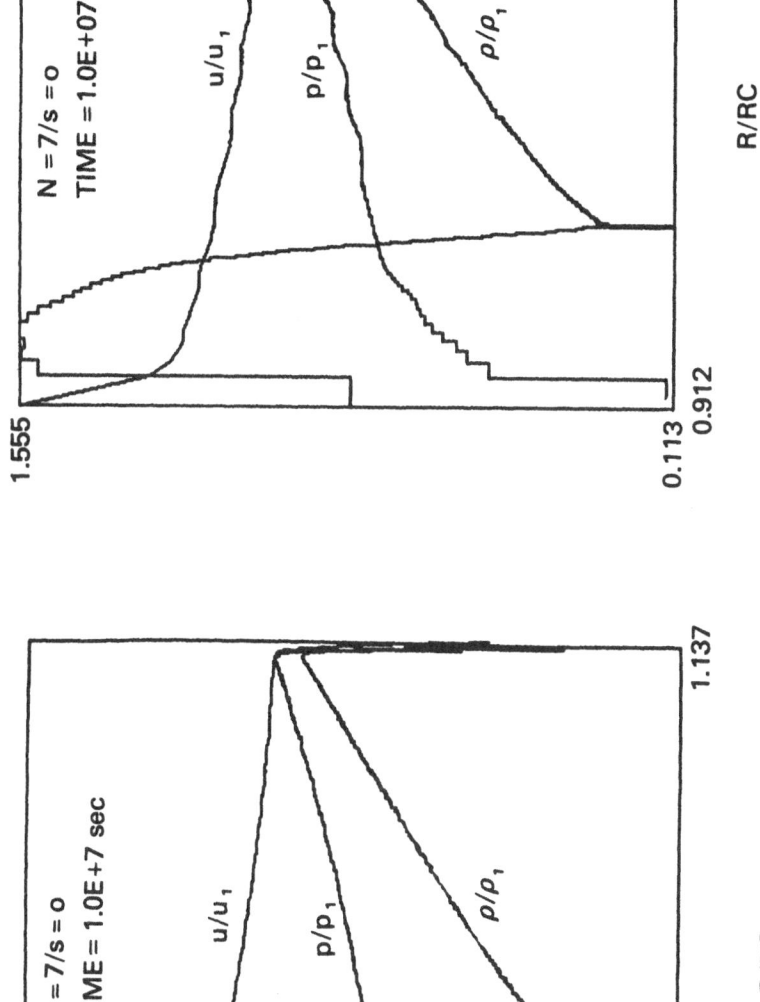

Figure 2 – Same problem as in Figure 1 except that the artificial viscosity coefficient has been reduced by a factor of four. The oscillations are of numerical origin, as discussed in the text.

Figure 1 – Velocities, pressures and densities between the shocks, normalized to shock values derived from the Chevalier Solution are plotted versus radius normalized to the similarity contact radius.

4. Numerical Instabilities (Non-Physical?) in the Calculations

Finally, we call attention to a significant numerical problem that can plague calculations of this kind. Figure 2 shows the pressure profile produced at 1.0e+07 seconds for the problem described above. The amplitude grows large at later times. The only difference between this run and the one shown previously is that the artifical viscosity coefficient has been reduced by a factor of four. The oscillations in the pressure and velocity profiles originate at the direct (outer) shock front and travel as acoustic waves toward the contact surface. The wavelength increases inward in response to the increasing sound speed. Consequently, the disturbance is resolved over many cells except very near the shock. Persistence of well-resolved disturbances is usually a sign that they have a physical rather than a numerical origin. However, we have done runs with different resolution and/or viscosity and have shown that, among other things, the wavelength decreases proportional to cell size. Apparently, frequency is just the frequency at which new cells are encountered by the shock.

We thank W. I. Newman and R. A. Chevalier for useful discussions. This work has been supported by the Los Alamos National Laboratory which is operated for the United States Department of Energy by the University of California.

5. References

Chavelier, R. A. 1976, Astrophy. J., vol.207. p.872.
Chevalier, R. A. 1977, Ann. Rev. Astron. Astrophys., vol.15, p.175.
Chevalier, R. A. 1982, Astrophys. J., vol.258, p.790.
Falk, S. W. and Arnett, W. D. 1973, Astrophys. J. Letters, vol. 180 p.L65.
Falk, S. W. and Arnett, W. D. 1977, Astrophys. J. Supp., vol.33, p.515.
Jones, E. M., Smith, B. W., and Straka, W. C., 1981, Astrophys. J., vol.249, p. 185.
Sedov, L. I. 1959, Similarity and Dimensional Methods in Mechanics, (New York:Academic Press).
Wilke, M. D. 1982, Los Alamos National Laboratory Report LA-9182 (Thesis).

X-RAY EMISSION FROM SUPERNOVA REMNANTS IN A CLOUDY MEDIUM

Christopher F. McKee
Departments of Physics and Astronomy
University of California, Berkeley, CA 94720

ABSTRACT

Young SNRs expand to much larger radii in a cloudy ISM than in a homogeneous medium, and they can have large variations in the pressure. The collision between supernova ejecta and an ambient cloud can result in an expanding high pressure region (a "secondary blast wave"). Observations of MSH 15-52 can be accounted for in this manner. X-ray emission from both young and older SNRs can provide an important probe for inferring the structure of the ISM.

1. INTRODUCTION

The interstellar medium is observed to be quite inhomogeneous, with densities ranging from those in molecular clouds ($n \geq 10^3$ cm^{-3}) to HI clouds and shells ($n \sim 10^{1-2}$ cm^{-3}) to warm medium ($n \sim 10^{-0.5}$ cm^{-3}, $T \sim 10^4$ K) to hot interstellar medium (HIM; $n \sim 10^{-2.5}$ cm^{-3}, $T \sim 10^{5.7}$ K). SNRs are thought to generate the HIM and pressurize the ISM (Cox and Smith 1974, McKee and Ostriker 1977), so to a significant extent SNRs govern the inhomogeneity of the ISM.

The cloudy structure of the ISM strongly influences the evolution and appearance of SNRs. The nature of the clouds encountered by the SNR depends on its age and type. Young SNRs, which are dominated by the ejecta, may have to contend with circumstellar material left by the pre-supernova star, as in the case of the quasi-stationary flocculi in Cas A (Peimbert and van den Bergh 1971) and possibly the "jet" in the Crab (Blandford et al. 1982). Massive type II SN may occur in a wind blown bubble (Castor, McCray, and Weaver 1975) such as those observed around Wolf-Rayet stars (Chu 1981); the low density in the bubble could allow the ejecta (plus shocked circumstellar gas) to penetrate to large radii, of order 20 pc. Outside the bubble, the strong ionizing radiation from the pre-supernova star tends to destroy the clouds and homogenize the medium out to the Stromgren radius. Massive type II SN tend to occur in stellar associations, where the combined effect of many

J. Danziger and P. Gorenstein (eds.), Supernova Remnants and their X-Ray Emission, 87–97.
© 1983 by the IAU.

stellar winds and SN is to produce a very large low density region, a
"superbubble" (Bruhweiler et al. 1980). On the other hand Type I SN and
relatively low mass or high velocity Type II's--which together probably
comprise the majority of SN--are likely to occur in the typical ISM with
the wide range of inhomogeneities described above.

2. YOUNG SNRs: THERMALIZATION OF THE ENERGY

Young SNRs are approximately in a state of free expansion and are
characterized by having most of their energy in kinetic form. The
conversion of a significant fraction of this energy into thermal energy
is accomplished by shock waves, one propagating outward into the ambient
medium and one (the reverse shock) propagating inward through the ejecta
(Ardavan 1973, Kahn 1974, McKee 1974). If the ambient medium and/or the
ejecta is cloudy, then the thermalization process and hence the onset of
the Sedov-Taylor phase of blast wave evolution is delayed.

The evolution of a young SNR is particularly simple if both the
ejecta and the ambient medium are homogeneous. Let

$$R_A = \left(\frac{3M_E}{4\pi\rho_A}\right)^{1/3} = 2.1\left(\frac{M_E/M_\odot}{n_A}\right)^{1/3} \text{ pc} \tag{1}$$

be the radius at which the swept up mass equals the ejected mass; ρ_A is
the mass density, and n_A the baryon number density, in the ambient gas
(for cosmic abundances, n_A is 1.4 times the hydrogen density). One can
readily show that the reverse shock traverses much of the ejecta,
thermalizing it in the process, by the time at which $R = R_A$. This is
confirmed by numerical calculations (Gull 1975), which show that the
thermal energy in the swept up gas has reached about half its final
value of 0.71 E when $R = R_A$; the thermal energy of the ejecta never
amounts to more than a few percent of E.

2.1 Homogeneous Ejecta Expanding into a Cloudy Medium

In the more realistic situation of a cloudy ambient medium (but
with uniform ejecta) we distinguish the cloud density ρ_{Ac} from the
intercloud density ρ_{Ai}, etc. The radius R_{Ai} at which the swept-up
intercloud mass equals the ejected mass is given by equation (1) with ρ_A
replaced by ρ_{Ai}; for $n_{Ai} \sim 3 \times 10^{-3} \text{ cm}^{-3}$ as in the HIM, this gives
$R_{Ai} \simeq 13(M/M_\odot)^{1/3}$ pc. Usually the clouds, which occupy a fraction
$f_{Ac} \ll 1$ of the volume, have most of the mass ($\rho_{Ac} f_{Ac} > \rho_{Ai}$), so that

$$R_{Ac} = (3 M_E/4\pi \rho_{Ac} f_{Ac})^{1/3} \tag{2}$$

is smaller than R_{Ai}.

There will be an outgoing shock in the intercloud medium and a
reverse shock in the ejecta. The energy of the ejecta can be
thermalized by the reverse shock and by collisions with the ambient

clouds. For the moment, assume that the intercloud medium is
sufficiently tenuous that the ambient clouds dominate the thermalization
process, and neglect evaporation and stripping of the clouds.
Approximate the ejecta as being in a thin spherical shell with a column
density of baryons N_E. (Chevalier 1982 has considered ejecta of
variable density interacting with a uniform ambient medium.) Collision
with a cloud of column density N_{Ac} results in a fraction $N_{Ac}/(N_{Ac} + N_E)$
of the initial kinetic energy being converted to heat. Now the
probability of intersecting a cloud in an interval dR along a given line
of sight is dR/λ_{Ac}, where λ_{Ac} = (density of ambient clouds x the cross
section of clouds)$^{-1}$ is the cloud mean free path. In terms of λ_{Ac} the
cloud column density is $N_{Ac} = n_{Ac}f_{Ac}\lambda_{Ac}$. Neglecting adiabatic expansion
losses, we find that the increase in thermal energy E_{th} is

$$\frac{dE_{th}}{dR} \approx \frac{E}{(R_{Ac}^3/3R^2) + \lambda_{Ac}} \tag{3}$$

with the aid of equation (2). Solution of equation (3) shows that the
energy of the ejecta is entirely thermalized (E_{th} = E) at a radius
$R_{Ac} + \lambda_{Ac}$ to within 20%. The intercloud medium will thermalize the
ejecta via the reverse shock at R_{Ai}, so altogether the thermalization
will be complete at

$$R_{therm} \approx \min[(R_{Ac} + \lambda_{Ac}), R_{Ai}]. \tag{4}$$

This radius marks the end of the "young" stage of SNR evolution.

As an example, consider an SNR in a three phase ISM. For
$M_E = 1.4 M_0$, and with the ISM parameters given by McKee and Ostriker
(1977), we find R_{Ac} = (2 pc, 5 pc) and λ_{Ac} = (88 pc, 12 pc) for the cold
cloud cores and warm cloud envelopes, respectively. The mean free path
of 88 pc for the cold cores is sufficiently large to ensure that they
are ineffective at thermalizing the ejecta. Rather, it is the warm
envelopes ($R_{Ac} + \lambda_{Ac}$ = 17 pc) together with the HIM (R_{Ai} = 15 pc) which
thermalize the ejecta. For an ejected mass $M_E = 10 M_0$, R_{Ac} and R_{Ai} are
about twice as large but λ_{Ac} is unchanged; the warm clouds would
thermalize the ejecta at $R_{therm} \approx$ 22 pc.

Evaporation (Cowie and McKee 1977, Balbus and McKee 1982) and/or
hydrodynamic stripping (Nulsen 1982) of the clouds could accelerate the
thermalization somewhat by increasing the intercloud density and thereby
driving R_{Ai} closer to R_{Ac}. The dynamical effects of evaporation and
stripping in young SNRs are limited by two factors, however: the short
time available prevents a steady evaporative flow from being
established, and the evaporated or stripped gas will tend to be blown
behind the cloud, reducing its interaction with neighboring ejecta which
did not hit the cloud. The evaporation will generally be suprathermal,
with the hot electrons freely penetrating the surface of the evaporating
cloud. The time to establish steady flow is of order $(a^2\mu/\Gamma)^{1/3}$, where
Γ and μ are the heating rate and mass per particle (Balbus and McKee
1982); in the shocked HIM, this is 1700 $a_{pc}^{2/3}T_{18}^{1/6}$ yr, where a_{pc} is

the cloud radius in parsecs and T_{18} is the intercloud temperature normalized to 10^8 K. A freely expanding SNR would reach a radius of order 10 pc in this time. The heated cloud material will emit X-rays at a temperature $8 \times 10^5 \, t_{yn_1}/T_{18}^{1/2}$ K, where t_y is the heating time in years. The numerical simulations of SNR evolution in a cloudy medium by Cowie, McKee, and Ostriker (1981) focussed on the later stages of the evolution and did not include these effects.

2.2 Cloudy Ejecta

Young SNRs like Cas A are considerably more complicated than the model above because the ejecta are also inhomogeneous. SN ejecta can fragment in several ways: (1) Due to Rayleigh-Taylor instability at the interface between the core and envelope of the exploding star (Falk and Arnett 1973, Chevalier 1975, 1976). The resulting clumps should have anomalous abundances. A significant fraction of the mass of the ejecta should remain in an intercloud medium. (2) Due to Rayleigh-Taylor instability at the interface between the ejecta and the swept up ISM (Gull 1973). If the development of this instability is limited by viscosity (Cowie 1975) only a small fraction of the mass of the ejecta may be involved. (3) Due to interaction with an embedded pulsar. We shall not consider this case here.

How is the thermalization radius R_{therm} estimated in equation (4) altered by clumping of the ejecta? The analysis proceeds exactly as in §2.1 except that the column density of the ejecta N_E may be increased by the clumping. Define the ejecta covering factor C_E as the fraction of the sky covered by ejecta as seen from the site of the explosion. If ω_{Ec} is the density of ejecta clouds and σ_{Ec} their cross section, and if the clouds are randomly distributed in direction, then the "optical depth" of the clouds is $\tilde{\tau}_E = \int \omega_{Ec} \, \sigma_{Ec} \, dR$ and $C_E = 1 - \exp(-\tilde{\tau}_E)$. The mean value of the ejecta column density is

$$ N_E = \frac{M_E}{4\pi R^2 m_p C_E} = \frac{1}{3} \frac{n_{Ac} f_{Ac} R_{Ac}^3}{R^2 C_E} . \tag{5} $$

We then recover equation (3) with R_{Ac} replaced by $R_{Ac} C_E^{-1/3}$. A similar argument shows that R_{A1} in equation (4) should be increased by the same factor, so that altogether

$$ R_{therm} \simeq \min[(\lambda_{Ac} + R_{Ac} C_E^{-1/3}), \, R_{A1} C_E^{-1/3}]. \tag{6} $$

This result will differ significantly from that for homogeneous ejecta, equation (4), only if $C_E \ll 1$. It is difficult to estimate C_E theoretically, but it is unlikely to be very small. Once the clouds are pressurized by the reverse shock, it is possible that σ_{Ac} will remain about constant and $\tilde{\tau}_E$ will fall off as R^{-2}; however, this occurs at $R \sim O(R_{A1})$ and cannot have a large effect by R_{therm}. If the shocks which compress the clouds are radiative, very high cloud densities can be attained, but the compression will be primarily one dimensional and will not significantly decrease $\tilde{\tau}$.

2.3 Comparison with Observation

A number of techniques have been developed to differentiate young SNRs from those in the Sedov phase, in which the dynamics are governed by the swept up ISM. Several observations suggest the young stage lasts well beyond the canonical estimate of $2(M_E/M_0)^{1/3}$ pc appropriate for a uniform ISM. By measuring the pressure in optical filaments in a number of SNRs in the LMC, Dopita (1979) inferred that the thermal energy E_{th} increased rapidly with SNR radius for $R \lesssim 15$ pc. This effect has been confirmed in a broader sample by Blair, Kirshner, and Chevalier (1981), but they pointed out that this procedure would underestimate E_{th} if the optical filaments were supported by magnetic pressure. A further complication in the interpretation is introduced by the fact that young SNRs are far from isobaric (see §3): the optical filaments may be shocked clouds in low pressure regions of the remnant, whereas the same clouds exposed to the mean SNR pressure would be heated to X-ray temperatures. In fact X-ray observations provide a better indication of the mean pressure, but their interpretation is complicated by the effects of the enhanced emissivity of the ejecta. Bearing these problems in mind, note that an approximate integration of equation (3) (which is more illuminating than the exact expression) gives

$$E_{th}/E \simeq R^3 / (R_{Ac}{}^3 + R^2 \lambda_{Ac}) \qquad (R < R_{therm}) \qquad (7)$$

to within 30%. For the values of R_{Ac} and λ_{Ac} in a three phase ISM quoted in §2.1, this is qualitatively consistent with the observed trend, although the predicted range of variation of E_{th}/E for $R > 5$ pc is smaller than the observed range.

With a complete sample of SNRs, it is possible to determine their dynamics statistically. If SNRs occur at a rate $\dot{N}$, then the number of SNRs smaller than R is $N(<R) = \dot{N}t(R)$, where $t(R)$ is the remnant age. If $R \propto t^m$, then $N(<R) \propto R^{1/m}$. Applying this technique to radio-selected SNRs in the LMC, Clarke (1976) found $N(<R) \propto R$ for $R < 25$ pc, which implied that either the remnants were freely expanding or the sample was incomplete; he favored the latter. Mathewson (1983) and Mills (1983), using SNR samples based on recent radio, optical, and X-ray observations, have substantiated the result that the LMC SNRs are freely expanding for $R < 20$ pc. This is consistent with equations (4) or (6) provided most of these SNRs are expanding in a very low density medium. This result is remarkable because $N(<R)$ tends to be dominated by the oldest SNRs, which are in the densest surroundings and are most luminous.

3. X-RAY EMISSION FROM YOUNG SNRS

3.1 Pressure Variations and Emissivity

Density variations in the ejecta and the ISM lead to large pressure variations in young SNRs, and these density and pressure variations

strongly affect the appearance of the SNR. To emit X-rays, shocked gas
in an SNR must be in the correct temperature range. Describe the
pressure p of the radiating gas by the characteristic density
$\rho_p \equiv p/v_E^2$; then the temperature of the radiating gas is

$$T = 260 \ (\rho_p/\rho_0)(v_{E9}^2/x_t) \ \text{keV} \tag{8}$$

where ρ_0 is the preshock density, x_t is the number of particles per
baryon (= 1.64 for a fully ionized gas of cosmic abundances), and
$v_{E9} = v_E/(10^9 \ \text{cm s}^{-1})$. The lowest pressure in the SNR is in the
unshocked ejecta, where $p \simeq 0$. In the shocked intercloud gas between
the outgoing and reverse shocks, the pressure is about $\rho_{Ai}v_E^2 = 7 \times 10^{-9}$
$(n_{Ai}/0.003 \ \text{cm}^{-3}) \ v_{E9}^2 \ \text{dyne cm}^{-2}$; a gas of cosmic abundances must be
initially less dense than $0.5(n_{Ai}/0.003 \ \text{cm}^{-3}) \ v_{E9}^2 \ \text{cm}^{-3}$ in order to be
shocked to a temperature above 1 keV by this pressure. Ejecta clouds
interacting with the ambient intercloud medium are subjected to this
pressure also, but ambient clouds of density ρ_{Ac} overtaken by ejecta of
density ρ_E are subjected to a greater pressure, amounting to $\rho_E v_E^2$ for
$\rho_E \ll \rho_{Ac}$.

The highest pressures in the remnant occur in the collision between
an ejecta cloud and an ambient cloud. Shocks are driven into the two
clouds at velocities v_{Ecs} and v_{Acs}, respectively, such that
$v_{Acs} + v_{Ecs} \simeq v_E$ and $p \simeq \rho_{Ac}v_{Acs}^2 \simeq \rho_{Ec}v_{Ecs}^2$. The pressure is $\rho_p v_E^2$
(see eq. 8), where

$$\rho_p^{-1/2} \simeq \rho_{Ac}^{-1/2} + \rho_{Ec}^{-1/2} \ ; \tag{9}$$

more crudely, we have $p \simeq \min(\rho_{Ac},\rho_{Ec}) \ v_E^2$. Both clouds will emit
X-rays, provided the density ratio is not too large. The luminosity of
the shocked region in each cloud is governed by the abundances (McKee
1974; Long, Dopita, and Tuohy 1982), temperature, and volume emission
measure,

$$VEM = \int n_e^2 \ dV \propto \rho^2 v_s \propto \rho^{3/2}p^{1/2} \ , \tag{10}$$

where we have assumed the shock has not crossed the cloud and that the
shock is fast enough ($v_s >$ few hundred km s^{-1}) that it is non-radiative.
Note that the initially denser gas strongly dominates the volume
emission measure and, being cooler, is a stronger source of soft
X-rays.

3.2 Secondary Blast Waves

The impact of ejecta, whether homogeneous or cloudy, upon an
ambient cloud creates a localized region of high pressure which expands
outward. In some cases the expansion is analogous to a blast wave, and
we term it a secondary blast wave; it will manifest itself as an X-ray
bright spot. Consider an ejecta cloud which is small compared to the
ambient cloud it strikes. The energy $(1/2)M_{Ec}v_E^2$ of the ejecta cloud is
dissipated in a time of order $t_d = m_p N_{Ec}v_E/\rho_p v_E^2$, where ρ_p is given by

equation (9). This corresponds to a column of shocked ambient cloud $N_{Acs} = n_{Ac}v_{Ac}s_{td}$, or

$$N_{Acs} = [1 + (\rho_{Ac}/\rho_{Ec})^{1/2}]N_{Ec} . \tag{11}$$

If the stopping length N_{Acs}/n_{Ac} is large compared to the size a_{Ec} of the ejecta cloud (which requires $\rho_{Ec} \gg \rho_{Ac}$), then this energy is distributed in the wake of the decelerating ejecta cloud. On the other hand, if the stopping length is $\lesssim a_{Ec}$, then the energy is initially confined to a volume of order a_{Ec}^3 and a secondary blast wave is produced. The evolution of the blast wave is complicated by the fact it occurs near the interface between the ambient cloud and the adjacent intercloud medium, but it passes through the same stages as a standard blast wave. Energy injection, corresponding to the SN explosion, occurs during the deceleration of the ejecta cloud. Next comes the free expansion phase, which lasts much longer in the intercloud medium than in the ambient cloud. Once the ejecta are thermalized in the cloud, mass and energy are blown from the cloud into the surrounding intercloud medium until the pressures are balanced. If the sound speed in the intercloud medium is small compared to v_E, then the pressure in the blast wave when the swept up intercloud mass equals the ejecta mass will be large compared to the intercloud pressure, and the blast wave will enter the Sedov-Taylor phase. The shocked cloud has a disk-like shape in this phase: the area corresponds to the radius of the blast wave in the intercloud medium, but the depth is smaller by $(\rho_i/\rho_{Ac})^{1/2}$. The non-equilibrium X-ray luminosity (Hamilton, Sarazin, and Chevalier 1982) is reduced from that due to a spherical blast wave in a medium of density ρ_{Ac} by $(\rho_i/\rho_{Ac})^{1/4}$.

3.3 MSH 15-52

The SNR MSH 15-52 appears to be a middle-aged remnant with a radius of 17 pc at a distance of 4.2 kpc. However, it was recently discovered to contain a pulsar with a spin down age of only 1600 yr (Seward and Harnden 1982). If this is the true age of the SNR, a possibility discussed by Seward et al. (1982), then it must still be in the free expansion phase: 17 pc/1600 yr = 10^4 km s^{-1}. As discussed in §2 above, this is possible provided the density of the interstellar medium is low, particularly if the ejecta are clumped so that their covering factor C_E is less than unity.

Direct evidence for the effect of cloudy ejecta may be provided by the two small bright X-ray sources near the rim of the SNR (Seward et al. 1982). The brighter one has a radius of 0.1 pc, a temperature of about 1 keV, an X-ray luminosity of 2 x 10^{34} erg s^{-1}, and a pressure estimated as (3 x 10^{-7}, 1.5 x 10^{-8}, 1.5 x 10^{-7}) dyne cm^{-2} based on (X-ray, optical, IR) observations. Since the IR and optical filaments may be supported by magnetic pressure (Hollenbach and McKee 1979, Blair et al. 1981), the pressure in the X-ray filament is more indicative of the total pressure in the region. This pressure is 400/E_{51} times greater than the mean pressure in an SNR with R = 17 pc and

$E = 10^{51} E_{51}$ erg. If we neglect the possibility of energy transport by jets such as those in SS433 (Begelman et al. 1980), such a localized region of high pressure in the remnant can be accounted for only by the impact of highly clumped ejecta upon an ambient cloud. For example, if we are viewing the collision just after the ejecta has been decelerated so that the column density of shocked gas in the ambient cloud is given by equation (11), then the observations can be accounted for by an ejecta cloud of radius 0.1 pc and density 0.33 cm^{-3} striking a larger ambient cloud of density 60 cm^{-3} at 10^4 km s^{-1}. The ejecta will be shocked to T > 100 keV and will be effectively invisible; the integrated effect of many such collisions might produce a detectable hard X-ray source, however. Pressure balance ensures that the ambient cloud will be shocked to T ∼ 1 keV; its X-ray luminosity will agree with the observed value if the non-equilibrium emissivity of Hamilton et al. (1982) is used [$\Lambda(0.2-2$ keV$) \simeq 1.5$ x 10^{-22} erg cm^3 s^{-1}]. An ejecta cloud with these properties would have decelerated significantly in passing through an intercloud medium with $n_{Ai} \gtrsim 3$ x 10^{-3} cm^{-3}, and it would have evaporated relatively quickly (Balbus and McKee 1982). Both problems would be alleviated if the ejecta cloud had a comparable mass to that assumed above, but 10 times the density, and if we were observing the collision after a blast wave had developed in the ambient cloud. Note that the time scale for variability could be as short as 0.1 pc/10^4 km s^{-1} ∼ 10 yr. Diffuse X-ray emission from this region of the SNR could be the result of many smaller or older impacts such as this.

4. MIDDLE AGED SNRs

     When the swept up interstellar mass exceeds the ejected mass, then the dynamics and appearance of the SNR are governed by the shocked ISM, and, as for young SNRs, are strongly affected by the inhomogeneity of the ISM. The structure of the ISM may in turn be determined by SNRs to a large extent (Cox and Smith 1974, McKee and Ostriker 1977). Here we shall briefly discuss the effect of an inhomogeneous ISM on the X-ray emission of SNRs.

     In a two-phase ISM (Field, Goldsmith and Habing 1969), cold clouds (T ∼ 10^2 K, $n_H \sim 10^{1.5}$ cm^{-3}) are embedded in a warm intercloud medium (T ∼ 10^4 K, $n_H \sim 10^{-0.5}$ cm^{-3}). The filling factor of the clouds is small (∼ few percent), so SNRs evolve as Sedov-Taylor blast waves in the warm medium. Clouds shocked to the mean pressure of the blast wave reach a temperature

$$T_c \simeq 2.2 \text{ x } 10^6 \text{ } E_{51} \text{ } [n_H(R/20 \text{ pc})^3]^{-1} \text{ K.} \tag{12}$$

For $n_H \gg 1$ cm^{-1}, the shocked clouds emit X-rays only in the early stages, R ≲ 15 pc. Once T_c drops to about 5 x 10^5 K, the shocked clouds cool and produce optical emission lines, as in the case of the Cygnus Loop. X-ray emission from the shocked intercloud medium follows the normal result for Sedov blast waves (e.g., Hamilton et al. 1982) and dominates the X-ray luminosity once the shocked clouds are too cool to emit X-rays. The SNR becomes radiative and forms a dense shell at $R_c \simeq 20\ E_{51}{}^{0.29}\ n_0{}^{-0.41}$ pc, where n_0 is the intercloud density; very few such shells have been seen, which argues against the pervasiveness of the two phase ISM.

In a three-phase ISM the warm medium is itself clumped and most of space is occupied by the hot interstellar medium ($T \sim 10^{5.7}$ K, $n_H \sim 10^{-2.5}$ cm^{-3}). SNRs expand to large radii in this medium, $R \gtrsim 100$ pc. Warm clouds are shocked to X-ray temperatures ($T \gtrsim 2 \times 10^6$ K) only for $R \lesssim 30\ E_{51}{}^{1/3}$ pc. Two different three phase models have been proposed. In one, there is little mass transfer between warm or cold clouds and the HIM (Cox 1979). SNRs expand as Sedov-Taylor blast waves until they are slowed by cloud drag at large radii. The expansion velocities are extremely large, $v \simeq 4000\ E_{51}{}^{1/2}(R/20\ \mathrm{pc})^{-3/2}$ km s^{-1}, which should be readily observable in galactic SNRs. The age of an SNR is correspondingly short, $t \sim 5000$ yr at $R \sim 30$ pc. The shocked HIM is ineffective at radiating X-rays because of its low density, so X-ray emission from an SNR essentially ceases for $R \gtrsim 30\ E_{51}{}^{1/3}$ pc. It is difficult to construct a self-consistent, steady-state model for the ISM of this type (McKee 1982).

On the other hand, in the three phase model proposed by McKee and Ostriker (1977) mass transfer between the clouds and the HIM is assumed to be efficient. They focussed on thermal evaporation of clouds, but turbulent stripping of the clouds (Nulsen 1982) may also contribute to the mass transfer. Under typical conditions, the model predicts that an SNR at $R = 20$ pc in a three phase ISM should be similar in mean density and age to a Sedov SNR in a medium with $n \simeq 0.2$ cm^{-3}. Thereafter the density of evaporated gas declines as $R^{-5/3}$. X-ray emission is produced by evaporated cloud gas as well as by shocked clouds, and it should be observable out to radii well in excess of 30 pc. The temperature of the evaporated gas decreases slowly, $T \propto R^{-4/3}$, whereas that of the shocked cloud drops as R^{-3}. The resulting spectrum has been calculated by Cowie et al. (1981) with the crude approximation of ionization equilibrium. Observational tests of the model have been discussed by McKee (1982).

5. CONCLUSIONS

Young SNRs are often observed to have clumpy ejecta and to be expanding into an inhomogeneous medium. There are two major effects of this inhomogeneity: the transition from a young SNR, dominated by the ejecta, to a middle-aged SNR, dominated by swept up ISM, occurs at a much greater radius than in the homogeneous case; and there can be large deviations from pressure equilibrium. The relative importance of clouds and intercloud medium in thermalizing the ejecta and thereby terminating the young stage was analyzed in §2. The observation that SNRs in the LMC follow an approximately linear number-radius relation for $R \lesssim 20$ pc implies that either (1) virtually all SNRs in the LMC occur in localized regions of low density, such as wind-blown bubbles, or (2) most of the volume of the LMC is filled by very low density gas, as in three phase models of the ISM. The second possibility is more likely.

In contrast to middle-aged SNRs, which are approximately isobaric, young SNRs have shock-heated gas ranging in pressure from (intercloud density) x v_E^2 to (cloud density) x v_E^2. There are correspondingly large spatial variations in the X-ray emissivity. The collision of an ejecta cloud with an ambient cloud can result in a secondary blast wave, a remote, miniature version of the original explosion. MSH 15-52, a relatively large SNR with a young pulsar, can be accounted for by highly clumped ejecta expanding into a low density medium and producing X-ray hot spots upon collision with ambient clouds.

The observational problem of determining the evolution of SNRs in later stages, when the swept up ISM is dominant, is to a large extent equivalent to determining which of the various ISM models—two phase, three phase with or without evaporation, or other—is correct. Studies of young SNRs may provide a valuable new tool in this effort.

ACKNOWLEDGMENTS

I wish to thank Roger Blandford and Charles Kennel for a number of valuable conversations on MSH 15-52, David Eichler for a helpful discussion on young SNRs, and the Institute for Theoretical Physics for providing the opportunity to have these discussions. This research is supported in part under NSF grant AST 79-23243.

REFERENCES

Ardavan, H.: 1973, Ap.J. 184, p. 435.
Balbus, S.A., and McKee, C.F.: 1982, Ap.J. 252, p. 529.
Begelman, M., Sarazin, C., Hatchett, S., McKee, C., and Arons, J.: 1980, Ap.J. 238, p. 722.
Blair, W.P., Kirshner, R.P., and Chevalier, R.A.: 1981, Ap.J. 247, p. 879.
Blandford, R., Kennel, C., McKee, C.F., and Ostriker, J.P.: 1982, Nature (in press).
Bruhweiler, F.C., Gull, T.R., Kafatos, M., and Sofia, S.: 1980, Ap.J. (Letters) 238, p. L27.
Castor, J., McCray, R., and Weaver, R.: 1975, Ap.J. (Letters) 200, p. L107.
Chevalier, R.A.: 1975, Ap.J. 200, p. 698.
Chevalier, R.A.: 1976, Ap.J. 207, p. 872.
Chevalier, R.A.: 1982, Ap.J. 258, p. 790.
Chu, Y.H.: 1981, Ph.D. Thesis, U.C. Berkeley.
Clarke, J.N.: 1976, M.N.R.A.S. 174, p. 393.
Cox, D.P.: 1979, Ap.J. 234, 863.

Cox, D.P., and Smith, B.W.: 1974, Ap.J. (Letters) 189, p. L105.
Cowie, L.L.: 1975, M.N.R.A.S. 173, p. 429.
Cowie, L.L., and McKee, C.F.: 1977, Ap.J. 211, p. 135.
Cowie, L.L., McKee, C.F., and Ostriker, J.P.: 1981, Ap.J. 247, p. 908.
Dopita, M.A.: 1979, Ap.J. Suppl. 40, p. 455.
Falk, S.W., and Arnett, W.D.: 1973, Ap.J. (Letters) 180, p. L65.
Field, G.B., Goldsmith, D.W., and Habing, H.J.: 1969, Ap.J. (Letters)
    155, p. L149.
Gull, S.F.: 1973, M.N.R.A.S. 161, p. 47.
Gull, S.F.: 1975, M.N.R.A.S. 171, p. 263.
Hamilton, A., Sarazin, C., and Chevalier, R.: 1982, Ap.J. (in press).
Hollenbach, D., and McKee, C.F.: 1979, Ap.J. Suppl. 41, p. 555.
Kahn, F.D.: 1974, in "The Interstellar Medium," ed. K. Pinkau
    (Dordrecht:Reidel), p. 235.
Long, K., Dopita, M., and Tuohy, T.: 1982, Ap.J. 260, p. 202.
Mathewson, D.S.: 1983, this volume, p. 553.
McKee, C.F.: 1974, Ap.J. 188, p. 335.
McKee, C.F.: 1982, in "Supernovae: A Survey of Current Research," M.
    Rees and R. Stoneham, eds. (Dordrecht: Reidel), p. 433.
McKee, C.F., and Ostriker, J.P.: 1977, Ap.J. 218, p. 148.
Mills, B.Y.: 1983, this volume, p. 563.
Nulsen, P.E.J.: 1982, M.N.R.A.S. 198, p. 1007.
Peimbert, M., and van den Bergh, S.: 1971, Ap.J. 167, p. 223.
Seward, F.D., and Harnden, F.R.: 1982, Ap.J. (Letters) 256, p. L45.
Seward, F.D., Harnden, F.R., Murdin, P., and Clark, D.H.: 1983, Ap.J.
    (in press).

DISCUSSION

RADHAKRISHNAN: Would you comment on the high ambient densities for the
historical SNRs which were mentioned this morning?

MCKEE: For SNRs high above the plane, such as Tycho, a circumstellar
rather than interstellar origin for this gas appears more reasonable.
Chevalier has argued against this possibility by showing that the
observed evolution is more consistent with a uniform ambient medium than
a wind. Perhaps this problem could be overcome if the wind, which would
come from the red giant precusor of the SN, were confined by the
pressure of the ambient medium.

FEDORENKO: Is it possible to associate the observed turbulence in Cas A
with multiple shocks produced by propagation of the expanding envelope
in a cloudy medium?

MCKEE: Such shocks undoubtedly contribute to the turbulence, as do the
Rayleigh-Taylor instability in the ejecta (Gull 1973) and cloud-cloud
collisions.

NON-EQUILIBRIUM IONIZATION X-RAY EMISSION FROM SUPERNOVA REMNANTS

J. Michael Shull
Joint Institute for Laboratory Astrophysics, University of
Colorado and National Bureau of Standards, Boulder, Colorado
80309

I. INTRODUCTION

X-ray spectra of young supernova remnants (SNR's) are perhaps the
most spectacular examples of hot, line-emitting astrophysical plasmas.
Heated to temperatures of 1 to 10 keV and enriched with the heavy
element products of stellar nucleosynthesis, the plasma inside these
SNR's emits prodigiously in lines of O, Ne, Mg, Si, S, Ar, Ca, and Fe.
Theoretical models of this emission provide measures of the plasma
temperature and density, elemental abundances, and the degree of approach to ionization equilibrium. Thus, astrophysicists are offered
the opportunity to test their understanding of the supernova explosion, its interaction with the interstellar medium, and the nucleosynthetic processes which enrich our galaxy with heavy elements.

However, the practical task of modelling the SNR X-ray spectra is
fraught with technical difficulties, primarily questions of the internal hydrodynamics and ionization state of the SNR. In this review, I
will describe recent X-ray spectral observations of young remnants and
will summarize the current state of non-equilibrium ionization modelling.

II. OBSERVATIONAL SUMMARY

Many young remnants (t < 1000 years) observed by HEAO-2 exhibit
the shell-like morphology predicted for SNR's in the adiabatic (Sedov)
phase. However, the existence of filled-shells such as the Crab
Nebula and Vela, and the patchiness of the emission in otherwise
spherical remnants suggests that departures from pure Sedov structure
are important. For example, a physically plausible case has been made
for the contribution of reverse-shocked ejecta to the X-ray emission
(McKee 1974; Gull 1975; Chevalier 1982).

The HEAO-2 spectrometers (SSS = Solid State Spectrometer, FPCS =
Focal Plane Crystal Spectrometer) observed prominent emission lines

J. Danziger and P. Gorenstein (eds.), Supernova Remnants and their X-Ray Emission, 99–107.

from He-like ionization stages of many heavy elements, as well as
other lines of Fe, O, and Ne. The SSS, with effective area 100 cm^{-2}
and energy resolution 160 eV from 0.6 to 4.5 keV (Becker et al.
1980a,b), demonstrated that the strongest lines arise from Si and S
(see Fig. 1). However, Ne, Mg, Ar, and Ca are also detected, together
with a blend of Fe L-shell lines between 0.8 and 1.4 keV. The FPCS,
with effective area 2-3 cm^{-2}, resolved the He-like "triplet" lines
(R = resonance, F = forbidden, and I = intercombination) of O VII and
Ne IX, provided important measurements of the H-like Lα and Lβ, and
resolved several lines of Fe XVII in Puppis A (Winkler et al. 1981).

Preliminary coronal ionization equilibrium analyses of the SSS
data (Becker et al. 1980a,b) required two temperature components to
fit the spectra: a hard component with kT = 5-10 keV to fit the continuum, and a soft component with kT ≈ 0.5 keV to fit the H-like to
He-like line ratios. These ionization equilibrium models for Tycho
suggested that Si and S were overabundant by factors of 10, but that
Fe was underabundant by a factor 0.15. Even stranger were the derived
abundance enhancements of Ar (×35) and Ca (×76). However, all of
these abundances are suspect because of the possibility of substantial
departures from ionization equilibrium.

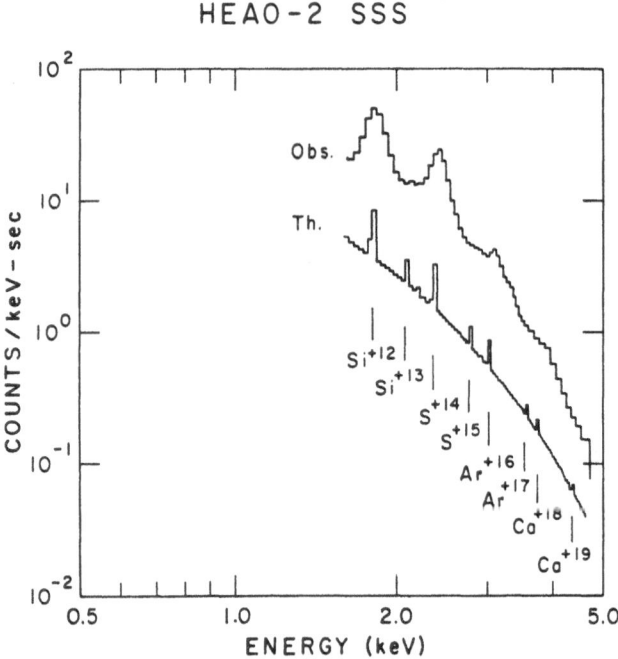

Fig. 1.   HEAO-2 SSS spectrum of Cas A (marked Obs.), plus theoretical
          spectrum for plasma in ionization equilibrium at 3 × 10^6 K
          (marked Th.). Positions of He-like and H-like emission lines
          of various heavy elements are marked. The observational line
          widths are instrumental; theoretical lines are binned in
          46 eV intervals.

III. THEORETICAL MODELS OF NON-EQUILIBRIUM EMISSION

     The application of ionization equilibrium (IE) models to young
SNR X-ray spectra is questionable for four major reasons: (1) the SNR
ages (400 to 1000 years) are comparable to the collisional ionization
times for He-like and H-like ions in low-density plasma; (2) the dramatically different temperatures (4 keV and 0.5 keV) required to fit
the continuum and lines, respectively, suggest that the plasma may be
in a transient, ionizing state; (3) the observed ratio G = (F + I)/R
of forbidden plus intercombination to resonance line intensity of He-
like O VII and Ne IX in Puppis A (Winkler et al. 1981) is lower than
that predicted for IE and characteristic of an underionized plasma
(Pradhan and Shull 1981); and (4) a comparison of S-line equivalent
widths to the Fe Kα line energy in Tycho (Pravdo et al. 1980) suggests
departures from equilibrium.

     Itoh (1977, 1979) showed that non-ionization equilibrium (NIE)
may have a significant effect on the emissivity of young remnants.
The physical effect is easy to understand: if the shocked plasma has
had insufficient time to ionize Si and S to their equilibrium state
at kT = 3-8 keV, then the fractional abundance in He-like stages and
the resulting emission line strengths will far exceed their values in
equilibrium. (The elements ionize "up the ladder" until they encounter the large ionization barrier from Li-like to He-like stages, at
which point they accumulate and radiate strongly in the He-like triplet lines.) In terms of the plasma emissivity, the situation is very
much like dropping an ice cube into a bath of hot water!

     Ionization equilibrium models of hot, optically thin, low density
plasmas (Raymond and Smith 1977; Shull 1981a) are relatively straight-
forward to construct. One simply solves for the steady-state ionization balance among competing rates of collisional ionization and
radiative plus dielectronic recombination, then computes the plasma
emission in the continuum and lines. The dominant processes which
contribute to this emission are:

     Continuum: 1) bremsstrahlung (free-free emission)
                  2) radiative recombination continua
                  3) two-photon continua (from H- and He-like ions)

     Lines: 1) electron collisional excitation (bound states)
                  2) radiative recombination cascade
                  3) dielectronic recombination satellite lines

Of these processes, heavy elements dominate the line emission while
H and He normally dominate the free-free emission. The electrons
which radiate are donated either by H and He, or in metal-enriched
plasmas from C, N, O, etc. Because the HEAO-2 SSS is not sensitive
to CNO line emission below 0.6 keV, the absolute abundances of heavy
elements are ambiguously determined. If CNO are enhanced in the
ejecta relative to H and He by a factor greater than about 100, then

the continuum emission is metal-dominated. Thus, in practice, the quoted abundances of Si, S, etc. are actually determined relative to an assumed cosmic abundance of CNO.

Non-equilibrium ionization models are far more complicated, because they require the specification of the past ionization history of every parcel of emitting plasma. While this ionization history has little effect on the free-free continuum, the emission line strengths are quite sensitive to the dominant ion stage at a given temperature. For young SNR's, non-equilibrium modelling involves the coupling of the gas hydrodynamics (density n and temperature T as a function of radius r) to a time-dependent ionization code and a spectral emissivity code. As a first step, most modellers have assumed an adiabatic Sedov interior solution to specify n and T. Only later have the effects of reverse shocked ejecta been considered.

In the Sedov solution (Sedov 1959; Taylor 1950), the radius, velocity, and temperature of the outer periphery are given by

$$R_s(t) = (4.97 \text{ pc})(E_{51}/n_o)^{1/5} t_3^{2/5} \tag{1}$$

$$V_s(t) = (1950 \text{ km s}^{-1})(E_{51}/n_o)^{1/5} t_3^{-3/5} \tag{2}$$

$$kT_s(t) = (4.54 \text{ keV})(E_{51}/n_o)^{2/5} t_3^{-6/5} \quad , \tag{3}$$

where E_{51} is the explosion energy in units 10^{51} ergs, n_o (cm^{-3}) is the ambient hydrogen number density assuming $\rho_o = 1.4 \text{ m}_H n_o$, and t_3 is the remnant age in units 10^3 years. By dividing the remnant into 100 shells, one may follow the thermal and ionization history of each parcel of the remnant interior by integrating rate equations for ionization, recombination, and the first law of thermodynamics (Shull 1982). Once the temperature, density, and ionization structure are known, one may compute the emergent X-ray spectrum by summing the shell emissivities, using a spectral code such as that described by Raymond and Smith (1977). The calculations to be described here used a recent spectral code (Shull 1981a) which incorporates the most accurate available collision strengths, particularly for ions in the He-like isosequence (Pradhan, Norcross, and Hummer 1981). Similar non-equilibrium ionization calculations have been made by Gronenschild (1979) and Hamilton, Sarazin, and Chevalier (1983).

The parameters of the non-equilibrium Sedov models may be expressed in a number of equivalent forms. In perhaps the simplest version, one specifies the ambient hydrogen number density (n_o), the explosion energy (E_{51}), the remnant age (t), and the metal abundances. The first three parameters determine the remnant's outer radius, velocity, and temperature. The metal abundance set is varied to fit the observed X-ray emission. Although it would be valuable to treat all parameters as free variables, we have fixed $E_{51} = n_o = 1$ for our standard models and varied only t and the metal abundances

(Hamilton, Sarazin, and Chevalier 1983 have computed a grid of models in which the scaling parameter, $E_{51} n_0^2$, is allowed to vary as well). For ease in comparing with equilibrium models of constant temperature, we will designate the non-equilibrium models by their immediate post-shock temperature T_s [see eq. (3)] rather than t.

Consider Sedov models with kT_s = 7.2 keV. The ionization structure and X-ray emission differ markedly in equilibrium and non-equilibrium. Because of the significant time for Si and S to ionize up to their equilibrium state, the He-like ion fractions are far greater in the non-equilibrium case. The overabundance of He-like ions in this hot plasma results in strong emission in the n = 2→1 lines, as well as a decrease in the ratio G = (F + I)/R of forbidden plus intercombination to resonance lines (Pradhan and Shull 1981) dictated by the increase in the relative importance of collisional excitation over recombination.

These non-equilibrium ionization enhancements in line emissivity affect both the derived metal abundances and ejecta masses. Figure 2 compares equilibrium and non-equilibrium X-ray spectra for a remnant with kT_s = 7.2 keV. The differences in He-like line strengths and He-like to H-like line ratios are apparent in the non-equilibrium case, as are the differences in Fe L-shell emission lines near 1 keV. A fit of SSS data on Tycho's remnant to this illustrative model (Shull 1981b, 1982) yielded the abundances in Table 1. In comparison with two-component ionization equilibrium models (Becker et al. 1980a), the "one component" non-equilibrium ionization model gave slightly higher chi-square per degree of freedom, but resulted in perhaps more believable abundance enhancements for Ar, Ca, and Fe. Iron is no longer underabundant, while the Ca and Ar enhancements are more in agreement with those of other elements. The Si and S abundances are affected by less than a factor of 2. More elaborate fitting will be described by Szymkowiak et al. (1983).

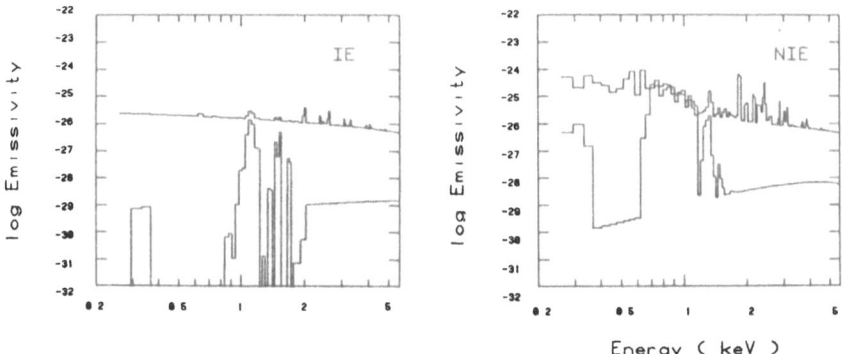

Fig. 2. X-ray emissivities (ergs cm^3 s^{-1} eV^{-1}) of Sedov remnant in ionization equilibrium (IE) and non-ionization equilibrium (NIE) at time t = 680 yr (kT_s = 7.2 keV, E_{51} = n_0 = 1). Abundances are those in Table 1. The two curves in each plot represent total emissivity and emissivity of Fe.

Table 1.   Model Elemental Abundances for Tycho[a]

Element	Equilibrium[b]	Non-Equilibrium[c]
Ne	1.	0.37
Mg	0.1	2.0
Si	6.0	7.6
S	13.5	6.5
Ar	34.6	3.2
Ca	76.	2.6
Fe	0.15	2.1

[a]Abundances relative to Solar Values, determined
from χ^2-fitting to HEAO-2 SSS data. H, He, and
CNO are assumed to be in solar abundance.

[b]Becker et al. (1980a), two-temperature component, ionization equilibrium model.

[c]Shull (1981b), single-velocity, non-ionization-
equilibrium model of SNR blast wave with immediate post-shock temperature, kT_s = 7.2 keV,
explosion energy 10^{51} ergs, and ambient H-
density n_o = 1 cm^{-3}.

Figure 3 illustrates the changes in broadband X-ray emissivities
produced by non-equilibrium ionization. The overabundance in He-like
ion stages results in an enhanced emissivity in the 1-3 keV range.
Supernova ejecta masses determined from equilibrium emissivities may
therefore overestimate the amount of emitting plasma by significant
factors (see Gorenstein, Seward, and Tucker 1983 for an explicit example of the application of these non-equilibrium emissivities to
ejecta masses in Tycho).

IV.   SUMMARY

Preliminary Sedov models indicate that Ar, Ca, and Fe abundances
are strongly affected by non-equilibrium ionization. The fact that
Fe, while not glaringly overabundant in Tycho, is at least present
in significant amounts is satisfying. Silicon and sulfur remain the
dominant contributors by mass to the X-ray emitting plasma. Likewise,
the ejecta masses inferred for Tycho are lower in non-equilibrium
models. Each of these inferences is important for ascertaining the
nature of the progenitor of the supernova.

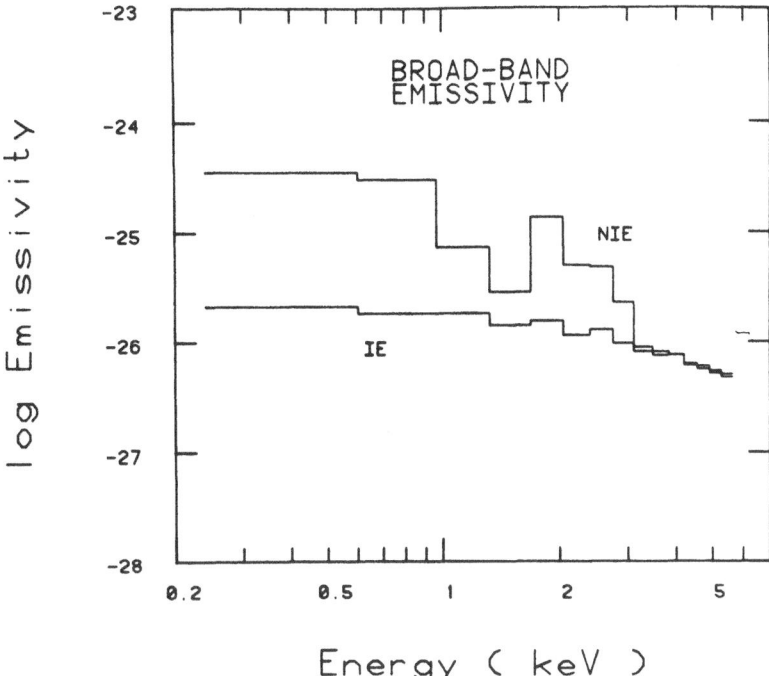

Fig. 3.   Broadband spectral emissivities (ergs cm^3 s^{-1} eV^{-1}) in ion-
          ization equilibrium and non-ionization equilibrium, averaged
          over 360 eV wide binds. Model parameters same as in Fig. 2.

These abundances, however, should still be regarded with some
skepticism, since a number of questions remain to be answered concerning the nature of the SNR hydrodynamics and plasma:

1.   What is the degree of ion-electron temperature equilibration?

2.   How far does the SNR depart from the idealized Sedov structure?

3.   How much mixing has occurred between ejecta and shocked interstellar gas?

4.   How certain are the atomic rates that go into the modelling?

5.   Can one exclude the possibility of heavy element depletion
into grains (or cold clumps)?

Some work has been done in these areas (see Sarazin, Hamilton,
and Chevalier 1983). The real answers, however, will most likely
await future X-ray spectrometers with higher spectral and spatial
resolution.

This work was supported by National Science Foundation grant
AST80-19960 through the University of Colorado.

REFERENCES

Becker, R.H., Holt, S.S., White, N.E., Boldt, E.A., Mushotzky, R.F.,
    and Serlemitsos, P.J.: 1980a, Astrophys. J. (Letters) 235, p. L5.
Becker, R.H., Holt, S.S., White, N.E., Boldt, E.A., Mushotzky, R.F.,
    and Serlemitsos, P.J.: 1980b, Astrophys. J. (Letters) 237, p.
    L77.
Chevalier, R.: 1982, Astrophys. J. 258, p. 790.
Gorenstein, E., Seward, F., and Tucker, W.: 1983, in IAU Symp. #101,
    this volume, p. 1.
Gronenschild, E.: 1979, Ph.D. Thesis, Utrecht.
Gull, S.F.: 1975, Monthly Notices Roy. Astron. Soc. 161, p. 47.
Hamilton, A.J.S., Sarazin, C.L., and Chevalier, R.A.: 1983, Astrophys.
    J. Suppl., in press.
Itoh, H.: 1977, Pub. Astron. Soc. Japan 29, p. 813.
Itoh, H.: 1979, Pub. Astron. Soc. Japan 31, pp. 429 and 541.
McKee, C.F.: 1974, Astrophys. J. 188, p. 335.
Pradhan, A.K., Norcross, D.W., and Hummer, D.G.: 1981, Astrophys. J.
    246, p. 1031.
Pradhan, A.K. and Shull, J.M.: 1981, Astrophys. J. 249, p. 821.
Pravdo, S.H., Smith, B.W., Charles, P.A., and Tuohy, I.R.: 1980,
    Astrophys. J. (Letters) 235, p. L9.
Raymond, J.C. and Smith, B.W.: 1977, Astrophys. J. Suppl. 35, p. 419.
Sarazin, C.L., Hamilton, A.J.S., and Chevalier, R.A.: 1983, in IAU
    Symp. #101, this volume, p. 109.
Sedov, L.I.: 1959, Dimensional Methods in Mechanics (New York:
    Academic Press).
Shull, J.M.: 1981a, Astrophys. J. Suppl. 46, p. 27.
Shull, J.M.: 1981b, in Proc. Symp. on X-Ray Astronomy and Spectroscopy
    (Goddard Space Flight Center, Oct. 1981), NASA Tech. Memo. 83848,
    p. 107.
Shull, J.M.: 1982, Astrophys. J. 262, p. 308.
Szymkowiak, A.E., Shull, J.M., Hamilton, A.K.S., and Holt, S.S.: 1983,
    in preparation.
Taylor, G.I.: 1950, Proc. Roy. Soc. London A 201, p. 159.
Winkler, P.F., Canizares, C.R., Clark, G.W., Markert, T.H., Kalata,
    K., and Schnopper, H.W.: 1981, Astrophys. J. (Letters) 246, p.
    L27.

DISCUSSION

COX: Shouldn't we be calling these models NEI rather than NIE? Also,
when you model remnants, do you fix E_{51} or derive it?

SHULL: NEI is probably appropriate, but NIE is analogous to non-LTE.
I fix the SNR energy at 10^{51} ergs to avoid many model calculations.
Hamilton et al. (1983) have let the scaling parameter $E_{51}n_o^2$ vary.

DICKEL: What is the physical basis for the prominence of He-like and even-Z ionization stages?

SHULL: The odd-even variation in Z is just cosmic abundances (nuclear binding). The prevalence of He-like stages results from the "hang up" in the ionization at the large barrier between Li-like and (closed shell) He-like stages.

BISNOVATYI-KOGAN: Did you take into account excited states of different elements in the calculations?

SHULL: No. At the low interstellar densities, all ions are in their ground state.

PRAVDO: Can you comment on our finding that your model for Tycho violates the high-energy data by a large factor, while our fit (which is consistent with all the data) results in elemental abundances higher by a factor of 10.

SHULL: In your fit, you must lower the continuum temperature to about 4.6 keV, which then raises the line fluxes and metal abundances. I assumed kT_S = 7.2 keV for the post-shock temperature (quoting your published HEAO-1 results). Evidently you have changed your mind about the temperature needed to fit the high energy data. Perhaps the Maxwellian tail of the electron energy distribution is depleted, or perhaps the relative normalization between the two data sets is uncertain.

CANIZARES: A plea to modellers: could you produce a table of a dozen or so line emissivities for various temperatures in ionization equilibrium? These could be used to compare the various models and separate effects due to atomic physics from those due to astrophysics.

SHULL: I agree with your idea. However, the strong He-like lines on which many of the SSS abundance fits are based, probably have the best-determined collision strengths (20%) -- Pradhan, Norcross, and Hummer (1981). The Fe L-shell lines near 1 keV are more uncertain.

GRINDLAY: Can you comment on the relative importance of a cloud-filled ambient medium for non-ionization equilibrium effects?

SHULL: I suspect that high-density clouds would be nearer to ionization equilibrium. Separating clouds from intercloud matter might be difficult without both spatial and spectral resolution. Also, the importance of the cloud component might not appear until the remnants were older and larger, owing to the delayed effect of evaporation.

X-RAY LINE EMISSION FROM SUPERNOVA REMNANTS AND MODELS FOR NONEQUILIBRIUM
IONIZATION

Craig L. Sarazin, Andrew J.S. Hamilton, and Roger A. Chevalier
Department of Astronomy
University of Virginia

ABSTRACT

An extensive grid of nonequilibrium ionization models for the X-ray
spectra of adiabatic supernova remnants (SNRs) is described. The models
are compared to the SSS spectra of remnants in the LMC, the Tycho SNR,
and SNR 1006. In Tycho, we show that the observed spectrum requires
significantly enhanced abundances of Si and S, and that this conclusion
is independent of the detailed ionization and thermal structure in the
remnant. We find that the SSS spectrum of SNR 1006 can be fit reasonably by a thermal emission model with abundances of about one half solar.
In this model, the weak line emission results from the very low ionization state in the remnant, and not because the X-ray emission is non-
thermal. We argue that the failure to detect strong Fe line emission
in young Type I SNRs poses a severe problem for models of Type I SN,
which predict that most of the ejecta be iron. Finally, the results of
UV observations of a star behind SNR 1006 are mentioned; these observations show that the remnant contains a large amount of rapidly moving,
cold iron.

NONEQUILIBRIUM IONIZATION MODELS

An extensive grid of nonequilibrium ionization models for the X-ray
spectra of adiabatic supernova remnants (SNRs) has been calculated. The
calculations assume a Sedov solution for the hydrodynamics of the blast
wave, and integrate the ionization equations for the plasma behind the
shock. A considerable effort was made to ensure that all processes
which can affect the ionization and emission of a highly underionized
or equilibrium plasma were included and accurately represented. The
details of the atomic physics are given in Hamilton, Sarazin, and
Chevalier 1982 (hereafter Paper I).

The Sedov solution is completely characterized by three parameters,
which may be taken to be the age of the remnant t, its conserved energy
E, and the ambient interstellar hydrogen density n_0. However, the shape

109

J. Danziger and P. Gorenstein (eds.), Supernova Remnants and their X-Ray Emission, 109–111.

of the X-ray spectrum of a remnant is determined by just two parameters, the shock temperature T_S, and the collisional timescale parameter $\eta \equiv n_0^2 E$ (see Paper I). This second parameter characterizes the rate at which the plasma relaxes to ionization equilibrium.

While previous calculations have been limited to isolated values of η, our calculations give a complete grid of models for the range $49 \leqslant \log \eta \leqslant 53$ and $6.25 \leqslant T_S \leqslant 8.25$ (cgs units), covering much of the astrophysically interesting domain of values for these parameters.

Because the heating rate of electrons due to plasma instabilities in the shock is uncertain, two grids of models have been calculated, in which the electrons are either assumed to be in equipartition with ions everywhere, or to be heated by Coulomb collisions with ions behind the shock.

The resulting grid of model spectra is given in Paper I. In addition to graphs of the spectra, we give contour plots for many important plasma diagnostics as a function of the two model parameters, T_S and η. For those requiring more information, the detailed model spectra are available on fiche or on a computer tape from the authors.

These model spectra may be extended to other physical models using the scaling relations given in Hamilton and Sarazin (in this volume, and 1982).

COMPARISON TO OBSERVED SPECTRA

The model spectra have been used to fit the SSS spectra of the three LMC remnants N132D, N63A, and N49 (Clark et al. 1982). These remnants are old enough that nonequilibrium effects are not terribly important. The remnants can be fit with abundances that are about 1/2 of solar. The required supernovae energies are $E_{SN} \sim 2 \times 10^{51}$ ergs, and these high energies may explain the unusually high surface brightness of these remnants.

The Tycho SSS spectrum shows extremely strong X-ray line emission (Becker et al. 1980a). The fit to this remnant requires very high abundances of Si and S, as compared to solar. Several temperature components are necessary, suggesting that the ejecta in this remnant are clumpy. We have shown that the conclusion about the high abundances of Si and S is independent of the detailed ionization or thermal structure in these models; no solar abundance plasma can produce lines as strong as those seen in Tycho.

The SSS spectrum of SNR 1006 is relatively featureless (Becker et al. 1980b), and it has been suggested that the X-ray emission is nonthermal (Reynolds and Chevalier 1981). We find that the SSS spectrum can be fit by a thermal model in which the weak line emission results from a very low nonequilibrium ionization state.

In either SNR 1006 or Tycho, the mass of iron which has been heated to X-ray emitting temperatures is limited to $\leqslant 0.05$ $M_\odot$.

PROBLEMS WITH TYPE I SNRs

These results suggested at least two problems with the structure of remnants of Type I SN. First, the measurements of the proper motion of the remnants suggest that they are adiabatic (e.g., the paper by Strom in this volume). However, the high abundances deduced from the X-ray spectra suggest that the ejecta are dynamically important.

Second, observations and theories for Type I SN suggest that they eject $\sim M_\odot$ of iron (e.g., the paper by Kirshner in this volume). However, very little hot iron is observed in the X-ray spectra. One possiblity is that the bulk of the iron in the historical Type I SNRs is cold. Evidence supporting this suggestion comes from a recent UV spectrum of a star behind SNR 1006 (Wu et al. 1982), which shows that this remnant contains a large mass of rapidly expanding, cold iron.

REFERENCES

Becker, R.H., Holt, S.S., Smith, B.W., White, N.E., Boldt, E.A., Mushotzky, R.F., and Serlemitsos, P.J.: 1982a, 235, p. L5.
Becker, R.H., Szymkowiak, A.E., Boldt, E.A., Holt, S.S., and Serlemitsos, P.J.: 1980b, Ap. J. (Lett.), 240, P. L33.
Clark, D.H., Tuohy, I.R., Long, K.S., Szymkowiak, A.E., Dopita, M.A., Mathewson, D.S., and Culhane, J.L.: 1982, Ap. J., 255, p. 440.
Hamilton, A.J.S., and Sarazin, C.L.: 1982, preprint.
Hamilton, A.J.S., Sarazin, C.L., and Chevalier, R.A.: 1982, Ap. J. Suppl., in press.
Reynolds, S.P., and Chevalier, R.A.: 1981, Ap. J., 245, p. 912.
Wu, C.-C., Leventhal, M., Sarazin, C.L., and Gull, T.R.: 1982, preprint.

THE EFFECT OF SNR STRUCTURE ON NON-EQUILIBRIUM X-RAY SPECTRA

A.J.S.Hamilton and C.L.Sarazin
University of Virginia, Charlottesville, Virginia 22903

A technique is presented for characterizing the ionization structure and consequent thermal x-ray emission of a SNR when non-equilibrium ionization effects are important. The technique allows different theoretical SNR models to be compared and contrasted rapidly in advance of detailed numerical computations. In particular it is shown that the spectrum of a Sedov remnant can probably be applied satisfactorily in a variety of SNR structures, including the reverse shock model advocated by Chevalier (1982) for Type I SN, the isothermal similarity solution of Solinger, Rappaport and Buff (1975), and various inhomogenous or 'messy' structures.

1. NON-EQUILIBRIUM SPECTRAL FORMATION

The ionization structure of a SNR is solved (cf. Hamilton et al. 1983) by integrating the time-dependent collisional ionization equations in Lagrangian gas elements behind the shock front:

$$\frac{Dn(X^i)}{n_e Dt} = C(X^{i-1},T_e)n(X^{i-1}) - \{C(X^i,T_e)+\alpha(X^{i-1})\}n(X^i) + \alpha(X^i,T_e)n(X^{i+1}) \tag{1}$$

Here $n(X^i)$ is the density of ion X^i, n_e and T_e are the electron density and temperature, and $C(X^i,T_e)$ and $\alpha(X^i,T_e)$ are respectively ionization and recombination rate coefficients out of and into ion X^i.

Equations (1) imply that the ionization structure of a Lagrangian gas parcel depends on its <u>ionization time</u> τ,

$$\tau = \int n_e Dt \tag{2}$$

(the integration being started from the moment the gas parcel is shocked), and on the thermal history of the gas parcel.

J. Danziger and P. Gorenstein (eds.), Supernova Remnants and their X-Ray Emission, 113–118.

The thermal emission from a gas parcel dM is proportional to $dE = n_e dM$, and depends otherwise on the composition, ionization structure and electron temperature of the gas parcel. For a spherically symmetric SNR it is useful to define

$$E = E(<r) = \int^r n_e dM_r \qquad (3)$$

Now <u>suppose</u> that the temperature dependence of ionization and emission rates could be ignored. Then the ionization structure at any point would depend only on the ionization time at that point, and the spectrum of the entire SNR would depend entirely on the variation of the ionization time τ with E through the remnant (the contribution of any element scaling linearly with its abundance). One might expect rates to be insensitive to temperature in a strongly ionizing plasma, in which kT_e exceeds the ionization potential of dominant ions so that Boltzmann factors are unimportant (a notable exception is lines excited through inner shell processes, such as the 7keV line of iron in an ionizing plasma; such lines remain temperature sensitive and are useful as temperature diagnostics).

The main source of temperature variation in the rate coefficients is the Boltzmann factor. The effect of thermal history on ionization structure can therefore crudely be accounted for by incorporating a Boltzmann factor into ionization times, defining therefore modified ionization times τ_χ:

$$\tau_\chi = \int \exp(-\chi/kT_e) n_e Dt \qquad (4)$$

where χ is an ionization potential. Obviously $\tau \equiv \tau_0$.

2. COMPARISON OF SNR MODELS

Fig.1 shows graphs of modified ionization times τ_χ and temperature T (assuming $T_e = T_{ion}$) versus E for:
a) the Sedov solution
b) the isothermal similarity solution of Solinger, Rappaport and Buff (1975);
c) the outer and
d) inner shocks in the self-similar reverse shock solution advocated by Chevalier (1982) for Type I SN, i.e. an ejected density profile $\rho \propto r^{-7}$ moving into a uniform density medium.

Consider first the Sedov solution. Fig.1 indicates that the degree of ionization, as characterized by the ionization times, advances inward from the shock front, reaches a maximum, then declines toward the center of the SNR. At early times when Boltzmann factors are unimportant ($\chi \ll kT$) the maximum ionization occurs at a point where $E = .11E_{total}$, corresponding to an interior mass fraction of .30, or a fractional radius of .89. Precisely this behaviour can be seen explicitly in the detailed ionization curves for early models traced by Itoh (1979) and Gronenschild and Mewe (1981). Later on, when Boltzmann factors become

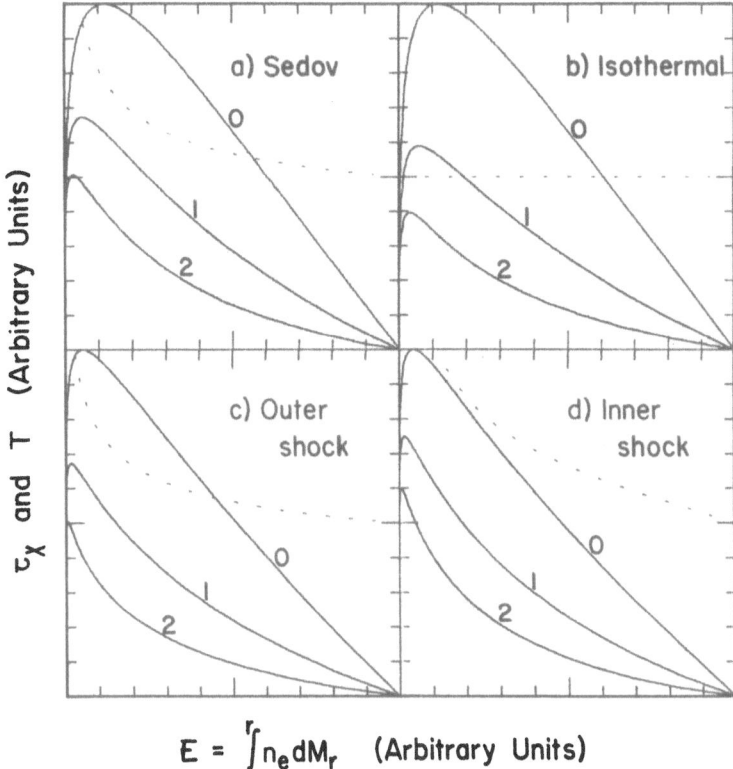

Figure 1. Modified ionization times τ_χ (————) and temperature T (— — —) versus E for various SNR models. τ_χ curves are labelled with values of χ/kT_{shock}. The shock front in each case is to the right.

important, the maximum degree of ionization moves relatively inward.

The ionization curves for the other SNR models shown in Fig.1, when suitably normalized, look remarkably similar to the Sedov curves. Moreover the runs of temperature are quite similar, except that the isothermal model lacks a high temperature inner region. The inference is that the resulting x-ray spectra should be quite similar. The major differences will occur in temperature sensitive emission, i.e. in general at the harder end of the spectrum. For example the isothermal spectrum will be relatively weaker in hard x-rays than the Sedov spectrum, but the soft x-ray spectrum will be almost indistinguishable from a Sedov spectrum.

The normalization of the curves in Fig.1 is fixed by three parameters: a characteristic ionization time, a characteristic temperature, and a total emission integral. Table 1 shows the normalizations used for each model in Fig.1.

Table 1. Normalization of quantities in Fig.1.

Model	$\dfrac{E_{total}}{n_e M}$ [1,2]	$\dfrac{\tau_{max}}{n_e t}$ [1,3]	$\dfrac{T_{inner}}{T_{outer}}$ [4]
Sedov	2.07	1.37	–
Reverse {Outer	1.92	1.79	–
shock {Inner	1.58	3.25	.353
Isothermal	1.38	.92	–

[1] Here n_e is the 'electron' density in the ambient medium; $n_e \simeq 1.23 n_H$ if n_H is the ambient hydrogen number density.

[2] M is the total mass of shocked material. In the reverse shock model the outer shock accounts for .67 of the mass M.

[3] τ_{max} is the maximum value of τ. t is the age.

[4] Ratio of post-shock temperatures in the reverse shock model assuming equal molecular weights.

3. INHOMOGENEITIES

The technique described can be used to consider emission from inhomogeneities distributed for example in some statistically homogenous way through a SNR (see Hamilton and Sarazin 1983 for further details). If the density contrast between clumps and ambient medium is not too large, then clumps can be shocked to x-ray emitting temperatures, and non-equilibrium ionization may be important.

The characteristic ionization times and temperatures of the clumped and unclumped components, having density contrast κ, are obviously related by

$$\tau_{clumped} \simeq \kappa \tau_{unclumped}$$

$$T_{clumped} \simeq \kappa^{-1} T_{unclumped}$$

(5)

which is a constraint on possible combinations of spectra. A further constraint follows from requiring that the filling factor of the clumped component be less than unity.

4. STEADY STATE APPROXIMATION

In a steady state approximation in which mass enters a blast wave at a constant rate, and the density and temperature profiles behind the shock remain constant, then

$$\frac{dE}{d\tau} = \frac{n_e \, dM}{n_e \, dt} = \frac{dM}{dt} = \text{constant}$$

(6)

i.e. τ varies linearly with E behind the shock front. Reference back
to Fig.1 shows that a linear increase of τ occurs through most of the
emitting region in each model, suggesting that a steady-state is a
good approximation there.

In the steady state approximation the emission from a line in any
ionizing ion stage is just

$$\text{Emission} = \frac{\text{Excitation rate coefficient}}{\text{Ionization rate coefficient}} \times \frac{dE}{d\tau} \qquad (7)$$

5. CONCLUSIONS

A simple technique has been presented for characterizing the
ionization structure and consequent thermal x-ray emission of a SNR
when non-equilibrium ionization effects are important.

It has been shown that the (soft, at least) x-ray spectrum of a Sedov
remnant is probably a good approximation for
-each of the outer and inner shock waves in the self-similar reverse
 shock model advocated by Chevalier (1982) for Type I SN;
-the isothermal model of Solinger, Rappaport and Buff (1975);
-statistically uniform clumpy SNRs (= 2-component spectrum).
Much of the reason for the agreement is the approximate validity of
a steady-state approximation in most of the emitting region. Sedov
spectra will presumably be satisfactory approximations wherever this
is the case.

A particular spectrum (for a given composition) is essentially
chracterized by three parameters, a characteristic ionization
timescale, a characteristic temperature, and an emission measure. The
first two parameters fix the shape of the spectrum, while the emission
measure fixes the absolute level of the observed flux.

Finally it should be noted that there are many cases where a Sedov
spectrum does not apply, notably where there are large systematic
density gradients (e.g. a circumstellar wind) in the ambient medium.

REFERENCES

Chevalier,R.A. 1982. Ap.J.,258,790.
Gronenschild,E.H.B.M. and Mewe,R. 1981. Astron.Astrophys.Supp.,48,305.
Hamilton,A.J.S. and Sarazin,C.L. 1983. Preprint.
Hamilton,A.J.S., Sarazin,C.L. and Chevalier,R.A. 1983. To be published
     in Ap.J.Supp.
Itoh,H. 1979. Pub.Astron.Soc.Japan,31,541.
Solinger,A., Rappaport,S. and Buff,J. 1975. Ap.J.,201,381.

DISCUSSION

WINKLER: If the Sedov model approximately reproduces the results of so
many other models including the isothermal model, then could one instead
use the even simpler isothermal model as the basis for approximating a
more complex reality?

HAMILTON: Recall the isothermal model is isothermal in Eulerian
coordinates, not Lagrangian coordinates. However, I think you could get a
good approximation on the ionization structure this way if you wanted to.

THE EVOLUTION OF YOUNG SUPERNOVA REMNANTS

A.C. Fabian[*], W. Brinkmann[+], G.C. Stewart[*]
*Institute of Astronomy, Madingley Road, Cambridge CB3 OHA, UK
†Max Planck Institut für Extraterrestrische Astrophysik,
  8046 Garching bei München, F.D.R.

1. INTRODUCTION

Einstein X-ray observations of the young supernova remnants
Cassiopeia A (Murray *et al.* 1980) and Tycho (Seward, Gorenstein and
Tucker 1982) indicate that the swept-up mass does not much exceed that
of the observed ejecta. The initial density distribution of the ejecta
and surrounding material is then important in determining the X-ray
structure and evolution. Some aspects of this behaviour have been
dealt with in previous numerical (e.g. Gull 1973; Itoh 1977; Jones,
Smith and Straka 1981) and analytical (e.g. Chevalier 1982a,b) studies.
We present here results obtained from numerical models covering a
wider range of initial conditions. In particular, we consider the
effect of a constant stellar wind from the progenitor star on the
expansion of the remnant. We have previously suggested that variable
mass loss from SN1006 may explain its warm filled interior (Fabian,
Stewart and Brinkmann 1982).

2. THE NUMERICAL MODELS

We have modelled the hydrodynamic evolution of our expanding
supernova remnant using a simple one-dimensional Lagrangian programme
following the prescription given in Richtemeyer and Morton (1967).
Mass and energy are conserved to better than a few percent throughout
the runs. The ejecta are distributed with a uniformly increasing
velocity out to a radius of 5×10^{16} cm and the initial temperature is
2×10^4 K everywhere. The radii of the shells of surrounding matter
increase in size logarithmically out to $\gtrsim 10^{20}$ cm. The count rate
emissivity expected in the Einstein High Resolution Imager (HRI) has
been computed for some and the ionization equilibrium of oxygen and
sulphur have been calculated for a few others using the rates given
by Shull and Van Steenberg (1982).

Fig. 1 shows the evolution of the inner and outer shocks from a
2.35×10^{33} gm, 7.673×10^{50} ergs explosion into either a constant

119

J. Danziger and P. Gorenstein (eds.), Supernova Remnants and their X-Ray Emission, 119–124.
© *1983 by the IAU.*

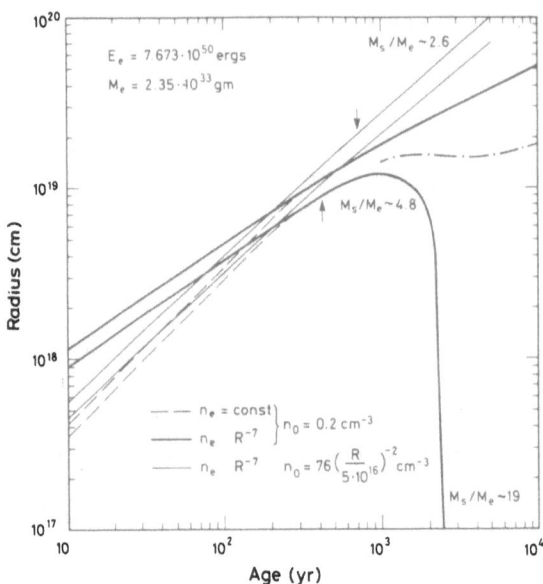

Fig. 1. Evolution of the inner and outer shocks from a low mass supernova of ejecta mass and energy M_e, E_e, into a surrounding medium of
density n_0 which is either constant or representative of a wind. The
solid curves are for an initial density distribution of ejecta, $n_e \propto R^{-7}$
over the outer $\sim 3/7$ of the ejecta mass. Arrows indicate the point at
which the reverse shock has reached the constant density interior. Note
that the dashed and heavy solid pairs of lines merge at ~ 200 yr. The
contact discontinuity is indicated by the chain line.

density (0.2 particles cm^{-3}) or constant mass-flux wind. Two of the
models have an outer density profile over the outer $\sim 3/7$ of the mass
proportional to (radius)$^{-7}$ to mimic the structure of a white dwarf
(cf. Chevalier 1982b). The detailed behaviour of the most complex
model is shown in fig. 2 and agrees with Chevalier's similarity
solution (1982a). We note that the reverse shock reaches the centre of
the remnant when the ratio of swept-up to ejecta mass (M_s/M_e) is ~ 19,
independent of the structure of the ejecta. This ratio is about twice
the value inferred from Chevalier's analytical solution (1982b), which
assumes that the ratio of the thermal pressures just behind the outer
shock to that just within the inner one, α, is $\sim 0.3 - 0.4$. The
numerical results at both early and late (Sedov) stages (the central
pressure in this last case) give α in this range but it is reduced to
~ 0.1 as the reverse shock travels back toward the centre. No simple
analytical result seems possible at this stage.

The almost free expansion of a constant density 15 $M_\odot$, 5.3 x 10^{51}
ergs supernova into a wind of mass-loss rate 5 x 10^{-4} $M_\odot$ yr^{-1} and
velocity of 10^8 cm s^{-1} is shown in fig. 3. The bubble blown by such a

wind after 10^5 yr in a constant density interstellar medium would be
∿ 20 pc in size and its edge is reached by the remnant in ∿ 10^4 yr.
We note that the shock temperature is reduced by the effect of the
wind velocity. The average mass loss rate (and thus total mass lost)
from a star is likely to be much lower than that which we use here.
Consequently the later expansion in a large bubble may be even more
rapid than indicated in fig. 3.

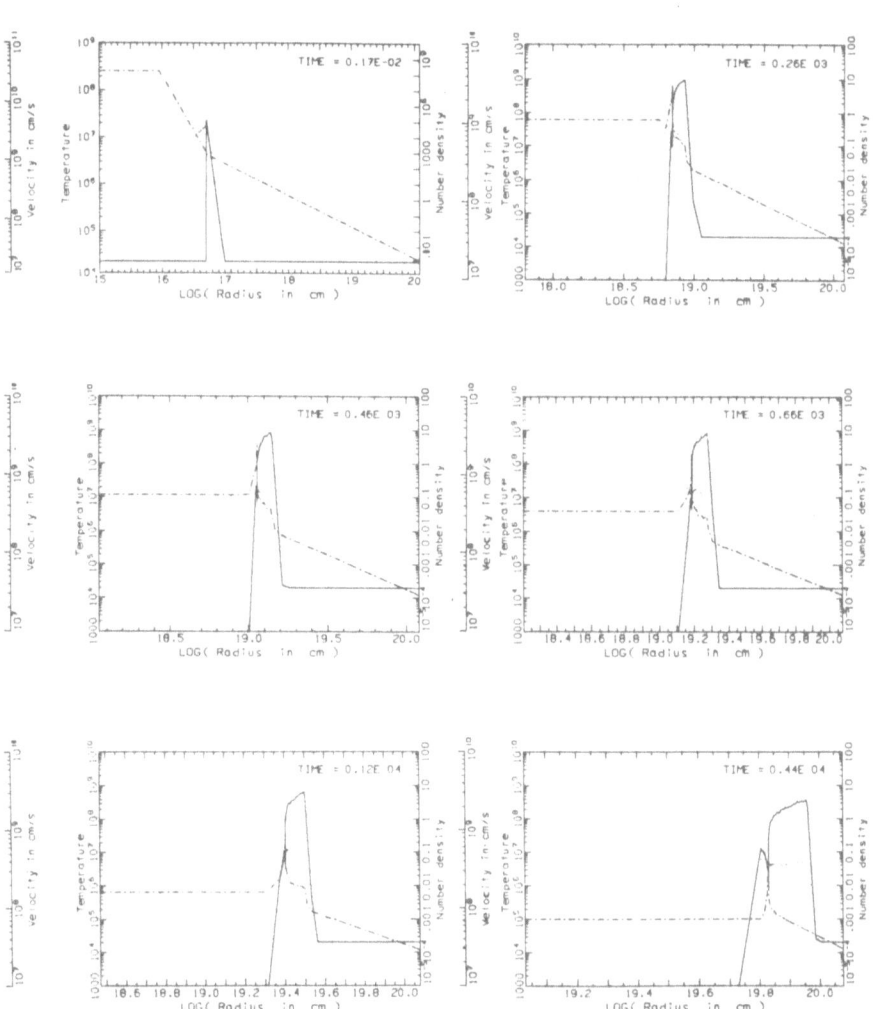

Fig. 2. Velocity (dots), density (chain) and temperature (solid)
evolution of the model shown as a fine solid line in fig. 1. The time
is given in yr.

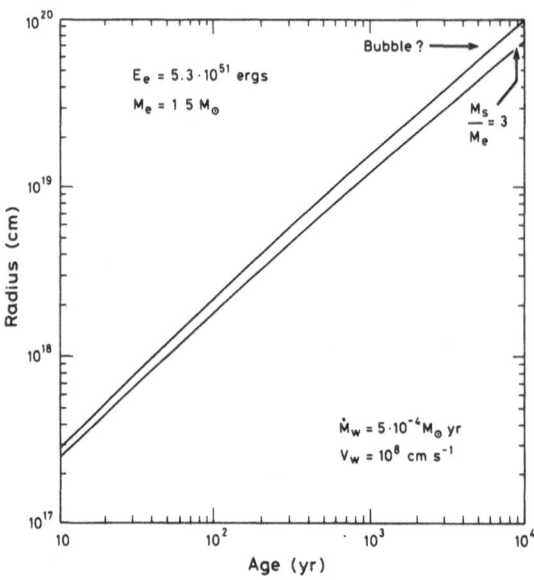

Fig. 3.   Evolution of a 15 $M_\odot$, 5.3 x 10^{51} erg supernova in a stellar wind. About 45 $M_\odot$ have been swept up in $\sim 10^4$ yr. A possible size of bubble blown by such a wind is indicated

We show the HRI emissivity of a 1 $M_\odot$, 10^{51} erg explosion in a constant density surrounding medium of density 20 cm^{-3} in fig. 4.   The outer blast wave dominates the X-ray appearance of the remnant after a few hundred years; the reverse shock only seriously contributes well before that time and becomes 'invisible' as it parts company with the contact discontinuity (which continues to be 'visible', although it is mainly an artefact of the initial steep density discontinuity).

3.   CONCLUSIONS

We emphasise (see Chevalier 1982b) that observed deceleration (e.g. from proper-motions) need not mean that the remnant is close to any Sedov-phase (fig. 1). The ejecta may instead have a steep outer density profile. Successive mass elements of the ejecta suffer a deceleration on being 'reverse'-shocked, but are then carried along by inner (and more massive) neighbours.

Thick reverse-shocked regions ($\Delta R/R \gg 0.1$) should be difficult to observe. When the reverse-shock starts to travel inward relative to the centre of explosion, the low inner density and tendency for pressure equilibrium mean that it does so rapidly. The shock temperature then varies roughly as $(time)^{9/5}$ and reaches $\sim 10^9$ K at the centre of our models. A Sedov phase begins when $M_s/M_e > 5$ and is only firmly established when the centre has been reached by the reverse shock ($M_s/M_e > 19$).

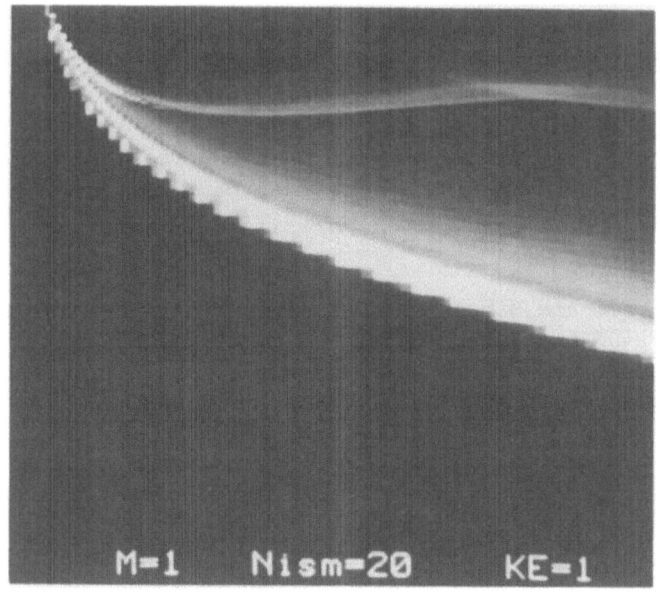

Fig. 4. HRI X-ray count-rate emissivity of a 1 $M_\odot$, 10^{51} erg explosion
in a high density surrounding medium (N_{ism} = 20 cm^{-3}). Radius and
time increase linearly downward and to the right up to $\sim$ 5 pc and
$\sim$ 1000 yr, respectively. The dynamic range of features shown is a
factor of 256. The broad structure lies just behind the outer shock;
the inner thinner line is the contact discontinuity. The effects on
this of a central bounce (at $\sim$ 300 yr) and of the coarse binning are
evident.

We note that the ratio of swept-up mass to shocked ejecta (M_s/M_r) is
always near unity in the early phases. Measurements of $M_s/M_r \sim$ 1 do
not indicate that free expansion will soon end. We confirm that X-ray
bright reverse-shocked matter is likely to be close to ionization
equilibrium (Itoh 1977). Simple checks may be made by comparing the
ionization/recombination time (given a deduced density) with the
remnant age.

    Finally we note that winds from massive stars may prevent a simple
Sedov phase from occurring.

ACKNOWLEDGEMENTS

    ACF thanks the Royal Society for financial support.

REFERENCES

Chevalier, R.A.: 1982a, Astrophys.J., 258, 790.
Chevalier, R.A.: 1982b, Astrophys.J., 259, L59.
Fabian, A.C., Willingale, R., Pye, J.P., Murray, S.S. and Fabbiano, G.:
    1980, Monthly Notices Roy. Astron. Soc., 193, 175.
Fabian, A.C., Stewart, G.C., and Brinkmann, W.: 1982, Nature, 295, 508.
Gull, S.F.: 1975, Monthly Notices Roy. Astron. Soc., 171, 263.
Itoh, H.: 1977, Publ. Astron. Soc. Japan, 29, 813.
Jones, E.M., Smith, B.W., and Straka, W.C.: 1981, Astrophys. J., 249,
    185.
Richtemeyer, R.A., and Morton, K.W.: 1967, Difference Methods for
    Initial Value Problems, Interscience, New York.
Seward, F.D., Gorenstein, P., and Tucker, W.: 1982, preprint.
Shull, J.M., and Van Steenberg, M.: 1982, Astrophys. J. Suppl., 48.

DISCUSSION

Winkler: You mentioned that material behind the reverse shock is in or very near ionization equilibrium. Why is that; is it due to high density?

Fabian: Yes.

Hamilton: Would you not expect the X-ray emission from the reverse shock to be greatly enhanced by the enhanced heavy element abundances?

Fabian: Only when the thickness of the reverse-shocked material is small, such that the temperature is low and line emission important.

Dickel: Is there enough density to see γ-rays from the super-hot reverse shock?

Fabian: No - pressure equilibrium gives very small density.

SPHERIZATION OF THE REMNANTS OF ASYMMETRICAL SN EXPLOSIONS IN A UNIFORM MEDIUM

G.S. Bisnovatyi-Kogan
Space Research Institute
Moscow

S.I. Blinnikov
Institute of Experimental
and Theoretical Physics
Moscow

The "snow-plough" approximation is applied to the problem of the propagation of the shock wave from a nonsymmetrical supernova explosion in a uniform medium. The spherization of the shape of the remnant is obtained. The results are in good agreement with the observed properties of Cas A.

Optical, X-ray and radio observations show that the shape of the projection of SNR's on the plane of the sky is close to a circle [1]. The deviations from a circle may be connected with asymmetry of the explosion, nonsymmetrical support from a central pulsar (for young SNR's) and nonuniformity of the interstellar matter. In the first and second cases we may expect a shape close to an ellipse, as observed in the Crab Nebula.

Asymmetry of the explosion is to be expected when the magnetorotational mechanism is producing the SN (Fig. 1). The transformation of the rotational energy into the energy of the explosion is so efficient as to give an energy output $\sim 10^{51}$ ergs [2,3].

We consider here the evolution of the remnant of an axisymmetrical (nonspherical) explosion in a uniform medium. The exact solution of the problem needs two-dimensional nonstationary hydrodynamical calculations, which is very complex. The problem may be simplified by taking into account the fact that during the propagation of the blast wave through the interstellar medium the main part of the mass is collected in a thin layer. This layer is treated within an approximation which may be called 1.5-dimensional hydrodynamics.

Let the density of the outer medium be ρ_0, and its pressure p_0 be insignificant, so that the strong wave approximation holds:

$$p_0 << p_i ; \tag{1}$$

here p_i is the pressure which is taken uniform across the cavity inside the shock, $p_i = p_i(t)$. We consider at the initial time $t = t_0$ an axisymmetric form of the SNR:

J. Danziger and P. Gorenstein (eds.), Supernova Remnants and their X-Ray Emission, 125–130.

$$t = t_o ; \quad r = r_o(\theta) \tag{2}$$

and an axisymmetric velocity field

$$v_f = v_{fo} \ (\theta, \ r_o(\theta)) \ , \tag{3}$$

r and θ being spherical coordinates.

There are different approximations for solving this problem. The simplest one is Kompaneetz approximation [4], where not only $p_i = p_i(t)$, but also the pressure along the shock $p_{sh} = p_{sh}(t)$. In this case the velocity of the shock is also position-independent

$$v_{sh} = (\frac{\gamma+1}{2} \frac{p_{sh}}{\rho_o})^{\frac{1}{2}} , \tag{4}$$

and is directed perpendicular to the shock front. Thus, all the matter on the shock moves with the same velocity and spherization occurs. The energy conservation law determines the time dependence of all quantities. The spherization of a flat disk in this model is shown in Fig. 2.

For a better approximation it is necessary to take into account the dependence $v(\theta, t, r(\theta,t))$ and the existence of a tangential velocity averaged across the shock.

The "snow-plough" model developed in plasma physics [5] is adequate for solving this problem. Let λ be a Lagrangian coordinate along the shock front (see Fig. 3); μ the surface density in Lagrangian coordinates; u_z, u_ω the average velocity components of the matter in the layer (in cylindrical coordinates ω, z); and D_n the normal component of the velocity of the shock relative to the interstellar medium. The set of equations includes the laws of mass and momentum conservation,

$$\frac{D\mu}{Dt} = \frac{\gamma+1}{2} \rho_o \omega \ (\frac{Dz}{Dt} \frac{\partial\omega}{\partial\lambda} - \frac{D\omega}{Dt} \frac{\partial z}{\partial\lambda}) , \tag{5}$$

$$\vec{u} = (\frac{D\omega}{Dt} , \frac{Dz}{Dt}) , \tag{6}$$

$$\frac{D(\mu \vec{u})}{Dt} = p_i \omega \ (- \frac{\partial z}{\partial\lambda} , \frac{\partial\omega}{\partial\lambda}) , \tag{7}$$

the integral energy conservation law

$$E_o = \frac{\pi p_i}{\gamma-1} \int_0^{\lambda_{max}} \omega^2(\lambda) \ \left| \frac{\partial z}{\partial\lambda} \right| \ d\lambda + \pi \int_0^{\lambda_{max}} \mu(u_\omega^2 + u_z^2) d\lambda \tag{8}$$

and a condition on the shock front

$$D_n = \vec{D}\cdot\vec{n} = (\gamma+1) u_n/2 = (\gamma+1) \ \vec{u}\cdot\vec{n}/2 . \tag{9}$$

The equations (5)-(9) have been solved for the following initial conditions:

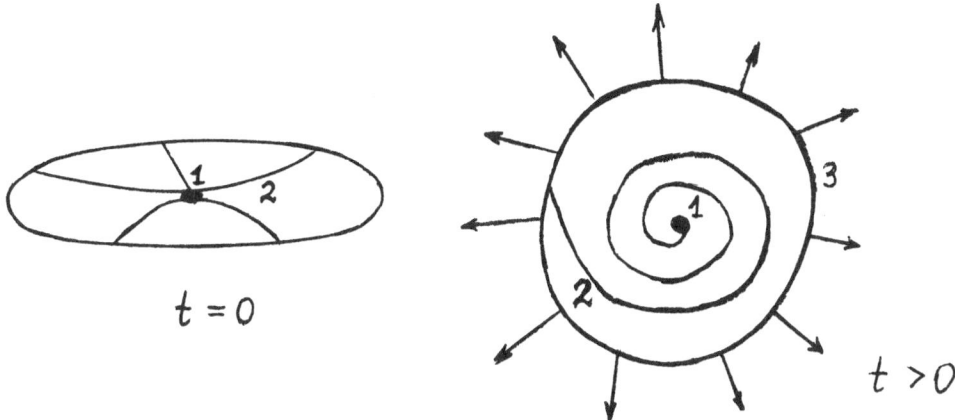

Fig. 1. Schematical model of magnetorotational supernova explosion.
1. neutron star; 2. flattened envelope with twisted magnetic
field; 3. shock wave.

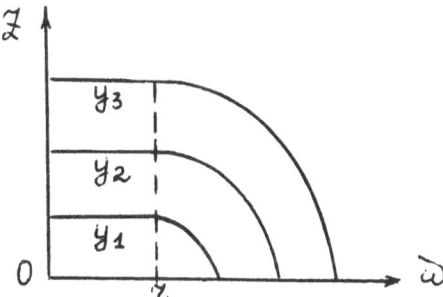

Fig. 2. The spherization of a flat disk in Kompaneetz model $[4]$, y = time
coordinate

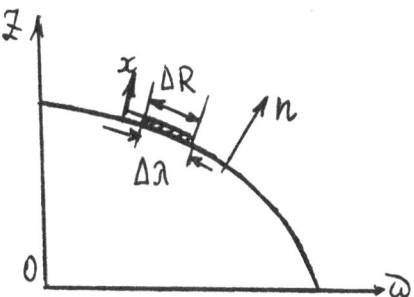

Fig. 3. For the derivation of the equations (5)- (9).

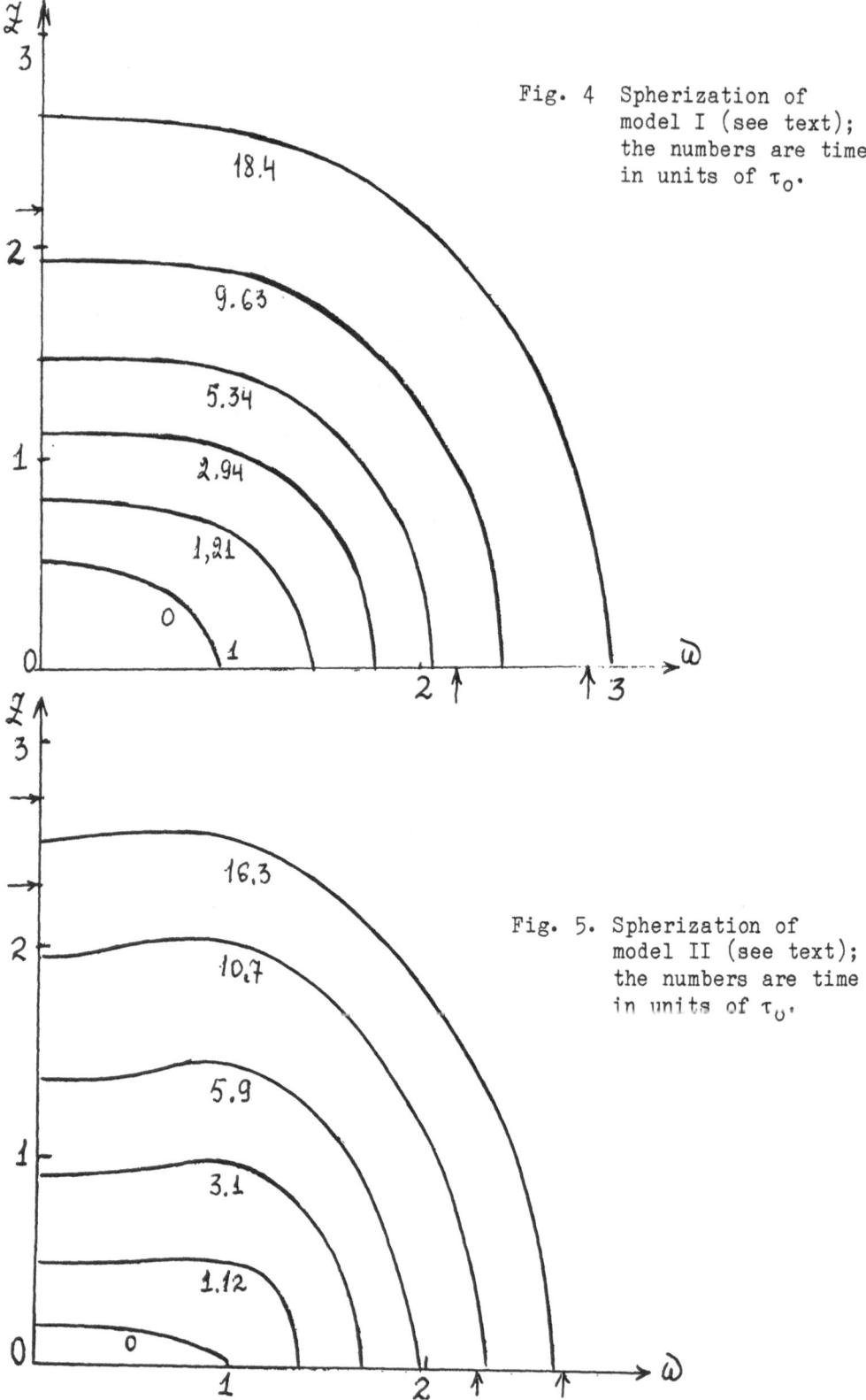

Fig. 4   Spherization of
         model I (see text);
         the numbers are time
         in units of τ_0.

Fig. 5. Spherization of
         model II (see text);
         the numbers are time
         in units of τ_0.

$t=0$: $u(\lambda) = u_0(1-b\cos 2\lambda)$, $\lambda(t=0) = \theta$ = spherical colatitude,

$\omega = R_e \sin \lambda$, $z = R_p \cos \lambda$, σ_0 = const.   (10)

The additional dimensional parameters E_0 - the total energy of the explosion -, ρ_0, M_0 and the adiabatic index γ determine the problem. In nondimensional calculations only the following values must be given at $t = t_0$:

u_e/u_p, E_{kin}/E_0, M_0/M_H, R_e/R_p,   (11)

where $M_H = \frac{4}{3}\pi\rho_0 R_e^2 R_p$ is the swept-up mass added to the initial mass of the shock M_0. We have calculated the variants with:

	γ	M_0/M_H	E_{kin}/E_0	R_e/R_p	u_e/u_p
model I	1.2	1.76	0.64	2	2
model II	1.2	4.47	0.47	5	2

The form of the shock wave for these models is shown in Figs. 4,5. The time of spherization is about ten characteristic times τ_0; for $E_0 = 10^{50}$ ergs, $\rho_0 = 10^{-24}$ g/cm^3, $M_0 = 1M_\odot$ $\tau_0 \approx 600$ years in both cases. Large differences of a factor 2 to 3 between σ_e and σ_p are still preserved. This may explain the observations of Cas A [6], where sphericity exists together with differences in surface brightness. The time τ_0 may be much smaller when spherization occurs during the propagation of the shock wave through the extended envelope of the pre-SN star with $\sigma_0 \gg 10^{-24}$ g/cm^3, as it could have been in Cas A. An extended version of this work is given in [7].

REFERENCES

1. Shklovskyi, I.S., 1976, Sverchnovyie Zwezdy, Nauka, M.
2. Bisnovatyi-Kogan, G.S., 1970, Astron. Zh., 47, 813.
3. Ardelyan, N.V., Bisnovatyi-Kogan, G.S., Popov, Yu.P., 1979, Astron. Zh., 56, 1244.
4. Kompaneetz, A.S., 1960, Doklady Ak.Nauk USSR, 130, 1001.
5. Leontovich, M.A., Osovetz, S.M., 1956, Atom. Energia, 3, 81.
6. Fabian, A.C., Willingale, R., Pye, J.P. et al., 1980, Mon. Not. RAS, 193, 175.
7. Bisnovatyi-Kogan, G.S., Blinnikov, S.I., 1981, preprint ITEF-124, Moscow; Astron. Zh., 1982.

DISCUSSION

McKEE: Is the swept up gas also in a thin shell? If so, and if its mass
is dominant, could you give a physical explanation of why the surface
density varies by a factor of several over the surface of the remnant
when it has become spherical?

BISNOVATYI-KOGAN: It is explained by the nonspherical velocity field
given by the initial conditions. At the time of spherization there is a
2-3 times difference in the surface density. During further, almost
spherical, expansion this difference decreases and goes to zero with
time. Therefore, we may expect to observe a nonuniform surface density
together with an almost spherical shape only in rather young SNR.

COX: Doesn't the spherization depend almost entirely on the assumption
of homogeneity?

BISNOVATYI-KOGAN: Yes, it depends on it very strongly. The almost
spherical shape of many of the observed SNR indicates that the
interstellar medium in their vicinity is rather uniform.

X-RAY STUDY OF THE CRAB NEBULA AND THE CRAB AND VELA PULSARS

F. R. Harnden Jr.
Harvard-Smithsonian Center for Astrophysics
Cambridge, MA U.S.A.

INTRODUCTION

The Crab Nebula has been intensely studied by X-ray astronomers ever since its discovery as the first, optically identified X-ray object (Bowyer et al. 1964); and a large majority of X-ray experiments during the past two decades have observed the Crab, seeking not only the answers to scientific questions but also assurance that the instruments' calibrations were understood. It is therefore no surprise that, following its launch in 1978 November, the Einstein X-ray Observatory too had the Crab Nebula on its list of mandatory targets.

The Einstein telescope, as had other pioneering experiments preceeding it, held the promise of new discoveries concerning this frequently observed object. Einstein for the first time brought X-ray imaging techniques to bear on questions regarding the spatial structure of the nebula and its famous 33 millisecond pulsar. As part of the effort to understand the relationship between the nebula and the pulsar, the only other such pair of objects known at the time of Einstein's launch, the Vela supernova remnant and the Vela pulsar, were also selected for intensive study.

Because the Vela supernova remnant (SNR) is so much closer and older (distance ~ 500 pc, age ~ 10,000 yr) than the Crab SNR (distance ~ 2 kpc, age ~1900 yr.), the Vela SNR has a much larger angular extent than does the Crab: nearly 5 degrees as compared to ~3 arcmin. This made study of the Vela SNR with the 1-degree field-of-view Einstein telescope tedious, requiring nearly 40 separate pointings to cover the entire extent of the X-ray shell. These observations were carried out, but will be discussed elsewhere. This paper centers mainly on the data for the two pulsars, and also includes brief discussions of the nebular structure surrounding each of them.

131

J. Danziger and P. Gorenstein (eds.), Supernova Remnants and their X-Ray Emission, 131–138.
© 1983 by the IAU.

OBSERVATIONS

All four <u>Einstein</u> focal-plane instruments (see Giacconi <u>et al.</u> 1979 for a description of the observatory) were used to observe the Crab and Vela SNR's, but only the High Resolution Imager (HRI) and Imaging Proportional Counter (IPC) data are discussed here. Table I gives some details of the observing program. The HRI, with its 4 arc second resolution, was used primarily to study detailed spatial structure, but the IPC with its greater sensitivity to low-surface-brightness features was also useful in studying the outer regions of the Crab's X-ray emission.

The possibility of "burning out" the IPC by exposure to the intense flux of the Crab dictated that the IPC observation of the Crab be postponed until the end of the mission. The IPC did survive this exposure; but the 1000 per second counting rate saturated the 125 per second telemetry capability, rendering temporal analysis of the pulsar data virtually impossible. Temporal analysis of the Crab pulsar has therefore been based entirely on HRI data, whereas both HRI and IPC temporal analyses have been carried out for the Vela pulsar.

TABLE I

EINSTEIN Imaging Observations

Crab Nebula			Vela Pulsar Region		
Exp.Time	Date	Notes	Exp.Time	Date	Notes
45760	1979 Mar 3	H	31460	1978 Nov 29	H,2,3
20784	1979 Sep 14	H	5607	1978 Dec 18	I
5795	1981 Mar 27	H,1-25	12259	1979 May 31	H
8083	1981 Mar 27	I,2,1-25	1626	1979 Oct 25	I,1-3
2583	1981 Mar 27	I,2	18333	1979 Oct 31	H
1024	1981 Mar 27	I	17035	1980 Apr 27	H
			2038	1980 Jun 5	I,1-26
			1100	1980 Jun 6	I,1-35

Notes:
  H   observed with HRI
  I   observed with IPC
  1-n   offset from target by n arcmin
  2   Al filter inserted in X-ray path
  3   Observed with HRI-2 (instead of HRI-3)
Exp.Time in seconds (uncorrected for deadtime)

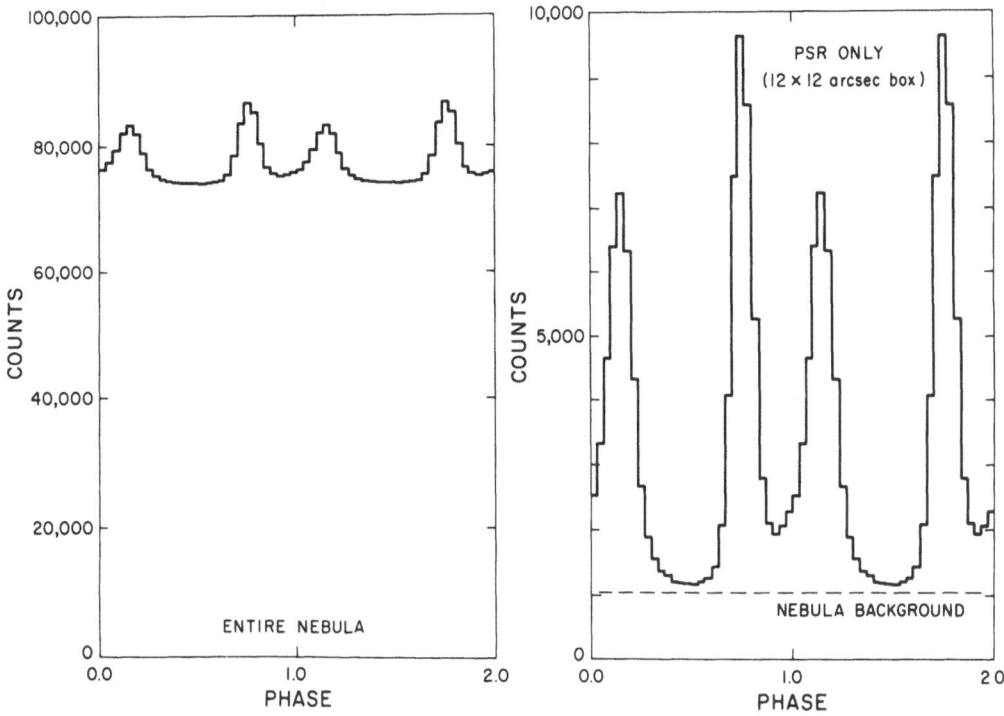

Figure 1: Light curve of the Crab Nebula X-ray data folded at the pulsar period of 33 msec. and plotted twice on each slide of the graph; left, all emission from the entire nebula; right, photons from the pulsar itself. Each curve has been corrected for a phase-dependent telemetry deadtime.

THE CRAB PULSAR

Temporal analysis for the Crab pulsar 1979 March data has yielded the light curve shown in Figure 1. An image made during the lowest phase interval, the next to last one of the 10 frames shown in Figure 2, reveals that a point source contribution from the pulsar is not statistically significant during this interval. The "3 σ" HRI count rate upper limit during this "off phase" is 0.22 counts per second.

This rate can be interpreted as an upper limit to the surface temperature of the neutron star; provided that the emission is assumed to be blackbody radiation. If a 10 km radius is assumed, an upper limit of ~ 2.5 million K results. Such a temperature is consistent with so-called cooling curves based upon a conventional equation of state for nuclear matter (cf. Nomoto and Tsuruta 1981). The corresponding 0.1 to 4.5 keV X-ray luminosity upper limit is 2.6×10^{34} erg sec^{-1}, but the use of such upper limits to constrain neutron star cooling curves seems rather ambiguous in view of the many possible heating mechanisms which could alter such curves (cf. Helfand et al. 1980).

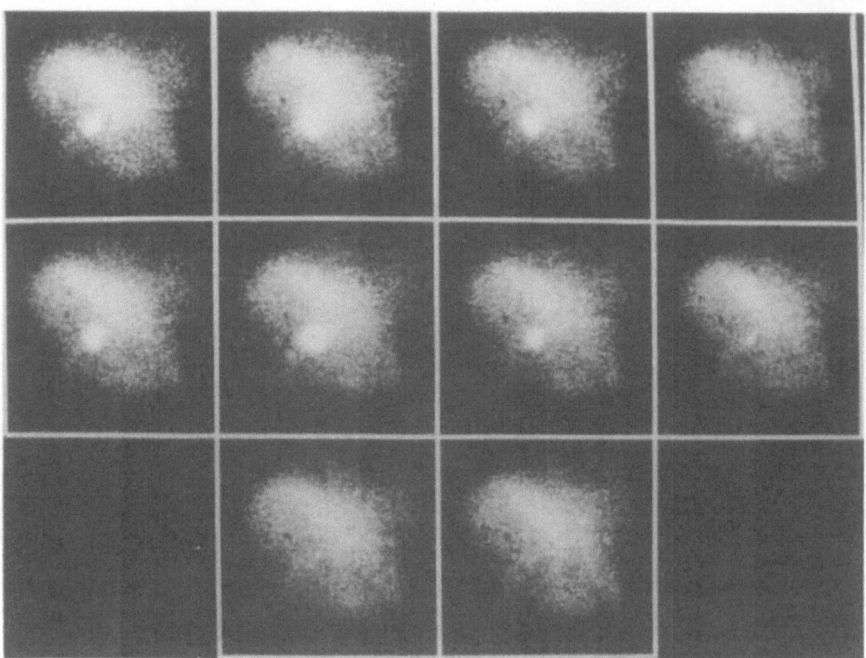

Figure 2: HRI images of the Crab Nebula and its pulsar.
March 1979 data from a 20,000 second exposure have been
folded at the pulsar period (33 msec.) and imgaged in 10
phase intervals. The main pulse occurs in the second
interval, the interpulse falls in the sixth, and the
pulsar is faintest in the ninth image.

THE VELA PULSAR

Because of previous reports of variability (Smith and Pounds 1975)
and earlier reports of possible X-ray pulsations (Harnden and Gorenstein
1973), it was important to observe the Vela pulsar (PSR 0833-45) several
times. However, all Einstein observations of this pulsar (see Table I)
are consistent with a constant flux. Furthermore, none of the data
exhibits pulsations at the 89 millisecond radio-optical-gamma-ray
period. The most sensitive upper limit (95 per cent confidence, from
IPC data) on pulsations is 3.2×10^{-13} ergs cm^{-2} sec^{-1} for the 0.2 to
4.0 kev passband. (Note that intercomparisons of fractional limits from
experiments with different fields of view can be misleading since there
is complicated spatial structure in the X-ray emission from the vicinity
of the pulsar — see below.)

The Vela pulsar is a relatively strong X-ray source despite the
absence of pulsations. It appears "unresolved" to the IPC (~1 arcmin
resolution), at a flux level of 3.9×10^{-11} erg cm^{-2} sec^{-1} in the 0.2 to
4.0 keV bandpass. The emission from this source is decidedly
non-thermal: the IPC spectrum is well fit by a power law model but

produces unacceptably high chi-square values for attempts to fit it to
with a thermal model. This is in contrast to the spectra of nearby,
bright, diffuse emission regions, which are well fit by the thermal
model but reject the power law. One such region, nearly a degree in
extent, partially overlaps the position of the pulsar and may account
for the extended source observed at the pulsar position by previous,
non-imaging experiments. On the basis of the spectral differences,
however, it seems unlikely that the pulsar is directly responsible for
this diffuse emission. Indeed, there are at least half a dozen other
localized, diffuse emission regions, throughout the 5 degree remnant,
which can characterized as knots and filaments.

Yet there is another scale to the pulsar source. With the HRI,
what was unresolved in the IPC becomes resolved into a point-like
component surrounded by a small nebula. The unresolved HRI source
accounts for ~1/3 to 1/2 of the emission, with the remainder coming
from a centrally condensed nebula of ~80 arc second extend.

STRUCTURE OF THE CRAB NEBULA

The shape of the X-ray Crab Nebula can be seen in Figure 2. Except
for the point-like pulsar, which contributes only ~ 5% of the
time-averaged image, there are no sharp features. The most intense
emission region is, like the optical wisps, to the northwest of the
pulsar. When subjected to a maximum entropy deconvolution procedure,
this region tends to coalesce into knot-like features, but the reality
of such features is not convincing, even though one such "knot"
coincides with the position of a 17th magnitude "stellar" object.
(Optical observations of this and other stellar objects within the
nebula are underway.)

SHOCKWAVE SHELL AROUND THE CRAB?

The Einstein data may be capable of answering this important
question, but the intricate analysis required has not yet been
completed. Two effects, one instrumental and the other astrophysical
make the detection of a possible, low-surface-brightness, blast wave
shell very difficult.

Scattering from surface imperfections of the telescope mirror makes
even a point source appear to have extended emission surrounding it at a
low level. The possibility of surface degradation between the preflight
calibration and the inflight observations necessitates caution in
applying the instrumental response function.

Scattering from interstellar dust grains has long been discussed as
a scientifically interesting effect which would produce a halo around a
point X-ray source. In the case of the Crab Nebula, untangling such an
effect (by subtle differences in spectral dependence) from mirror

scattering is non-trivial. A simple comparison of three celestial sources, the Crab, 4U1658-48 (also known as GX339-4) and GX13+1, merely demonstrates that mirror or grain scattering can indeed spatially mimic what one might expect from a blast wave shell.

Another difficulty in detecting a Crab X-ray shell is the extreme intensity of the synchrotron emission from the 3 arcmin X-ray nebula. A simulated experiment (F. Seward, private communication) of placing the prominent shell of SN 1006 around the Crab revealed that such a shell would be undetectable in the Crab even if the scaled surface brightness were increased by an order of magnitude.

In concluding, it can be noted that although the Crab Nebula has long been recognized to be quite unique among SNR's, this has not prevented observers and theoreticians alike from using it as the archetype for type I supernova. This was perhaps understandable in the past, due to the lack of other such objects (except for Vela which is quite different in many respects); but *Einstein* observing programs carried out by guest observers and consortium members have now discovered other SNR's with compact objects (e.g., RCW 103, G 109, and MSH 15-52), as well as "Crab-like" SNR's in the LMC. The very fact that this Symposium has occurred is a hopeful sign that new progress in the understanding of SNR may lie just around the corner.

REFERENCES

Bowyer, S., Byram, E.T., Chubb, T. A., and Friedman, H. 1964, Science, 146, p 912.
Giacconi, R., et al. 1979, Ap.J., 230. p 540.
Harnden, F.R., Jr. and Gorenstein, P., 1973, Nature, 241, p 107.
Helfand, David J., Chanan, Gary A., and Novick, R. 1980, Nature, 283, p 337.
Nomoto, Ken'ichi and Tsuruta, S., 1981, Ap.J. (Letters), 250, L 19.
Smith, A. and Pounds, K.A. 1977, Nature, 265, p 121.

DISCUSSION

KIRSHNER Is it correct to say that the point response function is a function of the incident energy, the position in the field and possibly even time? If that's true, can you give an idea whether it will be possible to place a firm limit on or possibly detect the shell around the crab that has been suggested in the past?

HARNDEN The unavoidable problem is the existence of three separate physical processes which produce similar observed features: (1) a blast wave shell, (2) interstellar dust grain scattering and (3) image degradation due to instrumental scattering from mirror-surface imperfections. No, it is not correct to say that the point response of the Einstein IPC is a function of time. The dependence upon incident energy and position is well understood and can be accurately modeled. Certainly a formal detection (or upper limit determination) will be possible, but it does require detailed analysis which has not yet been completed. With the Crab (whose spectrum is well determined) the major uncertainty is the possible existence of a dust grain halo; with sources whose spectra are not well determined, that uncertainty further complicates the analysis.

HELFAND The use of the minimum of the X-ray pulse profile to establish a neutron star temperature limit is a reasonable first approximation, but, in fact, the surface thermal conductivity is expected to be very nonuniform as a result of the strong magnetic field. Thus, the blackbody surface emission should be modulated, peaking at the nonthermal pulse peaks, if these represent the star's magnetic poles. Can you use the IPC/HRI rate as a function of pulse phase to get some spectral information? What about MPC/HRI?

HARNDEN Use of the IPC temporal information is hopelessly complicated by the very high count rate, which saturates the available telemetry, so I don't think the IPC data could help. The HRI rate is quite high, too, but the deadtime is well understood. I'm not sure what the MPC deadtime is as a function of pulsar phase, but if it can be accurately enough determined, perhaps the MPC/HRI ratio might shed some light on this question.

PRAVDO Is the temperature of the thermal X-rays near (within 1 degree) the Vela pulsar, the same as that of the X-rays in farther away regions of the SNR?

HARNDEN The Vela spectral analysis hasn't been completed yet so I can only give you a preliminary answer. As we saw in the slide of the 5-degree remnant, there are several localized regions of enhanced X-ray emission within the large remnant: the pulsar is one of these, as is a region ~ 20 arcmin to the north of the pulsar. If we fit a thermal plasma model to the IPC spectral data from these relatively intense regions, we can get acceptable fits with "reasonable" values of temperature for all regions (except for the region within ~ 3 arcmin radius of the pulsar, where a power law model is required to obtain an acceptable Chi-square value). These "reasonable" values definitely vary from one region to another. In fact, the intensity and the temperature appear to be correlated. The east-west filamentary structure which passes ~ 20 arcmin north of the pulsar has a temperature within the range of variation of the other bright spots, i.e., it's not particularly noteworthy except for its proximity to the pulsar.

DANZIGER   On deep exposure direct plates, one has seen in the immediate
vicinity of the Vela pulsar filamentary nebulosity belonging to the SNR.
One also sees emission from diffuse nebulosity of a type characteristic
of the SNR at the precise position of the pulsar. It could be either
foreground or background material.

HARNDEN   I think the X-ray data are consistent with that optical
picture, namely foreground or background thermal emission from the blast
wave are present at the pulsar position as elsewhere. But none of the
really pronounced regions of thermal X-ray emission coincides with the
pulsar, the nearest are being ~ 20 arcmin away.

PRESUPERNOVA EVOLUTION OF 8-12 M☉ STARS AND THE CRAB NEBULA'S PROGENITOR

Ken'ichi Nomoto
Dept. of Earth Science and Astronomy, College of General
Education, University of Tokyo, Meguro-ku, Tokyo   153
and
Max-Planck-Institut für Physik und Astrophysik
Institut für Astrophysik, Garching bei München

ABSTRACT

The elemental abundances of the Crab Nebula are compared with the presupernova models of 8-12 M☉ stars. The small carbon abundance of the Crab is consistent with only the star whose main-sequence mass was 8-9.5 M☉. More massive stars contain too much carbon in the helium layer and smaller mass stars do not leave neutron stars. A scenario for the Crab Nebula's progenitor is proposed.

1. INTRODUCTION

The comparison of the elemental abundances of the young supernova remnant with the nucleosynthesis in the evolving stellar interior gives an important clue to identify the progenitor star of the supernova remnants. Such a comparison has been made mostly between the massive star models (M $\gtrsim$ 15 M☉) and the oxygen-rich supernova remnants.

However, little work has been done for 8-12 M☉ stars although most of Type II supernovae originate from this mass range. In this paper, I summarize the features of the presupernova evolution of 8-12 M☉ stars based on the recent calculations (Nomoto 1981). In contrast to more massive stars, 8-12 M☉ stars could eject only a small amount of heavy elements. Thus it is interesting to compare these models with the abundances of the Crab Nebula, because the Crab has been known to contain little heavy elements.

2. PRESUPERNOVA EVOLUTION OF 8-12 M☉ STARS

The evolution of the helium core of 8-12 M☉ stars were calculated from the helium burning phase. The mass of the initial helium core (i.e., mass interior to the H-burning shell, $M_{H,0}$), the corresponding main-sequence masses (M), and the mass interior to the He-burning shell (M_{He}), and the C-burning shell (M_C) at the presupernova stages are summarized in Table. In Figures 1-4, shown are the chemical evolution of

139

J. Danziger and P. Gorenstein (eds.), Supernova Remnants and their X-Ray Emission, 139–143.
© 1983 by the IAU.

these stars. (Chemical compositions of these models are seen in Nomoto 1982).

$M/M_\odot$	$M_{H,0}$	M_{He}	M_C	Evolution
$\sim$ 12	3.0	1.62	1.60 .. off-center Si flash $\big\}$ $\Rightarrow$ $\big\{$ Fe core collapse
$\sim$ 10	2.6	1.45	1.43 .. off-center Ne flash $\big\}$ (photodisintegration)
$\sim$9.5	2.4	1.375	1.34 $\big\}$ no Ne ignition $\Rightarrow$ $\big\{$ O+Ne+Mg core collapse
$\sim$ 9	2.2	1.375	1.28 $\big\}$ (electron capture)

Since the core masses (M_C in particular) of these stars are close to the Chandrasekhar limit, the effect of electron degeneracy on the evolution is quite significant. Following are the characteristics of the evolution of 8-12 $M_\odot$ stars, which are very sentitive to the stellar mass:

1) Off-Center Flash: When the core (Si core of the 12 $M_\odot$ star and O+Ne+Mg core for M $\stackrel{<}{\sim}$ 10 $M_\odot$) becomes semi-degenerate, the central region is cooled down by the neutrino emission and temperature inversion appears. As a result, Si (for 12 $M_\odot$ star) or Ne (for 10 $M_\odot$ star) ignites off-center as seen in Figures 1 and 2. The Ne burning grows into a very strong flash. However, it is not strong enough to eject the H-He layer, i.e., weaker than the Ne flash of the 10 $M_\odot$ star computed by Woosley et al. (1980) because of our smaller M_C. Although the further evolution is very complicated, the 10 and 12 $M_\odot$ stars finally evolve into the Fe core collapse (Woosley et al. 1980).

2) Formation of Degenerate O+Ne+Mg Core: For stars of M $\stackrel{<}{\sim}$ 9.5 $M_\odot$, the core mass, M_C, is smaller than the critical mass of 1.37 $M_\odot$ for Ne ignition (Boozer et al. 1973). Therefore, the O+Ne+Mg core cools down without Ne burning and becomes strongly degenerate (Figures 3 and 4).

3) Dredging-Up of the He Layer: For the 9.5 $M_\odot$ and 9 $M_\odot$ stars, the core is more centrally condensed than for M $\stackrel{>}{\sim}$ 10 $M_\odot$ so that the He layer is more extended. When the 9.5 $M_\odot$ reaches the end stage of Figure 3, the He-layer expands greatly and the penetration of the surface convection zone into the He-layer starts. Such a dredging-up of the He-layer starts earlier for the 9 $M_\odot$ star (Figure 4) and proceeds until the hydrogen shell burning is ignited.

3. COLLAPSE OF THE O+Ne+Mg CORE OF 8-10 $M_\odot$ STARS

After the dredging-up of the He layer, both 9 and 9.5 $M_\odot$ stars evolve as follows, which is common to 8-10 $M_\odot$ stars.

1) The H/He double shell burnings form the C+O layer above the O+Ne+Mg core. The temperature of the C+O layer is too low to ignite the carbon shell flash. Then the evolutionary track in the central density, ρ_c, and temperature, T_c, merges into the same track as of the degenerate C+O core of 4-8 $M_\odot$ stars because of the same core growth rate.

2) When M_{He} exceeds 1.375 $M_\odot$, the electron captures on ^{24}Mg and ^{20}Ne

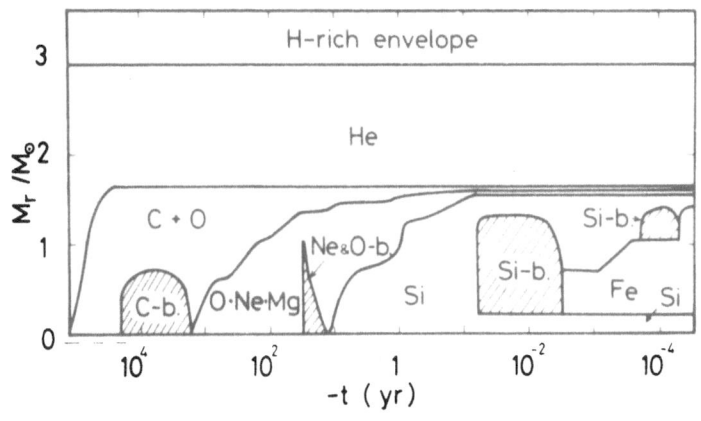

Fig. 1 Chemical Evolution of the 12 M$_\odot$ Star. Time is measured from the ignition of the central Si flash. Shaded regions are in convective equilibrium. (Convective He, C, and Ne burning shells are omitted from the figure.)

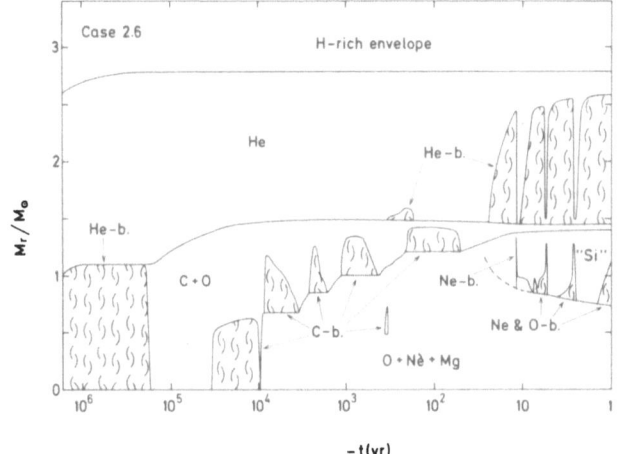

Fig. 2 Chemical Evolution of the 10 M$_\odot$ Star. Time is measured from the end stage of the computation. Curled regions are the convection regions. Off-center Ne flash is ignited.

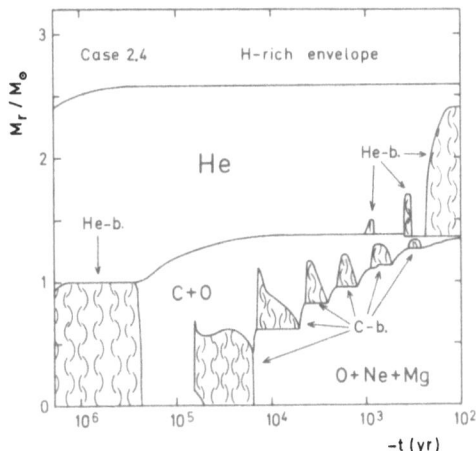

Fig. 3

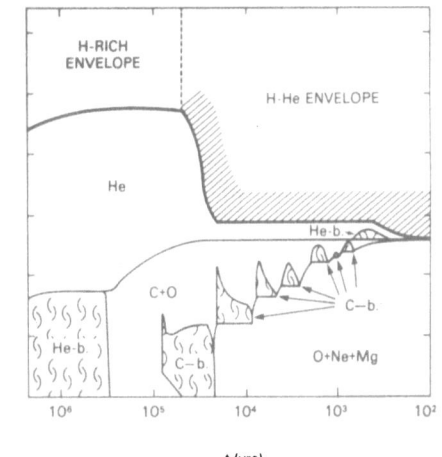

Fig. 4

Same as Fig. 2 but for the 9.5 M$_\odot$ star (Fig. 3) and 9 M$_\odot$ star (Fig. 4) where the degenerate O+Ne+Mg core is formed. Shaded region indicates the surface convection zone penetrating into the He-layer.

start to reduce the number of electrons per baryon, Y_e. Then the core
contracts rapidly. When ρ_c reaches 2.5×10^{10}g cm^{-3}, the oxygen deflagration is ignited and incinerates the material into nuclear statistical
equilibrium (NSE) elements. However, electron captures are sufficiently
rapid for the core to continue to collapse (see also Miyaji et al. 1980).

3) Preliminary calculation of the further collapse has been carried out
up to the neutrino trapping density by adopting the equation of state by
Wolff (1982) and the electron capture rates by Fuller et al. (1982). At
$\rho_c = 3 \times 10^{11}$g cm^{-3}, entropy per nucleon and Y_e at the center are s =
1.2 k and $Y_e = 0.35$, the average Y_e over 0.8 $M_\odot$ is 0.43, and the mass of
the NSE core behind the oxygen deflagration front is 0.52 $M_\odot$.

4) The hydrodynamical behavior of the core bounce and shock propagation
might be somewhat different from the iron core of massive stars: The
shock wave could be strengthened by the steep density gradient near the
He-burning shell at $M_r = 1.375$ $M_\odot$ (see Hillebrandt 1982) and might eject
only the He layer to leave a neutron star behind.

4.   CARBON ABUNDANCE OF THE EJECTED HELIUM LAYER

        The final fate of the 8-12 $M_\odot$ star is probably Type II supernova
explosion which leaves a neutron star behind. If the neutron star mass
of 1.4 $M_\odot$ is assumed, ejected remnant should not contain much heavy
elements as seen from M_{He} in Table. As discussed in the next section,
the chemical composition of the He layer of these stars is particularly
important to identify the Crab Nebula's progenitor:

1) For stars of M $\gtrsim$ 9.5 $M_\odot$, the carbon abundance of the helium layer is
so high as $X_C \sim 0.03$-0.05 because the He shell burning is so active as
to develop a convective shell (see Figures 2 and 3). Abundances of ^{14}N
and ^{16}O are $X_N \sim X_O \sim 0.005$. (Hereafter X denotes the mass fraction.)

2) For M $\lesssim$ 9 $M_\odot$, the carbon abundance is not large because the dredging-
up of the He layer prevents the convective He burning shell from developing (Figure 4). Such a mixing of H/He layer could produce a He-rich
envelope if the mass loss would reduce the stellar mass down to $\sim$ 3.5 $M_\odot$
before the dredge-up. The resultant composition in the He-rich envelope
would be $X_C \sim X_N \sim X_O \sim 0.004$ for $M_{H,0} = 2.2$ $M_\odot$ model.

5.   COMPARISON WITH THE ABUNDANCES OF THE CRAB NEBULA

        According to the recent optical, IUE and IR observations (Henry and
MacAlpine 1982; Davidson 1979; Davidson et al. 1982; Dennefeld and
Andrillat 1981), the Crab Nebula has a helium overabundance of 1.6 $\lesssim$
$X_{He}/X_H \lesssim 8$, relatively small oxygen abundance of $X_O \sim 0.003$, and the
carbon-to-oxygen ratio of 0.4 $\lesssim X_C/X_O \lesssim 1.1$. The small oxygen abundance
implies that only the hydrogen-rich envelope and the helium layer above
the helium-burning shell was ejected and the lower layer must have formed
the neutron star. The small carbon abundance is inconsistent with the
stellar models for M > 9.5 $M_\odot$ while it is consistent with M $\sim$ 9 $M_\odot$.

(Since all 8-12 M$_\odot$ models show ^{14}N overabundance in the He layer, ^{14}N abundance determination is also important. Fesen and Kirshner, 1982, suggested that ^{14}N may be overabundant; see, however, Dennefeld and Pequignot 1982, this conference.)

Based on the above discussion, the 8-9.5 M$_\odot$ star has been proposed to be the Crab Nebula's progenitor. The scenario of the post main-sequence evolution is as follows (see Nomoto 1982 and Nomoto et al. 1982 for details): During the red supergiant phase, the star lost 5-6 M$_\odot$ of its H-rich envelope by a stellar wind ($\dot{M} \sim 10^{-6} M_\odot y^{-1}$). During the C-burning, a He-rich envelope with $X_C/X_O \lesssim 1$ was formed by the mixing between the H-rich and He layer. After the C-burning, the O+Ne+Mg core was formed and finally collapsed by the electron captures. The bouncing shock wave was not strong enough to eject the core, but it was strong enough to eject the loosely bound He-rich envelope. The weak shock wave and the composition of the ejected matter was consistent with many of the observed features of the Crab Nebula. The suggested abundance variations among the filaments could be due to the partial mixing of the ejected material with the previously lost circumstellar material.

I would like to thank Drs. W.M. Sparks, R.A. Fesen, T.R. Gull, S. Miyaji, and D. Sugimoto for discussion on the Crab's progenitor (see Nomoto et al. 1982). Part of the calculation was carried out at Max-Planck-Institut. It is a pleasure to thank Prof. R. Kippenhahn for hospitality, Dr. W. Hillebrandt and Mr. R.G. Wolff for discussion on the core collapse, and H. Nomoto for preparation of the manuscript and figures.

REFERENCES

Boozer, A.H., Joss, P.C., and Salpeter, E.E. 1973, Ap.J. 181, 393.
Davidson, K. 1979, Ap.J., 228, 179.
Davidson, K., Gull, T.R., Maran, S.P., Stecher, T.P., Fesen, R.A., Parise, R.A., Harvel, C.A., Kafatos, M. and Trimble, V.L. 1982, Ap.J., 253, 696.
Dennefeld, M., and Andrillat, Y. 1981, Astron. Ap., 103, 44.
Fesen, R.A., and Kirshner, R.P. 1982, Ap.J., 258, 1.
Fuller, G.M., Fowler, W.A., and Newman, M.J. 1982, Ap.J. Suppl., 48, 279.
Henry, R.C., and MacAlpine, G.M. 1982, Ap.J., 258, 11.
Hillebrandt, W. 1982, Astron. Ap., 110, L3.
Miyaji, S., Nomoto, K., Yokoi, K., and Sugimoto, D. 1980, Publ. Astron. Soc. Japan, 32, 303.
Nomoto, K. 1981, in Fundamental Problems in the Theory of Stellar Evolution, ed. D. Sugimoto et al. (Reidel), p.295.
Nomoto, K. 1982, in Supernovae: A Survey of Current Research, ed. M.J. Rees and R.J. Stoneham (Reidel), p.205.
Nomoto, K., Sparks, W.M., Fesen, R.A., Gull, T.R., Miyaji, S., and Sugimoto, D. 1982, Nature, in press.
Wolff, R.G. 1982, private communication.
Woosley, S.E., Weaver, T.A., and Taam, R.E. 1980, in Type I Supernovae, ed. J.C. Wheeler (Univ. of Texas), p.96.

THE CRAB NEBULA'S JET

Robert Fesen and Theodore Gull
Laboratory for Astronomy and Solar Physics
NASA/Goddard Space Flight Center
Greenbelt, Maryland

Using deeply exposed Schmidt plates, van den Bergh (1970) discovered a peculiar jetlike feature extending about 100" beyond the Crab Nebula's northern boundary. The feature roughly coincided with a spur in the outer intensity contour map of the Crab's continuum emission (5200-6400A) made by Woltjer (1957) from plates taken by Baade with the 200 inch. However, little if any continuous synchroton emission is visible in the jet's region on Scargle's (1970) deep continuum photograph. More recent imaging of the jet by Chevalier and Gull (1975) and Wyckoff et al. (1976) using narrow passband interference-filters, show it slightly better than van den Bergh's plates, and indicate it is most visible in [O III] 5007A emission. Spectroscopy of the jet's southern portion by Davidson (1978, 1979) confirmed that the [O III] 4959A, 5007A lines are the brightest optical emissions in the jet and suggested a low radial velocity of about 100 km s^{-1}.

We have recently obtained a very deep image of the Crab Nebula using an [O III] interference-filter and the Carnegie Image Tube Direct camera attached to the 0.9m telescope at Kitt Peak. The image was digitally enhanced using a PDS microdensitometer and processed with a Comtal Display System at the GSFC. Figure 1 shows a "flattening" histogram transform for the [O III] image of the Crab, permitting simultaneous viewing of both the very bright and very faint [O III] emission features. Figure 2 is an enlargement of the nebula's northern region to show the jet in greater detail.

In our [O III] image, the jet appears as a wide (≈ 45"), yet strikingly collimated filamentary feature extending about 75" above the main nebula's northern outer boundary. The jet appears brighter along its sharply defined and parallel eastern and western sides, suggesting a tube-like structure rather than a gaseous sheet or filled cylinder. At a resolution of 2", the jet's emission appears clumpy, consisting of several small discrete condensations. The feature's western side is especially straight and sharply defined in our image. At its northern end, the jet appears to terminate

145

J. Danziger and P. Gorenstein (eds.), Supernova Remnants and their X-Ray Emission, 145–148.
© *1983 by the IAU.*

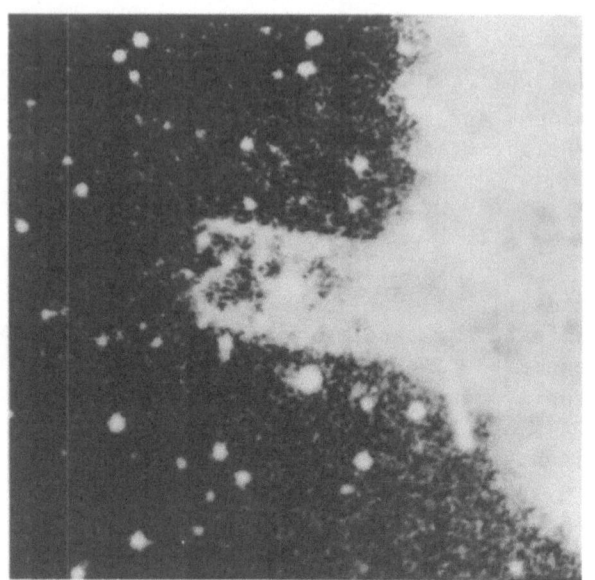

Fig. 1

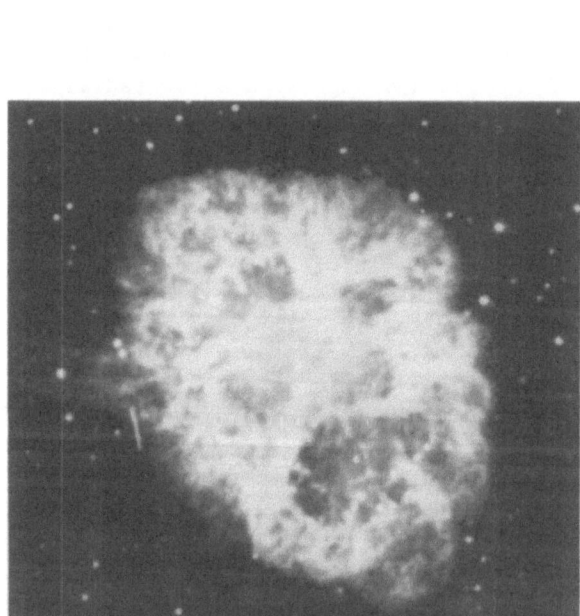

Fig. 2

Figure 1:   Deep [O III] image of the Crab Nebula which has been digitally enhanced. The short E-W streak just east of the base of the jet was caused by the motion of the minor planet Holmia, accidentally recorded as it moved in front of the nebula.

Figure 2:   Enlargement of the northern area of nebula showing the jet's structure in greater detail.

abruptly and evenly. Despite its orderly morphology, however, the
jet surprisingly does not project back to either the Crab's center
of expansion or its pulsar. Instead, it is inclined to the
northwest by about 15°.

No other jetlike or extended [O III] feature is visible on our
photograph with a brightness exceeding about 1/4 that of the
northern jet. Indeed, with the exception of the jet, the Crab
Nebula has very well defined boundaries (see Figure 1). The
remnant's maximum minor and major axes dimensions are 335" x 435"
using the [O III] image and 315" x 420" using an Hα + [N II]
photograph. Our measured [O III] dimensions are slightly larger
than those quoted by van den Bergh probably due to deeper imaging as
well as a 1.5% expansion from 1970-1981. At a distance of 2 kpc
(Trimble 1973), our [O III] photograph indicates the Crab Nebula's
physical dimensions are 3.2 pc x 4.2 pc.

The origin of the Crab Nebula's jet is unknown. Potential
theories for its formation must provide answers as to why it is the
only filamentary extension outside an otherwise well bounded nebula,
and why it exhibits such an organized structure in a largely chaotic
and amorphous appearing supernova remnant? Because of its
appearance, it is not clear that the jet has undergone the same
physical processes which shaped the rest of the Nebula. Also,
though its structure suggests an energetic origin, one cannot even
be sure that the jet is directly related to the AD 1054 event since
it is not in radial alignment with either the Nebula's center of
expansion or pulsar.

There are several possibilities for its origin. 1) the jet
consists of some unusually high velocity material ejected by the
1054 explosion, 2) it is the result of a violent but separate event
from the 1054 SN, 3) it represents some post-supernova input of
energy by the Crab pulsar and/or plasma instabilities within the
remnant, or 4) it could represent a tubular structure of material
lost by the pre-supernova star during its red giant phase (Blandford
et al. 1982). However, clearly more information regarding the jet's
physical properties (e.g. proper motion, temperature and density
estimates, and the presence of synchrotron radiation) must be
obtained before we can understand the jet's structure and origin.

We thank K. Davidson for useful discussions and D. Klinglesmith
for assistance with the digital image processing.

REFERENCES

Blandford, R. D., Kennel, C. G., McKee, C. G., and Ostriker, J. P.:
    1982, preprint.
Chevalier, R. A. and Gull, T. R.: 1975, Astrophys. J. 200, pp. 399-
    401.

Davidson, K.:   1978, Astrophys. J. 220, pp. 177-185.
Davidson, K.:   1979, Astrophys. J. 228, pp. 179-190.
Scargle, J. D.:   1970, Pub. A. S. P. 82, pp. 388-392.
Trimble, V.:   1973, Pub. A. S. P. 85, pp. 579-585.
van den Bergh, S.:   1970, Astrophys. J. 160, pp. L27.
Woltjer, L.:   1957, Bull. Astr. Inst. Netherlands, 13, pp. 301-311.
Wyckoff, S., Wehinger, P. A., Fosbury, R. A. E., and McMullan, D.:
    1976, Astrophys. J. 206, pp. 254-256.

HIGH OXYGEN ABUNDANCE IN A BRIGHT FILAMENT OF THE CRAB NEBULA

Daniel Péquignot
Observatoire de Meudon, Meudon, France

Detailed stationary photoionization models with homogeneous chemical composition were constructed for a bright filament of the Crab Nebula centered ~42" W-SW of the pulsar. So far this filament is the only one which has been observed in the ultraviolet (Davidson et al, 1982), in the visual (Miller, 1978; Davidson et al, 1982) and in the near-infrared (Dennefeld and Péquignot, 1982), so that numerous constraints and meaningful tests of the models exist. On the other hand the computer program used in the calculations includes recently introduced atomic physics processes (charge exchange reactions and "low-temperature" di-electronic recombinations) and was tested and calibrated by means of models of the high-excitation planetary nebula NGC 7027 whose line spectrum shares some characteristics with that of the Crab Nebula.

The two final models obtained fit all important features of the observed spectrum, in particular the critical ratio [OIII] 4363/5007, within observational uncertainties: these models differ in the assumed shape of the ionizing continuum and can be considered as the two most extreme assumptions yet compatible with currently available continuum measurements in the ultraviolet and X-ray range.

The main features of the gas density distribution are as follows: (1) In order to explain the strength of [NeV] 3426 (and CIV 1549 as well), the thermal gas pressure of the high-excitation zone must be several times less than in the bulk of the filament; therefore this high-excitation zone has a low density and comprises much of the volume of the filament. (2) In order to explain the strength of the "photon-counting" line HeII 4686, the high-excitation zone must substand a larger solid-angle, as seen from the source, than the low- or inter-mediate-excitation zones (emitting, e.g. [OI], [OII] and a fraction of [OIII]); this is in harmony with point (1) above if the filaments are cylinder-like.

The inferred elemental abundances are quite similar in both models and differ remarkably from previous determinations based on model calculations in that not only helium but the heavy elements C, N, O, Ne, S

149

J. Danziger and P. Gorenstein (eds.), Supernova Remnants and their X-Ray Emission, 149–151.

as well are markedly <u>overabundant</u> with respect to hydrogen so that, when the abundances are given in number <u>per nucleon</u>, they appear fairly <u>typical of Population I.</u>

This finding allows one to solve an old dilemma concerning the progenitor of the supernova. Until now the abundances (per nucleon) of the heavy elements, notably oxygen (the most easily accessible), were believed to be no more than one third the cosmic value (e.g. Davidson, 1979) so that the progenitor should have been a Population II and there-fore very low mass star which would have expelled a large amount of helium with at most a <u>moderate</u> enrichment of heavy elements. This did not match any known scenario of supernova explosions and the Crab Nebula, the closest and best known supernova remnant, appeared as an extraordinary object that could not be attributed any reasonable status. In fact the situation was so uncomfortable that the recent proposals for the explosion involved a ~10 M_0 Population I star (e.g. Chevalier, 1977, Nomoto, 1983, Nomoto et al. 1982) and therefore simply <u>ignored</u> the underabundance of heavy elements in spite of repeated claims of Davidson and others.

Another important result of our calculations is that the abundance of <u>nitrogen</u> is certainly <u>not</u> more than normal with respect to oxygen so that there is no trace of the well-known enrichment which characterizes the CNO bi-cycle. This implies that the gas forming the filament has been somehow further processed at high temperatures favorable to helium burning. Since oxygen and carbon are roughly normal (the latter less accurate) while neon may be slightly overabundant (~ factor 2), this material should not have been processed beyond the very beginning of shell helium-burning and negligible mixing with inner C-rich layers should have occurred. The small amount of hydrogen presumably results from mixing with more external (unprocessed) layers of the star. The scenario recently proposed by Nomoto et al (1982) based on a ~ 9.5 M_0 progenitor accounts for the abundances of helium and oxygen but predicts an overabundance of nitrogen. A way to preserve the basic features of their attractive scenario may be to accept that, contrary to their expectation, the passage of the supernova shock wave burned a large fraction of nitrogen at least in some specific layers of the H-He en-velope. Our result would then suggest that very little mixing occurred during or after the explosion.

A discussion of the photoionization models and of the reasons why previous analyses reached a different conclusion about abundances is given by Péquignot and Dennefeld (1982).

REFERENCES

Chevalier, R.: 1977, "Supernovae", D. Schramm, ed., p. 53, Reidel, Dordrecht, Holland.
Davidson, K.: 1979, Astrophys. J. <u>228</u>, 179.

Davidson, K., Gull, T.R., Maran, S.P., Stecher, T.P., Fesen, R.A.,
    Parise, R.A., Harvel, C.A., Kafatos, M., Trimble, V.L.: 1982,
    Astrophys. J. 253, 696.
Dennefeld, M., Péquignot, D.: 1982, Astron. Astrophys., submitted.
Miller, J.S.: 1978, Astrophys. J. 220, 490.
Nomoto, K.: 1983, this volume, p. 521.
Nomoto, K., Sparks, W.M., Fesen, R.A., Gull, T.R., Miyaji, S.,
    Sugimoto, D.: 1982, Nature, submitted.
Péquignot, D., Dennefeld, M.: 1982, Astron. Astrophys., submitted.

DISCUSSION

WOLTJER: What is changed if you allow for some shock or other collisional heating in addition to photoionization energy input?

PEQUIGNOT: The answer to your question depends at once on the place
where the complementary heating would be effective and on the problem
that this heating would be supposed to solve:
1. An extra-heating of the highest excitation zone would mainly amplify
CIV which, in the present schematic models, tends indeed to be small
(although still within the observational uncertainties). Keeping in mind
our procedure of convergence based on [NeV], a smaller contrast in thermal pressure between the high- and low-excitation zones would follow.
2. An extra-heating of the high- or intermediate-excitation zone would
oblige the consideration of even higher elemental abundances since our
final models were obtained by enforcing that the temperature-dependent
line ratios (notably [OIII] and [OII]) be in agreement with observation.
Also the collisionally excited lines would grow stronger relative to Hβ:
since, with photoionization alone, they may tend to be slightly too
large, we must conclude that the supplementary heating is moderate.
3. An extra-heating of the low-excitation zone would lead, in the
framework of our rough models, to a smaller optical thickness of the
filament without other significant consequences.

    Now if the extra-heating is thought as an ad hoc alternative to
explain the collisional high-excitation lines without considering, as we
did, an ad hoc gradient of thermal pressure, then I believe that major
difficulties would have to be faced. In effect [NeV] 3426 is perhaps the
best indicator of pure photoionization by hard radiation since the collisional excitation of this line is perfectly effective at moderate temperature while any collisional ionization of Ne^{+3} to Ne^{+4} requires very
high temperatures which would entail extremely large (and not observed)
consequences on other lines such as CIV 1549. In a word the pressure
gradient seems virtually unescapable provided that [NeV] has really been
detected, although its exact value is still controversial.

THE PECULIAR X-RAY MORPHOLOGY OF THE SNR G292.0+1.8 : EVIDENCE FOR AN ASYMMETRIC SUPERNOVA EXPLOSION

D.H. Clark[1,3] and I.R. Tuohy[2,3]

1   Rutherford Appleton Laboratory
2   Mount Stromlo and Siding Spring Observatories
3   Guest Investigator, Einstein Observatory.

A high resolution X-ray image from the Einstein Observatory of the young supernova remnant G292.0+1.8 (MSH11-54), previously noted as peculiar in terms of its spectral and morphological properties at optical and radio wavelengths, also shows an unusual X-ray morphology. Instead of a limb-brightened X-ray shell characteristic of most SNRs, the remnant consists of a central bar-like feature superposed on an ellipsoidal disc of approximately uniform surface brightness. We attribute the bar emission to a ring of oxygen-rich material ejected in the equatorial plane of a massive rotating progenitor, and the uniform disc component to emission from material with roughly cosmic composition heated by the accompanying blast wave. This interpretation provides observational support for the rotating precursor model of a Type II supernova discussed by Bodenheimer and Woosley.

1. INTRODUCTION

The supernova remnant (SNR) G292.0+1.8 has attracted considerable attention recently since it is only the second galactic SNR (after the fast moving knots in Cas A) to show an optical spectrum dominated by oxygen emission. The oxygen-rich material in these SNRs is believed to be alpha-processed ejecta from the relatively recent explosion of a massive ($\sim$25 $M_\odot$) star. Thus the study of such remnants offers the potential of probing both the nucleosynthetic processes in Population I stars and the dynamics of Type II supernova explosions.

At radio wavelengths, G292.0+1.8 (MSH11-54) shows a bright central peak of about 2 arcminutes extent, surrounded by a faint plateau of approximately 10 arcminutes diameter (Lockhart et al. 1977). No radio shell is present. The optical counterpart (Goss et al. 1979) consists of faint and irregular nebulosity visible only in the emission lines of oxygen and neon. Murdin and Clark (1979) found the velocity dispersion of the oxygen and neon rich material to be in excess of 2000 km s^{-1}, consistent with G292.0+1.8 being a young (<2000 years) remnant in which

153

J. Danziger and P. Gorenstein (eds.), Supernova Remnants and their X-Ray Emission, 153–158.
© *1983 by the IAU.*

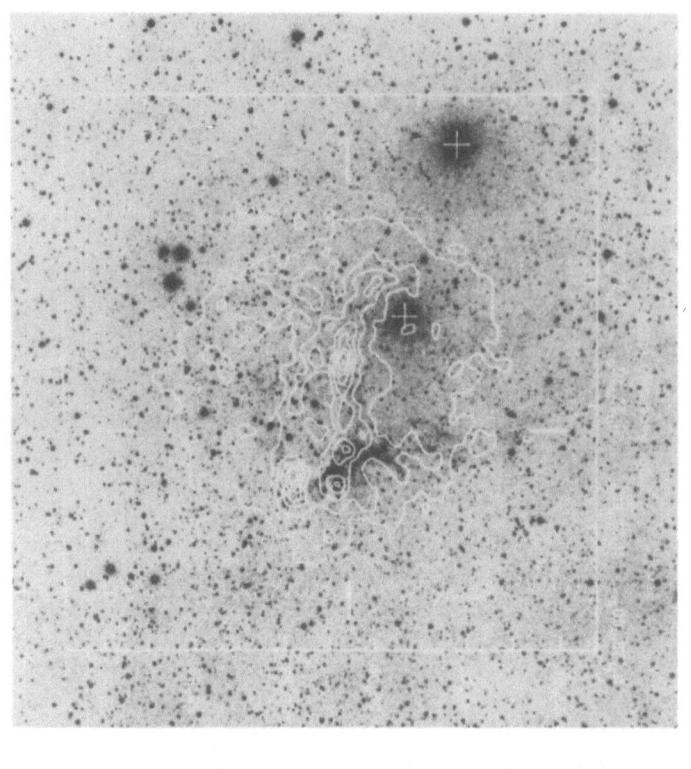

Fig. 1b Smoothed X-ray contour diagram of
G292.0+1.8 overlaid on an [OII] λ3727 exposure
taken at the prime focus of the AAT by Murdin
and Malin. The two crosses denote SAO stars
used for alignment purposes, while the intersection of the four edge marks defines the
centroid of the radio peak measured by Lockhart
et al. (1977). Figures 1a and 1b are to a
similar scale.

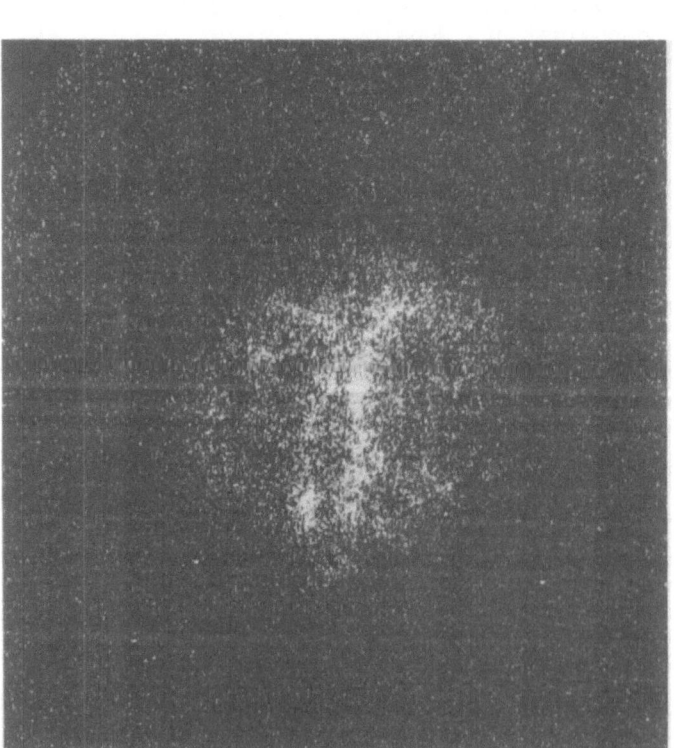

Fig. 1a X-ray image of G292.0+1.8 recorded by
the High Resolution Imager on the Einstein
Observatory. The number of counts per 2 arcsecond square pixel ranges from 0 to 5. North
is at the top and East is to the left.

the supernova ejecta has not yet been contaminated by swept-up interstellar material. In this paper we report the first X-ray image of
G292.0+1.8 which we interpret in terms of an asymmetric Type II supernova
explosion (see also Tuohy, Clark and Burton 1982).

2. OBSERVATIONS AND RESULTS

The X-ray image of G292.0+1.8, obtained using the High Resolution
Imager on the Einstein Observatory (Giacconi et al. 1979), is displayed
in Figure 1a. Figure 1b shows the X-ray data as a smoothed contour map
overlaid on an [OII] λ3727 photograph taken at the prime focus of the
Anglo-Australian Telescope (Murdin and Malin, private communication).
The intersection of the four edge marks defines the centroid of the radio
peak measured by Lockhart et al. (1977).

The X-ray emission from G292.0+1.8 is confined to an ellipsoidal
"disc", punctuated by a bright central feature bisecting the disc in an
E-W direction. The emission from this central "bar" is irregular, and
furthermore, shows bifurcated structure having a mean separation of ∿ 1
arcminute. The dimensions of the disc are 8 arcminutes by 6.5 arcminutes,
with the N-S elongation being almost perpendicular to the central bar.
Away from the bar, the disc emission has a relatively uniform surface
brightness without limb-brightening (see also Figure 3 in Tuohy, Clark
and Burton 1982). The X-ray image shows no evidence of a point source
that could be associated with a neutron star.

The correlation between the X-ray and optical features in Figure 1b
is not striking; the two dominant hotspots in the X-ray bar lie away
from any significant optical nebulosity. The brightest [OII] nebulosity
is located at the eastern extremity of the X-ray bar, and coincides with
two small knots in the X-ray image. A spur of X-ray emission running to
the south appears to be associated with a similar southerly extension of
the [OII] emission. The centroid of the radio peak lies near the bright
[OII] nebulosity and is approximately midway between the two hotspots in
the X-ray bar. Thus it is probable that the peaked radio component,
which shows evidence for E-W extension, is associated with the X-ray bar,
while the broad radio plateau referred to earlier can be ascribed to the
X-ray disc.

The X-ray image accurately delineates the extent of G292.0+1.8 for
the first time. The angular size and the distance to the remnant of
≥ 3.7 kpc (Caswell et al. 1975) imply a mean linear diameter of ≥ 7.8 pc.
This diameter is consistent with the young age of the remnant (≤ 1600
years) inferred from optical velocity data (Murdin and Clark 1979).

3. DISCUSSION

The X-ray structure of G292.0+1.8 reported here is unique amongst
the ∿ 50 SNRs for which X-ray maps have now been obtained. Neither of

the two principle morphological classification criteria are met, namely
shell-like structure or centrally peaked (Crab-like) X-ray emission.
Initially it was thought that G292.0+1.8 *was* Crab-like on the basis of
the centrally condensed radio structure (Lockhart et al. 1977, Goss et
al. 1979). Doubt was cast on this classification by Weiler (1978) and
van den Bergh (1979), and also by the failure to observe a synchrotron
X-ray spectrum (Clark, Tuohy and Becker 1980). Thus the cumulative
optical and X-ray data now rule out any close relationship between
G292.0+1.8 and the Crab Nebula.

A plausible explanation of the unusual X-ray morphology of
G292.0+1.8 follows from the recent work of Bodenheimer and Woosley
(1980, 1982) and Woosley and Weaver (1981,1982). They have discussed
firstly the collapse of a non-rotating 25 $M_\odot$ star, and find that the
outward moving shock produced by the "bounce" of the iron core at nuclear
density travels only part way out before being extinguished by infalling
material and by the photo-disintegration of ^{28}Si. The net result is that
a supernova explosion does not accompany stellar collapse. However, when
stellar rotation is invoked, the impeding effect of the centrifugal
motion on the infalling oxygen-rich envelope leads to rotational braking.
Inertial overshoot of this material occurs, followed by an outward radial
bounce, which, powered by explosive nucleosynthesis, results in the
ejection of mass in the *equatorial* plane. The effect of the equatorial
expulsion of oxygen would be to form an expanding ring of ejecta. Such
a ring, if viewed nearly edge-on at an angle of $\sim 10^{\circ}$, could mimic the
oxygen-rich nebulosity and the bifurcated X-ray bar of G292.0+1.8.

Velocity mapping of isolated portions of the optical nebulosity,
which is concentrated along the X-ray bar, show complex motions (Danziger
and Goss, private communication). It is impossible to model these in
isolation. However, a preliminary assessment of velocity data covering
the complete remnant (Murdin and Clark, in preparation) show a general
expansion commensurate with the equatorial ejection model. Some deviation from a simple planar ring model would not be unexpected in view of
the evidence for a distorted ring of ejecta in the oxygen-rich SNR,
1E0102.2-7219 (Tuohy and Doptia, 1982, 1983).

The origin of the uniform X-ray disc remains to be understood.
Woosley and Weaver (1982) have considered the explosin of a 25 $M_\odot$
rotating precursor further, and find that while the oxygen-rich material
is ejected equatorially, a portion of the kinetic energy is likely to be
shared with the outer (un-processed) envelope of the star, creating a
nearly spherical shockwave. Ejection of some of this envelope material,
together with matter swept up from the interstellar medium, would lead
to the formation of the observed X-ray disc in G292.0+1.8. Presumably as
the remnant evolves further and more material is swept-up, a limb-
brightened shell will develop. The lack of limb-brightening at this
time might also be explained in part by a radial decrease in interstellar
density away from the site of the explosion resulting from pre-supernova
mass loss.

Markert, Canizares and Winkler (1981) have recently obtained direct evidence for a ring of X-ray emitting material in another oxygen-rich SNR, Cas A, from an analysis of the velocity structure of the S XV line. These authors attribute the X-ray ring in Cas A to either the asymmetric ejection mechanism of Bodenheimer and Woosley described above, or alternatively, to the effect of a spherically symmetric supernova explosion interacting with inhomogeneous interstellar material that is fortuitously arrayed to the line of sight. This latter explanation now appears unlikely in view of the evidence for a similar X-ray ring in a second oxygen-rich SNR, G292.0+1.8.

The occurrence of asymmetric Type II supernova explosions is given further support by the optical velocity mapping of N132D by Lasker (1980). Lasker finds the inner oxygen-rich material in this remnant to be expanding as an inclined ring with a velocity of 2250 km s^{-1}. Similar velocity mapping of the young oxygen-rich SNR in the Small Magellanic Cloud by Tuohy and Doptia (1982,1983) has also provided compelling evidence for an asymmetric Type II supernova explosion.

Our interpretation of the X-ray structure of G292.0+1.8 has two implications. Firstly, the two spatial components should have distinctly different compositions; the bar component should be dominated by oxygen, neon, etc., as with the optical remnant, while the disc emission should be characteristic of roughly cosmic material. Secondly, X-ray velocity differences of over 2000 km s^{-1} should exist within the bar material of G292.0+1.8, analogous to those detected in Cas A.

REFERENCES
Bodenheimer, P.B., and Woosley, S.E.: 1980, Bull. Am. Astron. Soc.,
      12, 833.
Bodenheimer, P.B., and Woosley, S.E.: 1982, Astrophys. J. in preparation.
Caswell, J.L., Murray, J.D., Roger, R.S., Cole, D.J., and Cooke, D.J.:
      1975, Astron. Astrophys. 45, 239.
Clark, D.H., Tuohy, I.R., and Becker, R.H.: 1980, Monthly Notices Roy.
      Astron. Soc. 193, 129.
Giacconi, R. et al.: 1979, Astrophys. J. 230, 540.
Goss, W.M., Shaver, P.A., Zealey, W.J., Murdin, P., and Clark, D.H.:
      1979, Monthly Notices Roy. Astron. Soc. 188, 357.
Lasker, B.M.: 1980, Astrophys. J. 237, 765.
Lockhart, I.A., Goss, W.M., Caswell, J.L., and McAdam, W.B.: 1977,
      Monthly Notices Roy. Astron. Soc. 179, 147.
Markert, T.H., Canizares, C.R., and Winkler, P.F.: 1981, Talk presented
      at Colgate University, April 1981.
Murdin, P., and Clark, D.H.: 1979, Monthly Notices Roy. Astron. Soc.
      189, 501.
Tuohy, I.R., Clark, D.H., and Burton, W.M.: 1982, Astrophys. J. (Letters)
      260, L65.
Tuohy, I.R., and Doptia, M.A.: 1982, Astrophys. J. (Letters), in press.
Tuohy, I.R., and Dopita, M.A.: 1983, this volume, p. 165.
van den Bergh, S.: 1979, Astrophys. J., 234, 493.

Weiler, K.W.: 1978, Mem. Soc. Astr. Italiana, 49, 545.
Woosley, S.E., and Weaver, T.A.: 1981, Ann. N.Y. Acad. Sci. 375, 357.
Woosley, S.E., and Weaver, T.A.: 1982, Proc. NATO Conference on
      Supernovae, Cambridge, England, June–July 1981, p. 79.

DISCUSSION

DICKEL: One of the most important characteristics of Cas A is the
intensity variation in the filaments which are quite detectable over a
few years. If a second epoch optical photograph does not exist, one
should be taken soon.

TUOHY: At present only one high quality optical photograph of
G292.0+1.8 exists, and that is the [OII] $\lambda3727$ exposure taken at the
prime focus of the Anglo-Australian Telescope in 1980 by Murdin, Clark
and Malin. This plate would be suitable for comparison with later epoch
measurements.

THE KINEMATICS OF THE SNR G292.0+1.8

R. Braun[1], W.M. Goss[2], I.J. Danziger[3], A. Boksenberg[4].
[1]Sterrewacht Leiden, The Netherlands
[2]Kapteyn Astronomical Institute, Groningen, The Netherlands
[3]European Southern Observatory, Garching bei München, F.R.G.
[4]Royal Greenwich Observatory, Herstmonceux Castle, Hailsham, U.K.

ABSTRACT

Optical velocity field mapping of G292.0+1.8 in the [OIII] λ5007 Å line has been carried out using the IPCS with the 3.6 m ESO telescope at La Silla. Our data are not consistent with the suggestion that the [OIII] emitting material in the western portion of this remnant is concentrated in an expanding ring. The existing data on G292.0+1.8 suggests that only the brightest portion of a thick shell of ejecta with high velocity spurs is observed. The expansion centroid, size, velocity and age of this SNR are derived.

INTRODUCTION

The SNR G292.0+1.8 was first found to be center-filled on the basis of radio observations at 21 cm (Lockhart et al. 1977), and similar structure has been recently observed at 843 MHz by Caswell (private communication) using the MOST telescope with an angular resolution of 43" arc. Optical nebulosity was found in its vicinity (Goss et al. 1979) which showed strong oxygen and neon lines but virtually none of hydrogen, sulfer or nitrogen. Together with the high velocity dispersion (> 2000 km s^{-1}) of this material (Murdin and Clark 1979) a young SNR was indicated in which the optical emission was dominated by ejecta which had not yet experienced appreciable contamination by the ISM. To better understand the nature of this source optical velocity field mapping was carried out using the 3.6 m ESO telescope. Since this work began, the Einstein SSS spectrum (Clark et al. 1980) has suggested that the X-ray emission also arises within ejecta, while the HRI image (Tuohy et al. 1982) has been interpreted as evidence for an expanding ring of ejecta.

OBSERVATIONS

A total of 15 spectra were obtained on 14-15 March 1980 using the IPCS detector with the 3.6 m ESO telescope and Boller and Chivens spectro-

J. Danziger and P. Gorenstein (eds.), Supernova Remnants and their X-Ray Emission, 159–164.
© *1983 by the IAU.*

graph at La Silla. The spectra were centered on $\left[\text{OIII}\right]$ $\lambda5007$ Å with resolution 170 km s^{-1} determined by the 2" arc slit width. The slit had length 2!9 arc with a pixel separation of 2!06 arc. The slit orientation was east–west and each of the 15 spectra was displaced in declination (separation 10" – 20" arc) to define a grid containing most of the conspicuous emission features. The calibrated spectra were aligned into a 'data cube' (on the basis of stellar images and night sky lines) from which maps at fixed velocity were produced after subtracting the stellar images. Absolute velocities are accurate to within about ±20 km s^{-1} and positions within about ±2" arc.

Three prominent features are apparent in these maps. The first is a fairly smooth component extending over the entire mapped field. This is centered at $V_{LSR} \approx 0$ km s^{-1} and corresponds to the feature described by Murdin and Clark (1979). Secondly, an arc of emission is prominent at all velocities between about –700 to +700 km s^{-1} extending from the NE to the west central portion of the field. This feature is illustrated by the maps at V_{LSR} = –570 and +285 km s^{-1} shown in Figures 1a and 1b. Finally a southern spur is prominent at large positive velocities between about +700 and +1200 km s^{-1}. This feature is illustrated by the map at V_{LSR} = +850 km s^{-1} in Figure 1c.

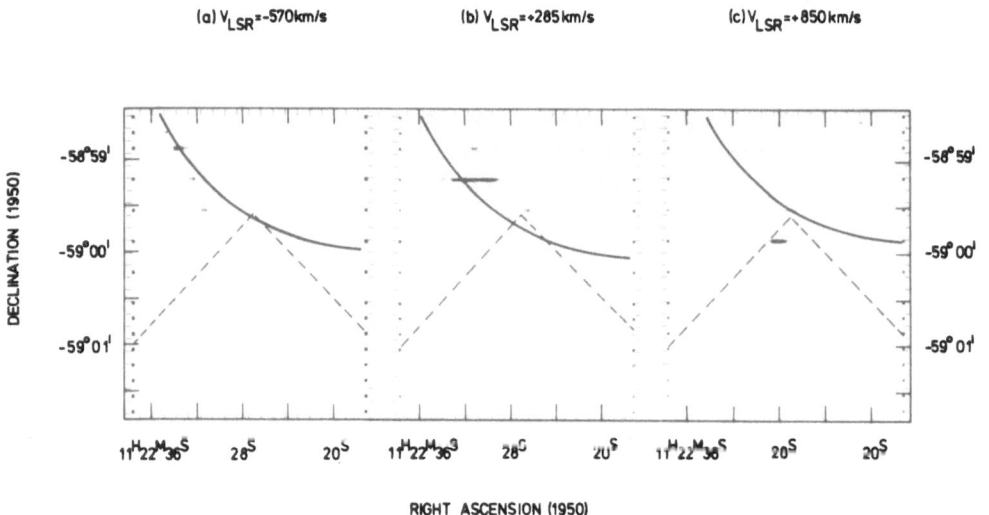

Figure 1. $\left[\text{OIII}\right]$ $\lambda5007$ Å emission maps of a portion of G292.0+1.8 in a 142 km s^{-1} interval centered at the indicated velocity. Shaded squares along the right and left hand edges indicate the positions of the spectra. The grey-scale is linear beginning from the 5σ level. The curves are discussed in the text.

DISCUSSION

To understand the significance of these features it is instructive
to consider the $\overline{[OII]}$ λ3727 Å interference filter photograph of this
region obtained at the prime focus of the AAT (Murdin, private communication) shown in Figure 2. A portion of the same photograph also appears
in Tuohy et al. (1982). Extended emission is indeed apparent over a
large region. This emission is not centered on the SNR and may well not
be physically related. The arc of emission sampled in our spectra is
also prominent but is seen to have an extension to the west beyond the
region sampled. This feature curves to the north again before disappearing. There are also weak filaments up to 3.5 arc north of this 'southern
rim' as well as a suggested extension of the rim to the NE. The southern
spur is also apparent and is seen to extend to about 4' arc south of the
rim.

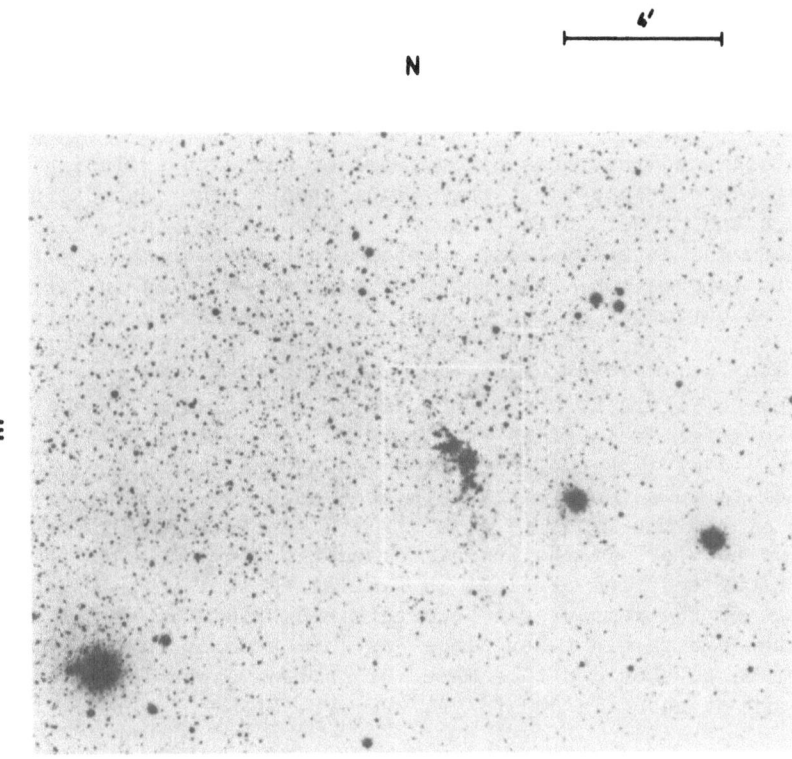

Figure 2. $\overline{[OII]}$ λ3727 Å interference filter photograph of the region of
G292.0+1.8 obtained at the prime focus of the AAT. The rectangle indicates the field of our velocity mapping study while the cross indicates
the assumed expansion center.

Another valuable comparison is that with the X-ray image of Tuohy et al. (1982). The southern rim of oxygen line emission coincides with a series of prominent X-ray hot spots and these extend symmetrically even farther to the NW. (The radio map of Lockhart et al. (1977) also shows a weak extension at this position.) Other X-ray features include a weak counterpart to the optical southern spur and a prominent spur extending to the SW. It is perhaps noteworthy that the center of curvature of the southern rim is easily compatible with a radial extrapolation from the X-ray spurs.

What type of physical entity could give rise to the observed structure? The suggestion that the oxygen-rich material is concentrated in an expanding ring is not compatible with the kinematic data. The [OIII] emission features along the NE segment of the southern rim seen in Figures la and lb clearly show that both strong red and blue-shifted emission are observed, contrary to what one would expect from the approaching or receding edge of a ring. However, the continuous southern rim feature is very suggestive of a portion of an expanding shell; while the two spurs appear to be instances of prominent breakout of high velocity ejecta beyond this region of more orderly expansion.

To investigate this possibility, all of our data excluding the region of the southern optical spur (indicated by the dashed line in Figures la,b and c) were used to model the expanding shell. A spheroid was fitted to the data (by least squares) to represent the positional and velocity information. The expansion center was fixed in position by the center of curvature of the southern rim as determined from both the [OII] emission of Figure 2 and the X-ray map of Tuohy et al. (1982) giving $\alpha(1950) = 11^h22^m18^s\pm2^s$, $\delta(1950) = -58°57'25"\pm15"$. This position is indicated in Figure 2. The rest velocity was fixed at $V_{LSR} = 0$ km s^{-1} since line-of-sight velocities at $\ell^{II} = 292°$ are within 0 ± 25 km s^{-1} out to distances, d of 10 kpc. The derived radius and expansion velocity with respect to this center are R=2!6 arc and V=2200 km s^{-1}. If we assume a distance of 5.4 kpc (Goss et al. 1979) this corresponds to a radius R=4.2 (d/5.4 kpc) pc. If the further assumption is made that the physical dimension along the line of sight is similar to that in the plane of the sky, the age of the remnant can be determined. Since negligible contamination of the ejecta has taken place (as indicated by the spectrum), it is appropriate to consider this material in free expansion, i.e. R=VT; giving an age of about T=1800 (d/5.4 kpc) yr.

The compatability of the model with the data is illustrated by the histogram in Figure 3a, where volume emissivity E_v (defined as the number of observed samples in a spheroidal shell divided by the volume of the partial shell which was sampled) is plotted against normalized distance ($D_n = 1$ corresponds to a sample lying on the solution spheroid). A single well-defined peak is observed with a halfwidth of about 0.9 (d/5.4 kpc) pc as opposed to the flat curve expected for data samples distributed randomly in position and velocity. The distribution of the southern spur emission features with respect to the model is illustrated in the same way in Figure 3b. A significant component is found associated with the

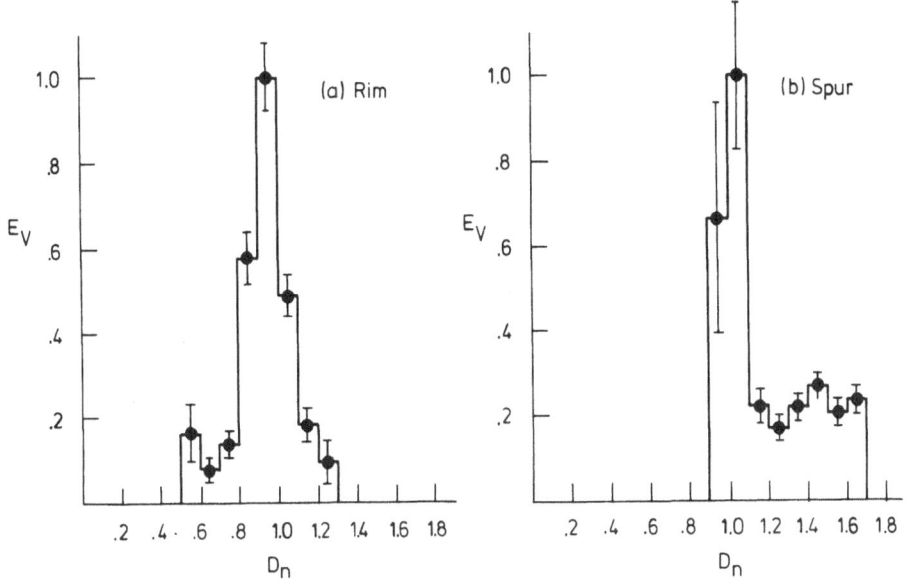

Figure 3. Histogram of volume emissivity as a function of normalized distance from the assumed expansion center for $[OIII]$ $\lambda 5007$ Å emission features.

shell as well as the expected high velocity component which shows quite uniform filling. The model spheroid is also included as the solid curve in Figures 1a,b and c for direct comparison with the data.

This kinematic model is limited by the incomplete sampling. Future observations of the fainter features shown in Figure 2 would lead to a better determination of the expansion center as well as to a more precise knowledge of the total extent of the SNR. However, it is doubtful that the estimated age of the remnant would change dramatically since experimentation with movement of the assumed expansion center has shown that the inferred age of the remnant remains fixed to within about 10% of the quoted value.

The degree of consistency of this interpretation with the kinematic data gives strong support to its applicability. The implication is that there is a northern portion of this more extended remnant that remains to be studied. This explains in a natural way the peculiar radio and X-ray morphology observed.

Such asymmetry in brightness of opposite sides of a SNR is certainly not uncommon. In radio SNR it is quite common to encounter asymmetries of a factor of 10 or more. In particular, a remnant as young as Kepler's SNR (T=380 yr) shows a northern rim which is more than 10 times brighter in both the radio (Gull, 1975) and the X-ray (White and Long, 1982) than

the southern counterpart. A similar degree of asymmetry would be sufficient to explain the absence of corresponding structure in the present
radio and X-ray observations of G292.0+1.8.

CONCLUSION

A consistent interpretation of the SNR G292.0+1.8 is possible if
the 'southern rim' feature seen in [OII], [OIII], X-ray and radio
observations is identified with part of an expanding shell of ejecta of
finite thickness and the two spurs with high velocity ejecta which has
broken out of this region of more orderly expansion. While this interpretation is suggested by the morphology, it receives strong support
from the systematic velocities of the [OIII] emitting material discussed
here. The center of curvature of this rim corresponds to $\alpha(1950)$ =
$11^h22^m18^s\pm2^s$, $\delta(1950)$ = $-58°57'25"\pm15"$, while the derived parameters of
a least-squares-fit spheroid with respect to this center are velocity
$V=2200$ km s^{-1}, radius R=4.2 (d/5.4 kpc) pc halfwidth $\Delta R=0.9$
(d/5.4 kpc) pc and age T=1800 (d/5.4 kpc) yr.

ACKNOWLEDGEMENTS

We wish to thank Paul Murdin and David Malin for allowing us to use
their unpublished AAT [OII] plate of G292.0+1.8. R.B. was supported
during the course of this work by The Netherlands Ministry of Education
and Science through a Netherlands Government Scholarship.

REFERENCES

Clark, D.H., Tuohy, I.R., and Becker, R.H.: 1980, M.N.R.A.S. 193, 129
Goss, W.M., Shaver, P.A., Zealey, W.J., Murdin, P., and Clark, D.H.: 1979,
     M.N.R.A.S. 188, 357
Gull, S.F.: 1975, M.N.R.A.S. 171, 237
Lockhart, I.A., Goss, W.M., Caswell, J.L., and McAdam, W.B.: 1977,
     M.N.R.A.S. 179, 147
Murdin, P., and Clark, D.H.: 1979, M.N.R.A.S. 189, 501
Tuohy, I.R., Clark, D.H., and Burton, W.M.: 1982, Ap. J. 260, L56
White, R.L., and Long, K.S.: 1982, preprint

DISCUSSION

DOPITA: The oxygen rich SNR in the SMC shows evidence for a highly distorted expanding ring. Your analysis covers only part of the SNR and you
have rejected a portion of the data. Can you therefore firmly reject the
hypothesis of an expanding twisted ring to explain the dynamics?

BRAUN: Any form of simple expanding ring is ruled out by our observations.
It is of course usually possible to construct an arbitrarily complex
variation of some preferred model to explain an observation. This does not
seem justified in this case where a reasonable and simple alternate explanation is sufficient.

A DYNAMICAL STUDY OF THE YOUNG OXYGEN-RICH SUPERNOVA REMNANT IN THE
SMALL MAGELLANIC CLOUD

Ian R. Tuohy & Michael A. Dopita

Mount Stromlo and Siding Spring Observatories
Australian National University

We present a velocity map of the young oxygen-rich supernova remnant
(1E0102.2-7219) in the Small Magellanic Cloud, obtained with the Anglo-
Australian Telescope. The velocity structure is complex, and implies a
high degree of asymmetry during the Type II supernova explosion. Our
data can be modelled geometrically in terms of a severely distorted ring
of oxygen-rich ejecta. This result, together with the evidence for
expanding rings in similar remnants, suggests non-spherical ejection to
be an intrinsic characteristic of Type II supernovae. We have also
obtained two-dimensional spectroscopy of the diffuse halo of emission
which partially surrounds 1E0102.2-7219. The halo exhibits the high
excitation line of HeII λ4686, and is either a fossil HII region created
by a UV flash accompanying the supernova, or alternately, is being excited
by intense UV radiation from the remnant itself. It is the first clear
association of a high excitation region with a supernova remnant.

1. INTRODUCTION

In a recent X-ray survey of the Small Magellanic Cloud (SMC),
Seward and Mitchell (1981) reported the discovery of an intense soft
X-ray emitting supernova remnant, 1E0102.2-7219. The optical counterpart
was found by Dopita, Tuohy and Mathewson (1981) and consists of a series
of filaments (~ 24 arcsec diam.) emitting only in the forbidden lines of
oxygen and neon. A faint outer halo, separated from the filaments by a
dark annular region, partially surrounds the remnant. In this paper we
report the results of a detailed two-dimensional spectroscopic study of
both the oxygen-rich filaments and the diffuse outer halo. A more complete discussion of this work is given by Tuohy and Dopita (1982).

2. OBSERVATIONS AND RESULTS

The observations were made in June 1981 using the RGO spectrograph on
the Anglo-Australian Telescope. A total of 11 long slit spectra were
taken over the full area of the SNR and part of the halo, resulting in a

J. Danziger and P. Gorenstein (eds.), Supernova Remnants and their X-Ray Emission, 165–169.

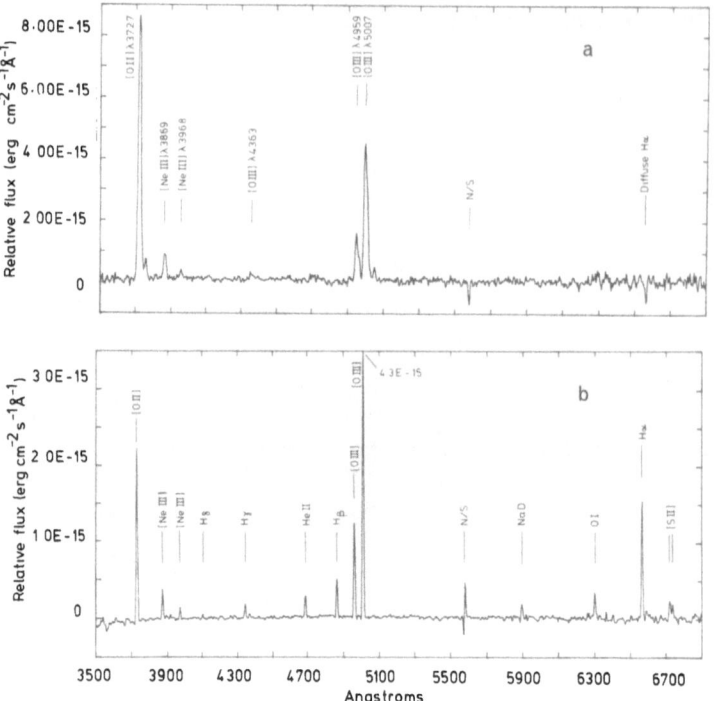

Fig. 1   (a)   A representative sky-subtracted and calibrated spectrum of
               a filament in 1E0102.2-7219.
         (b)   Spectrum of the diffuse halo which partially surrounds
               1E0102.2-7219.   Note the He II λ4686 line.

data set consisting of a spatial grid of 58 x 11 pixels, each with 2040
spectral elements spanning the wavelength range 3000 to 7000 Å
(resolution ~ 8 Å FWHM).   A typical spectrum of a single filament is
shown in Figure 1a.

2.1 Velocity Structure

The velocity of the oxygen-rich material ranges from -2500 to +4000
km s^{-1}, relative to the local SMC standard of rest.   Narrow band images
have been constructed from the spatial grid of spectra by summing wavelength elements over velocity intervals centered on the [OII] λ3727 line
(the [OIII] λ5007 line was not suitable because of velocity overlap from
the λ4959 component).   The total [OII] emission from the remnant in the
velocity range -2500 to +4000 km s^{-1} is shown in Figure 2a.   The [OII]
image agrees well with the [OIII] discovery image which covers a
restricted velocity interval, but has higher angular resolution (Tuohy and
Dopita 1982).

Figure 3 depicts a mosaic of 12 narrow band images, each equivalent
to a velocity range of 520 km s^{-1}.   Comparison of the individual images

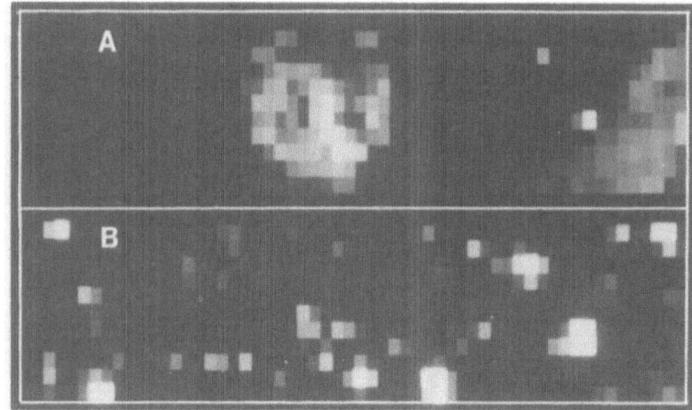

Fig. 2 An [OII] image of 1E0102.2-7219 (and N76 to the west) obtained by summing emission in the velocity range -2500 to +4000 km s^{-1}. A continuum image (b) centered near 4000 Å has been subtracted to cancel the stellar component. Individual pixels measure 3.0 arcsec (N-S) and 2.3 arcsec (E-W).

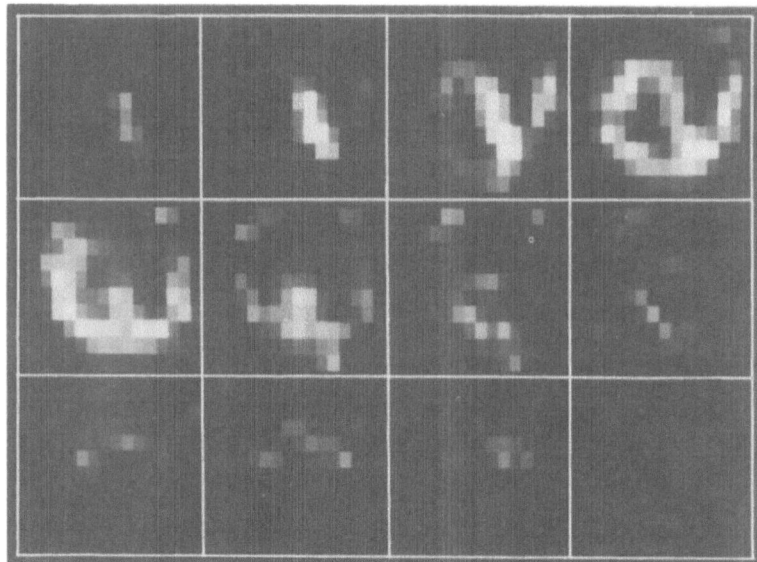

Fig. 3 A mosaic of 12 images, each corresponding to a velocity range of 520 km s^{-1}. The sequence begins at -2600 km s^{-1} (top left) and ends at +3640 km s^{-1} (bottom right). The stellar component (Figure 2b) has been subtracted. Pixel sizes are the same as Figure 2.

shows that the velocity structure of the SNR is both highly ordered and
complex. In particular, it is not consistent with a simple expanding
shell of material. We have attempted to model the complex motions of the
filaments in terms of an expanding ring of ejecta. Evidence for such
rings has been found in three other oxygen-rich SNRs: N132D (Lasker
1980), Cas A (Markert, Canizares and Winkler 1981) and G292.0+1.8 (Tuohy,
Clark and Burton 1982, Clark and Tuohy 1983). For 1E0102.2-7219 however,,
the ring must be grossly distorted. We have allowed for this distortion
by applying a sin 3θ deformation of varying amplitude to a simple planar
ring, as summarized in Figure 4. While our mathematical model is
idealized, approximate, and the uniqueness of the solution is not established, we find nevertheless that it gives an acceptable representation
of the gross properties of the data.

The complex velocity structure of 1E0102.2-7219 provides compelling
evidence for a high degree of asymmetry during the supernova explosion
(assuming that the oxygen emission does in fact trace out the ejection
pattern). As noted above, similar evidence has been found for three other
members of the class (N132D, Cas A and G292.0+1.8), suggesting that

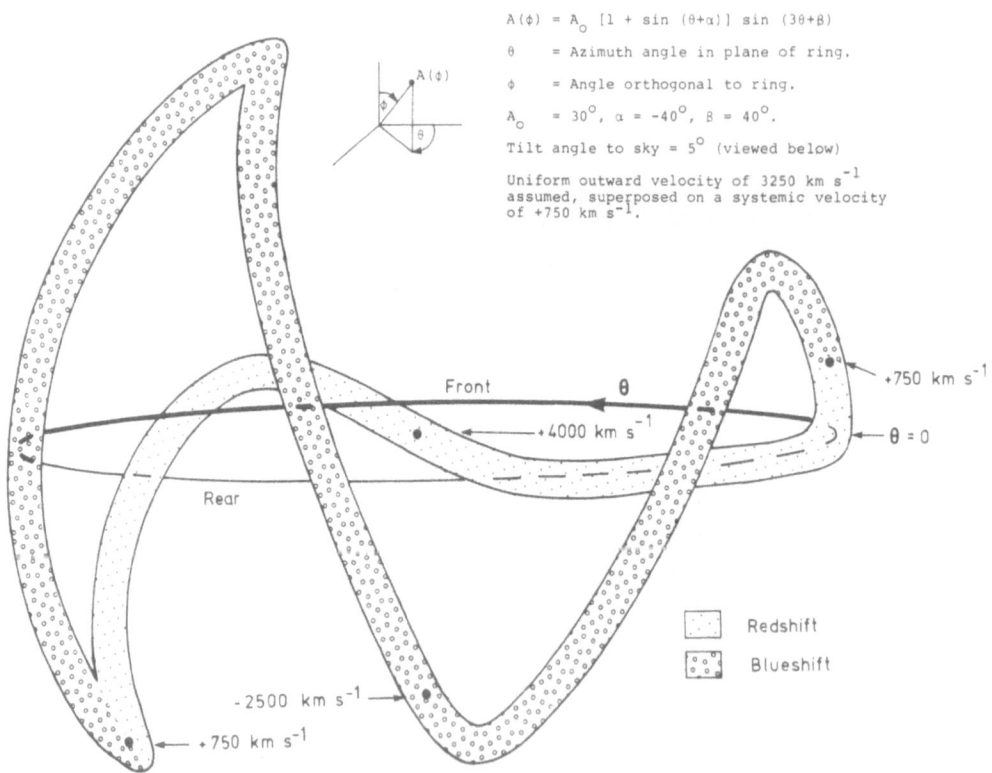

Fig. 4   A schematic representation of the distorted ring model of
         1E0102.2-7219.

non-spherical ejection may be an intrinsic property of Type II supernovae. We note that the ejection of a planar ring of oxygen-rich
material can be understood in terms of the collapse of a massive
($\sim$25 M$_\odot$) <u>rotating</u> pre-supernova star (Bodenheimer and Woosley 1980,
Woosley and Weaver 1982). The ejection of a distorted ring (if our
modelling of 1E0102.2-7219 is valid) would appear to require an extension
of this model; e.g., allowance for magnetized plasma instabilities.

2.2   The Halo Emission Region

    A sky-subtracted spectrum was also obtained of the diffuse halo
which is prominent to the east of 1E0102.2-7219. The halo spectrum
(Figure 1b) has a typical Case B Balmer decrement of Hα:Hβ = 2.9:1.0,
but the most unusual feature is the presence of the high excitation
He II λ4686 line. There are two possible explanations for this unique
association of a high excitation halo with an SNR. First, the halo may
be a fossil HII region created by a UV flash accompanying the supernova.
Second, modelling of radiative shocks in the oxygen-rich filaments by
Dopita, Binette and Tuohy (1982) shows that a sufficient specific
intensity of UV radiation (principally [OVI] λ150.1) can escape from the
SNR itself and ionize the surrounding medium. We note in conclusion
that HII regions have been detected in the vicinity of other oxygen-rich
SNRs (Tuohy and Dopita 1982), suggesting that these may also be associ-
ated with the supernova events (or remnants).

REFERENCES

Bodenheimer, P.B., and Woosley, S.E.: 1980, Bull. Am. Astron. Soc.,
        12, 833.
Clark, D.H., and Tuohy, I.R.: 1983, this volume, p. 153.
Dopita, M.A., Tuohy, I.R., and Mathewson, D.S.: 1981, Astrophys. J.
        (Letters), 248, L105.
Dopita, M.A., Binette, L., and Tuohy, I.R.: 1982, in preparation.
Gull, S.F., and Daniell, G.J.: 1978, Nature, 272, 686.
Lasker, B.M.: 1980, Astrophys. J., 237, 765.
Markert, T.H., Canizares, C.R., and Winkler, P.F.: 1981, Talk presented
        at Colgate University, April 1981.
Seward, F.D., and Mitchell, M.: 1981, Astrophys. J., 243, 736.
Tuohy, I.R., Clark, D.H., and Burton, W.M.: 1982, Astrophys. J. (Letters)
        260, L65.
Tuohy, I.R., and Dopita, M.A.: 1982, Astrophys. J. (Letters), in press.
Woosley, S.E., and Weaver, T.A.: 1982, Proc. NATO Advanced Study
        Institute on Supernovae, R. Stoneham and M. Rees, Cambridge,
        England, June-July, 1981, p. 791.

RADIO SUPERNOVAE

K. W. Weiler[1], R. A. Sramek[2], J. M. van der Hulst[3], N. Panagia[4]

[1]National Science Foundation, Washington, D. C. 20550
[2]NRAO - VLA, P. O. Box O, Socorro, NM 87801
[3]Neth. Found. for Radio Astron., Dwingeloo, The Netherlands
[4]Inst. for Radio Astron., Via Irnerio 46, 40126 Bologna, Italy

Three supernovae have so far been detected in the radio range shortly after their optical outbursts. All are Type IIs. A fourth supernova, a Type I, is being monitored for radio emission but, at an age of approximately one year, has not yet been detected. For two of the supernovae, extensive data are presented on their "light curves" and spectra and models which have been suggested in the literature are discussed.

I. INTRODUCTION

There are four examples which yield essentially all of our knowledge concerning the radio properties of supernovae. These are listed in Table 1. The first three (SN1970g, SN1979c, and SN1980k) were all Type II and are the only supernovae which have ever been detected in the radio range. The last, the Type I supernova in NGC4536, is recent and, although optically bright, has not yet been detected at 6 cm wavelength to a limit of less than 0.06 mJy (1σ) at an age of approximately one year. A lack of detection is, of course, never definitive so that a search for radio emission continues.

Table 1: Radio Supernovae

Object	Galaxy	SN Type	Optical Max.
SN1970g	M101	II	1 Aug. 1970
SN1979c	M100 (NGC 4321)	II	23 Apr. 1979
SN1980k	NGC 6946	II	3 Nov. 1980
SN1981b (undet.)	NGC 4536	I	5 Mar. 1981

In order to put the properties of radio supernovae in perspective, they are compared with several well known galactic supernova remnants in Table 2. The columns are self-explanatory except for, perhaps, column 8 which gives an example of how strong each source would be if it were placed at a standard distance of the galactic center (10 kpc). Examination of Table 2 shows that young Type II supernovae are likely to be intrinsically very bright and compact objects. From column 9 it is

171

J. Danziger and P. Gorenstein (eds.), Supernova Remnants and their X-Ray Emission, 171–176.

apparent that the detected radio supernovae are from one to more than two orders of magnitude stronger than the brightest supernova remnant known, Cassiopeia A.

Table 2: Comparisons

SOURCE	S_{6cm} Peak (Jy)	DIST. (kpc)	S_{6cm}^{c} (Jy)	$SIZE^{c}$ (arcsec)	$SIZE^{c}$ (pc)	F Peak (W Hz^{-1})	S_{6cm}^{10kpc} Peak (Jy)	RATIO to Cas A
Cas A		2.8	900	300	4	$8*10^{17}$	71	1
3C10		5	30	600	14	$8*10^{16}$	8	0.1
Crab		2	660	360	3.5	$3*10^{17}$	26	0.4
3C58		8	29	480	18.5	$2*10^{17}$	19	0.3
SN1970g	$\approx$.005	$7*10^{3}$	undet	$3.6*10^{-3}$[a]	$1.2*10^{-1}$[a]	$3*10^{19}$	2,500	35
SN1979c	.008	$16*10^{3}$	.007	$4.1*10^{-4}$[a]	$3.2*10^{-2}$[a]	$2*10^{20}$	20,500	290
SN1980k	.003	$5*10^{3}$	.001	$6.2*10^{-4}$[a]	$1.5*10^{-2}$[a]	$8*10^{18}$	650	9
SN1981b	$\underline{<}.00006$[b]	$20*10^{3}$	undet	$1.1*10^{-4}$[a]	$1.1*10^{-2}$[a]	$\underline{<}3*10^{18}$	$\underline{<}240$	$\underline{<}3$

[a] Assumes an average expansion velocity of 10^{4} km s^{-1}
[b] One sigma
[c] In April 1982

  SN1970g had only five significant detections (Marscher and Brown, 1978) so that little information is available on its radio light curve and spectrum and the Type I supernova in NGC4536 has not yet been detected. Thus, we will concentrate on the new results for SN1979c and SN1980k. A more detailed discussion of their general properties has recently been presented by Weiler et al. (1981, 1982).

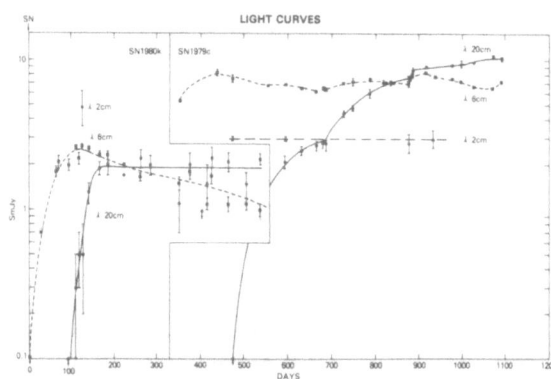

Figure 1: The radio "light curves" for SN1979c and SN1980k. All points are taken with the VLA. The "age" in days for both objects is counted from the date of maximum optical light.

II. RESULTS

The available information on the radio observations of SN1979c (α (1950) = $12^h20^m26.^s71$, δ (1950) = $+16^\circ04'29."5$) and SN1980k (α (1950) = $20^h34^m26.^s68$, δ (1950) = $+59^\circ55'56."5$) is shown in Figure 1. The observations at 20 cm (1.5 GHz), 6 cm (5 GHz) and 2 cm (15 GHz) are all taken with the VLA[1]. Since their initial detections, both SN1979c and SN1980k have been monitored approximately once per month and additions to the light curves are still being observed.

The main features of the light curves are their extremely sharp turn-on, the delays between 6 and 20 cm, the differing delays before turn-on for SN1979c and SN1980k, and the irregular, "bumpy" form for the well determined light curve of SN1979c. It is also interesting to note that SN1979c shows higher peak flux densities at lower frequencies while SN1980k shows the reverse.

The radio spectra of both SN1979c and SN1980k are steepening with time as is seen in Figure 2. The rate of steepening, however, appears to be slowing for both sources and an asymptotic approach to a limit of $\alpha \sim -0.5$ - -0.7 appears possible.

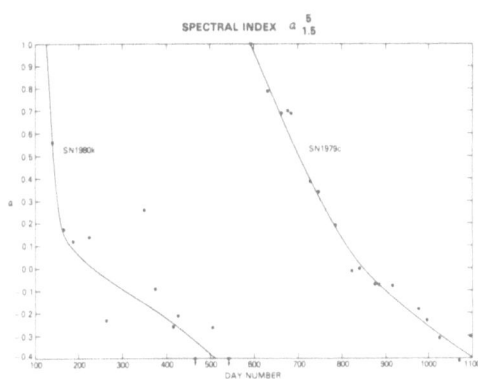

Figure 2:   Spectral index (S $\propto \nu^{+\alpha}$) between 20 cm and 6 cm plotted as a function of time.

III. DISCUSSION

The sharp turn-on of the radio light curves and the time delay in the turn-on between 6 and 20cm can most easily be described by an optical depth effect. This has been discussed at greater length by Weiler et al. (1982).

Although a detailed discussion of models for the emission mechanism of radio supernovae will be deferred to a future paper, there are two general problems which must be addressed in any successful description:

1) how the relativistic synchrotron particles are generated, and
2) how the radio radiation escapes through the great amount of matter believed to be in the photosphere of a Type II supernova.

An older model, that of impulsive creation of relativistic electrons expanding adiabatically from a single outburst (van der Laan, 1966) is inconsistent with the data. The model predicts a slow turn-on at any frequency and significantly higher maxima at higher frequencies, neither of which are observed in SN1979c. The higher maxima at higher frequencies are observed in SN1980k, but the slow turn-on is not.

Three new models have recently been advanced which, although they lack the details necessary for a quantitative comparison, deserve further consideration. All three can permit the sharp turn-on of the radio radiation as an optical depth effect. However, they differ significantly in their generation and distribution mechanisms for relativistic particles.

1) The Shklovskii Model (Shklovskii, 1981) consists of a young plerion (a young Crab Nebula - Weiler and Panagia, 1980) which is formed by the pulsar remaining after the Type II supernova explosion. The pulsar's magnetic field and relativistic synchrotron particles "leak" through the filaments of the supernova shell to regions of sufficiently low optical depth to be visible at radio wavelengths. This situation is illustrated in Figure 3.

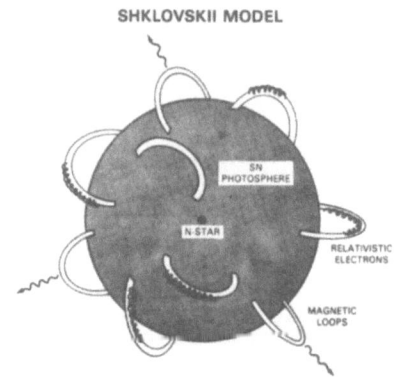

Figure 3: Author's (KWW) visualization of the Shklovskii Model.

2) The Pacini and Salvati Model (Pacini and Salvati, 1981; see also Pacini and Salvati, 1973 and Brown and Marscher, 1978) consists similarly of a young plerion formed by the remaining neutron star but this "mini-Crab" remains near the center of the supernova. With proper input parameters, the model can match the observed properties of the radio supernovae. However, the

problem remains of transporting the radiation through the
expectably massive supernova shell. This can be solved if the
shell quickly breaks into a net of dense filaments which provide
low density paths to the central regions. This situation is
illustrated in Figure 4.

PACINI & SALVATI MODEL

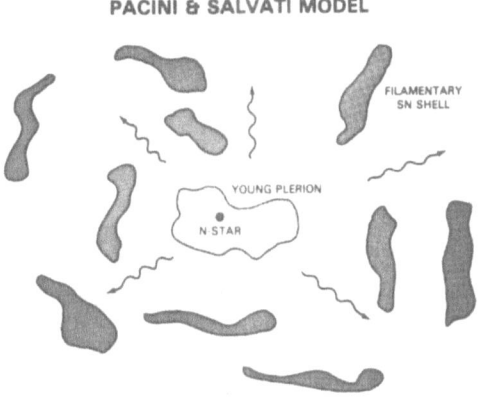

Figure 4: Author's (KWW) visualization of the Pacini and Salvati Model.

3) Whereas the previous two models are broadly related to plerionic
   supernova remnants, the Chevalier Model (1981) involves
   mechanisms resembling those of shell-type supernova remnants.
   The shock wave from the supernova explosion expands into the
   interstellar cocoon, formed by mass loss during the last stages
   of stellar evolution, and accelerates relativistic particles.
   This model can be made to fit the observed properties of SN1980k
   very well, but has difficulty describing the relatively constant
   level of radio emission still observed for SN1979c.

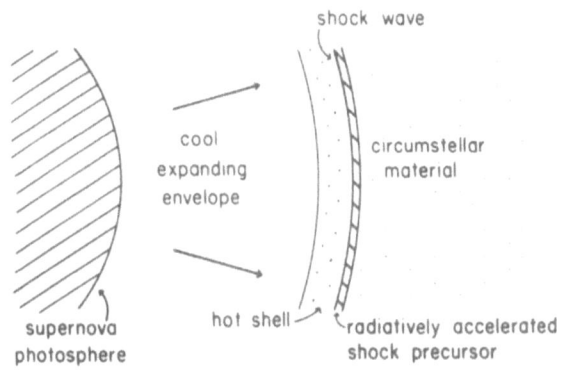

Figure 5: Chevalier Model (Chevalier, 1981)

M. Salvati (1983) has suggested that the flat light curve of SN1979c can be explained by a combination of Models 2 and 3. Initially, the shock wave in the circumstellar material predominated as in the Chevalier model, but more recently the central plerion has become visible through the cooling supernova shell as in the Pacini and Salvati model. The plerion emission is thus compensating for the expected decline in the shock-excited shell emission.

[1]The VLA is operated by the National Radio Astronomy Observatory of Associated Universities Inc. under contract to the National Science Foundation.

REFERENCES

Brown, R.L., Marscher, A.P.   1978, Astrophys. J. 220, 467.

Chevalier, R.   1981, Astrophys. J. 251, 259.

Laan, H. van der   1966, Nature 211, 1131.

Marscher, A.P., Brown, R.L.   1978, Astrophys. J. 220, 474.

Pacini, F., Salvati, M.   1973, Astrophys. J. 186, 249.

Pacini, F., Salvati, M.   1981, Astrophys. J. Lett. 245, L107.

Salvati, M. 1983, this volume, p. 177.

Shklovskii, I.S.   1981, Sov. Astron. Lett. 7(4), 263.

Weiler, K.W., Panagia, N.   1980, Astron. Astrophys. 90, 269.

Weiler, K.W., Hulst, J.M. van der, Sramek, R.A., Panagia, N.   1981, Astrophys. J. Lett. 243, L151.

Weiler, K.W., Sramek, R.A., Hulst, J.M. van der, Panagia, N.   1982, "Supernovae: A Survey of Current Research," M.J. Rees and R. J. Stoneham (eds.), D. Reidel Publishing Co., Dordrecht, Holland, p. 281.

RADIO AND X-RAY EMISSION FROM NEWLY BORN REMNANTS

Marco Salvati
Istituto di Astrofisica Spaziale - CNR
C.P. 67   00044 Frascati   Italy

ABSTRACT

Radio and X-ray observations of SN 1979c and SN 1980k offer a unique opportunity of monitoring the transition from supernovae to remnants. By means of the two-frequency radio light curves, we test the hypothesis that these objects are surrounded by circumstellar matter, originated in a pre supernova wind, and derive the relevant parameters. Then we use the absorption-corrected light curves to test the various proposed models. SN 1980 k appears to be powered by a canonical shock, while SN 1979c is a good plerion candidate. An optical pulsar could still be detected at its location.

1. INTRODUCTION

While the physical connection between Supernovae and Supernova Remnants has always been obvious, until very recently these two phenomena were dealt with in rather independent contexts. The reason was the lack of direct experimental information on the intermediate phases, through which an exploding star develops into an expanding remnant. This gap now begins to be bridged because of the detection of radio emission at the sites of SN's 1970g, 1979c, and 1980k, and of X-ray emission at the site of SN 1980k. In all cases the detection first occurred only a few months after maximum light; the underlying objects, then, granted their kinship to the classical SNR's, must be in a very early, embryonic stage, characterized by length and time scales several orders of magnitude less than the conventional ones. It should be noted that the association of the X-ray flash from SN 1980k with a SNR-like phenomenon is model dependent (Canizares et al., 1982). On the other hand, the evidence collected about the radio emission strongly suggests that the basic physical mechanisms involved here are the same which we invoke for the classical SNR's.

A complete discussion of the embryonic stage of SNR's should take into account the numerous observations carried out at the sites of recent Supernovae, 10 to 100 years after maximum light, which however have given

177

J. Danziger and P. Gorenstein (eds.), Supernova Remnants and their X-Ray Emission, 177–182.
© 1983 by the IAU.

only upper limits (de Bruyn, 1973; Brown and Marscher, 1978; Cowan and Branch, 1982); also, there is at least one member of a new class of objects which show many similarities to the ones under discussion, but whose parent explosions, if any, have gone undetected (Kronberg et al., 1981). Here, instead, we will limit our analysis to SN's 1979c and 1980k, which are by far the best observed cases and the only ones to constrain significantly the theory. Most of the emphasis will be placed on the assessment of the competing models referred to in the following, and on the dis cussion of the near-future behavior to be expected of the sources. The very extensive radio data have been obtained by Weiler et al. (1983), who generously allowed their use in advance of publication. General information on the Supernovae themselves, as well as reference to related works, can be found in Panagia et al. (1980) and Panagia (1982), respectively.

2. POSITION OF THE PROBLEM

Upon inspection of the two-frequency light curves obtained by Weiler et al. (1983) one notices several distinctive features, such as the different switch-on time at different frequencies, the rapid onset followed by a plateau or a slow decline, the steady variation of the observed spec tral index. The most obvious interpretation is that both sources have ex perienced a steady decline of optical thickness in the radio range. The required opacity is very likely due to free-free absorption in a shell of circumstellar matter, with a higher density than the general interstellar medium. In the case of SN 1979c, independent evidence for the existence of such matter is provided by UV emission lines (Panagia, 1982; Fransson et al., 1982). Further evidence can be drawn from the fact that embryo SNR's appear to be associated with Type II Supernovae, whose progenitors are believed to be red giants with a dense, low-velocity wind. The wind parameters which are obtained under this hypothesis are quite reasonable (see below), and may be looked at as an a posteriori justification.

As for the energy supply in magnetic fields and relativistic parti-cles, two basic models have been proposed. It is conceivable that the new ly born remnants are a small scale version of classical shells like Tycho or Kepler (Chevalier, 1981, 1982): the equipartition energy deduced from the radio observations is only 1 percent of the shock-driven turbulent energy. It is equally conceivable that the newly born remnants are a small scale version of the so-called plerions, of which the Crab Nebula is the prototype (Pacini and Salvati, 1981): this approach assumes that a pulsar is the primary energy source, and attributes the observed radia tion to a bubble of relativistic fluid (electrons and fields) inflated by the pulsar within the supernova material. Note that the pulsed radio emission is expected to be negligible, even at the shortest pulsar pe-riods; but the rotational energy can be made larger than the equiparti-tion energy, and there are no difficulties in stockpiling the latter in-side the bubble.A crucial problem with the plerion model is the enormous opacity of the supernova ejecta (Bahcall et al., 1970); this is overcome if the relativistic fluid bulges out of the stellar debris (Shklovskii, 1981), or if an effective filamentation process begins operating at sufficiently early times; the observed filamentary structures in classical remnants only suggest that the time scales of the process are shorter than several hundred years.

The shell model and the plerion model are most easily distinguished by means of their time behavior: it is obvious that in the latter case the continuing activity of the pulsar prevents a rapid decline of the lu minosity, until the rotational energy reservoir becomes depleted; this feature will form the basis of the following analysis (Panagia, private communication). We assume: a) that the opacity is due to CSM, and the CSM was provided by a steady presupernova wind: hence the density scales as $n \propto r^{-2}$, and the optical depth as $\tau \propto r^{-3}$; b) that the dynamical influen ce of a possible pulsar is negligible, and the radius of the ejecta fol lows a self-similar law as discussed by Chevalier (1982): for the sake of definiteness, we set the envelope profile exponent equal to 12, so that $R \propto t^{0.9}$, and $\tau \propto t^{-2.7}$; c) that in a given object the intrinsic spectral index α is time independent. A certain value of α defines the intrinsic flux ratio at the observed frequencies; then, comparison with the measured ratio gives the value of τ at that time. Since the time de pendence of τ is determined by our assumptions to within a normalization factor, A, the only free parameters are A and α, which we find by ordi nary χ^2 techniques. A is related to the CSM density and to the presuper nova mass loss rate; more important still, its knowledge allows us to correct the observed flux at a fiducial frequency, and to compare its in trinsic time behavior with model predictions.

The light curve to be expected from a shell-type source is taken from Chevalier (1982), with a minor correction to relate the typical elec tron energy to the shock velocity. In the plerion case, one must distin guish between two alternatives: the low field, slow pulsar case, where the synchrotron lifetime is longer than the expansion time; and the high field, fast pulsar case, where the opposite is true. The predicted light curves are taken from Pacini and Salvati (1973), corrected for a non-li near expansion. One has

$$F \propto t^{-(0.1+1.4\alpha)} \nu^{-\alpha} \qquad \text{shell}$$
$$F \propto t^{(0.15-0.85\alpha)} \nu^{-\alpha} \qquad \text{plerion, } \nu < \nu_b \qquad (1)$$
$$F \propto t^{(0.85-0.85\alpha)} \nu^{-\alpha} \qquad \text{plerion, } \nu > \nu_b$$

The break frequency which divides the two plerionic regimes evolves as

$$\nu_b \propto t^{-2} B^{-3} \propto t^{0.55} \qquad (2)$$

Its normalization coefficient contains information on the bubble magne tic field, hence, indirectly, on the rotational energy loss and initial period of the pulsar. Additional information is provided by the X-ray ob servations, and by the early portions of the radio light curves, where only single-frequency measurements are available, and the optical depth has to be extrapolated from later times.

3. RESULTS AND DISCUSSION

When the outlined procedure is applied to SN 1979c, one finds a satisfactory fit for the τ-vs-time curve; if t is measured in days, and the
optical thickness is referred to $\nu = 5$ GHz, the best fit parameters are

$$A = 4.6 \ 10^6 \qquad\qquad\qquad \alpha = 0.67 \qquad\qquad\qquad (3)$$

A is easily transformed into a density profile and a presupernova mass
loss rate

$$n = C \ r^{-2} \quad cm^{-3} \qquad\qquad C = 9.0 \ 10^{37} \ \nu_{o9}^{1.5} \quad cm^{-1}$$

$$\qquad\qquad\qquad\qquad\qquad\qquad\qquad\qquad\qquad\qquad\qquad (4)$$

$$\dot{M} = 3.5 \ 10^{-5} \ \nu_{w6} \ \nu_{o9}^{1.5} \quad M_o \ yr^{-1}$$

where ν_{o9} is the velocity of the ejecta at a reference time $t_o = 3 \ 10^6$ s,
in units of 10^9 cm s^{-1}; and ν_{w6} is the wind velocity in units of 10^6 cm
s^{-1}.

The problems with this particular object arise when one tries to fit
the corrected light curve to the theoretical models of Eq.1. The curve,
either corrected or uncorrected, exhibits statistically significant irre
gularities which cannot be reproduced by any power law; the best which
can be done is to account for the general trend, and under this respect
the best performing model is a plerion with a fast pulsar inside, $\nu > \nu_b$.
The fast pulsar condition implies that the spectral index unperturbed by
synchrotron losses is equal to $\alpha - 0.5 = 0.17$: spectra as flat as this
are peculiar to plerionic remnants (Weiler and Panagia, 1980). Furthermore, the requirement that ν_b be smaller than 1.5 GHz for at least 1200
days is a severe constraint on the pulsar period. The exact upper limit
is a sensitive function of the plerion volume, that is, of the ejecta ve
locity and of the bubble filling factor; our best guess is between a few
and several milliseconds, which implies an average optical magnitude <
< 22 according to the canonical scaling (Pacini, 1971). Since ν_b is an
increasing function of time, one expects a change in spectral slope and
time slope, as prescribed by Eq.1; later still, when the pulsar initial
lifetime is elapsed, the energy supply will start to decline, and the
source will be switched off with the expansion time scale. If both events
were actually observed, one could evaluate separately the pulsar period
and the plerion volume.

Now we turn to SN 1980k, and in analogy with Eqs.3 and 4 we find

$$A = 5.6 \ 10^4 \qquad\qquad\qquad \alpha = 0.38 \qquad\qquad\qquad (5)$$

$$C = 9.9 \ 10^{36} \ \nu_{o9}^{1.5} \quad cm^{-1}$$

$$\qquad\qquad\qquad\qquad\qquad\qquad\qquad\qquad\qquad\qquad\qquad (6)$$

$$\dot{M} = 3.8 \ 10^{-6} \ \nu_{w6} \ \nu_{o9}^{1.5} \quad M_o \ yr^{-1}$$

With the given α, a satisfactory fit to the corrected light curve is obtained with the shell model: apart from opacity effects, and at variance

with SN 1979c, the radio flux from SN 1980k is declining at a substantial rate, and the plerion model could work only if the pulsar power input were declining as well. A Crab-like pulsar rotating at the breakdown period, $P \sim 0.5$ msec, would indeed have a lifetime of about one month, but would also be so energetic that some exotic and unmistakable manifestations should be observed.

A further argument in favor of the shock interpretation is that classical shell-type remnants have a spectral index distribution with a median value of precisely 0.4. It is then natural to interpret the X-ray emission from SN 1980k in analogy with classical remnants, i.e. as a thermal emission from shock-heated material (Chevalier, 1982). It might be noted that a similar shock and a similar emission process is unavoidable in SN 1979c, which is surrounded by an even thicker CSM cocoon; X-ray observations were carried out at comparable post-outburst times, with negative results (Palumbo et al., 1981). If the expansion velocity is assumed to be the same, and if the small difference due to the slightly discrepant ages is neglected, the luminosity should scale as the τ coefficient A; however, the result exceeds the measured upper limit by almost an order of magnitude. A possible remedy, may be only a partial one, is the assumption that v_o had a lower value in SN 1979c.

The same considerations can be repeated for the radio emission: the shock system surrounding SN 1979c is likely to be the site of instabilities and energization processes as well as SN 1980k, so that the plerion emission must be contaminated by a shell emission. The scaling law for a universal value of α is

$$F \propto t^{-(0.1+1.4\alpha)} \nu^{-\alpha} D^{-2} A^{0.25(3+\alpha)} v_o^{0.25(13+19\alpha)} \tag{7}$$

and the shell component at, say, t = 800 d is comfortably smaller than the observed total, especially if the effect of a different v_o is included. A qualitative support to this picture is found by extrapolating the τ curve and by deriving the corrected light curve even at very early times, for which only high-frequency measurements are available. SN 1980k nicely follows the best-fit power law which describes the later, two-frequency measurements. SN 1979c, instead, has an intrinsic flux definitely higher than the best fit, and the discrepancy decreases smoothly with increasing time until it disappears around day 600. It is certainly disturbing that the higher intrinsic flux, if real, would conspire with opacity in such a way as to give an almost constant observed flux. But it is nonetheless appealing to interpret the early-time excess as a shell component, which then faded and became dominated by a time-steady plerion component.

If the analogy between the two objects were complete, one would expect in the near future a flattening of the SN 1980k light curve, in correspondence with the taking over of a plerion. Until then, however, the place to search for an extragalactic optical pulsar would be the site of SN 1979c.

We gratefully acknowledge very interesting discussions with N. Panagia, F. Pacini, and R. Bandiera.

REFERENCES

Bahcall, J.N., Rees, M.J., and Salpeter, E.E.: 1970, Astrophys. J. 162, p.737
Brown, R.L., and Marscher, A.P.: 1978, Astrophys. J. 220, p.467
Canizares, C.R., Kriss, G.A., and Feigelson, E.D.: 1982, Astrophys. J. 253, p.L17
Chevalier, R.A.: 1981, Astrophys. J. 251, p.259
Chevalier, R.A.: 1982, preprint
Cowan, J.J., and Branch, D.: 1982, Astrophys. J. 258, p.31
de Bruyn, A.C.: 1973, Astron. Astrophys. 26, p.105
Fransson, C., Benvenuti, P., Gordon, C., Hempe, K., Palumbo, G.G.C., Panagia, N., Reimers, D., and Wamstekers, W.: 1982, NORDITA preprint 82/24
Kronberg, P.P., Biermann, P., and Schwab, F.R.: 1981, Astrophys. J. 246, p.751
Pacini, F.: 1971, Astrophys. J. 163, p.L17
Pacini, F., and Salvati, M.: 1973, Astrophys. J. 186, p.249
Pacini, F., and Salvati, M.: 1981, Astrophys. J. 245, p.L107
Palumbo, G.G.C., Maccacaro, T., Panagia, N., Vettolani, G., and Zamorani, G.: 1981, Astrophys. J. 247, p.484
Panagia, N.: 1982, Proc. 3rd IUE European Conf., to be published
Panagia, N., et al.: 1980, M.N.R.A.S. 192, p.861
Shklovskii, I.S.: 1981, Pis'ma Astron. Zh. 7, p.479 (Sov. Astron. Lett. 7, p.263)
Weiler, K.W., and Panagia, N.: 1980, Astron. Astrophys. 90, p.269
Weiler, K.W., Sramek, R.A., van der Hulst, J.M., and Panagia, N.: 1983, this volume, p. 171.

PARTICLE ACCELERATION AND RADIO EMISSION OF THE SUPERNOVAE REMNANTS AT
DIFFERENT STAGES OF THEIR EVOLUTION

V.N. Fedorenko
A.F.Ioffe Institute of Physics and Technology, Leningrad, USSR

In this Paper, I consider physical processes, governing
relativistic electrons in SNRs.

a) SNRs at the age $t > 10^2$ yr. I argue that the shock wave
acceleration faces some difficulties. Then I show that the temporal
evolution of the SNRs radio emission can be accounted for without
involving the acceleration.

b) SNRs at the age $t < 10^2$ yr. I associate the lack of radio emission
at this stage (Brown and Marscher, 1978) with the weakness of the
magnetic field.

c) I infer that the most efficient particle acceleration and radio
emission of the SNRs should occur at the stage $t \sim 10^2$ yr.

a) Relativistic electrons in the SNRs at the stage $t > 10^2$ yr.

These are just the SNRs we observe. The relative roles of adiabatic
deceleration, acceleration and leakage of particles are still unclear.
Shklovskii (1960) proposed a model, incorporating only adiabatic
deceleration. The model qualitatively describes the secular evolution
of SNR radio emission, but quantitatively disagrees with observations.
This is why some models were suggested, incorporating particle
acceleration either by the hydromagnetic turbulence (Chevalier et al.,
1978), or by the shock front (Bell, 1978; Blandford and Ostriker,
1978). In this Section I would like to emphasize some difficulties,
related to these mechanisms.

The shock acceleration mechanism requires efficient scattering of
the particles on the small-scale hydromagnetic turbulence. A rough
estimate of the characteristic acceleration time is (Toptygin, 1980):
$\tau_a \sim D/U^2$, where D and U are the diffusion coefficient and shock
velocity respectively. The problem is to determine the value of D in
the SNRs. If the SNR shock wave propagates in an unperturbed ISM (see
Fig. 1), it is necessary to amplify the level of the ISM turbulence in
the upstream region, since otherwise $\tau_a \gg t$ (Toptygin, 1980).

183

J. Danziger and P. Gorenstein (eds.), Supernova Remnants and their X-Ray Emission, 183–186.

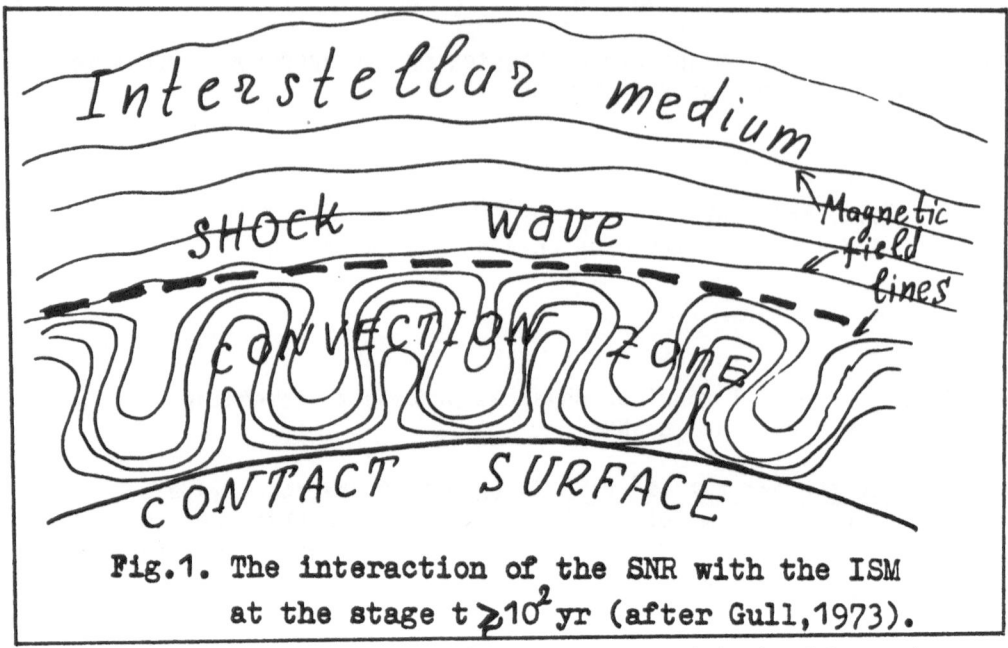

Fig.1. The interaction of the SNR with the ISM at the stage t $\gtrsim 10^2$ yr (after Gull,1973).

Figure 1. The interaction of the SNR with the ISM at the stage t > 10^2 yr. (after Gull, 1973).

For this purpose, Bell (1978) proposed the self-generated turbulence mechanism. However, as shown by Fedorenko (1981), to amplify the turbulence, a rather stringent condition must be satisfied: $\Gamma\, t_{pass} \gg 1$, Γ is the instability growth rate, and $Q_{pass} \sim D/U^2$ is the plasma element passage time in the upstream region. This results in very low values of the energy of the electrons, which experience the increased scattering: $E \ll 10^3 m_e c^2$. Thus, the radio-emitting electrons cannot be accelerated by this mechanism. The situation can be radically changed, if SN explodes into the region of the presupernova stellar wind (Chevalier, 1982). As in the solar wind, we might expect the continuous m.h.d. turbulence spectrum to be present with a typical scale length $L_0 \sim$ few pc. Then, estimations show (Fedorenko, 1982), that the acceleration condition $\tau_a \ll t$ is fulfilled. However, the existence of the pre-supernova stellar wind is still questionable, especially for the SN I (Chevalier, 1982).

According to Chevalier et al. (1978), the power spectrum of the relativistic electrons is produced by the joint action of the adiabatic deceleration and the Fermi acceleration processes. However, it was pointed out by Fedorenko (1981a), that in such a model the spectral index depends on many model parameters. It seems unlikely that they naturally combine to give $\gamma = 2.0 \div 2.6$.

These considerations induced me to investigate the possibility of a model without particle acceleration. This was done in Fedorenko (1981a). The main assumptions were: 1) spherical expansion with $R(t) \propto t^\alpha$, R is the SNR radius, and $2/5 < \alpha < 1$; 2) relativistic particles with initial power spectrum are decelerated due to expansion; acceleration and leakage were not considered; 3) magnetic field B evolving as $B \propto R^{-\beta}$; $3/2 < \beta < 2$, which assumes turbulent amplification - see Gull (1973), and Fig. 1. The particle distribution function is governed by a kinetic equation, in contrast to the model of Shklovskii (1960). Varying the parameters α and β enabled me to fit such a model to the observational data. As in the Shklovskii model, particle power spectrum is conserved during the SNR expansion, but additional consideration of the self-generated turbulence enabled me to explain the flattening of the Cas-A spectrum (see Fedorenko 1979, 1981). The same effect provided particle confinement in the SNRs.

Evidently, at the stage of $t > 10^2$ yr., particle acceleration is either absent, or weak compared to the adiabatic deceleration. Therefore, the acceleration of particles in SNRs should occur at the stage $t < 10^2$ yr.

b) <u>Relativistic electrons in the SNRs at the epoch with $t < 10^2$ yr.</u>

The radio observations of such objects in our Galaxy are absent. Brown and Marscher (1978) establish very low limits of the radio emission for nearly 50 SNRs with ages 1 yr < t < 30 yr in the external galaxies. According to Brown and Marscher, this result may be connected either with the lack of accelerated particles, or with the weakness of the magnetic field. Simple considerations of Fedorenko (1981a) seem to clear up the situation. According to Gull (1973), the turbulent amplification of the ISM magnetic field at the shock front (see Fig. 1), occurs at the stage $t < 10^2$ yr., due to convective motions. Maximum magnetic energy corresponds to the epoch with $t \sim 10^2$ yr (see Fig. 2).

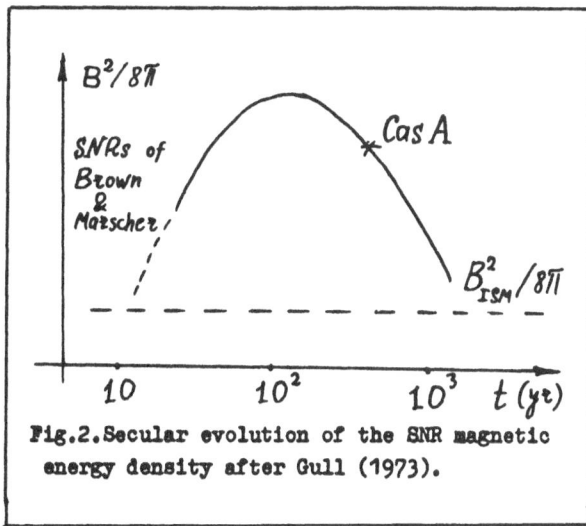

Fig.2. Secular evolution of the SNR magnetic energy density after Gull (1973).

Figure 2. Secular evolution of the SNR magnetic energy density after Gull (1973).

Therefore, the SNRs of Brown and Marscher correspond to the epoch with rather weak magnetic field and have low radiation fluxes. It is also possible that these SN explode into a very dilute ISM, such as the "coronal" phase in our Galaxy (McCray and Snow, 1979). Thus, in this case, the SNR magnetic field and radio-emission will be very weak. Nevertheless, three "radio supernovae" with ages t ~ 1 yr. were detected in the external galaxies (Weiler et al., 1981). This phenomenon might be connected with the pulsar activity at this stage. Probably, "radio-supernovae" cease at the age of few years (Weiler et al., 1981).

c) Scenario of the SNR evolution

The considerations mentioned above, lead us to the following inference: the most efficient pumping of the SNRs with accelerated particles occurs at the stage t ~ 10^2 yr. Probably, acceleration takes place at the shock front, the scattering rate being increased compared to the stage t > 10^2 yr (see Sec. a). This is compatible with the model of Gull (1973). According to him, at the stage t ~ 10^2 yr the shock wave makes contact with the convection zone, so the latter may be efficiently filled with the accelerated particles (see Fig. 1). At t > 10^2 yr the convection zone is removed from the shock region causing the injection rate of particles to be decreased. At this stage, adiabatic deceleration of the captured particles is dominant. Therefore, the maximum of the radio emission of the SNRs should occur at t ~ 10^2 yr. Remember that all of this refers to the SNRs without pulsars.

REFERENCES

Bell, A.R. 1978, MNRAS, 182, 443.

Blanford, R.D. and Ostriker, J.P. 1978, Ap.J. (Letters), 221, L29.

Brown, R.L. and Marscher, A.P. 1978, Ap.J., 220, 467.

Chevalier, R. 1978, Ap.J. (Letters), 225, L27.

Chevalier, R. 1982, Ap.J. (Letters), 259, L85.

Fedorenko, V.N. 1979, Astron. ZH. (USSR), 56, 1235. Translated in Soviet Astronomy, 23, 700.

Fedorenko, V.N. 1981a, Astron. ZH. (USSR), 58, 790. Translated in Soviet Astronomy, 25, 451.

Gull, S. 1973, MNRAS, 161, 47.

McCray, R. and Snow, T.P. Jr. 1979, Ann. Rev. Ast. and Ap., 17, 213.

Shklovski, I.S. 1960, Soviet Astronomy, 4, 335.

Toptygin, I.N. 1980, Space Science Reviews, 26, 157.

Weiler, K.W., Van der Hulst, J.M., Sramek, R.A., and Panagia, N. 1981, Ap.J. (Letters), 243, L151.

THE EVOLUTION OF SUPERNOVA REMNANTS AS RADIO SOURCES

R. Cowsik & S. Sarkar *
Tata Institute of Fundamental Research, Homi Bhabha Road,
Bombay 400005, India.
*now at Department of Astrophysics, University of Oxford,
South Parks Road, Oxford OX1 3RQ, England.

ABSTRACT

The acceleration of relativistic electrons by hydromagnetic turbulence in shell-type supernova remnants (SNRs) is examined within the framework of previous studies of their structural evolution through interaction with the interstellar medium. The predicted evolution of the synchrotron radio emission by the electrons is in agreement with a wide variety of observations.

1. INTRODUCTION

Gull (1973, 1975) has suggested that the relativistic electrons and magnetic field in young SNRs are generated during the transition from free expansion to the adiabatic phase of evolution, thus avoiding the severe expansion energy losses associated with an origin in the supernova explosion itself. His hydrodynamical calculations show that the excitation of a Rayleigh-Taylor instability in the decelerating ejecta results in the formation of a shell-shaped convection zone, in which $\sim 1\%$ of the blast energy is transferred through turbulence into magnetic field, relativistic electrons and hydromagnetic waves.

We treat the relativistic electrons as a collisionless plasma coupled to the magnetic field and to the thermal plasma through collective interactions. The transport equation governing the electron energy spectrum is analytically solved and the transport coefficients estimated from the hydrodynamic calculations. The evolution of the synchrotron radio spectrum can then be followed. We present a brief summary below of the results that are detailed elsewhere (Cowsik & Sarkar, 1982).

J. Danziger and P. Gorenstein (eds.), Supernova Remnants and their X-Ray Emission, 187–192.
© 1983 by the IAU.

2.   RADIO EMISSION FROM SNRs

2.1   Evolution of the Electron Energy Spectrum

The relativistic electrons undergo adiabatic energy changes due to the time variation of the magnetic field and the confinement volume, as well as stochastic energy changes (2nd order 'Fermi' accelertion) by scattering against magnetosonic waves. In addition, gyro-resonant interactions with Alfven waves rapidly isotropize the particle trajectories, thereby ensuring both the continuation of stochastic acceleration as well as the spatial confinement of the electrons (see Kulsrud, 1979).

The electron spectrum naturally evolves towards a power law shape from any steep form at injection under the combined effects of such convection and diffusion in energy. Figure 1 illustrates the flattening of the spectrum with time, t, for monoenergetic injection with $E = E_0$, both impulsively at $t = t_0$, and continuously from t_0 onwards. The power law shape, becomes better defined in the latter case, but the spectrum evolves more slowly than for impulsive injection. The corresponding synchrotron spectra evolve in a similar manner.

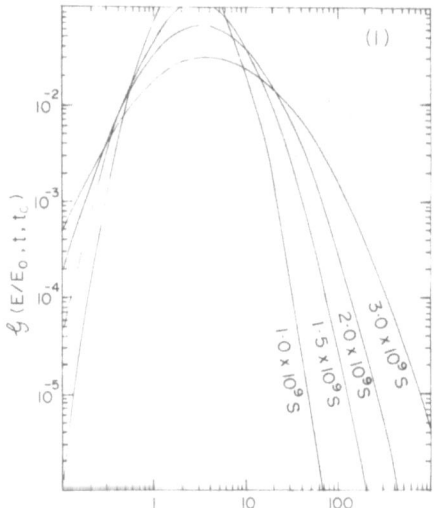 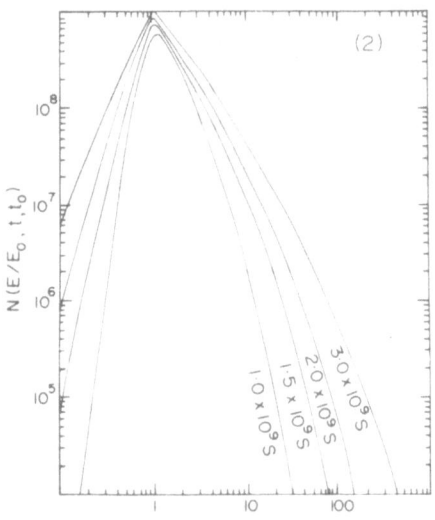

Fig.1. Evolution of the electron energy spectrum corresponding to, (1) Impulsive injection of 1 electron, and (2) continuous injection of 1 electron/s.

2.2   The 'Turn-on' of Radio Emission

We find that $\sim$ 30-100 yr after the supernova explosion the build up of the magnetic field is sufficiently rapid that the concommitant betatron acceleration overcomes the effect of expansion energy losses. The correlated increase of the field and electron energies thus leads to a sudden turn on of the radio emission, with peak luminosity being reached at

$$\tau \sim 100 \text{ yr} \cdot (n_0/1 \text{ cm}^{-3})^{-1/3} \cdot (E_{tot}/10^{51} \text{ erg})^{-1/2} \cdot (M_{ej}/10^{33} \text{ gm})^{5/6},$$

where n_0 is the ambient interstellar density, E_{tot}, the total explosion energy, and, M_{ej}, the ejected mass.

Just such a scenario is suggested by the failure of attempts to detect radio emission from the locations of many extragalactic supernovae that have occurred between a few years to almost a century ago. In Figure 2, we show the predicted evolution of the radio luminosity at the various search frequencies together with the observational limits. Note that these upper limits are grossly violated by both a backward extrapolation, of the observed luminosity-diameter relationship for older, galactic SNRs ($L \sim D^{-2}$; Milne, 1979) and a model that assumes the magnetic flux and total number of relativistic electrons in a young SNR such as, Cassiopeia-A to have been conserved at earlier epochs ($L \sim D^{-5}$, Shklovskii, 1968).

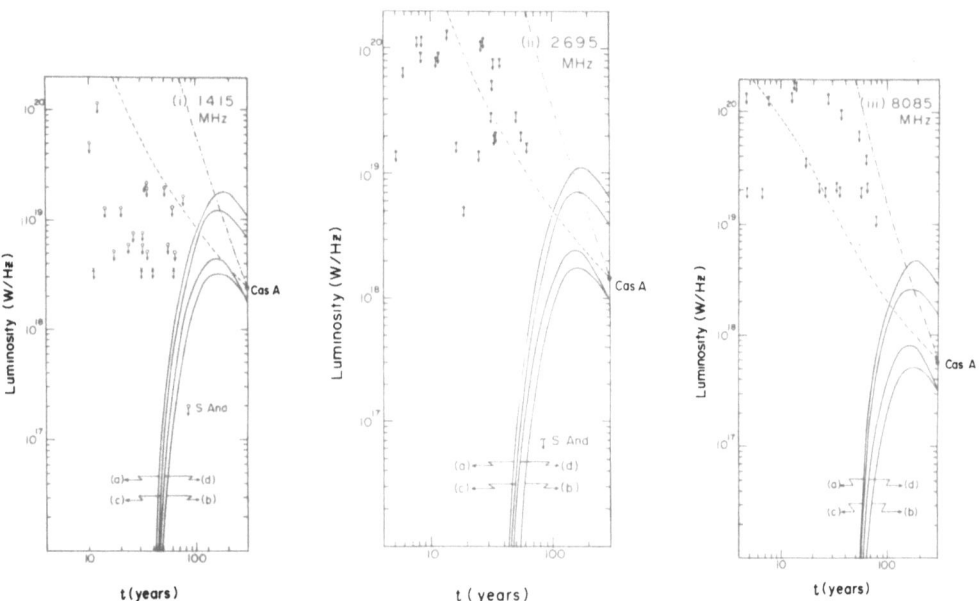

Fig.2. Evolution of the monochromatic synchrotron luminosity (solid lines) at different frequencies. Observational upper limits are derived from Spencer & Burke (1973, ⊥), de Bruyn (1973, ♀), Brown & Marscher (1978, ♀) and Ulmer et al. (1980, ⨎). The labels a-d refer to different choices of the initial conditions. The empirical evolution law of older galactic SNRs (broken lines) and a particle number-cum-magnetic flux conserving model for Cassiopeia-A (dot-dashed lines are shown for comparison.

2.3   The Supernova Remnant Cassiopeia-A

The many detailed observations that have been made of Cas A, the youngest galactic SNR, allow several tests of the theoretical model. The chaotic, cellular structure of its radio shell (Bell, Gull & Kenderdine, 1975), with its turbulent internal velocity field (Bell, 1977), as well as the strong anticorrelation between the polarized and unpolarized emission (Dickel and Greisen, 1979) do argue strongly in favour of small-scale turbulence as the origin of the radiating electrons and magnetic field.   Independent observational evidence for the turbulent enhancement of the magnetic field has also been presented (Cowsik and Sarkar 1980).   Moreover, the good spatial correlation between the radio emission and the soft X-ray emission from the ejecta (Fabian et al., 1980) demonstrates that the radio shell is well within the outer interstellar shock front.

A departure from spherical symmetry is evident in both the radio and X-ray structures suggesting that the expansion has been decelerated more on the eastern side, presumably due to a higher local interstellar density (Dickel and Greisen, 1979).   Our model then predicts that the radio spectrum should have evolved further and become flatter on this side, in accordance with tentative observational evidence for such a variation of the spectral index across the remnant (Rosenberg, 1970). A similar assymetry is perhaps also evident in the distribution of flux variations across the remnant, with most of the brightening features being located on the western side (Dickel and Greisen, 1979), consistent with the above suggestion that it is 'dynamically' younger.

A fit to the the spectum of the total radio emission that takes into account such a distribution of 'ages', or equivalently acceleration rates, in the remnant is shown in Figure 3. We have assumed that the compact features in the radio shell are the primary sites of low energy electron injection and that the electrons leak out of them in an energy independent manner into the surrounding extended regions.   The average spectral index of the compact features would then be the same as that of the total emission as is, in fact, observed (Rosenberg 1970).   The spectral shape of the total emission can then be fitted by summing together the compact feature spectra.   The average rates of decay of the compact and extended features would however be different, as shown in Figure 4.   A weighted average of the two in accordance with their respective contribution (1 : 2) to the total flux is in reasonable accord with observations.

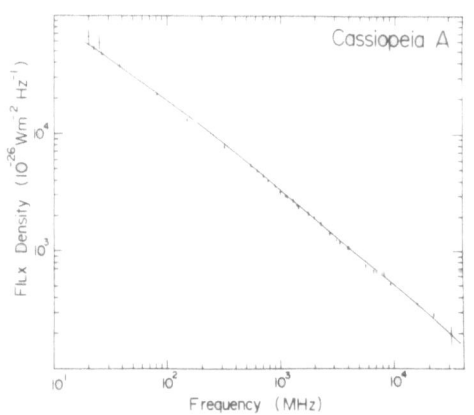

Fig.3   Fit to the radio spectrum of Cas A, with data from Baars et al. (1977).

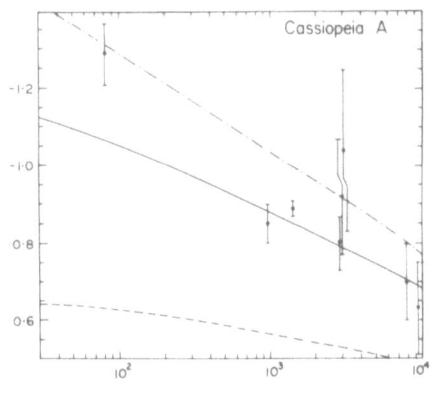

Frequency (MHz)

Fig. 4 The expected decay rate of compact (broken line) and extended (dot-dashed line) features in the radio shell, and of the total flux (solid line) of Cas A. Data are from Baars et al. (1977).

Finally we note that any turbulence initiated at the present epoch in Cas A would generate radio emission with a steep spectrum which would contribute to the total flux primarily at low frequencies. This contribution would decrease rapidly with time as the spectrum flattens. The recently observed 'flare' at 38 MHz (Read 1977) could be due to such an event.

2.4   The collective properties of galactic radio SNRs

In old SNRs such as IC 443 and the Cygnus Loop, the radio emission is from the compressed interstellar field in the associated radiative shock waves (Duin and van der Laan, 1975). Most galactic SNRs however appear to be the adiabatic phase of evolution (Clark and Caswell 1976), so that their radio emission is presumably still of internal origin. Their surface brightnesses are however much higher than would be expected if the radio emission simply decayed with the adiabatic expansion of the convection zone, following the Rayleigh-Taylor instability phase. Moreover, the spectral index appears to continue flattening with increasing diameter, while the total (minimum) energy in particles and field does not show any appreciable decrease. Taken, together these observations imply a mild, ongoing stochastic acceleration of particles throughout the adiabatic phase, counteracting the energy losses due to expansion.
We have examined two possible sources of such energy inputs; sporadic regeneration of turbulence by interactions with interstellar 'clouds', and absorption of magnetic dipole radiation from a central, spinning neutron star that has failed to become a pulsar (Radhakrishnan and Srinivasan, 1980). A semi-quantitative study suggests that the general trends in the collective properties (e.g. $\Sigma - D$) relationship) can be reproduced, and the large scatter ascribed to differences in the local environments of SNRs in the inhomogeneous interstellar medium. Further since the total energy in electrons (of up to a few GeV) does not decrease appreciably in the expansion, SNRs might, after all, contribute significantly to the galactic cosmic rays.

3. CONCLUSIONS

(a) The sudden emergence of SNRs as radio sources several decades after the explosion is explained. (b) Their subsequent evolution with the complex spectral and temporal changes as well as several structural details exemplified by the remnant Cas A follows naturally. (c) Their collective properties such as the Σ-D relationship are reproduced and these provide evidence for a mild ongoing acceleration throughout the adiabatic phase. (d) When the SNRs finally merge into the interstellar medium they contain enough relativistic electrons to contribute significantly to galactic cosmic rays.

REFERENCES

Baars, J.W.M., Genzel, R., Paulini-Toth, I.I.K. and Witzel, A.: 1977, Astron.Astrophys., 61, p.99.
Bell, A.R. : 1977, Monthly Notices Roy.Astron.Soc., 179, p.573.
Bell, A.R., Gull, S.F. and Kenderdine, S.: 1975, Nature, 257, p.463.
Brown, R.L. and Marscher, A.P. : 1979, Astrophys.J., 220, p.467.
Clark, D.H. and Caswell, J.L.: 1976, Monthly Notices Roy.Astron.Soc., 174, p.267.
Cowsik, R and Sarkar S.: 1980, Monthly Notices Roy.Astron.Soc., 191, p.855.
Cowsik, R. and Sarkar, S.: 1982, Monthly Notices Roy.Astron.Soc., to be published.
de Bruyn, A.G.: 1973, Astron.Astrophys., 26, p.105.
Dickel, J.R. and Greisen, E.W.: 1979, Astron.Astrophys., 75, p.44.
Duin, R.M. and van der Laan, H.: 1975, Astron.Astrophys., 39, p.33.
Fabian, A.C., Willingale, R., Pye, J.P., Murray, S.S. and Fabbiano, G.: 1980, Monthly Notices Roy.Astron.Soc., 193, p.175.
Gull, S.F.: 1973, Monthly Notices Roy.Astron.Soc., 161, p.47.
Gull, S.F.: 1975, Monthly Notices Roy.Astron.Soc., 171, p.263.
Kulsrud, R.: 1979, in 'Particle Acceleration Mechanisms in Astrophysics', A.I.P. No.56, p.13.
Milne, D.K.: 1979, Aust.J.Phys., 32, p.83.
Radhakrishnan, V. and Srinivasan, S. : 1980, J. Astrophys.Astron., 1, p.25.
Read, P.L. : 1977 : Monthly Notices Roy.Astron.Soc., 181, p.63.
Rosenberg, I. : 1970 : Monthly Notices Roy.Astron.Soc., 151, p.109.
Shklovskii, I.S. : 1968, 'Supernovae', Wiley-Interscience, London.
Spencer, J.H. and Burke, B.F. : 1973, Astrophys.J., 185, L83.
Ulmer, M.P., Crane, P.C., Brown, R.L. and van der Hulst, J.M. : 1980, Nature, 285, p.151.

OPTICAL PROPERTIES OF SUPERNOVA REMNANTS

I. J. Danziger

European Southern Observatory,
8046 Garching bei München, West Germany

INTRODUCTION

Over the past decade there has been an increasing tempo of detailed optical and UV spectroscopy of SNR leading to comparisons with prevailing models of radiative shocks.

In these endeavours there have been reports of great success, for example with IC 443 (Fesen and Kirshner 1980) and SNR's in the LMC in the framework of shocked cloudlets (Dopita 1979), of moderate success, for example with the Cygnus Loop (Miller 1974), and of serious discrepancies, for example with Cygnus (Fesen et al. 1982), Vela (Danziger 1982) and RCW 86 and RCW 103 (Leibowitz and Danziger 1982). Since a new generation of models will appear in the very near future, if not at this symposium, and will be discussed in detail by those who have done this work, it seems more profitable to discuss other more general areas.

SOME INDIVIDUAL SNR

Reference is made to Table 1, where one finds a highly selective listing of SNR's from our Galaxy and from external galaxies. The emphasis here has been placed on objects where there has been some evidence for abundance effects (i.e. variations from solar-type abundances that may not be part of a galactic abundance gradient). Note that objects such as Tycho and SNR 1006 and similar objects in the Magellanic Clouds have not been discussed, even though, with an emission spectrum due only to hydrogen, they are certainly not normal SNR's. Objects above the double line in Table 1 are those in which I believe it is impossible at this stage to conclude that increased abundances of some heavy elements are not present. Below the double line there are objects where there is room for doubt. Indeed the last group of 3 are probably only representative of a bigger group which are not carefully defined here.

The outstanding characteristic of this table is that with the exception of the Crab, oxygen overabundances always appear in high velocity ($\sim$ thousands of km/sec) filaments, and therefore in presumably

J. Danziger and P. Gorenstein (eds.), Supernova Remnants and their X-Ray Emission, 193–203.
© *1983 by the IAU.*

young objects (≤2000-3000 years). Nitrogen overabundances occur in low
velocity filaments, but in objects that can be extremely young as well as
of moderate age. These properties have led to the formulation of models
involving some form of low velocity mass loss of nitrogen-enriched material
prior to the SN explosion. The oxygen-enriched material is apparently
consistent with the remnants formed from material in the interior of a $25M_\odot$
star prior to collapse and explosion.

Table 1. Some Properties of SNR

SNR	RV Km/s	Abundances	Comment
CRAB	>1200 - 1500 (6000?)	Variable He N, O, Ne, S Low x 2-3	
CasA HVF QSF	5000 ~100	O Var. S, Ar, Ca N + He	
LMC N132D	2200	O	similar to CasA?
G292.0 + 1.8	1600	O	
LMC 0540-69.3	1500	O	
SMC 0102-72.3	2000	O	
NGC 4449	3500	O	
Puppis A	300	N + He O?	WN Star?
W 50	Small	N	
Kepler	300	N	Dense, [FeII] strong
3C58	100	N	AD1181
LMC 0525-66.0	Small?	N	
RCW 86	200	N (Variable?)	AD185?
RCW 103	300	N	
RCW 89	Small?	N	[FeII] strong

Comments are made on individual objects:

Crab. An analysis using photoionisation models and recent optical (Fesen and Kirshner 1982) and UV (Davidson et al. 1982) data has been made by Henry and MacAlpine (1982). There is no large inconsistency with earlier published work and the carbon abundance seems normal. These authors give a warning that "the strong [SII] lines originate in the H° zone, and their strength is indicative of a large volume of gas which is relatively unionized, instead of an overabundance of sulphur". This might serve as a warning for the interpretation of some of the spectroscopic features of Cas A.

Cas A. While the oxygen overabundance in the high velocity knots seems certain, one could be more sceptical about S, Ar, Ca. If the ionisation structure of sulphur is difficult to model, this is surely true also of calcium. The forbidden and allowed transitions of CaII are seen in great strength in some but not all much older and more conventional SNR's, where one would be less willing to interpret it as an effect of overabundance in the absence of more sophisticated models and where oxygen is not apparently overabundant. The case for an overabundance of argon would be enhanced if it could be measured against another ion of similar excitation and ionisation properties.

LMC N132D. Danziger and Dennefeld (1976) first suggested this was identical to Cas A. Recently I have become less convinced of this exact identity. There are two major differences. The quasi-stationary flocculi of Cas A enriched in nitrogen are not apparent in N132D. The high velocity oxygen knots are very distinct and separate in N132D even at a distance of 55 kpc. If Cas A (or G292.0 + 1.8, the other oxygen rich SNR in the galaxy) was at the distance of the LMC, with our method of observing them through apertures 2 - 4 arcseconds in diameter, the resulting spectra would resemble the spectrum of the object in NGC 4449, i.e. oxygen lines with broad wings and no separate identity. This difference may be reflecting a difference in the environments of the two SNR's rather than in the natures of the exploding stars.

The other oxygen rich objects in the Magellanic Clouds have broad oxygen lines. However SMC 0102 - 72.3 has its own peculiar dynamical pattern referred to later. An unidentified line near 4905 in N132D might be blue shifted [OIII]5007, indicative of velocities as high as 6000 km/sec.

G292.0 + 1.8. This SNR has recently been imaged at X-ray wavelengths by Tuohy et al. (1982) who suggest the morphology supports a model of an expanding ring of material ejected from the exploding star. An analysis of the motions of the optically visible material by Braun, Goss and Danziger (1983, reported later in this symposium) shows that the oxygen-rich material does not follow the pattern of an expanding ring.

Puppis A. This object has the strongest [NII] lines of any known SNR, and in some filaments the [NII]/Hα = 20. [NI] 5200 is also strong, reinforcing the conclusion that we are seeing an abundance effect. The [NII] electron temperatures are normal. The nitrogen line strengths vary from filament to filament. Although Danziger et al. (1982) have proposed that the precursor star might be a WN star, it is not yet clear whether the nitrogen enrichment occurred before the SN explosion or as a result of ejection at the time of the explosion. In its radio-optical properties it resembles a young SNR.

W50. Kirshner and Chevalier (1980) concluded that the remnant was not excited by particle beams from SS433, and certainly the [NII] electron temperature obtained from the [NII] 6584/5755 ratio for this SNR is normal (Danziger, unpublished material).

Kepler. Recently Leibowitz and Danziger (1982) published detailed spectra of the various filaments. It was not possible to find a published model to fit the spectra. One reason is that the indicated density is very high $Ne > 10^4$ cm^{-3}. Indeed plausible quantitative arguments were made that in some parts of the filament one is currently seeing a shock front "eating" its way into one of the dense clouds. Comparison with published models suggests that a successful model for Kepler will require at least a much denser medium for propagation of the shock, a higher shock velocity and an increased nitrogen abundance. The [FeII] lines are strong in Kepler, but we shall see later that this is probably a density effect.

3C58. It has been suggested by Kirshner and Fesen (1978) that 3C58 is similar to Kepler. It is, in the sense that high velocity filaments have not been detected in what are considered to be young remnants. However the [NII] lines are stronger and the densities are an order of magnitude greater in Kepler.

LMC 0525-66.0. This object is included here not because the [NII] lines and the nitrogen abundance are enhanced relative to conventional Galactic SNR's, because they are not, but because they are enhanced relative to other SNR's in the LMC (Dopita, Mathewson, Ford, 1977; Danziger and Leibowitz, 1982). The relative enhancement is not of the same order as is seen in Puppis A, posing for us the question: Are there any Puppis A type objects remaining to be discovered in the metal poor LMC or SMC? So far none have been found, which in itself may provide a clue as to what type of star can become a SN providing huge overabundances of nitrogen.

RCW 86, RCW 103, RCW 89. Ruiz (1981) analysed the spectra of filaments in RCW 86, and suggested that there were real variations of nitrogen abundance with position. Leibowitz and Danziger (1982) also analysed several positions with an advantage that electron temperatures were directly measurable for [OIII] rather than being inferred from [OII]/[OIII] ratios. Both found significant temperature variations although the Leibowitz and Danziger values were systematically and significantly lower. Because of these temperature fluctuations, one is inclined to doubt whether real nitrogen abundance variations are occurring. It is important to settle this question, because Ruiz has used the inferred variation as a sign of youth and consequent identification with the SN of AD 185. In any case if one is observing abundance effects in nitrogen in RCW 86, one must surely be

seeing it in RCW 103, where the [NII] lines on average are twice as strong
as observed in RCW 86 and comparable to those in Kepler. The [NII]/Hα ratio
is also variable.

RCW 89 is included here because of the strong [NII] lines shown in
Figure 1. The other interesting feature is the great strength of
the [FeII] 5159 line which will be discussed later.

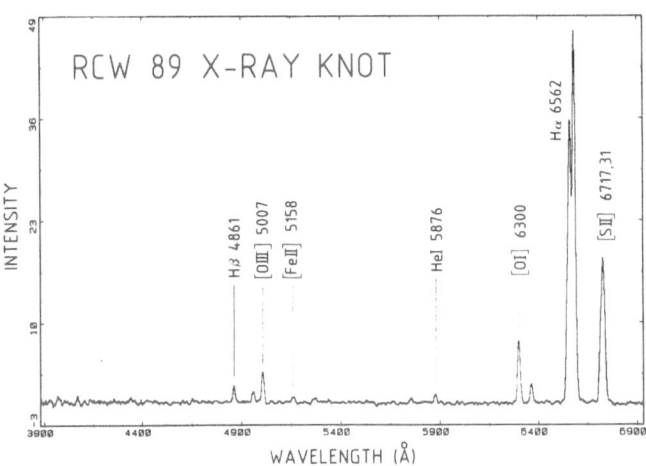

Figure 1. A spectrum of the optical knot in RCW89 showing
strong [NII] lines and a strong [FeII] 5159 line.

NITROGEN AND HELIUM ABUNDANCES

Table 2 lists average line ratios for SNR's in various galaxies.
Although this is a simplistic approach it contains valuable information.

Note the trend of decreasing [NII]/Hα towards less massive systems. A
similar trend is apparent for the [OIII] and [SII] lines. This supports
what is already known about the metallicity of these systems. Note also
especially the dispersion and in particular the very low dispersion in the
LMC. This suggests that [NII] lines if calibrated can give a very reliable
determination of the abundance, a point already made by Dopita (1977). It
also suggests that there is virtually no nitrogen abundance gradient nor
variation in the interstellar medium of the LMC. The larger dispersions
seen in the more massive galaxies are also consistent with what we know
from more detailed studies of HII regions and SNR's.

Table 2. Comparison of Average Line Ratios

	Galaxy	M31	M33	LMC	SMC
[OIII] 5007/Hβ	4.07 ± 2.75	2.63 ± 1.27	1.86 ± .83	1.70 ± 1.66	0.94
[NII] 6584/Hα	1.25 ± 0.63	0.83 ± .34	0.35 ± .14	0.21 ± .07	0.08
[SII] 6716,31/Hα	1.03 ± .36	1.03 ± .24	0.87 ± .23	0.72 ± .22	0.45

In Table 3 can be found a listing of ionised helium abundances obtained by Danziger and Leibowitz (1982) who have derived an average helium abundance $N(He^+)/N(H^+)$ using the HeI 4471, 5876 lines. For ten remnants in the LMC where we have a uniform set of data, we obtain $N(He^+)/N(H^+) \sim 0.079 \pm 0.013$. Also listed is the result for one SNR in the SMC obtained by Danziger and Dennefeld (1982).Since some models (Dopita 1977) show near coincidence of the H^+ and He^+ zones it may be that no correction for He^0 is necessary. This result would then support the results for the HII regions in the LMC of Peimbert (1975) and Shaver et al. (1982) who obtain values of 0.080 ± 0.005 and 0.083 respectively. Or the argument could be turned around to demonstrate that in real SNR the ionisation zones of the 2 species are on average very similar.

Table 3. Ionised Helium Abundances, $N(He^+)/N(H^+)$

	LMC
SNR	0.079 ± 0.013
HII (Peimbert)	0.080 ± 0.005 (0.084)
HII (Shaver et al.)	0.083 ± 0.008

	SMC
SNR	0.074 ± 0.017
HII (Peimbert)	0.077 ± 0.015 (0.081)
HII (Shaver et al.)	0.072 ± 0.008

THE [FeII] SPECTRUM

It was intended to look for abundance effects in iron in the same empirical way as we discussed for nitrogen and helium. However it soon became clear that excitation effects were present and important, and would need to be understood before one could say anything about abundances. Figure 2 shows a plot of the density sensitive line ratio [SII] 6731/6717 versus the relative strength of [FeII] 5159. This line was chosen because it is generally one of the strongest [FeII] lines in spectra of SNR's. The data come from published sources and include both Galactic and LMC SNR's. Preference has been given to photoelectric measurements. It is possible that selection effects exist in this data. However a strong correlation exists, suggesting that [FeII] 5159 increases with electron density. It is not surprising then that one could not show a systematic difference in the [FeII] 5159/Hβ ratio, between the Galaxy and LMC. Clearly, modelling which reproduces this effect is necessary before conclusions concerning iron abundances can be drawn.

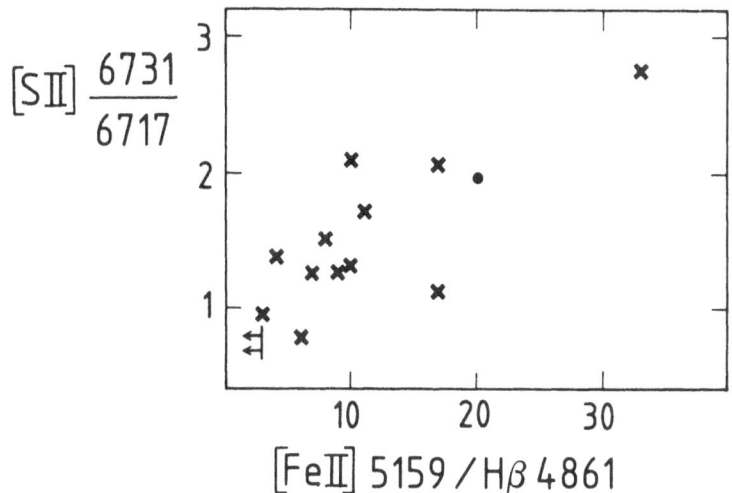

Figure 2. A plot showing the variation of [FeII]5159 with density indicated by the [SII]6731,17 doublet ratio. The dot is the point for the Herbig-Haro object HH-1 given by Dopita (1978). The point deviating most from the correlation results from a photographic measurement.

Since RCW 89 (bright X-ray knot) has a very strong [FeII] 5159 line, one might predict that the electron density is high ($Ne \sim 10^4$). Unfortunately our [SII] line measurements are of too low resolution to suggest whether a high density is present, but the spectroscopic results presented at this symposium by Murdin et al. (1983) do confirm this prediction.

CORONAL [Fe XIV]

Since the original report of the presence of [Fe XIV] 5303 in the spectrum of N49 by Danziger and Dennefeld (1976), I have been concerned about the interpretation of this feature. The resolved structure of the line seen in our original spectra and further elaborated by Murdin et al. (1978) shows components separated by 4 angstroms (or 240 km/s) in N49. This same structure can be seen in the spectra of N103B and probably N132D (Figures 3 and 4), where Danziger and Leibowitz (1982) have detected the feature for the first time. Thermal broadening alone at a temperature of $2 \times 10^6 K$ is not nearly sufficient to explain this structure. At present we know very little about the mass motion of the gas at these high temperatures, so this feature provides a possible means of learning more. One might also be tempted to speculate that [FeXIV] does not provide the only contribution to this feature, even though an alternative identification is not obvious.

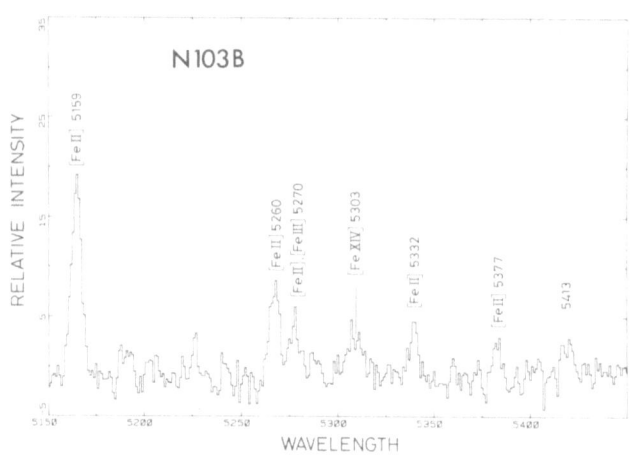

Figure 3. A spectrum of N103B showing the feature assumed to be [FeXIV]5303.

An attempt to detect spectroscopically [Fe XIV] 5303 in the LMC SNR N63A has not been successful. (Danziger and Dennefeld, 1982; Danziger and Leibowitz, 1982). The upper limit is about a factor 2 - 3 lower than a previously reported detection by Dopita and Mathewson (1979) using filter techniques. This applies to the main northern component of N63A and a separate cloudlet to the east.

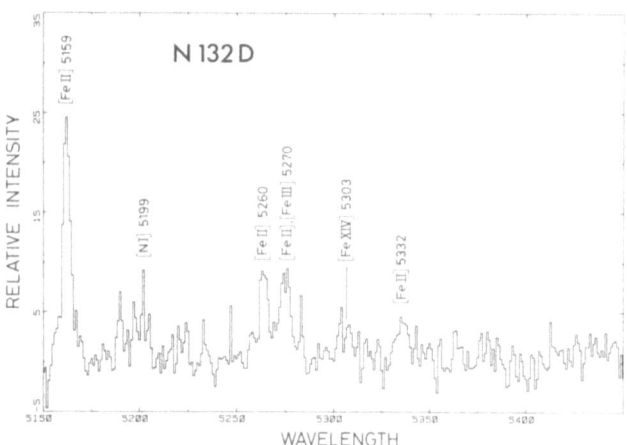

Figure 4. A spectrum of N132D showing a possible feature assumed to be [FeXIV]5303.

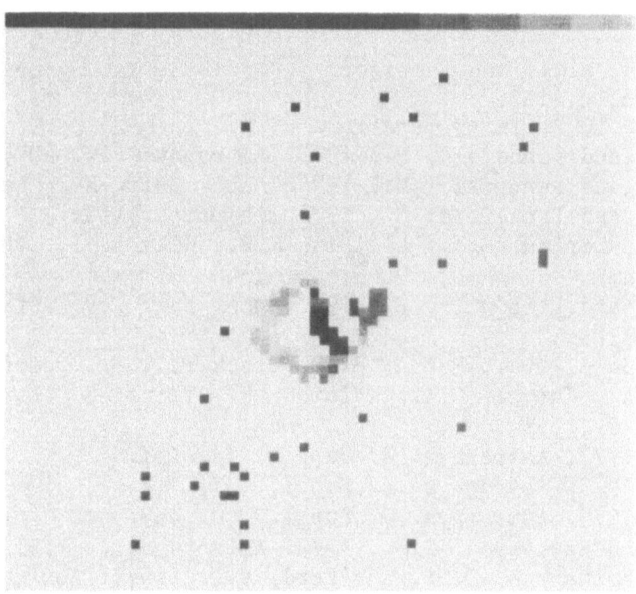

Figure 5. A grey scale map of all negative velocities in SMC 0102 - 72.3. The darkest levels represent velocities ~-2000 km/sec. The shell in this plot is ~30 arcseconds in diameter. North is to the top, east to the left.

OTHER OXYGEN RICH SNR

In spite of my earlier remarks concerning the lumpy filamentry structure seen in N132D, and the lack of a coherent expanding optical ring in G292.0 + 1.8, it is worth noting that SMC 0102 -72.3, the oxygen rich SNR in the SMC discovered by Dopita, Tuohy and Mathewson (1981), has a distinctive coherent velocity pattern unlike those two. Figure 5 is a velocity map of 0102 - 72.3 shown as grey scale levels. This map results from TAURUS Imaging Fabry-Perot observations of [OIII] emission made at La Silla by Atherton, Taylor, Boksenberg, Baade and myself. This map shows all velocities that are negative relative to the systemic velocity (i.e. approaching material). The largest velocities of approach (shown by the darkest grey levels) tend to be near the centre of the object. What is most striking is the continuity of material expanding in various directions, and maintaining a certain cohesion over projected distances greater than 5 parsecs. While older SNR contain elongated filaments of some considerable dimensions that must be formed as a result of instabilities in the interstellar medium, it is worth examining whether in the case of these younger SNR the apparent toroidal structures are the imprint of this SN explosion itself and possibly of the structure of the exploding star.

Much of the new work reported here results from a fruitful collaboration with E. Leibowitz to whom I am grateful.

REFERENCES

Braun, R., Goss, W.M., and Danziger, I.J., 1983. IAU Symposium 101, this volume, p. 159.
Danziger, I.J., 1982. in preparation.
Danziger, I.J. and Dennefeld, M., 1976. Astrophys. J., 207, 394.
Danziger, I.J. and Dennefeld, M., 1976. Publ. astr. Soc. Pacific, 88, 44.
Danziger, I.J. and Dennefeld, M., 1982. in preparation.
Danziger, I.J., Dopita, M.A., Elliott, K.H., Wilson, I., 1982. in preparation.
Danziger, I.J. and Leibowitz, E., 1982. Mon. Not. Roy. astr. Soc., in press.
Davidson, K., Gull, T.R., Maran, S.P., Stecker, T.P., Fesen, R.A., Parise, R.A., Harvel, C.A., Kafatos, M., Trimble, V.L., 1982. Astrophys. J., 253, 696.
Dopita, M.A., 1977. Astrophys. J. Suppl., 33, 437.
Dopita, M.A., 1978. Astrophys. J. Suppl., 37, 117.
Dopita, M.A., 1979. Astrophys. J. Suppl., 40, 455.
Dopita, M.A. and Mathewson, D.S., 1979. Astrophys. J., 231, L147.
Dopita, M.A., Mathewson, D.S., and Ford, V.L., 1977. Astrophys. J., 214, 179.
Dopita, M.A., Tuohy, I.R., Mathewson, D.S., 1981. Astrophys. J., 248, L105.
Fesen, R.A. and Kirshner, R.P., 1980. Astrophys. J., 242, 1023.
Fesen, R.A. and Kirshner, R.P., 1982. Astrophys. J., 258, 1.
Henry, R.B.C. and MacAlpine, G.M., 1982. Astrophys. J., 258, 11.
Kirshner, R.P. and Chevalier, R.A., 1980. Astrophys. J., 242, L77.
Kirshner, R.P. and Fesen, R.A., 1978. Astrophys. J., 224, L59.

Leibowitz, E., and Danziger, I.J., 1982. Mon. Not. Roy. astr. Soc.,
    in press.
Miller, J.S., 1974. Astrophys. J., 189, 239.
Murdin, P., Clark, D.H., Culhane, J.L., 1978. Mon. Not Roy. astr. Soc.,
    183, 79p.
Murdin, P., Seward, F.D., Harnden, R.F., 1983, this volume, p. 429.
Peimbert, M., 1975. Ann. Rev. Astron. Astrophys., 13, 113.
Ruiz, M.T., 1981. Astrophys. J., 243, 814.
Shaver, P.A., McGee, R.X., Newton, L.M., Danks, A.C., Pottasch, S.R.,
    1982. Mon. Not. Roy. astr. Soc., in press.
Tuohy, I.R., Clark, D.H., Burton, W.M., 1982. Astrophys. J., 260, L65.

DISCUSSION

CHEVALIER: I wish to comment on whether there are abundance effects in the
Cas A fast knots. There are many lines observed in the brightest knots:
three stages of ionization of O and Ar and two of S. The pattern of
observed elements is compatible with heavy element nucleosynthesis. Oxygen
and silicon group elements are present. It is unlikely that the pattern can
be explained by atomic physics effects.

DANZIGER: Arguments based on atomic physics seem more reassuring than
those based on possible complicated evolutionary nuclear physics scenarios.

DOPITA: Two points. First the [FeII] strength/density effect you mention
is well known in the Herbig-Haro objects which are also shock-excited but
somewhat denser on average than SNR. Have you compared the two sets of
data? Second, Binette, D'Odorico, Benvenuti and myself (Astron. Astrophys,
in press) give evidence for a galactic abundance gradient in
the [NII]/Hα ratio of SNRs. W50 fits very well on this correlation which
seems to argue that the strong [NII] lines in this object are the result of
position in the galaxy, rather than any abundance peculiarity intrinsic to
the SS433/W50 system.

DANZIGER: One of the points in the plot is for HH-1. However the Herbig-
Haro objects do not follow the same trend as SNR, and do not show any clear
correlation at all.

HIGH RESOLUTION X-RAY SPECTRA OF SUPERNOVA REMNANTS

C.R. Canizares[1,2], P.F. Winkler[3], T.H. Markert[1] and C. Berg[1]

[1]Department of Physics and Center for Space Research
Massachusetts Institute of Technology
[2]Alfred P. Sloan Research Fellow
[3]Department of Physics, Middlebury College

ABSTRACT
     We review results obtained with the Focal Plane Crystal
Spectrometer (FPCS) on the Einstein Observatory. Clear evidence is
found for departures from ionization equilibrium in the interior of
Puppis A. This comes from the observed weakness of the forbidden lines
relative to the resonance lines for the He - like triplets of O VII and
Ne IX. However, it is shown that this departure from equilibrium does
not alter our conclusion, based on previous FPCS results, that O and Ne
are overabundant relative to Fe. The spectrum of N132D shows strong
O VIII emission and very weak Fe emission, suggesting an even greater
O/Fe abundance enhancement than in Puppis A. In the Cygnus Loop, the O
to Ne abundance ratio is approximately solar; we have no information
about Fe. The O VII triplet shows clear evidence for departures from
ionization equilibrium in the Cygnus Loop. The spectrum of Tycho's SNR
contains lines from ionization stages of Fe XVII through Fe XXIII and
XXIV, indicating that a wide range of ionization conditions are
present. Cas A and Kepler's SNR show relatively less emission from the
higher ionization stages. For Tycho, we measured the strength of the
strong Si XIII lines, and we find that a many-fold overabundance of Si
relative to Fe is required regardless of the equilibrium state of the
emitting plasma (confirming the Solid State Spectrometer results). On
a separate topic, the completed analysis of X-ray Doppler shifts in Cas
A suggests that the emitting material is concentrated in a ring that is
inclined to the line of sight and is expanding at ~5000 km s^{-1}.

INTRODUCTION

     This paper will give a status report of the work on Supernova
Remnants (SNRs) that we have been carrying our with the X-ray data from
the Focal Plane Crystal Spectrometer on the Einstein Observatory. The
reasonably high spectral resolution of our instrument allows us to
apply plasma diagnostic techniques similar to those used to study
laboratory plasmas and the solar corona: we can measure individual
line strengths and take selected line ratios to deduce the physical
properties of the emitting material (e.g., see Winkler <u>et al</u>. 1982 and

J. Danziger and P. Gorenstein (eds.), Supernova Remnants and their X-Ray Emission, 205–212.
© *1983 by the IAU.*

references given below). Although we are limited by sensitivity and by the necessity of measuring lines one at a time, we have acquired data on most of the brighter SNRs. Our analysis is far from complete, but several interesting results have already emerged. First we will address questions of elemental abundances, which inevitably includes the topic of nonequilibrium ionization, and then we will review our data on the X-ray Doppler shifts in Cas A.

ABUNDANCES AND IONIZATION DISEQUILIBRIUM

Puppis A has by far the richest and brightest X-ray spectrum, an appropriate complement to its spectacular X-ray image (Petre et al. 1982, also, Petre et al. this conference). We have already published some of our analyses of the spectrum of the interior region, which undoubtedly contains a mixture of supernova ejecta and swept-up interstellar material (Winkler et al. 1981a, 1981b, 1982, Canizares and Winkler 1981). In a separate presentation, we give some new spectral results for the eastern bright knot, which appears to be a recently shocked interstellar cloud (Winkler et al., this meeting).

In our work on the interior of Puppis A, a major conclusion was that the O/Fe and Ne/Fe abundances are enhanced relative to cosmic abundances (Canizares and Winkler 1981). We most plausibly attribute this to enrichment by a type II SN in a 20-25 $M_\odot$ star. Our analysis was based on a comparison of the ratios of selected O, Ne, and Fe lines both to ratios computed theoretically for equilibrium plasma and to similar ratios observed in solar active regions. Here we explore further the question of ionization equilibrium. We now have good evidence for the departure from equilibrium of the emitting plasma, but we can also strengthen our previous conclusion that the abundance determinations are unaffected.

The evidence for departures from equilibrium comes from an examination of the relative strengths of the closely spaced triplet of lines emitted by He - like ions, such as O VII and Ne IX. The ratio G of the Forbidden ($F:1s^2 - 1s2s\,^3S$) plus Intercombination ($I:1s^2 - 1s2p\,^3P$) lines to the Resonance ($R:1s^2 - 2s2p\,^1S$) line depends on electron temperature and on the importance of recombination vs. collisional excitation for populating the excited states. Recombination cascades favor the ^3S and ^3P states over the ^1S state (because of the statistical weights). A plasma at equilibrium has a given contribution from each process, but in an ionizing plasma the recombination term is suppressed giving smaller F and I lines relative to the R lines (smaller G). Pradhan (1982) has calculated the line strengths and found values that agree very well with solar data (e.g., McKenzie and Landecker 1982). In contrast, for the Puppis A interior one would expect G ~ 1 for O VII in an equilibrium plasma at an electron temperature similar to our observed ionization temperature ($T_i \sim 2$-3 x 10^6K), but we see G $\simeq 0.32^{+0.18}_{-0.12}$ implying that the plasma is ionizing with $T_e > 5$ x 10^6K (See Figure 1). A similar result is found for Ne IX. This is the first clear evidence for non-equilibrium conditions in an SNR.

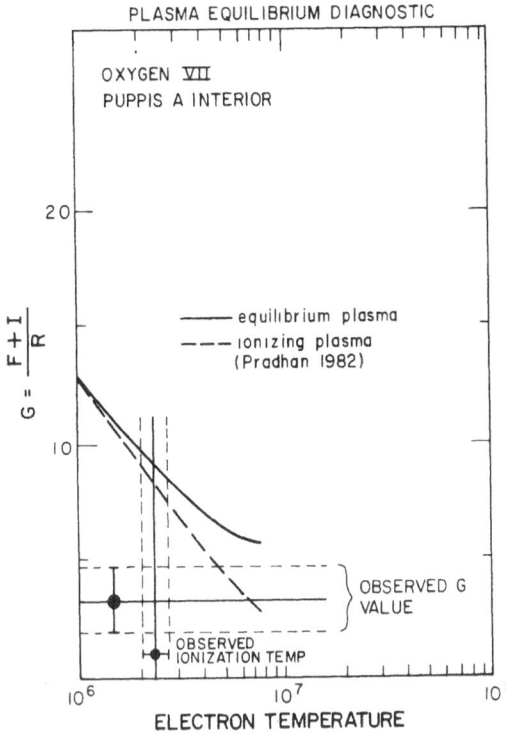

Figure 1: Plasma equilibrium diagnostic for the interior of Puppis A using the ratio Forbidden (F) + Intercombination (I) to Resonance (R) line strengths of O VII (see text). The curves show the calculated ratios from Pradhan (1982) for equilibrium plasma (solid) and for an ionizing plasma (dashed; the recombination terms are suppressed). The observed value with ± 1 σ errors is shown, as is the observed oxygen ionization temperature derived from O VII/O VIII line ratio (Winkler et al. 1981a). Clearly the plasma must be out of equilibrium with $T_e > 5 \times 10^6$K.

Although the establishment of ionization disequilibrium is important for the interpretation of SNR spectra in general, it has little effect on our deduced relative abundances. The reason for this is that the ratio of two lines of similar energy from different elements (e.g., O VIII Lβ at 755 eV to Fe XVII $2p^6$ - $2p^53d$ at 825 eV) depends on the relative abundances and the ionization fractions for each element but has no other significant dependence on T_e (e.g., see eq. 1 in Winkler et al. 1981a). Thus the abundances can be determined once the ionization fractions are known. The ionization fractions can be found from ratios of lines at approximately the same energy from different ionization stages of the same element. For the interior of Puppis A, we find that oxygen is 60% O VIII and 40% O VII (Winkler et al. 1981b; we can safely take the O IX fraction to be <10% and the fraction of stages below O VII to be zero). Iron is largely in the form of Fe XVII because the lines of Fe XVIII (~850 - 870 eV) and Fe XX (~960 - 970 eV) are very weak (Canizares and Winkler 1981). These values give the excess O/Fe and Ne/Fe abundance ratios independent of assupmtions about ionization equilibrium. Finally, to drive the point home, we have made a preliminary analysis of our Puppis A data using the models of Hamilton, Sarazin and Chevalier (1982) for non-equilibrium plasma in a Sedov phase SNR. Again, the loci of model parameters required to explain the O and Ne line ratios do not intersect those required by the O/Fe line ratios unless the O/Fe abundance is several times the solar value.

The remnant N132D in the LMC is known to contain O rich optical filaments (Lasker 1978) and to be very luminous in X-rays (Long, Helfand and Grabelsky 1981). Some of our FPCS data is shown in Figure 2. The spectrum is dominated by the O VIII Lα line -- nearly 10% of

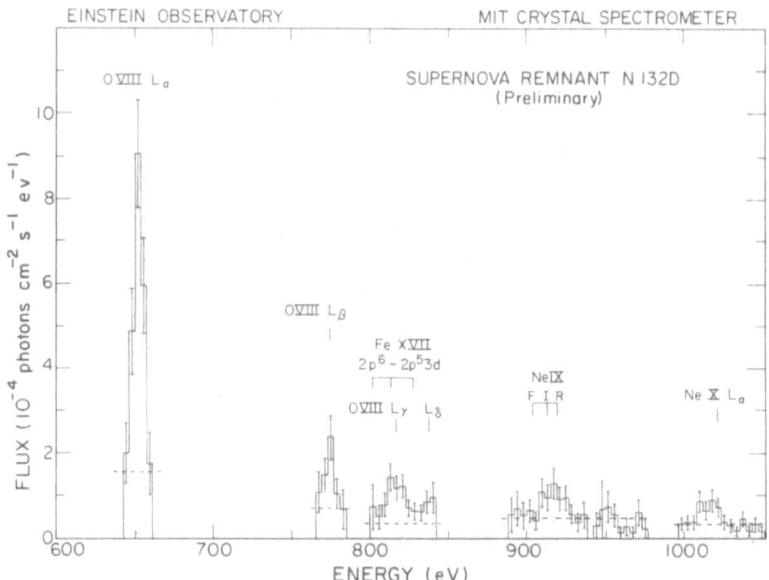

Figure 2:   A portion of the X-ray spectrum of N132D corrected for instrumental efficiency.

the total luminosity of the source is in this one feature. By comparison, the Fe XVII line is still weaker than in Puppis A; it may not even be detected (we are exploring the likely possibility that the emission around 826 eV is largely due to O VIII L_γ and Lδ). A large O/Fe excess throughout the remnant appears to be inevitable. The enhancement factor will come out of our full analysis. Because N132D may be younger than Puppis A (Lasker 1980), we are probably seeing less diluted ejecta and would expect a larger enhancement if the progenitors of the two SNRs are similar.

Our observations of the Cygnus Loop were confined to a bright region along its northern limb. Peter Vedder has measured fluxes for 6 lines of O VII, O VIII and Ne IX. Unfortunately we did not observe Fe XVII. The O VII/O VIII ionization temperature is ~3 x 10^6K. However, as in Puppis A, the F/R ratio for O VII is smaller than expected for ionization equilibrium implying that T_e must be higher and the plasma is still ionizing. This is firm evidence for the importance of nonequilibrium effects even for older SNRs like the Cygnus Loop as was first recognized by Itoh (1979 and references therein).

For the younger remnants Tycho, Kepler and Cas A, we have made systematic observations of the line complex around 1 keV (most of the lower energy lines we describe above are not detectable because of the large interstellar absorption). Although the lines in this region are too closely spaced to be fully resolved, we can separate the various line blends from ionization stages of Fe XVII through Fe XXIV. Our analysis is far from complete, but examination of this complex should be very important for establishing the degree of departures from equilibrium ionization (e.g., see Hamilton, Sarazin and Chevalier 1982). This will help constrain the abundances derived from SSS data (see Holt and Shull, this conference).

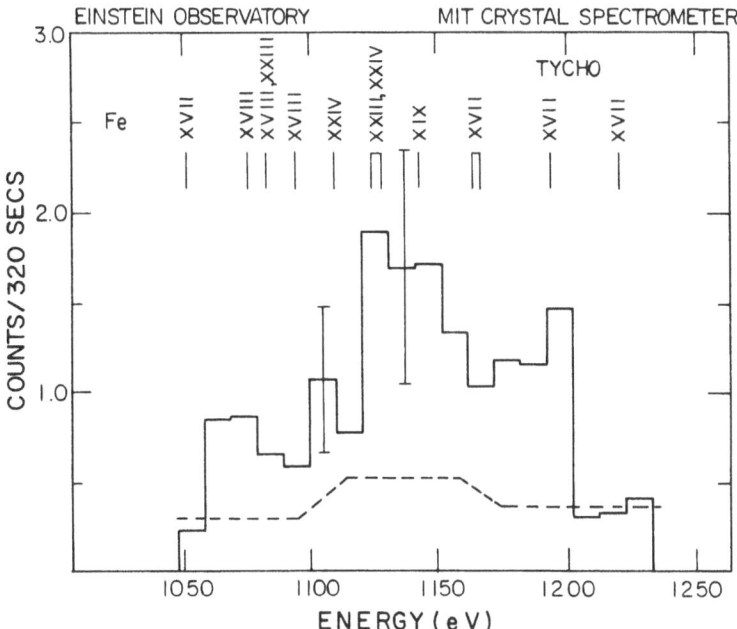

Figure 3: A portion of the X-ray spectrum of Tycho's SNR uncorrected for instrumental efficiency. Location of Fe lines seen in solar flare spectra have been marked (McKenzie et al. 1980).

Figure 3 shows the spectrum of Tycho around 1 keV. A preliminary analysis shows that Tycho has considerably more emission from the higher ionization stages of Fe XXIII and XXIV than do Kepler or Cas A. Note that this would not be reflected in the SSS fits, which give very similar temperatures for the line emitting material in all three SNRs; Becker et al. 1979, 1980a, 1980b. The Fe XVII lines are also present in Tycho, implying that a wide range of ionization is present.

In addition to the Fe lines and blends, for Tycho we have measured the strength of the Si XIII line that is so prominent in the SSS spectrum (Becker et al. 1980b). Even a qualitative examination confirms the SSS conclusion that Si must be overabundant in Tycho: none of the equilibrium models of Raymond and Smith (1977, 1979) or the nonequilibrium models of Hamilton, Sarazin and Chevalier (1982) can come close to reproducing the Fe and Si lines unless the Si abundance is very considerably enhanced (see also Shull, this conference).

X-RAY DOPPLER SHIFTS IN CAS A

We have recently completed our analysis of the X-ray Doppler shifts of Si and S lines from Cas A (Markert et al. 1983). The lines show broadening corresponding to ~5000 $km\ s^{-1}$ (FWHM) and a northwest/southeast asymmetry of 1820 ± 290 $km\ s^{-1}$. Clearly the bulk of the X-ray line emitting material has velocities comparable to those of the fast moving optical knots (van den Bergh 1971), giving a kinetic energy ~4 x $10^{51} erg\ s^{-1}$. The remnant is probably still in its free expansion phase with the X-ray line emission coming from reverse shocked stellar ejecta. The NW/SE asymmetry and the X-ray image (Fabian et al. 1980) can be reconciled with a model in which this emitting material partially fills a ring that is inclined to the line of sight and expanding at ~5000 $km\ s^{-1}$. The asymmetric range of optical radial velocities for the NW filaments (van den Bergh 1971) are also consistent with this picture. Such a ring may be the result of an inhomogeneity in the circumsource medium into which Cas A is expanding. An excess density could be caused by equatorial mass loss from the pre-supernova star, for example, which would drive the reverse shock faster and thus deeper into the dense stellar ejecta. An alternative explanation is that the ejecta themselves are preferentially located in the equatorial plane of the rapidly rotating pre-supernova star. This was suggested by Weaver and Woosley (1980). Similar ring-like geometries are seen in two other members of the class of oxygen rich SNRs, N132D (Lasker 1980) and G 292.0+1.9 (Tuohy, Clark and Burton 1982). It may be that this class shares common dynamics as well as elemental composition, and that it represents the remnants of massive stars.

REFERENCES

Becker, R.H., Boldt, E.A., Holt, S.S., Serlemitsos, P.J. and
    White, N.E. 1980, Ap. J. (Letters), 237, p. L77.
Becker, R.H., Holt, S.S., Smith B.W., White, N.E., Boldt, E.A.
    Mushotzky, R.F., and Serlemitsos, P.J. 1980, Ap. J. (Letters),
    235, p. L5.
Becker, R.H., Holt, S.S., Smith, B.W., White, N.E., Boldt, E.A.
    Mushotzky, R.F. and Serelemitsos, P.J. 1979, Ap. J. (Letters),
    234, p. L73.
Canizares, C.R. and Winkler, P.F. 1981, Ap. J. (Letters), 246, p. L33.
Fabian, A.c., Willingale, R., Pye, J.P., Murray, S.S. and Fabbiano, G.
    1980, M.N.R.A.S. 193, p. 175.

Hamilton, A.J.S., Sarazin, C.L. and Chevalier, R.A. 1982, preprint.
Itoh, H. 1979, Publ. Astron. Soc. Japan, 31, 541.
Lasker, B.M. 1980, Ap. J. 237, p. 765.
Lasker, B.M. 1978, Ap. J. 223, p. 109.
Long, K.S., Helfand, D.J. and Grabelsky, D.A. 1981, Ap. J. 248, p. 925.
Markert, T.H., Canizares, C.R., Clark, G.W. and Winkler, P.F. 1983,
    Ap. J. (in press).
McKenzie, D.L. and Landecker, P.B. 1982, Ap. J. 259, p. 372.
McKenzie, D.L., Landecker, P.B., Broussard, R.M., Rugge, H.R.,
    Young, R.M., Fledman, U. and Doschek, G.A. 1980, Ap. J. 241,
    p. 409.
Petre, R., Canizares, C.R., Kriss, G.A. and Winkler, P.F. 1982, Ap. J.
    258, p. 22.
Pradhan, A. 1982, Ap. J. (submitted).
Raymond, J.C. and Smith, B.W. 1979, Private communication.
Raymond, J.C. and Smith, B.W. 1977, Ap. J. Suppl. 35, p. 419.
Tuohy, I.R., Clark, D.H., and Burton, W.M. 1982, Ap. J. (Letters), 260,
    p. L65.
van den Bergh, S. 1971, Ap. J. 165, p. 457.
Weaver, T.A. and Woosley, S.E. 1980, Ann. N.Y. Acad. Sci., 336, 335.
Winkler, P.F., Canizares, C.R., Clark, G.W., Markert, T.H., Kalata, K.
    and Schnopper, H.W. 1981a, Ap. J. (Letters) 246, p. L27.
Winkler, P.F., Canizares, C.R., Clark, G.W., Markert, T.H., and
    Petre, R. 1981b, Ap. J. 245, p. 574.
Winkler, P.F., Canizares, C.R., Markert, T.H. and Szymkowiak, A.E.
    1982, in Supernovae: A Survey of Recent Research, Rees, M.J.
    and Stoneham, R.J., eds., (D. Reidel), 501.

DISCUSSION

DANZIGER: Are you prepared to make a quantitative statement about
Nitrogen in Puppis A?

CANIZARES: No. Unfortunately our instrument was quite insensitive
at the energies of the nitrogen lines and so we did not even attempt
a measurement. This will have to wait for AXAF.

TUOHY: There are presently 6 oxygen rich SNRs known in our galaxy and
in other galaxies. For the 4 SNRs of this group for which we have
sufficient velocity or spatial information, each shows evidence for an
asymmetric Type II supernova explosion. In particular, there is
evidence for expanding rings of ejecta in N132D, G292.0+1.89 and
1E0102.2-72.3, similar to the expanding ring of X-ray ejecta reported
for Cas A.

SARAZIN: You did not mention SN1006 where there is a claim of a possible detection of very strong oxygen emission by Galas et al. Do you have a good limit on oxygen lines from SN1006?

WINKLER: With the FPCS we searched for O VII and O VIII emission from the bright southwest limb of the SN1006 shell, and detected nothing. Scaling our upper limits to the entire remnant (using the Einstein HRI image) gives 3σ upper limits for both the O VII (561-574 eVs) and O VIII (654 eV) lines that correspond to about 1/2 the flux reported by Galas et al. Either the region observed with the FPCS is anomalously low in oxygen line emission, or the Galas et al. is too high. Galas et al. obtained their result from fits to low-resolution HEAO-1 data, and it is possible that a power-law-spectrum (which they did not investigate) may explain their data. Becker et al. and Toor have both reported successful power-law fits to the SN1006 spectrum.

RECENT RADIO STUDIES OF SNR

John R. Dickel
Department of Astronomy, University of Illinois

I. INTRODUCTION

SNR can generally be recognized as extended sources of continuum radio emission with non-thermal spectra located near the galactic plane. The emission is synchrotron radiation from relativistic electrons which have either been accelerated or trapped in the expanding shell and its associated shocks. Early lists of remnants (e.g. Milne 1970) were culled from general catalogs of radio sources and confirmed by several other kinds of evidence including the presence of shell structure, significant polarization, lack of recombination line emission, strong optical [S II] lines, and soft x-ray emission. While a few efforts to detect more data on faint remnants are continuing (e.g. Bonsignori and Tomasi 1979; Reich and Braunsfurth 1981), about 150 SNR are now known in the Milky Way and most studies have turned to detailed investigation of specific objects to determine their energy sources, emission mechanisms, and interactions with their surroundings. These studies have shown that while most remnants fall within two general categories, standard shell and filled center, there is no unique evolution within a class and irregularities in the local interstellar medium dominate any statistical properties of individual remnants. A few objects, in particular Cas A and CTB80, do not fit within either category.

Surveys of SNR in other galaxies of the local group are now producing significant results, although the required high resolution and sensitivity have generally limited detailed radio studies to remnants identified by optical or x-ray techniques. As well as providing some statistical information on SNR, the data may be used to indicate variations among different galaxies.

In the following sections we shall describe the kinds of observations currently being obtained and what they tell us about the objects. The data include continuum observations which can reach resolutions of about 1 arcsec using aperture synthesis techniques and also spectral line observations of the interstellar matter being encountered by the remnants.

213

J. Danziger and P. Gorenstein (eds.), Supernova Remnants and their X-Ray Emission, 213–219.

II.   SNR IN OUR GALAXY

A.   Standard-Shell Objects

1.   <u>Young SNR.</u> The radio emission from young supernova remnants
can be well characterized by the map of Tycho's SNR shown by Strom
(this volume). Although somewhat patchy, the basic structure is that
of a quite amorphous shell. The brightest emission arises just inside
a few optical filaments which probably delineate denser areas of the
interstellar medium which have just been encountered by the shock.
There is also faint radio emission with a very sharp edge lying a
short distance outside the main bright part of the shell suggesting
the presence of both a forward shock expanding into the interstellar
medium and a reserve shock propagating back into the ejectum. The
relativistic particles responsible for the radio emission have apparently been accelerated in situ by Rayleigh-Taylor instabilities at the
shock interfaces. This acceleration approximately balances losses due
to expansion as the overall flux density is decreasing only marginally. The gas has also been heated by the shocks as evidenced by a very
close association of x-ray and radio morphologies. As the shock enters into the concentrations seen in the optical recombination emission they should be heated further and undergo more particle acceleration to enter the x-ray and radio emitting regimes.

The overall magnetic field structure in the young remnants has a
net radial alignment although there must be much small scale turbulence as the net polarization of the emission even with the best resolution is only about 10%.

2.   <u>Older SNR.</u> The older remnants such as IC443 or the Cygnus
Loop are qualitatively different from the young objects. The radio
emission arises in thin sheets or filaments which show a one-to-one
correlation with optical features (e.g. Dickel and Willis 1980) and
there is little or no correlation with the x-ray morphology (Watson,
this volume). Recent radio observations are beginning to resolve individual filaments which appear to be the same size and shape as their
optical counterparts. After the interstellar medium has been heated
by the expanding shocks, thermal instabilities will produce cool,
dense sheets which can then break into filaments while remaining in
pressure equilibrium with their surroundings (Duin and van der Laan
1975). Increased radio emissivity is provided by the compression of
the relativistic particle frozen-in magnetic fields.

The structure and conditions in the old remnants are completely
dominated by the surrounding interstellar medium. This can be seen in
their very irregular shapes and also in the disordered magnetic fields
which are observed. The field orientations generally have quite cellular patterns and tend to merge into the general galactic background
with little discontinuity (e.g. Dickel and Milne 1976). Again, the
polarization percentages are somewhat small, indicating unresolved
structure, although they often reach 20 - 30%, somewhat larger than in

the young objects.

In a few instances we can observe the encounter of a remnant with a molecular or neutral hydrogen cloud. The spectral line data usually indicate that the cloud virtually stops the expansion, although in IC443 the presence of shocked molecular emission from within the remnant (DeNoyer and Frerking 1981) possibly suggests that the expanding shell may have overtaken a molecular cloud which is now being heated and evaporated.

3. The Relations Between Size and Brightness. Generally older, larger SNR appear to be fainter than the young ones, but there is tremendous scatter between individual objects. Remnants with the same diameters can have differences in surface brightness of over 10 to 1 (Milne 1979; Caswell and Lerche 1979) and single remnants can have sectors with radii differences greater than 3 to 1 but nearly the same surface brightness (Landecker et al. 1982). Most of the shells appear to be quite thin ($\Delta R/R$ typically 0.1 to 0.2) and there is no apparent relation between shell thickness and luminosity. Finally, as discussed above, there is a quantitative difference in the radio morphologies and also the sources of relativistic particles between the young and old remnants. The shock heating and acceleration in the young objects is very dependent upon the ejected mass and velocity plus irregularities in the ambient medium. The thermal-instability compression in old SNR will be controlled by such factors as the composition (affecting the cooling rate), the relativistic particle spectrum, the magnetic field strength and orientation in the surrounding medium, and the varying shock strengths. The transition from one phase to the other will depend critically upon the interrelationship of the various factors. We note that perhaps Puppis A may represent an intermediate stage; its radio, optical and x-ray structures are all rather filamentary but do not show good coincidence (Petre et al. 1982).

Not all of the above phenomena have been fully evaluated theoretically making it difficult to predict accurately the behavior and observed characteristics of a given remnant. Furthermore, ambient conditions can vary drastically around the galaxy and attempts to derive mean relations, such as those for surface brightness and diameter as a function of height above the galactic plane (Milne 1979; Caswell and Lerche 1979), remain fraught with large scatter. We conclude that there is neither an observational nor a theoretical basis for a unique surface brightness-diameter relation for SNR. The properties of individual remnants cannot be evaluated by this statistical approximation.

B. Crab-like SNR or Plerions

A complete review of the characteristics of these objects has been presented by Weiler (this volume). They include a filled distribution of emission from within the whole volume of the source, nearly uniform magnetic fields, and a flat radio spectrum. The power

for such objects is attributed to the spindown of a central neutron
star.

Recently several sources of this type have been discovered within
standard-shell SNR and a number of examples are presented in this volume by various authors. In general, the compact filled feature bears
little or no relation to the structure of the extended shell. This
supports the idea that the pulsar-like activity may be a transient
phase. An early occurence of such activity may mask the young shell
stage of the SNR while a later occurrence may stay independent of the
shell and decay before seriously affecting it. This leaves unresolved
the question of whether every supernova produces both an ejected SNR
and a neutron star.

C. Unclassifiable Objects

1. Cas A. At the current time, Cas A has several unique properties, although further detailed study of G292.0-1.8, which appears
similar optically (Tuohy, this volume), or other objects may reveal
additional sources of this kind. As well as being the brightest SNR
(by a factor of about 100 over Tycho's), Cas A contains numerous fine
scale components, many of which are unresolved with 1-arcsec resolution. The total remnant is expanding but the proper motions of individual features also show large random components (Tuffs, this volume). The overall flux density of the remnant is decreasing at about
0.7%/year (Dent, Aller, and Olsen 1974) and individual knots change
rapidly. They turn on quickly and then decrease with a mean 1/e lifetime of 48 years (Dickel and Greisen 1979). Presumably much of the
observed motion may be attributed to varying excitation and acceleration in the interacting wakes of rapidly moving optical filaments
rather than physical motion of the emitting material. Hopefully further monitoring will tell us in detail just how the object is changing
and whether we may be viewing a very irregular circumstellar envelope
which is slower than most young SNR to arrive near adiabatic equilibrium, perhaps an extreme case of multiple reverse shocks, or some
other phenomenon.

2. CTB80. This amazing non-thermal galactic radio source contains a filled central object with faint edge brightening (Dickel et
al. 1981). The brightest spot corresponds with what may be a point x-
ray souce (Becker, Helfand, and Szymkowiak 1982). The central feature
has an angular extent of about 1 arcmin and sits on a plateau which
appears to trail off almost like a wake toward the east for over 30
arcmin. Finally, three jets extend outward 40 arcmin from the core
toward the north and southwest without any apparent symmetry (Angerhofer et al. 1981). Very likely, multiple events are involved but the
cause of the jet-like structures remains a mystery.

III. SNR IN OTHER GALAXIES

Currently available radio telescopes now allow the detection and

even mapping of SNR in other galaxies within the local group but high
sensitivity and resolution measurements have so far been limited to
small selected regions. Therefore the most complete investigations
have relied upon identification of the SNR by other techniques. In
the Magellanic Clouds, Mills (this volume) has used the x-ray surveys
by the Einstein Observatory (Long, Helfand, and Grabelsky 1981; Seward
and Mitchell 1981). A plot of number versus diameter suggests that
most remnants are still in a nearly free-expansion phase with little
retardation by the surrounding interstellar medium and are thus younger than previously estimated. In M31, Dickel et al. (1982), using a
list of optical candidates based upon bright [S II] emission (D'Odorico, Dopita and Benvenuti 1980), tentatively conclude that the SNR in
that galaxy are on the average several times fainter at radio wavelengths than their counterparts in the Milky Way. Although we do not
know of any biases in the radio properties of SNR in our own galaxy
caused by optical or x-ray selection criteria, we are sampling only a
fraction of the total number of remnants in these other galaxies and
there could be serious deficiencies. Any conclusions must be treated
with caution until we are certain of an unbiased list of sources.

IV. CONCLUDING REMARKS

In this review we have deliberately avoided attempts to redo
various statistical analyses to determine such parameters as the supernova rate. Although a few new remnants have been found in the past
ten years, the basic numbers have remained unchanged and the detailed
investigations have revealed that many assumptions of uniformity and
unique evolutionary scenarios do not appear to be valid. Further progress will require considerably more observational and theoretical
work on just what governs the dynamics and emission processes in the
remnants. The new data are beginning to tell us exactly where the radio emission arises and how the varying conditions in the interstellar
medium affect the expansion. It is now important to collect such information for many remnants spanning a full range of ages and properties. Then we can determine where each remnant fits into the over
evolutionary pattern. On the theoretical side, we need more magneto-
hydrodynamic calculations on how structures of various sizes can form
and dissipate and the effects of shocks on regions with different
physical conditions.

Even with a more complete understanding of the physical processes
occurring in SNR, it will be difficult to improve statistical studies
because of the uncertainty in distances to most remnants. This limits
the determination of their true sizes and absolute luminosities. In
addition, measurement limits which depend upon size and brightness are
functions of position within the Galaxy, making it difficult to establish any true physical variations in galactic distribution. Both of
these problems can be avoided by studies of other galaxies whose SNR
are all at essentially the same distance from the observer. As mentioned above, we now have the means to acquire the large well-
calibrated data bases necessary for such studies, although the project

will require long term systematic surveys with major synthesis instruments. The prospect of obtaining such answers should certainly justify the effort.

   In summary, current radio studies of SNR in the Milky Way are providing us with exciting new details of the emission processes and dynamics of such objects in all stages of their evolution and we look forward to refined statistical analyses based upon complete samples of the SNR in external galaxies.

REFERENCES

Angerhofer, P. E., Strom, R. G., Velusamy, T., and Kundu, M. R. 1981. Astron. Astrophys., **94**, 313.

Becker, R. H., Helfand, D. J., and Szymkowiak, A. E. 1982. Ap. J., **255**, 557.

Bonsignori-Facondi, S. R. and Tomasi, P. 1979. Astron. Astrophys., **77**, 93.

Caswell, J. L. and Lerche, I. 1979. MN, **187**, 201.

DeNoyer, L. K. and Frerking, M. 1981. Ap. J. Letters, **246**, L37.

Dent, W. A., Aller, H. D., and Olsen, E. T. 1974. Ap. J. Letters, **188**, L11.

Dickel, J. R., Angerhofer, P. E., Strom, R. G., and Smith, M. D. 1981. Vistas in Astronomy, **25**, 127.

Dickel, J. R., D'Odorico, S., Felli, M., and Dopita, M. 1982. Ap. J., **252**, 582.

Dickel, J. R. and Greisen, E. W. 1979. Astron. Astrophys., **75**, 44.

Dickel, J. R. and Milne, D. K. 1976. Australian J. Phys., **29**, 435.

Dickel, J. R. and Willis, A. G. 1980. Astron. Astrophys., **85**, 55.

D'Odorico, S., Dopita, M., and Benvenuti, P. 1980. Astron. Astrophys. Suppl., **40**, 67.

Duin, R. M. and van der Laan, H. 1975. Astron. Astrophys., **40**, 111.

Landecker, T. L., Pineault, S., Routledge, D. and Vaneldik, J. F. 1982. Ap. J. Letters, in press.

Long, K. S., Helfand, D. J., and Grabelsky, D. A. 1981. Ap. J., **248**, 925.

Milne, D. K. 1970. Australian J. Phys., **23**, 425.

Milne, D. K. 1979. Australian J. Phys., **32**, 83.

Petre, R., Canizares, C. R., Kriss, G. A., and Winkler, P. F. 1982. Ap. J., **258**, 22.

Reich, W. and Braunsfurth, E. 1981. Astron. Astrophys., **99**, 17.

Seward, F. D. and Mitchell, M. 1981. Ap. J., **243**, 736.

DISCUSSION

R. P. KIRSHNER: Isn't it likely that remnants selected optically (like the M31 remnants) will be fainter, on the average, than remnants selected by radio means as in our galaxy?

S. VAN DEN BERGH: In our galaxy, the bias in favor of radio-bright SNR's can be avoided by looking only at objects with $90° < \ell < 270°$

because most all of these remnants are seen at both radio and optical
wavelengths.

W.P. BLAIR: It is interesting that even the early, low resolution
radio surveys had the sensitivity to detect objects similar to Cas A
or the Crab Nebula in M31, yet none have been found! This seems to be
another indication that things are different in M31.

J.R. DICKEL: There are some point sources in the 5C3 and Westerbork
surveys which could be such objects, but they have not yet been followed up optically. There are also a number of sources in these surveys which are clearly too bright to be any form of standard SNR and
are probably unassociated with M31.

NON-EQUILIBRIUM MODELLING OF THE STRUCTURE AND SPECTRA OF SHOCK WAVES

Michael A. Dopita and Luc Binette
Mount Stromlo and Siding Spring Observatories
Research School of Physical Sciences
Australian National University, Canberra.

1. INTRODUCTION

Our understanding of the structure of radiative shocks of modest ($v_S \lesssim 200$ km s^{-1}) velocity has improved greatly since the pioneering work of Cox (1972). This advance has been accomplished primarily by a more complete development of the physics of the shock front, its ionisation precursor and of the cool recombination zone. (Dopita 1976, 1977; Raymond 1979; Shull and McKee 1979; Shull, Seab and McKee, this conference; D'Odorico and Dopita, this conference). However, all models used for spectrum synthesis have so far involved one dimensional steady-flow hydrodynamics and most have covered only a very limited range in parameter space. This has meant that whilst they can be used to estimate shock velocities and metallicities of the (assumed) fully radiative shocks found in nearby SNRs, they are of limited applicability in more exotic types of objects.

In this paper, we discuss new models for three classes of more unusual object; the oxygen-rich Cass A type of filament, very high velocity shocks and/or thermal instabilities and shocks which are not yet fully radiative. These models have been generated using the general purpose modelling code MAPPINGS, (Binette 1982; Binette, Dopita and Tuohy 1983).

2. THE OXYGEN-RICH SNR

The fast-moving knots of Cass A have been intensively studied (Chevalier and Kirshner 1979), since the strong oxygen and neon emission lines and the absence of hydrogen and helium recombination lines suggest that here we are actually seeing the material thrown out from the core of a massive star at the time of the supernova event. If the spectra of such regions could be interpreted, we would therefore expect to derive a wealth of information on nucleosynthesis in massive stars and the role of these objects in the chemical evolution of galaxies.

J. Danziger and P. Gorenstein (eds.), Supernova Remnants and their X-Ray Emission, 221–230.
© *1983 by the IAU.*

Recent SNR searches in our Galaxy, (Goss et al. 1979) the Magellanic Clouds (Danziger and Dennefeld 1976, Mathewson et al. 1980, Dopita, Tuohy and Mathewson 1981) and in NGC 4449 (Kirshner and Blair 1980) have revealed several other objects of this type, and these show subtle but important spectral differences between them.

It has generally been assumed that the optical emission of this fast-moving material is the result of shock-heating (Peimbert and van den Bergh 1971, Lasker 1978, Chevalier and Kirshner 1978, 79; Goss et al. 1979; Murdin and Clark 1979, Kirshner and Blair 1980). However, interpretation of spectra has been hampered by a lack of theoretical modelling.

The work of Itoh (1981a,b) represents a major advance in this regard, and he was the first to describe the curious structure of a shock propagating through a pure oxygen plasma. We have recently applied our modelling code MAPPINGS to this problem, and the results are the subject of this section.

As far as the general shock structure is concerned the characteristics of our models (Fig. 1a,b) agree with the Itoh work. In the initial cooling zone, the electron temperature is much lower than the ion temperature (T_i), and T_e reaches a quasi-equilibrium level at which the rate

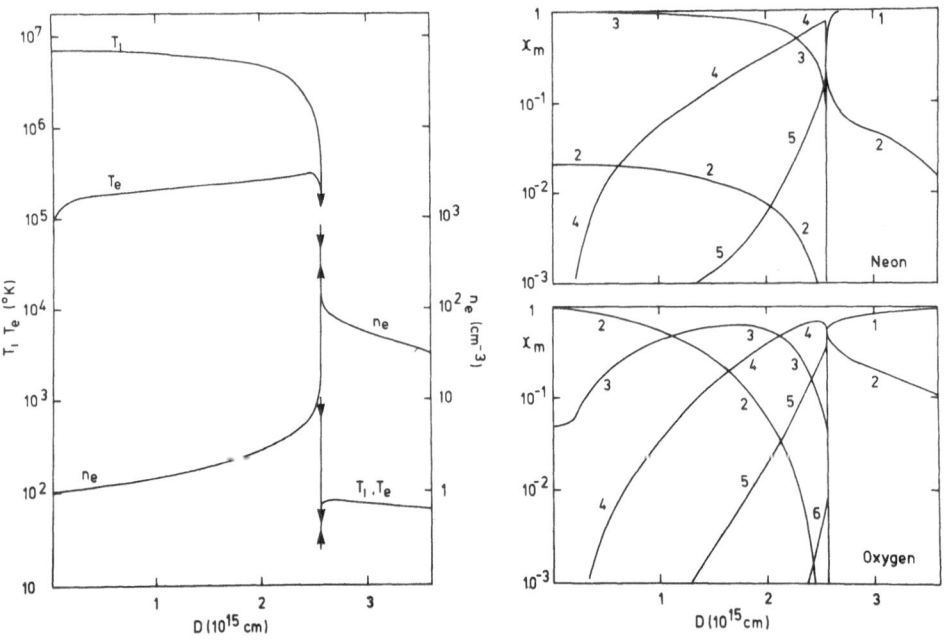

Figure 1a,b   The temperature (a) and ionisation (b) structure of the oxygen-rich shock model described in the text. Note how the ionisation state continually increases up to the region of temperature 'collapse'

of energy loss by collisional excitation of the ions matches the gain by collisional transfer of energy from the ions (Spitzer 1967). This temperature varies little through much of the cooling zone. If collective heating of the electrons by plasma instabilities is important (McKee and Hollenbach 1980), this result may be invalid, and then a model which sets $T_e = T_i$ would be more applicable. This can be done with MAPPINGS, but has not yet been investigated.

As T_e and T_i draw together, the approximately isobaric compression of the gas causes a rapid increase in volume emissivity of the plasma and a collapse in temperature. Because the heavy elements are such important coolants, the cooling time remains shorter than the recombination time until infrared cooling in fine structure lines is quenched, which occurs below $100^\circ K$. Thus, a high ionisation state is 'frozen' in at very low temperatures.

The initial recombination of the plasma is very rapid because of the high rates and densities prevailing at such low temperatures, but when the plasma has recombined sufficiently to see the UV radiation it produced while hot, photionisation heating becomes important in a very extended final recombination zone.

The major discrepancy between the models (Table 1) occurs in this recombination zone. At no point do we see the bounce-back to over $1000^\circ K$ in temperature which characterised the pure-oxygen Itoh models. Similarly, in the radiative precursor (Itoh 1981b) temperatures remain low. This is most likely due to the very important [CII] $\lambda 156\mu$ and

Table 1: Comparison of oxygen-rich shock models.

Parameter		Itoh	Dopita & Binette
Composition		Oxygen	Mixed[1]
Shock conditions			
Velocity	V	141.4 km s^{-1}	137.8 km s^{-1}
Pre-shock density	n_1	1.0 cm^{-3}	0.25 cm^{-3}
Post-shock ion temp.	T_i	7.2 x 10^6 $^\circ K$	7.2 x 10^6 $^\circ K$
Post-shock elect. temp.	T_e	10^4 $^\circ K$	10^4 $^\circ K$
Cooling length to $T_e < 10^4$ $^\circ K$		$10^{15}\left(\dfrac{n_1}{cm^{-3}}\right)^{-1}$ cm	$6.5 \times 10^{14}\left(\dfrac{n_1}{cm^{-3}}\right)^{-1}$ cm
Max T_e		$105000^\circ K$	$186000^\circ K$
Max T_e, T_i (recombination zone)		$\sim 2000^\circ K$	$86^\circ K$

[1] Inner 10 $M_\odot$ or 25 $M_\odot$ model; Woosley and Weaver (1980).
A mixture of C, O, Ne, Mg, Si, Ca, S and Ar.

[SiII] $\lambda35\mu$ lines, which dominate the cooling in these zones by very
large factors. This has the unfortunate consequence that shocks of a
more realistic composition give less good agreement with observations,
as in our models there is negligible flux in the [OI] $\lambda6300$, 63 Å lines
either in the shock structure or its photoionisation precursor, and the
[OIII] $\lambda4363/\lambda4959+5007$ Å ratio, predicted at 0.086, indicates a higher
temperature than observed. (Lasker 1978; Chevalier and Kirshner 1979;
Kirshner and Blair 1980).

More generally, we are not optimistic about the prospects for this
type of model to ever describe the observations in a satisfactory manner.
For example, the excitation as measured by the [OII] $\lambda3727,9$/[OIII] $\lambda5007$
ratio is an exceedingly sensitive function of temperature. For the model
of Table 1 it is 0.86, about the middle of the observational range.
However, a variation of shock velocity of only about 50 percent in either
direction is sufficient to drive the theoretical value away from the
observed range. It seems to us most improbable that objects with a
velocity dispersion of several thousand km s^{-1}, sometimes showing shears
of order a thousand km s^{-1} within an individual knot are excited by
shocks with such a narrow range of velocities. If the knots are excited
by the propagation of a radiative reverse shock into a cloudy ejecta,
then equating the ram pressures implies a cloud/intercloud density ratio
of order 100:1, which seems unreasonably high. In any case, such a
scenario would not predict velocity shear within a cloudlet, nor why the
fast-moving knots are only visible over comparatively short timescales
(Kamper and van den Bergh 1976).

An idea which seems promising to us is that the knots are not radiative, but are in fact collisionless shocks. If we immerse a cloud of
sufficiently small dimension into a lower density intercloud medium with
large relative velocity, the intercloud medium will pass through the
cloud more or less freely as a suprathermal gas. Since the stopping
timescale varies as the cube of the relative velocities (Spitzer 1967),
this effect is favoured at high velocity. The optical emission will then
result from the ionisation and heating by these suprathermal particles,
which will occur throughout the cloudlet. By analogy with X-ray ionised
regions, which are predominantly heated and ionised by suprathermal Auger
electrons, we expect a spectrum showing a mixture of ionisation states at
fairly modest equilibrium temperature. Such a physical state seems, from
the observational material, to be exactly what is required. We have
therefore undertaken to model this situation.

2. HIGH VELOCITY SHOCKS AND THERMAL INSTABILITIES.[1]

For none of the published steady-flow radiative shock models does the
velocity exceed 200 km s^{-1}. In part, the reason for this is that the
physics is computationally complex, but the major reason must be that
cooling timescales in the plasma become much longer than the timescales
relevant to SNR evolution. At galactic scales, however, it is possible
for a low density, high velocity shock to become radiative or for low

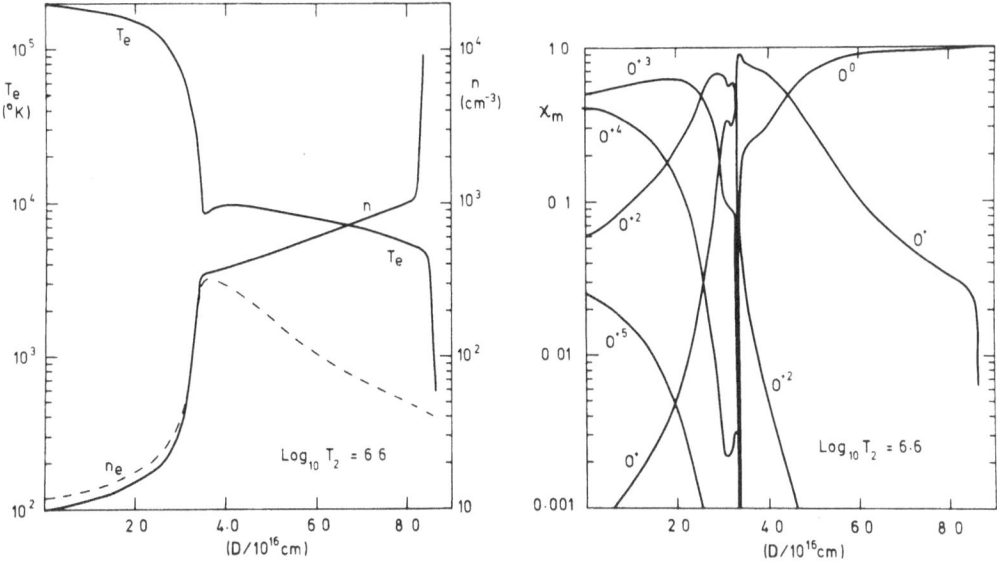

Fig. 2a,b   The temperature (a) and ionisation structure in oxygen (b)
            for a high velocity shock of initial Temperature
            log T_e = 6.6 $^\circ$K. Note how the cooling zone is followed by an
            extended region ionised by X-rays.

density hot plasma to cool by thermal instability in the condensation
mode (Field 1965; Mathews 1978). Such models may therefore be applicable to optical nebulosity associated with galactic jets, such as in Cen A
(Blanco et al. 1975; Osmer 1978; Graham and Price 1981) or to filaments
in the vicinity of giant elliptical galaxies such as NGC 1275 (Kent and
Sargent 1979) or M87 (Ford and Butcher 1979).

   To simplify the computation with MAPPINGS, we have divided the cooling
zone in which photoionisation is unimportant into an X-ray emission zone
with T_e > 2×10^5 and a cooler UV-optical emission zone. Since the first
zone determines the radiation field for photoionisation but is too highly
ionised to treat with our code, we have assumed that the plasma cools
isobarically and at collisional ionisation equilibrium, and computed the
resulting radiation field between 7.64 and 5000 eV using the Raymond-
Smith code. (Raymond and Smith 1977; Raymond, Cox and Smith 1976). For
the cooler region and the subsequent zones, we solve the full time
dependent hydrodynamic flow.

   The general characteristics of this are as follows (see Fig. 2a,b).
When the initial temperature exceeds 10^6 $^\circ$K, photoionisation effects
become important in the region normally associated with the recombination
of hydrogen. A narrow 'supercooled' zone appears where the ionisation

state is still fairly high so that photoionisation heating is not
sufficient to maintain the temperature. This is followed by a bounce-
back in temperature as the plasma adjusts to photoionisation equilibrium.
For high initial temperature of the plasma, there exists an X-ray
absorbing photoionised region which comes to dominate the total emission
for initial temperatures, T_i, in excess of $2x10^6$ °K. As Figure 2 shows,
this region has low ionisation state and modest temperature, and the
equilibrium is dominated by Auger electrons and the secondary ionisation
and heating they produce.

The emission from plasma cooling from $T_i > 2x10^6$ °K is therefore
rather similar to that computed for regions ionised by a power law non-
thermal spectrum. Indeed, the ionising radiation field computed can be
fairly well approximated by a power law of the form:

$$F_\nu = F_{\nu_c} (\nu_c/\nu)^{0.5} \qquad \nu \lesssim \nu_c$$

$$= F_{\nu_c} (\nu_c/\nu)^{2.5} \qquad \nu > \nu_c$$

Where ν_c is a critical frequency determined by the initial temperature
and the geometry of the hot plasma with respect to the cool. This result
demonstrates that the ability to model a emission-line region in a galaxy
or a QSO by a power law ionising spectrum does not necessarily imply a
non-thermal source for the ionisation.

4.   SHOCKS OF FINITE AGE

Consider the timescale for cooling of a shocked plasma. For a final
temperature of 10^4, this timescale (t_c) for 'cosmic' abundance is given
in our models by:

$$t_c = 870 \ V_{100}^2 \ /n_1 \quad yr$$

where V_{100} is the shock velocity measured in units of 100 km s^{-1} and n_1
is the pre-shock density in units of cm^{-3}. Thus, even for a typical
shock velocity of 150 km s^{-1} with an ambient density of 5 cm^{-3}, this
cooling timescale of 400 yr is uncomfortably long compared with typical
ages of SNR. When one considers that the timescale for full recombination is about ten times longer, and that magnetic pressure support
extends both of these timescales, it is clear that the steady flow
approximation is of dubious validity.

Shocks in SNR can then be characterised by an 'age parameter' which
can be defined as the ratio of shock age and cooling timescale. Consider
two clouds of different density shocked at the same epoch. If the intercloud pressure is the same, then it follows from the above expression for
t_c that the ratio of the age parameters is in the ratio of the square of
the pre-shock densities of the two clouds. Thus in a cloudy medium, the
densest cloudlets will become radiative first, and different filaments

will be characterised by different age parameters.

It has been shown elsewhere (Dopita, Binette, D'Odorico and
Benvenuti 1982 and D'Odorico and Dopita this conference) that neither
the pre-ionisation condition nor the shock velocity has much effect on
the visible spectra of SNR, provided that the shock velocity exceeds
100 km s^{-1}. Also, global spectrophotometry of SNR in external galaxies
shows that SNR shock spectra are primarily dependent on chemical
abundances. Thus, assuming that all filaments in an SNR have a given
chemical abundance suggests that all the spectrophotometric data within
a given SNR should be capable of being modelled simply by changing the
age parameter.

Figure 3 suggests that something of the sort is happening. In
RCW 103, the filaments are dense, and there is little evidence for
scatter in line ratios between filaments (the solid lines show the
approximate limits defined by global spectrophotometry of SNR in external
galaxies). However, both RCW 86 and IC 443 show evidence for systematic
spectral variations between filaments, and the measurements scatter along
a line at roughly 45°, showing that strong [NII] is correlated with
strong [SII]. Ruiz (1981) suggests that this is an abundance effect, but
this conclusion is not consistent with our abundance grids, which are
characterised by a 'saturation' in the relative intensity of the [SII]
lines at these abundances.

We have attempted to model this effect by radiative shocks of finite
age, generated by the simple expedient of 'sawing off' the recombination
zone when the required shock age is reached. Figure 4 shows the result
for RCW 86. The abundance set used in the MAPPINGS model was adjusted to
fit filament B1 of Ruiz (1981) which shows the weakest [SII] and [NII]
lines with respect to hydrogen, and is therefore assumed to be the most
envolved. The model was a fully self-consistent (Shull and McKee 1979)

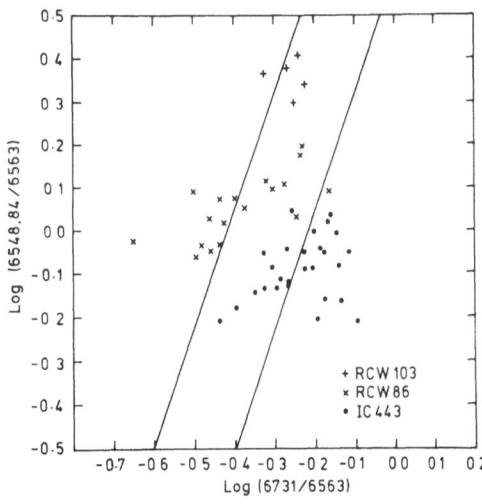

Fig. 3 Spectrophotometry of
individual filaments in three
galactic SNR, showing evidence for
variable age parameter (see text).
The data are drawn from D'Odorico
1974; Dopita, D'Odorico and
Benvenuti 1980; Ruiz 1981
Leibowitz and Danziger 1982 and
Fesen and Kirshner 1980.

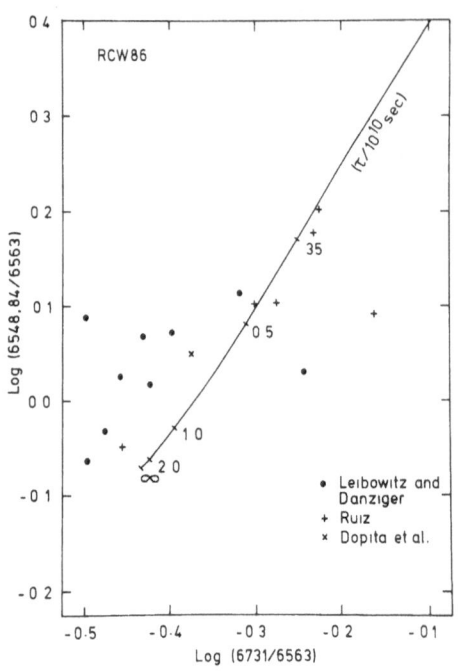

Fig. 4 Spectrophotometry of the individual filaments of RCW 86 fitted by models of varying age parameter. The pre-shock density, magnetic field and post-shock temperature were respectively 22 cm^{-3}, 2.2 μG and 180000 °K. A time of 10^{10} sec corresponds to an age parameter (see text) of 6.7.

plane-parallel radiative shock wave of finite age. As can be seen, all observed points lie within 0.15 dex of the computed age trajectory, which is well within modelling and observational errors. We thus conclude that RCW 86 shows evidence for varying age parameter in its filaments.

Curiously enough, the spectrophotometric results of Fesen, Blair and Kirshner (1982) on seventeen filaments in the Cygnus Loop cannot be as simply explained. They found a tight correlation between the intensity ratios of the [OII] and [OIII] lines with respect to Hβ, and a similar correlation between the [NeIII] λ3869 Å/H ratio and the [OIII]/Hβ ratio. Such a correlation is expected in shocks of varying age parameter, but we find that the slope of the expected correlation is wrong in the sense that the [OIII]/Hβ ratio increases more slowly from the [OII]/Hβ ratio. This remains true even if we suppose that the spectrophotometry refers to a 'time slice' of a shock only, as may be the case in shocks which have propagated through very thin sheets of denser gas.

We have only two possible explanations for these observations. It is possible that a few filaments may have unusual abundances. However enhancements of the oxygen abundance by a factor of about twenty are needed in some filaments, which would be very surprising in such an old remnant. An alternative explanation, which we have not yet investigated, is that the lower density filaments are photoionised by the UV radiation field produced by the denser ones. If this maintained a higher degree of ionisation in the cooling zone, the observations might be explicable.

[1]This section is the subject of a paper with I.R. Tuohy which has been submitted to Astrophysical Journal.

REFERENCES

Binette, L.: 1982, Thesis, Australian National University.
Binette, L., Dopita, M.A., and Tuohy, I.R.: 1983, Astrophys. J. (in press).
Blanco, V.M., Graham, J.A., Lasker, B.M., and Osmer, P.S.: 1975, Astrophys. J. (Letters), 198, L63.
Chevalier, R.A., and Kirshner, R.P.: 1979, Astrophys. J., 233, 154.
Cox, D.P.: 1972, Astrophys. J., 178, 143.
Danziger, I.J., and Dennefeld, M.: 1976, Astrophys. J., 207, 394.
D'Odorico, S.: 1974, 'Supernovae and Supernova Remnants', Ed. C. Cosmovici, D. Reidel, Dordrecht.
D'Odorico, S., and Dopita, M.A.: 1983, this volume, p. 529.
Dopita, M.A.: 1976, Astrophys. J., 209, 395.
Dopita, M.A.: 1977, Astrophys. J. Suppl., 37, 117.
Dopita, M.A., Binette, L., D'Odorico, S., and Benvenuti, P.: 1982, Astrophys. J. (in press).
Dopita, M.A., D'Odorico, S., and Benvenuti, S.: 1980, Astrophys. J., 214, 179.
Dopita, M.A., Tuohy, I.R., and Mathewson, D.S.: 1981, Astrophys. J. (Letters), 248, L105.
Fesen, R.A., Blair, W.P., and Kirshner, R.P.: 1982, Astrophys. J. (in press).
Fesen, R.A., and Kirshner, R.P.: 1980, Astrophys. J., 242, 1023.
Field, G.B.: 1965, Astrophys. J., 142, 531.
Ford, H.C., and Butcher, H.: 1979, Astrophys. J. Suppl., 41, 147.
Goss, W.M., Shaver, P.A., Zealey, W.J., Murdin, P., and Clark, D.H.: 1979, Monthly Notices Roy. Astron. Soc., 188, 357.
Graham, J.A., and Price, R.M.: 1981, Astrophys. J., 247, 813.
Itoh, H.: 1981a, Publ. Astron. Soc. Japan, 33, 121.
Itoh, H.: 1981b, Publ. Astron. Soc. Japan, 33, 521.
Kamper, K., and van den Bergh, S.: 1976, Astrophys. J. Suppl., 32, 361.
Kent, S.M., and Sargent, W.L.W.: 1979, Astrophys. J., 230, 667.
Kirshner, R.P., and Blair, W.P.: 1980, Astrophys. J., 236, 135.
Lasker, B.M.: 1978, Astrophys. J., 223, 109.
Leibowitz, E.M., and Danziger, I.J.: 1982, Monthly Notices Roy. Astron. Soc., in press.
McKee, C.F., and Hollenbach, D.J.: 1980, Ann. Rev. Astron. Astrophys., 18, 219.
Mathews, W.G.: 1978, Astrophys. J., 219, 413.
Mathewson, D.S., Dopita, M.A., Tuohy, I.R., and Ford, V.L.: 1980, Astrophys. J. (Letters), 242, L73.
Osmer, P.S.: 1978, Astrophys. J. (Letters), 226, L79.
Peimbert, M., and van den Bergh, S.: 1971, Astrophys. J., 167, 223.
Raymond, J.C.: 1979, Astrophys. J. Suppl., 39, 1.
Raymond, J.C., Cox, D.P., and Smith, B.W.: 1976, Astrophys. J., 204, 290.

Raymond, J.C., and Smith, B.W.: 1977, Astrophys. J. Suppl., 35, 419.
Ruiz, M.T.: 1981, Astrophys. J., 243, 814.
Shull, J.M., and McKee, C.F.: 1979, Astrophys. J., 227, 131.
Shull, J.M., Seab, C.G., and McKee, C.F.: 1983, this conference.
Spitzer, L. (Jr.): 1967, 'The Physics of Fully Ionised Gases' 2nd
          rev. edition, Interscience J. Wiley and Sons, N.Y.
Woosley, S.E., and Weaver, T.A.: 1980, "Nuclear Astrophysics" ed.
          C. Barnes, D. Clayton and D. Schramm, Cambridge Uni. Press.

DISCUSSION

PREITE-MARTINEZ: I do not understand how you can get a strong
recombination zone behind a shock with V = 141 km/s and n = 0.25.
According to fully HD computations the remnant should be in the
transition phase between adiabatic and radiative phases, with no
recombination region at all. Even more puzzling is the cool almost
neutral region behind a shock with $V_s \sim 10^3$ km/s. A remnant with such
a shock velocity is certainly in the adiabatic phase.

DOPITA: If you are referring to the oxygen-rich shocks in the first
part of your question, the answer is that oxygen is a very efficient
coolant. The model described in the text took only 35.8 years to
cool to 10^4 °K. With regard to the high velocity shocks, these
models do not apply to SNRs but to galactic jets and cooling pools
of hot gas.

FALLE: With reference to the spread of shock velocity necessary to
explain the observations, we have some evidence that a steady radiative
shock with a velocity greater than 100 km s^{-1} cannot exist. Would
this effect your results?

DOPITA: Probably.

THE STRUCTURE AND EMISSION OF A NON-RADIATIVE SHOCK

J.C. Raymond
Center for Astrophysics
60 Garden St.
Cambridge, MA 02138, USA

Faint filaments are observed a few arcmin outside the bright optical
filaments of the Cygnus Loop. They show nearly pure Balmer line
emission spectra, and they are interpreted as emission from non-
radiative shocks (1). Each neutral H atom passing through the shock
front emits on average about 0.1 Hα photon before it is ionized.
Since this radiation arises very close to the shock front, rather than
in an extended post-shock cooling zone (2), it can be used to study
the physics of the shock front itself. The structure of a shock poses
several important questions (3). There may be an electron thermal
conduction precursor ahead of the shock and there may be plasma
turbulence. The shock thermalizes 3/4 of the bulk velocity of the
incoming particles, so the ions initially have nearly all of the
thermal energy. The electron and ion temperatures can reach
equilibrium on the Coulomb collision time scale, but plasma turbulence
may bring them into equilibrium much more rapidly. The Coulomb
equilibration time scale is similar to the hydrogen ionization time,
so that the hydrogen line emission will depend on the nature of the
equilibration. The interpretation of the Hα line profile in terms of
the shock velocity also depends on this equilibration, so this
question is important for comparison of shock models with X-ray
spectra.

Models have been computed with the assumptions that the electrons and
ions each have Maxwellian distributions (possibly with different
temperatures) and that 90% of the pre-shock hydrogen is neutral.
Energy losses due to ionization and excitation of hydrogen and helium
were included, and great care was taken in the selection of atomic
rate coefficients. The assumption of Maxwellian velocity
distributions is questionable, but the available data do not yet
justify the great complexity of models which drop this assumption.

The observations available for detailed comparison with these models
are an IUE SWP spectrum, a blue spectrum, and an Hα profile of a
filament 5 arcmin outside the bright northeast portion of the Cygnus
Loop (4). The line profile gives a shock velocity of 170 km s^{-1} if

J. Danziger and P. Gorenstein (eds.), Supernova Remnants and their X-Ray Emission, 231–233.
© *1983 by the IAU.*

the electron and ion temperatures equilibrate instantly or about 150 km s^{-1} if the equilibration is more gradual. The observed relative intensities in the UV and optical are compared with models using these two assumptions in Table 1.

TABLE 1

Observed and Theoretical
Relative Intensities

	Observed	$T_e = T_i$	$T_e \neq T_i$
N V λ1240	1.0	1.0	1.0
C IV λ1550	.73	.78	.43
He II λ1640	.64	.12	.03
2-photon	11.	2.9	1.1
Hβ (Broad)	1.0	1.0	1.0
O III λ5007	.29	.04	.13
O II λ3727	.54	.002	.02
Ne V λ3420	.89	.13	2.1

Comparison of the models with the observations shows that the assumption of Coulomb equilibration of T_e and T_i gives much better fits to the optical forbidden line intensities, though the predicted [O II] intensity is still far too small. The He II λ1640 intensity predicted by either model is too small. This can be qualitatively explained by the conversion of He II λ256 photons into λ1640 and λ304 photons as the λ256 photons are scattered by He$^+$ in the emission region. The optical depth for resonant scattering of λ256 photons is about 1 at line center, and conversion of λ256 photons could enhance the intensity of λ1640 by a factor of three or four. A detailed radiative transfer calculation is required for more precise comparison of this line with the models.

The $T_e = T_i$ model predicts C IV and He II intensities relative to the N V lines in agreement with the observations. The two-photon continuum seems brighter than predicted, but the uncertainty in this measurement might be as large as a factor of three. All these aspects of the UV spectrum are more poorly accounted for by the slow equilibration model. On the other hand the absolute intensities of the C IV and N V lines are straining the limits of the $T_e = T_i$ model, and resonant scattering of these lines (5) presents a possible problem. These difficulties are less severe by a factor of four in the Coulomb equilibration model.

In conclusion, the present models do not fully account for the observations, and they do not answer the equilibration question. The filament observed is atypical both in its brightness and in the strength of its optical forbidden lines, so the more normal non-radiative shocks may be more amenable to modelling. The approximately equal intensities of the broad and narrow components of the Hα line show that any thermal conduction precursor present is not extensive enough to ionize much of the pre-shock hydrogen.

This work has been supported by NASA Grant NAS 5-5 and NASA contract NAGW 117 to the Smithsonian Astrophysical Observatory.

REFERENCES

1. Raymond, J.C., Davis, M., Gull, T.R., and Parker, R.A.R 1980, Ap. J., **238**, L21.
2. Chevalier, R.A., and Raymond, J.C. 1978, Ap. J., 225, L27.
3. McKee, C.F., and Hollenbach, D.J. 1980, Ann. Revs. Ast. and Ap., 18, 219.
4. Raymond, J.C., Fesen, R., and Gull, T.R. 1981, B.A.A.S., 13, 887.
5. Raymond, J.C., Black, J.H., Dupree, A.K., Hartmann, L., and Wolff, R.S. 1980b, Ap. J., 246, 100.

NEAR INFRARED SPECTROSCOPY OF GALACTIC AND MAGELLANIC CLOUDS SUPERNOVA REMNANTS

M. Dennefeld,
Institut d'Astrophysique de Paris

The spectral range redwards of about 7300 Angströms has up to now been rarely explored for any kind of object because of lack of adequate detectors. This situation is changing with the availability of silicon detectors whose response extends up to 1.1μ with reasonable quantum efficiency. We have used the ESO Reticon System behind the Boller and Chivens Spectrograph at the Cassegrain focus of the ESO 3.6m telescope to obtain spectra of several Supernova Remnants in the range 6000-10500 Å. The spectral resolution was 8 Å and the dual array allowed a clear sky-subtraction, indispensable because of the presence of numerous atmospheric OH bands in this region. Data reduction was done with standard techniques using wavelength calibration, flat-field correction, response curve determination through observations of standard stars and extinction correction. We present below results based on the first spectra obtained for the Galactic remnants RCW 86, RCW 103 and Kepler SNR and the Magellanic Cloud ones N 49 and N 63A. The Crab Nebula will be discussed separately at the end.

As already seen for the visible range (Danziger and Dennefeld, 1976), the spectra present striking similarities. The strongest lines are, as expected, the │S III│ lines 9069-9532 Å. Other typical lines are the │S II│ quadruplet around 10300 Å however faint because of the drop in sensitivity redwards of 10000 Å and the Paschen series of hydrogen. The │C I│ doublet 9827-9850 Å previously identified in the Cygnus Loop (Dennefeld and Andrillat, 1981) is detected with reasonable strength in all objects and opens interesting perspectives for carbon abundance determinations. Several new │Fe II│ lines are detected in this spectral range, among them the strongest one seen in the whole visible range λ 8616.9 Å from multiplet 13 F with strength about 5 % of H alpha. A detailed description of the iron spectrum has been given for Kepler SNR by Dennefeld (1982). Note that the iron spectrum of Kepler SNR is 3 to 4 times stronger than in the other studied remnants. Other interesting lines are │Ca II│ 7292 and the lines at 7379 and 7412 which are assigned to │Ni II│ as discussed later.

The │S III│ 9069-9532 to │S II│ 6717-6731 ratio shows a very well

J. Danziger and P. Gorenstein (eds.), Supernova Remnants and their X-Ray Emission, 235–239.
© *1983 by the IAU.*

defined correlation with |S II|/Hα for all objects available :
SNR, H II regions and Planetary Nebulae, about 60 objects in total. The
slope of the line is close to 1, indicating that |S II| is indeed a
better discriminator than |S III| between different types of objects.
The correlation deviates from the straight line for the SNR, as expected because photoionization is not the dominant ionization mechanism in
those objects.

Availability of the far red |S II| lines allows a model indepen-
dent reddening determination (Miller, 1968) by comparison of the obser-
ved |S II| 10300/|S II| 4070 ratio with the theoretical value. For the
objects under study, the reddening found was always compatible (within
the observational errors) with the one derived from the Balmer decrement under assumption of simple recombination. There is therefore no
evidence for collisional excitation of hydrogen. However this effort is
expected to be dominant only in low velocity shocks (Shull and Mc Kee,
1979) which are not represented in our sample.

An abundance determination for species like iron, calcium, carbon
or nickel, accessible from our spectra, requires a detailed comparison
with model predictions. The only models making specific predictions
(but only for |C I| and |Ca II|) are those of Raymond (1979). We note
that, on the average, the |Ca II| predictions fall short by an order of
magnitude, while those of |C I| are too large by a factor of 2 to 3 and
more so for galactic objects than for the Cloud ones (possibly indicating an abundance effect). However, a full appraisal of the data requires models including charge exchange reactions and updated atomic coefficient, with specific predictions for the lines observed here. Waiting
for such calculations to be made, we try meanwhile some empirical comparisons between specific line ratios, with emphasis on the possible
differences between Galactic objects on one hand and the Magellanic
Cloud ones on the other. These comparisons should however be taken with
caution, if only because of the small number of objects involved.

Starting with lines of neutral species, the ratio |C I|/|O I| gi-
ves an average value of 0.18 for Galactic objects and 0.05 for LMC. The
difference is significant (remembering the established underabundance
of oxygen in LMC) and points towards a lower C/O ratio in LMC than in
the Galaxy. This is in accordance with the results found for H II regions by Dufour et al. (1982).

Neutral carbon and singly ionized calcium both have ionization po-
tentials below 13.6 eV and should therefore partly coexist.
The ratios of |Ca II| to |C I| show, after correction for collisional
de-excitation of the |C I| line, an average value of 1.0 in the Galaxy
compared to 2.6 in LMC. This difference could be entirely accounted for
by the carbon deficiency in LMC (as suggested above and by the IUE observations of Benvenuti et al., 1980) and detailed calculations are required before a conclusion can be reached for calcium.

The situation is less clear as far as iron is concerned : for four objects, the | Fe II| 8616/Hα ratio shows a very similar value around 0.05, the exception being Kepler with a value about three times higher. There is no similar situation for the | Fe III|/Hβ ratio so that an abundance effect is excluded. More likely, the stronger | Fe II| spectrum in Kepler is due to a density effect as suggested by Danziger (this conference) and there is up to now no evidence of a large iron abundance difference between the Galaxy and LMC.

The line at 7379 has been a puzzle for a long time. Systematically seen in most SNR (Danziger and Dennefeld, 1976) it was tentatively ascribed to | Ni II| by Fesen and Kirshner (1980). However, it is only since the latest computations of atomic coefficients by Nussbaumer and Storey (1982) that this identification gains some support on the basis of a second | Ni II| line at 7412 Å detected in our spectra with an intensity compatible with the predictions for standard nebular conditions (see the discussion by Dennefeld, 1982). With this identification and the coefficients of Nussbaumer and Storey (1982) we calculate the Ni^+ /Fe^+ ionic abundance ratios to find, with very little scatter, 0.45 for the galactic objects and 0.18 for LMC. Under the reasonable assumption that so closely similar atoms as Fe and Ni should have very similar distributions in the post-shock region of a SNR, Ni^+/Fe^+ would reflect the total abundance ratio of those species and indicate a difference between the two galaxies. However, if one remembers that the cosmic abundance ratio of Ni to Fe is about 0.050 (Astrophysical Quantities) one sees that Ni^+/Fe^+ alone would already indicate a large overabundance of nickel with respect to iron.

Even if a number of people would like to see large quantities of nickel and preferably iron be produced in Supernovae, specially of type I, some of the hypotheses or parameters used might prove to be wrong before the conclusion be accepted. In order of decreasing likelihood : the ionic distributions of Ni and Fe could nevertheless be largely different, the abundance ratio could be modified by selective absorption on grains, a yet unknown mechanism might selectively enhance the | Ni II| lines, the atomic coefficients could still be wrong or the identification of the | Ni II| line still erroneous. The importance of these two elements for nucleosynthesis theories justifies some efforts to solve the question.

Finally, several filaments of the Crab Nebula have been observed in similar conditions. The surprising presence of the | C I| lines (Dennefeld and Andrillat, 1981) is confirmed in spectra which otherwise show large resemblances with those of shock-excited SNR. The strength of this | C I| is largely variable from filament to filament, its intensity going from equal or larger than | S III| 9532 to less than one tenth of the same line. The a-priori surprising presence of neutral carbon (ionization potential well below 13.6 eV) in a photoionized nebula such as the Crab could find some explanation in mechanisms as for example dielectronic-recombinations, as shown in the latest models by Péquignot (this conference) who calibrated them with the planetary Nebula NGC 7027 where | C I| is seen also (Péquignot and Dennefeld, 1982).

Ni II| lines are also seen in the Crab Nebula and analysis of Ni/Fe abundance ratio leads to the same problems as encountered above despite the different ionization mechanism (Dennefeld and Péquignot, 1982).

Finally we would like to emphasize the presence in one of the near IR spectra of the Crab Nebula of faint features attributed to high velocity filaments. The velocities are measured at +3380 and +4940 km/s respectively and are much higher than seen up to now. They are in fact much closer to what one would except from a young remnant of type II supernova.

REFERENCES

Benvenuti,P., Dopita,M. and D'Odorico,S.: 1980, Astrophys.J. 238, 601
Danziger,I.J. and Dennefeld,M.: 1976, P.A.S.P. 88, 44
Dennefeld,M. and Andrillat,Y.: 1981, Astron. Astrophys. 103, 44
Dennefeld,M.: 1982, Astron. Astrophys. 112, 215
Dennefeld,M. and Péquignot,D.: 1982, Astron. Astrophys. submitted
Dufour,R.J., Shields,G.A. and Talbot,R.J.: 1982, Astrophys.J. 252, 461
Fesen,R.A. and Kirshner,R.P.: 1980, Astrophys.J. 242, 1023
Miller,J.S.: 1968, Astrophys.J. 154, L 57
Nussbaumer,H. and Storey,P.J.: 1982, Astron. Astrophys. 110, 295
Péquignot,D. and Dennefeld,M.: 1982, Astron. Astrophys. submitted
Raymond,J.C.: 1979, Astrophy.J. Suppl. 39, 1
Shull,J.M. and Mc Kee,C.F.: 1979, Astrophys.J. 227, 131

DISCUSSION

SHULL: You indicated that |S II|/Hα ≃ 0.8 for N 63A and N 49. By ⌊S II| do you mean one or both of the doublet lines ?

DENNEFELD: In the diagram, I have always used the sum of the ⌊S II| lines 6717-6731 and the sum of the ⌊S III| lines 9069-9532 Å.

DOPITA: I was very pleased to see how your data discriminates between excitation mechanisms according to the ⌊S III|/|SII| ratio. Theoretical models predict this and also predict that a plot of |SIII|/|SII| against ⌊O III|/⌊O II| will very clearly define regions excited by shocks, thermal sources and non-thermal sources. The theoretical uncertainties in such a plot are caused principally by the uncertainty in the O^{++} + H → O^+ + H^+ charge exchange reaction. Have you plotted such a diagram, and if so, what is the result ?

DENNEFELD: Yes, I have done it. A relation exists, but the scatter is much larger than in the case of the relation |S III|/|S II| versus |S II|/ Hα. I am afraid that in the controversial cases (like bubble-type objects for example) this diagram taken alone will not make the decision because of the overlap between weak shocks and low-ionization H II regions. However a combination of several such diagrams might do.

SHULL: Concerning your request for the inclusion of Fe-lines in theoretical shock models, I would like to mention that Chris Mc Kee, Greg Seab and I have added Fe to our shock code (see poster display). The strongest lines of | Fe II| are at 1.26μ and 1.6μ, with strengths comparable to Hα.

DENNEFELD: I am glad to see the | Fe II| lines included. The strongest lines you mention are unfortunately just located at the sensitivity boundary of two very different observational techniques and might be difficult to observe.

I would also be happy to see the ⌊C I⌋, ⌊Ca II⌋ and |Ni II| lines included in model predictions for the corresponding abundance determination.

SPECTROPHOTOMETRY OF THE OPTICAL EMISSION FROM RCW103[1]

Maria Teresa Ruiz
Departamento de Astronomía, Universidad de Chile

ABSTRACT

Spectra for five different positions in RCW103 were obtained with the 1.5m and 4m telescopes of CTIO equiped with a SIT Vidicon detector.

From the observed Hα/Hβ ratio we found a variation of about 1.5 magnitudes in reddening for different filaments. The minimum value of Av found was 4.4 magnitudes implying a distance of 6.5 kpc.

Temperatures of about 10^5 K and densities of 70 cm^{-3} were found. Nitrogen is overabundant at least by a factor 3.

I INTRODUCTION

The discovery by Tuohy and Garmire (1980) of a compact X-ray source in the center of RCW103 gave special interest to the study of this object. Determination of its distance, age and possible interaction of the compact X-ray source with the gaseous filaments are very important.

Absorption measurements of HI and OH towards RCW103 give a lower limit to the distance of 3.3 kpc (Caswell et al. 1975). A distance of about 8 kpc is found from the Σ-D relation (Caswell et al. 1980). Westerlund (1969) finds a distance of 3.9 kpc assuming that the remnant is part of an OB association having 3 magnitudes of reddening. In 1982 Leibowitz and Danziger using the observed Hα/Hβ ratio for several filaments in RCW103 found a reddening of 4.5 magnitudes implying a distance of 6.6 kpc from the Sun.

II OBSERVATIONS

The spectra obtained with the 4m telescope of CTIO and the 40mm Vidicon tube covered the wavelength region between 5000 Å and

241

J. Danziger and P. Gorenstein (eds.), Supernova Remnants and their X-Ray Emission, 241–243.

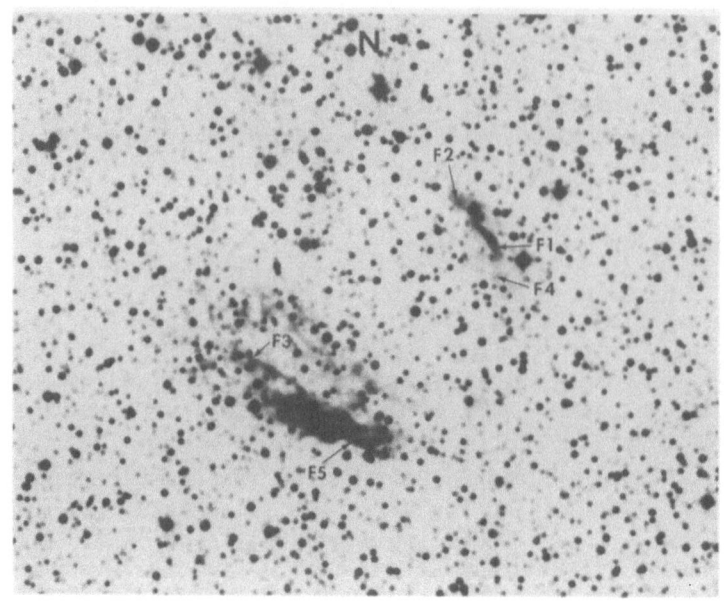

Figure 1. RCW103

6800 Å with 6 Å of resolution. At the 1.5m telescope the 16 mm Vidicon
was used covering the region between 4363 Å to 6800 Å with about 15 Å
of resolution and the regions between 3700 Å to 5100 Å and 6200 Å to
7300 Å with a resolution of about 10 Å.

        The slit was always kept E-W; its width was 1.8". The length
of the observed spectra was 13" for filaments F1, F3 and F4; 23" for F2
and 27" for F5. Figure 1 is a reproduction of RCW103 with the positions
of the observed filaments indicated.

III RESULTS

        Table 1 gives the line intensities corrected by reddening
using a normal reddening law and assuming a theoretical Hα/Hβ ratio of
3. The accuracy of the line strengths are about 15% for lines stronger
than Hβ and 30% for lines 1/2 Hβ. The values of Av for each filament
are indicated in Table 1, the difference between the Av for filament 2
and 5 must be due to internal reddening. The minimum Av found was 4.4
magnitudes indicating a distance of about 6.5 kpc from the Sun and a
galactocentric distance of about 4 kpc.

        Comparison of the observed line intensities of Table 1 with
models by Dopita (1977) and Shull and McKee (1979) indicates that the
temperatures in the filaments of RCW103 are of the order of 10^5 K, the

TABLE 1
Line Intensities in RCW103

Line	F1(Av=5.3)	F2(Av=4.4)	F3(Av=4.5)	F4(Av=5.0)	F5(Av=5.9)
[OII] 3726+3729	951				966
Hγ 4340	64				57
[OIII] 4363	73				
Hβ 4861	100	100	100		100
[OIII] 4959	87	38	185	118	136
[OIII] 5007	243	140	320	297	416
[NI] 5200	29		22		59
HeI 5876	7	21			
[OI] 6300	94	71	95	111	73
[OI] 6363	31	31	36	40	23
[NII] 6548	168	138	189	186	158
Hα 6563	300	300	300	300	300
[NII] 6584	492	394	534	555	520
[SII] 6717	115	181	158	157	136
[SII] 6730	137	197	192	204	176
HeI 7065	17	18			
[ArIII] 7135		45	31		26
[FeII] 7155	26	21			15
[CaII] 7293	43	39	12		
[OII] 7320+7330	62	105	112		74

densities about 70 cm^{-3} and there is an overabundance of N by a factor 3, S and O show smaller overabundance with respect to cosmic.

IV CONCLUSIONS

        The observed minimum value of Av=4.4 magnitudes is in agree-
ment with the Av=4.5 magnitudes found by Leibowitz and Danziger (1982)
implying a distance to RCW103 of 6.5 kpc, thus the remnant is at a galactocentric distance of about 4 kpc. The observed overabundances could
be explained as due to a galactic abundance gradient, although a small
N abundance variation between filaments seems to be present.

(1) A more detailed discussion of this work has been sent for publication to the Astronomical Journal.

REFERENCES

Caswell, J.L., Roger, A.S., Murray, J.D., Cole, D.J. and Coke, D.J.,
     1975, Astron. Astrophys., 45, p.p. 239-258.
Caswell, J.L., Haynes, R.F., Milne, D.K. and Wellington, K.J., 1980,
     M.N.R.A.S., 190, p.p. 881-889.
Dopita, M.A., 1977, Ap. J. Suppl., 33, p.p. 437-449.
Leibowitz, E.M. and Danziger, I.J., 1982, ESO preprint N°197.
Shull, J.M. and McKee, C.F., 1979, Ap. J., 277, p.p. 131-149.
Tuohy, I. and Garmire, G., 1980, Ap. J. (Letters), 239, p.p. L107-110.

NON-EQUILIBRIUM IONIZATION IN PUPPIS A

P.F. Winkler[1], C.R. Canizares[2,3], and B.C. Bromley[1]

[1]Department of Physics, Middlebury College
[2]Department of Physics and Center for Space Research
Massachusetts Institute of Technology
[3]Alfred P. Sloan Research Fellow

ABSTRACT

High resolution X-ray spectroscopy of the brightest knot of emission in the Puppis A supernova remnant shows that it is made up of ionizing plasma, far from equilibrium. Flux measurements in several X-ray lines enable us to determine the non-equilibrium conditions: electron temperature, ion populations, and time since the knot was heated by the supernova shock. Imaging and spectroscopic data from the Einstein Observatory together suggest that this knot is a cloud of density about 10 cm^{-3} which has recently been shocked to a temperature 7×10^6 K. Radio and optical data on the region appear consistent with this picture.

INTRODUCTION

A recurring theme of this meeting and of recent literature on supernova remnants (SNR) is the importance of non-equilibrium ionization (NEI) in interpreting X-ray spectra. The time scales for oxygen and heavier elements to achieve ionization equilibrium in shocked plasmas are often greater than the ages of young SNR. The X-ray spectrum, consisting primarily of lines from highly stripped ions, will be significantly different from that of an equilibrium plasma (e.g., as calculated by Raymond and Smith 1977 or Shull 1981) both because the ion populations in an NEI plasma are different from those at equilibrium, and also because in an ionizing plasma the emissivities of many lines are enhanced over their equilibrium values. These effects may reduce the extremely high heavy-element abundances that have been inferred from equilibrium-model fits to the X-ray spectra of several young SNR (Shull 1982 and references therein).

We present here the first clear measurements of NEI conditions in an SNR. These have been carried out for a bright knot of plasma located just behind the shock front in Puppis A, using data from the Focal Plane Crystal Spectrometer (FPCS) at the Einstein Observatory.

J. Danziger and P. Gorenstein (eds.), Supernova Remnants and their X-Ray Emission, 245–252.

Puppis A is extremely rich both in its X-ray line spectrum and in its spatial structure. We have already published analysis of the spectrum from the interior region of Puppis A (Winkler et al. 1981a, 1981b, 1982; Canizares and Winkler 1981; Canizares et al., this meeting). Here we will discuss new results from observations of the "bright eastern knot" of X-ray emission discussed by Petre et al. (1982 and this meeting). This knot, centered at $\alpha = 8^h 22^m 30^s$, $\delta = -42°48'$ (1950), has the highest X-ray surface brightness of any feature in Puppis A. This knot lies just inside the shock front, and both X-ray and radio maps show a steep intensity gradient separating it from the unshocked ISM to the east. Furthermore, the SNR shell appears to have been retarded where it meets the knot. These facts suggest that the knot is a recently shocked cloud with density ~10 cm^{-3}.

Several important energy bands of the spectrum from the bright eastern knot were scanned with the Einstein FPCS in a series of observations during November 1979 and June 1980. The 6' circular aperture of the FPCS, used in all these observations, defined a field which matched the size of the knot well. The composite spectrum of the eastern knot is shown in Figure 1, which may be compared with that from the interior of Puppis A in Figure 1 and 2 of Winkler et al. 1981a. In both spectra the predominant ions are O VII, O VIII, Ne IX, Ne X and Fe XVII; however, there are important differences in the relative strengths of lines from these ions, as we discuss below.

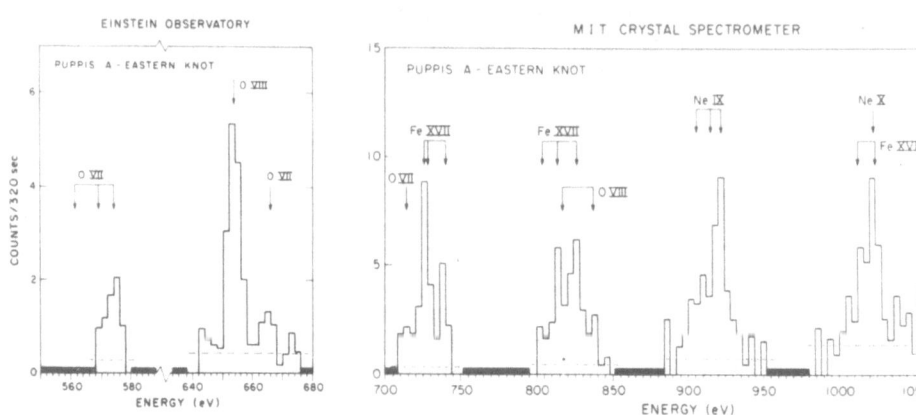

Figure 1. X-ray spectrum of the Puppis A bright eastern knot as observed with the Einstein FPCS. Count rates have not been corrected for instrumental efficiency.

PLASMA DIAGNOSTICS

Since the FPCS enables us to measure directly the flux in numerous spectral lines, we can use ratios of line strengths as diagnostics for several important properties of the emitting plasma. (See Winkler et al. 1981b, 1982 for a more complete discussion.) The flux in any line can be written as a product of several factors; e.g., for O VIII Lyman α:

$$F = \left(\frac{n_{O+7}}{n_O}\right) \times \left(\frac{n_O}{n_e}\right) \times \Omega_{L\alpha} \times e^{-E_{exc}/kT_e} \times e^{-\sigma_E N_H} \times \int n_e^2 dV \times \frac{1}{4\pi d^2}$$

$$\underset{\substack{\text{ion}\\\text{fraction}}}{} \quad \underset{\substack{\text{oxygen}\\\text{abundance}}}{} \quad \underset{\substack{\text{collision}\\\text{strength}}}{} \quad \underset{\text{temperature}}{} \quad \underset{\text{absorption}}{} \quad \underset{\substack{\text{emission}\\\text{integral}}}{} \quad \underset{\text{distance}}{}$$

Most of these factors involve quantities which are poorly known, but by taking line flux ratios we can eliminate many of the unknowns.

A. Electron Temperature and Column Density

The first diagnostic uses ratios of different lines from the same ion to determine the electron temperature T_e and the absorption column density N_H. In taking the ratio, all the factors in the flux equation cancel except for the third, fourth and fifth. If the collision strengths are known (we have used those from Shull 1981), then each ratio leads to an allowed region in the T_e-N_H parameter space.

Five different ratios can be formed from lines observed in the eastern bright knot of Puppis A, as listed in Table 1.

Table 1.   Puppis A - Eastern Knot

Ratio	Lines and Energies[a]	Observed Value
O VII BA	Heβ (666eV)/Heα (574eV)	0.45 ± 0.17
O VIII CA	Lγ (817eV)/Lα (654eV)	0.24 ± 0.08
Fe XVII BA	$2p^5 3d$(802-826eV)/$2p^5 3s$(725-739eV)	1.51 ± 0.36
Fe XVII CA	$2p^5 4d$(1011-1023eV)/$2p^5 3s$	0.19 ± 0.10
Fe XVII CB	$2p^5 4d$/$2p^5 3d$	0.29 ± 0.15

[a]See Winkler et al. (1983) for complete spectroscopic identification and line fluxes.

Some of the lines are blends only partially resolved by the FPCS, e.g., Fe XVII lines at 802, 813, and 826 eV are blended with O VIII Lγ and Lδ at 817 and 837 eV. In such cases, we have achieved deblending by fixing the relative strength of members of the same "multiplet" (e.g., O VIII Lδ relative to Lγ) at the theoretical value and varying the total strength of the different multiplets (e.g., O VIII vs. Fe XVII) relative to one another to obtain the best fit to the observed spectrum.

The five ratios yield quite a narrow region of intersection in the T_e-N_H parameter space. In Figure 2 we have plotted the 90 percent-confidence limits for all ratios except Fe XVII BA, which does not further restrict the region of intersection and is thus omitted for simplicity. The column densities allowed to the eastern knot are consistent with the range obtained by Winkler et al. (1981b) to the interior of Puppis A.

B. Ion Populations

A second diagnostic gives the relative population of different ions of the same element by comparing lines from these ions. We have done this for oxygen and neon in the Puppis A eastern knot. The ratio of the 666 eV O VII line to O VIII Lα at 654 eV indicates that

$$n_{O^{+6}}/n_{O^{+7}} = 0.5 \pm 0.2 \qquad (2a)$$

For neon the ratio of similar ions gives

$$n_{Ne^{+8}}/n_{Ne^{+9}} = 2.0^{+1.5}_{-1.0} \qquad (2b)$$

For a plasma at equilibrium any ion population ratio measures the temperature, but the oxygen and neon ratios lead to different temperatures 2.4×10^6 K and 3.2×10^6 K, respectively. Thus, these populations are incompatible with a plasma at equilibrium.

We can investigate NEI conditions by considering the transient ionization populations of oxygen and neon behind a shock. In a simple model we have assumed the sudden heating of electrons in a plasma cloud to a temperature T_e as a shock passes. The ion populations are then calculated as a function of time after the shock has passed for different values of T_e. It is convenient to use the parameter $n_e t$ as the independent variable since the ionization rate equations then become independent of electron density n_e.

Each measurement of an ion population ratio leads to an allowed band in the parameter space of T_e vs. $n_e t$, as shown in Figure 3. For our measurements of the oxygen and neon ratios (Eq. 2), the bands do not overlap at times long enough to be near equilibrium which reflects the different ionization temperatures. However, in a rapidly ionizing plasma at $T_e \gtrsim 6 \times 10^6$ K oxygen ionizes enough faster than neon to achieve the observed ionization structure in a region indicated in the figure. The shock passage must have been relatively recent: $n_e t \lesssim 10^3$ cm^{-3}yr.

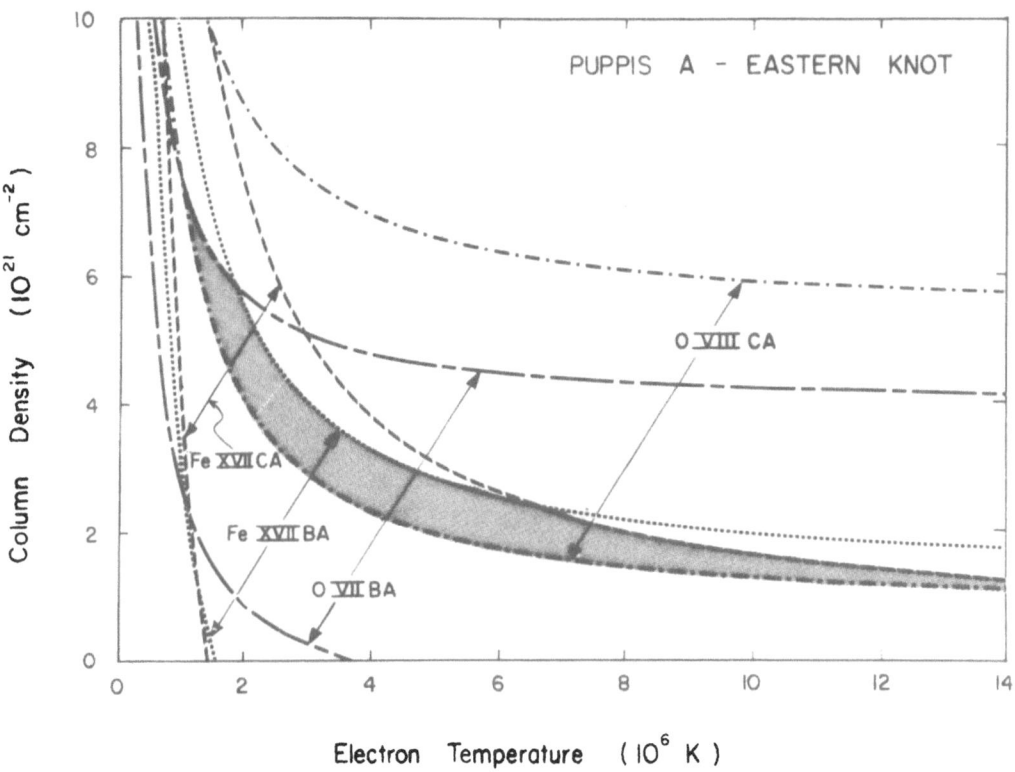

Figure 2. Allowed regions (90% confidence) in $T_e - N_H$ parameter space from four independent line ratios. Shaded area indicates region of overlap.

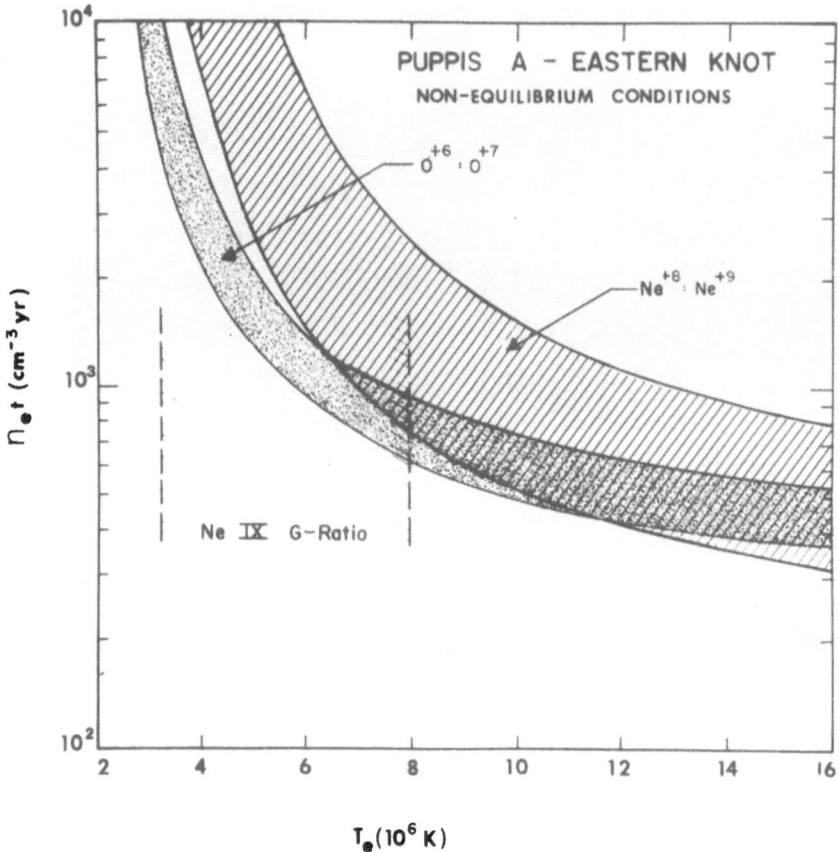

Figure 3.   Post-shock transient ionization conditions in bright
eastern knot of Puppis A as determined from O and Ne ion populations.
Only small triangle at $T_e \simeq 7 \times 10^6$K, $n_e t \simeq 10^3$ cm^{-3}yr is consistent
with all the data.

C. He-like Triplets

A third diagnostic that enables us to restrict further the NEI conditions is the ratio G between different members of the triplet of lines from helium-like ions (Canizares et al., this meeting, and references therein). From the eastern knot in Puppis A we observe G = 0.7 $\pm$ 0.2 for the Ne IX triplet. The calculations of Pradhan (1982) show that for an ionizing plasma (no recombination) an electron temperature T_e = 3 - 8 x 10^6 K would give G values for Ne IX consistent with our observations. (See Figure 1 of Canizares et al., this meeting, for details of the same method with the O VII G-ratio, which was not measured in the eastern knot.)

CONCLUSIONS

We can now combine information from all three diagnostics not only to conclude that the eastern bright knot of Puppis A has not reached ionization equilibrium, but also to obtain a relatively precise measurement of the NEI conditions there. When the restriction on T_e from the G-ratio is added to Figure 3, only a very small region of parameter space is allowed: T_e = 7$\pm$1 x 10^6 K, $n_e t$ = 1.0 $\pm$ 0.2 x 10^3 cm^{-3} yr. Furthermore, the narrow range for T_e restricts the allowed column density in Figure 2 to values $N_H \simeq$ 2 x 10^{21} cm^{-2}. The X-ray spectra indicate that the eastern bright knot in Puppis A has been shocked in the relatively recent past, that the electron temperature is high, and that rapid ionization is still taking place.

The measurement of NEI conditions in the eastern knot is a nice complement to the conclusion by Petre et al. (1982 and this meeting) that the knot is a recently shocked cloud of density $n_e \simeq$ 10 cm^{-3}. If the electron temperature results from shock heating, a shock velocity $v_s \simeq$ 700 km s^{-1} is required. Since the NEI conditions indicate that $n_e t \simeq 10^3 cm^{-3}$ yr, a plasma with $n_e \simeq$ 10 cm^{-3} must have been shocked about 10^2 years ago. The shock would have since traversed a characteristic distance $v_s t \simeq$ 0.1 pc, comparable to the sizes of bright features in the eastern knot.

We would like to thank Roger Chevalier, Tom Markert, Robert Petre, Anil Pradhan, Michael Shull, and Andrew Szymkowiak for stimulating discussions, at this symposium and elsewhere, on the material presented here. This research has been supported in part by NASA through contract NAS-8-30752 and grant NAG-8389, and by Research Corporation.

REFERENCES

Canizares, C.R., and Winkler, P.F. 1981, Ap. J. (Letters),
    <u>246</u>, L33.

Petre, R., Canizares, C.R., Kriss, G.A., and Winkler, P.F.
    1981, Ap. J., 258, 22.
Pradhan, A.K. 1982, Ap. J., 263, 477.
Raymond, J.C., and Smith, B.W. 1977, Ap. J. Suppl., 35, 419.
Shull, J.M. 1981, Ap. J. Suppl., 46, 27.
Shull, J.M. 1982, Ap. J., 262, 308.
Winkler, P.F., Canizares, C.R., Clark, G.W., Markert, T.H.,
    Kalata, K., and Schnopper, H.W. 1981a, Ap. J. (Letters),
    246, L27.
Winkler, P.F., Canizares, C.R., Clark, G.W., Markert, T.H., and
    Petre, R. 1981b, Ap. J., 245, 574.
Winkler, P.F., Canizares, C.R., Markert, T.H., and Szymkowiak,
    A.E. 1982, in Supernovae: A Survey of Current
    Research, Proceedings of the NATO Advanced Study
    Institute held at Cambridge, UK, June 29-July 10, 1981,
    M.J. Rees and R.J. Stoneham, eds. (Dordrecht: D. Reidel),
    501.

DISCUSSION

RAYMOND: Did you assume a single value of $n_e t$ for the entire feature?

WINKLER: Yes. Clearly this model is over-simplified, and we intend to refine it by averaging over material at different ages and distances behind the shock.

MURDIN: This spot is concident with Filament 37 in Puppis A which has the strongest [Fe XIV] line I've seen in an SNR, with strong spatial stratification of the ionic species. I think this fits nicely with your model of a shock wave encountering a blob of material.

WINKLER: Quite so. It's especially interesting that the optical [Fe XIV] filaments as mapped by Dopita et al. match the X-ray structure of the eastern knot in detail. This may enable us to model the iron ionization structure.

DETAILED X-RAY OBSERVATIONS OF THE CYGNUS LOOP

W. H.-M. Ku, K. Long, R. Pisarski, and M. Vartanian
Columbia Astrophysics Laboratory, Columbia University

ABSTRACT

High quality X-ray spectral and imaging observations of the Cygnus Loop have been obtained with three different instruments. The High Resolution Imager (HRI) on the Einstein Observatory was used to obtain arcsecond resolution images of select bright regions in the Cygnus Loop which permit detailed comparisons between the X-ray, optical, and radio structure of the Loop. The Imaging Proportional Counter (IPC) on the Einstein Observatory was used to obtain an arcminute resolution map of essentially the full Loop structure. Finally, an Imaging Gas Scintillation Proportional Counter (IGSPC), carried aloft by a sounding rocket last fall, obtained modest resolution, spatially resolved spectrophotometry of the Cygnus Loop. An X-ray map of the Loop in the energy of the O VIII line was obtained. These data combine to yield a very powerful probe of the abundance, temperature, and density distribution of material in the supernova remnant, and in the interstellar medium.

1. INTRODUCTION

The Cygnus Loop is one of the most well-studied supernova remnants (SNR) in our galaxy. Its close distance (770 pc) and large size (~3°) make the Cygnus Loop an ideal candidate for detailed studies. Observations a decade ago by Gorenstein et al. (1971), and Stevens and Garmire (1973) showed that the remnant was a copious source of subkilovolt X-rays whose spectrum was consistent with thermal emission from a thin hot plasma at a temperature of 3 million degrees. These observations also suggested the presence of oxygen line emission near 0.6 keV. A high resolution search with a Bragg crystal spectrometer (Stark and Culhane 1978) led to only an upper limit on the O VIII emission of 6% of the total X-ray flux. More recently Inoue et al. (1980) and Kahn et al. (1980) presented new evidence for the presence of oxygen line emission from the Cygnus Loop. These data, in addition to the detection of 5303 Å Fe XIV coronal line by Woodgate et al. 1974 constitute strong evidence for the presence of hot plasma.

The spatial structure of the Cygnus Loop has been mapped by several one-dimensional imaging systems (e.g., Stevens and Garmire

253

1973, Rappaport et al. 1974), and small aperature two-dimensional nonimaging detectors (Charles et al. 1975, Gronenschild et al. 1976) and has been found to be consistent in size with the 3° optical diameter of the remnant. The first true two-dimensional X-ray image of the Cygnus Loop was obtained by Rappaport et al. (1979) with a multiwire proportional counter (MWPC) at the focal plane of a grazing incidence telescope on a sounding rocket. Their picture suggests that the X-ray emission arises from a limb-brightened shell of hot gas which resulted from the expansion of a blast wave into an inhomogenous interstellar medium (ISM). Unfortunately their observation had only limited useful spectral information (Kayat et al. 1981).

2. INSTRUMENT AND OBSERVATIONS

We describe here observations of the Cygnus Loop made with three different X-ray imaging systems over the past three years: (1) An imaging gas scintillation proportional counter (IGSPC) with solid state spectrometer-type spectral resolution below a 1 keV and 12' spatial resolution carried aloft by a sounding rocket; (2) An imaging proportional counter (IPC) with modest spectral resolution (100% @ 1.5 keV) and 1'-2' spatial resolution operated on the Einstein Observatory; (3) A microchannel plate (HRI) with no spectral information but 4" spatial resolution operated on the Einstein Observatory. Table 1 summarizes the properties of the three instruments at 654 eV.

TABLE 1
Instrument Parameters

Instruments	F.O.V.	(Performance @ 654 eV)		
		$\Delta E(eV)$ (FWHM)	ΔS (FWHM)	B $c/ks/(mm)^2$
IGSPC/Rocket	150'	177	~12'	0.06
IPC/Einstein	60'	~1000	2'	0.5
HRI/Einstein	25'	------	4"	4.0

A Nike boosted Black Brant sounding rocket carrying an IGSPC at the focal plane of a grazing incidence telescope was launched on 17 November, 1981. The rocket pointed at the W limb of the Cygnus Loop and then scanned over to the NE limb- holding at each position for 120 s. Fig. 1 shows a copy of a POSS picture of the Cygnus Loop overlaid with the circular field of view of the IGSPC. The positions of the millimeter wide window support ribs relative to the centers of the two IGSPC pointings are sketched.

Fig. 2 shows a merged image of the two pointings corrected for exposure but uncorrected for ribs and vignetting. The log intensity scale in arbitrary units is shown to the right. The spatial scale is shown on the bottom. Large scale X-ray features observed by several groups prior to <u>Einstein</u> are clearly seen in Fig. 2. The bright NE, N, and W limbs are seen corresponding roughly to similar bright features in the optical and radio regimes.

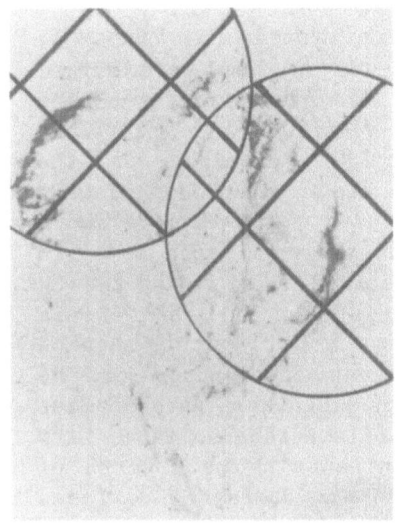

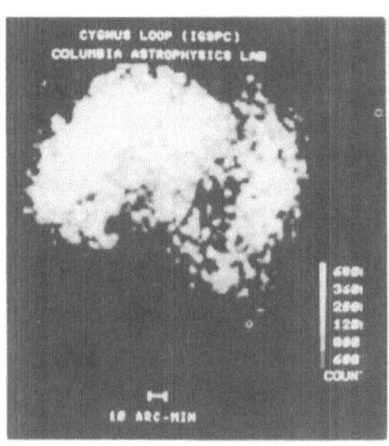

Fig. 1 - Overlay of the 150' IGSPC field of view on the POSS print of the Cygnus Loop.

Fig. 2 - The total spatial data from the two IGSPC pointings have been combined to form one image.

An arcminute resolution image of the total 0.1-4.5 keV X-ray emission observed by the IPC is plotted on top of a POSS print in Fig. 3. The composite image is formed by merging the central 35' x 35' region of 59 overlapping IPC fields (one field near center north is missing). Over all, the X-ray bright shell demonstrates remarkable symmetry after 18,000 years of expansion. The X-ray emission is centered on R.A. (1950.0) = $20^h49^m15^s$ and Dec (1950.0) = 30°51'30", with a radius of 84'. Except for the breakout near the south (away from the galactic plane), the shell is circular to within ±5'. The contrast seen is very high with the brightest feature on the NE and W limb ~300 times brighter than the diffuse X-ray background and ~100 times brighter than the faintest region within the circular shell. Faint X-ray emission ($\lesssim$1000 times the peak surface brightness) extends beyond the shell on the W and SW and extends ~25' beyond the bottom of this image (to δ = 28° 30'). Radial surface brightness profiles confirm this and also reveal considerable (>3:1) limb brightening - favoring the presence of adiabatic (rather than isothermal) blast waves (Solinger, Rappaport, and Buff 1975). IPC images formed in several different energy bands were examined and found to be extremely similar in general outline although, as we shall see later, spectral variations across the remnant do exist. Two point sources, ~ 10^{-3} times the total intensity of the Loop, are observed within the X-ray shell at α = $20^h46^m24^s$, δ = 31°31'43" and α

= $20^h43^m35^s$, δ = 30°32'20". Eleven other point sources- stars or
active galaxies- are seen just outside the shell. No point sources
are visible near the center of the remnant although enhanced emission
is seen 7' to the NW and NE of the center near enhanced optical emission. Any neutron star remnant must be colder than 9 × 10^5 K.

The brightest regions of the loop were mapped with the HRI.
Five pointings on the W limb and twelve on the NE limb were merged
together to form the image shown in Fig. 4. All of the features seen
in the IPC are reproduced clearly. In addition, several finer
features may be noted when the images are examined on the
subarcminute scale. One sees that while the shock is strong and well
defined in some regions, it is diffuse in others. The NE limb shows
a bright ~8' wide barlike feature with excess X-ray emission 3'-4' in
front ending in a fainter but still well-defined front. On the SW
limb, a very strong bow-shaped shock front is seen. A break across
the main shock front with a new shock forming ~5' in front is seen on
the W limb. The position of this new front corresponds to the
location of a faint optical filament found to exhibit a "pure" Balmer
spectrum, similar to the spectra of optical filaments associated with
the much younger Tycho SNR and in sharp contrast to the spectra of
the bright optical filaments in Cygnus which are dominated by lines
of [OIII]. Our observations support the suggestion of Raymond et al.
(1980) that these lines result from the collisional excitation of
cold ISM material swept up by a fast shock. The close
identification of this X-ray feature with the optical filament is
repeated for the Hβ + Hα + [NII] filament noted by Gull et al.
(1977) on the northern boundary of the Loop and the nonradiative
shock outside the NE limb studied by Raymond (this volume). Other
optically bright filaments are seen to
correspond to regions of enhanced X-ray emission in general but not
in detail. For example the bow shaped shock front has no bright
optical counterpart, while a bright optical filament on the NE limb
actually extends past the boundary of the X-ray emission. The
identification of X-ray features with radio emission is also generally good but poor in detail. We have compared the X-ray emission
with the 2.7 GHz map of Keen et al. (1974). The X-ray and optically
bright NE and W limb are seen clearly in the radio. Faint radio
emission is seen all around ~5' inside the X-ray emission except in
the SE quadrant. Weak radio emission is seen outside the W limb
similar to the optical and X-ray images. In the S and SW, the
breakout is seen more strongly in the radio than in the X-ray. The E
edge of the breakout is strong in all bands while the W edge is only
strong in the radio.

While the HRI provides the greatest detail on the spatial structure of the Cygnus Loop, it has no spectral information. The IPC has
some spectral capabilities which are just now being exploited. Temporal and spatial variations in the IPC detector gain have hampered
the analysis of the pulse height information for most sources. However the Cygnus Loop IPC observations benefit from several features
which improve the reliability of the spectral parameter estima-

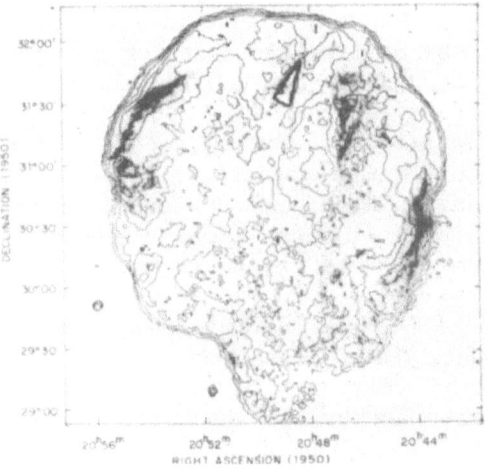 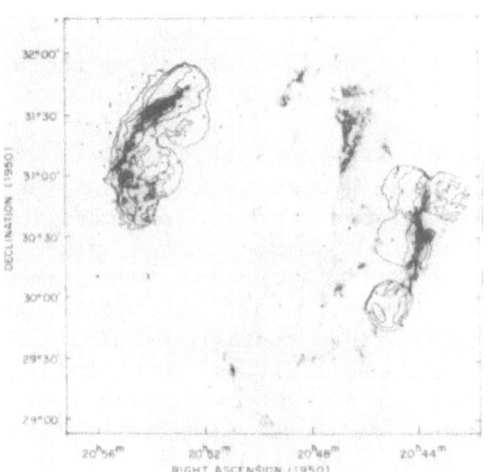

Fig. 3 - The central 35' × 35', 0.1-4.5 keV data from 59 IPC <u>Einstein</u> pointings merged in an exposure corrected manner and overlaid on a POSS image of the Cygnus Loop. Contours are set at equal log intervals (0.3).

Fig. 4 - The central 25' diameter data from 19 HRI <u>Einstein</u> pointings merged in an exposure corrected manner and overlaid on a POSS image of the Cygnus Loop. Contours are set at equal log intervals (0.3).

tions. 1) Because of the considerable overlap among observations, we can limit ourselves in most cases to examining the central IPC regions where the IPC's response is better mapped; for interesting off-axis regions (>20') we can check the spectra obtained in >2 pointings against one another. 2) Because the observations were often done alternately inside and outside the shell at near coincident time with similar gain and background conditions, we can use the outside fields for background subtraction. Using 6' radius cells, we summed all the IPC counts in a source image, subtracted the exposed normalized counts from the same counter region in a background field and fitted the spectra to a Raymond and Smith (1979) model with cosmic abundances. Both N_H the hydrogen column density and T the plasma temperature, were allowed to vary although due to the poor resolution of the IPC at low energies and the strong anticorrelation between N_H and T there is little leverage on the N_H parameter so we limited its variation to $1 - 10 \times 10^{20}$ atoms cm^{-2} (accomodating all X-ray values, and E(B-V) $\simeq 0.05 - 0.15$ ($2 - 8 \times 10^{20}$) reddening measurements (Raymond <u>et al.</u> 1981)). We also added an estimated 5% residual systematic error to the data before minimizing χ^2 measures of the fit of models to the data. A typical fit to the data is shown in Fig. 5. Most of the flux at the best fit temperature of 2.1×10^6 K is due to line emission from OVII, OVIII, NVI, NVII, and CVI with contributions from FeXVII. To simplify the interpretation of possible spectral gradients, we fixed log N_H at 20.6. When this was

done, the best fit temperature along the limb is found to be generally lower than that toward the center. The temperature is somewhat
higher in the NE compared to the W. The derived temperatures generally anticorrelate with intensity consistent with the relationship
for pressure equilibrium. Limb spectra are usually well fit while
central spectra have poor χ^2 suggesting the presence of
multitemperature components or nonequilibrium conditions. These
effects are in fact, expected to be more pronounced toward the
center. When N_H was allowed to vary between
$1 - 10 \times 10^{20}$ atoms cm^2 the temperature varied by a factor of 2
from 2.0-4.3×10^6 K- within the range of all previous X-ray
measurements. Comparisons of the IPC to HRI surface brightness over
the NE and W limbs yield ratios which are consistent with the
temperatures and column densities found from the IPC data. Some
evidence for moderate spectral variations across the shock front may
be seen. While IPC spectral fitting and comparisons of IPC/HRI
brightness ratios can yield useful estimates of the temperature variation across the remnant they do not permit detailed studies of the
distribution of temperature, density, and abundance gradients in the
Loop. The IGSPC provides better diagnostics. Fig. 6 shows the total
spectrum obtained by the IGSPC for the NE limb. The measured spectrum is markedly different from that obtained by the IPC and reflects
the presence of absorption near 0.28 keV by the carbon in the 2μ
polypropylene window used (unresolved by the IPC), line emission from
oxygen near 0.65 keV, and mirror cutoff above 1.5 keV. The fit (histogram) does not include detailed postflight calibration done recently at MSFC and is therefore preliminary. Nevertheless a
2×10^6 K plasma with cosmic abundance and log N_H = 20.65 reproduces
the general features of the spectrum quite well. The presence of
line emission near 0.65 keV is absolutely required to fit the data.
The W limb shows qualitatively the same features.

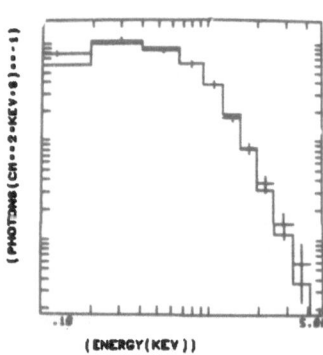

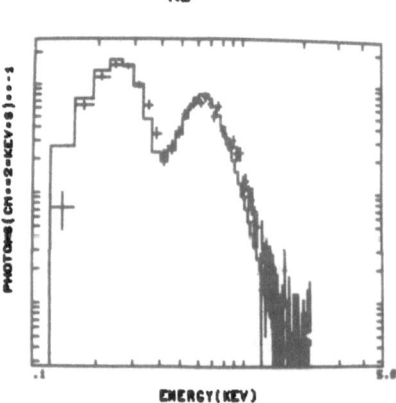

Fig. 5 - Typical IPC pulse
height spectrum of photons collected in 12' diameter circle
centered on α = $20^h44^m10^s$, δ =
29°41'.

Fig. 6 - Typical IGSPC pulse
height spectrum of photons collected in 150' diameter circle
centered on α = $20^h52^m10^s$, δ =
31°50'.

3. SUMMARY AND CONCLUSIONS

New X-ray observations of the Cygnus Loop clearly reveal the remarkable symmetry of the expanding SNR blast wave. Observations of strong limb brightening and a hotter interior suggest that the remnant is adiabatic, near the end of the Sedov stage, and approaching the radiative stage of evolution. Table 2 summarizes those measured and derived parameters relevant to the Sedov model as applied to the whole remnant. Despite the large scale uniformity, it is

TABLE 2
Cygnus Loop: Sedov Model

Distance:	$d = 770$ pc
Radius:	$R_x = 84' = 18.8$ pc (± 1 pc)
Temperature:	$T_x = 3 \times 10^6$ K
Limb temperature:	$T_s = 2 \times 10^6$ K
X-ray luminosity	$L_x = 9 \times 10^{35}$ ergs s^{-1} (0.1-4.0 keV)
ISM density:	$N_H = 4 \times 10^{20}$ atoms cm^{-2}; $n_H = 0.16$ cm^{-3}
Shock velocity:	$V_s = 400$ km s^{-1}
Initial SN energy:	$F_0 = 3 \times 10^{50}$ ergs
Age:	$t = 18,000$ yrs
Total mass:	$M = 100$ M$_\odot$

also clear that a range of different temperatures, densities, and abundances apply to different regions of the remnant. We believe these differences reflect differences in the ISM. The Cygnus Loop probably sits in a low density 0.1 cm^{-3}, warm ~10^4 K ISM. This ISM contains many cold $\lesssim 10^3$ K , ~1-10 pc clouds with densities $\gtrsim 10$ cm^{-3} which have been shocked by the blast wave. The compressed gas and magnetic field yield the radio synchrotron emission. Material heated and compressed by the passing shock cool and radiate in the optical. The general coincidence of X-ray, optical, and radio peaks suggest that evaporative heating of these clouds in the hot postshock region may be occurring. The breakout in the south may reflect the shock front encountering a hot, $\gtrsim 10^5$ K low density $\lesssim 0.01$ cm^{-3}, ~10 pc "tunnel" in the ISM. Further work on detailed studies of the temperature and density variations in the Loop and detailed comparisons between X-ray, optical, and radio fine structure may clarify the structure of the ISM.

However, it is already clear from these data that X-ray observations can provide an extremely detailed look at the ISM. In particular, spatially resolved spectrophotometry with the IGSPC on AXAF should provide an excellent diagnostic of the SN shock and the ISM.

The authors gratefully acknowledge the support of the National Aeronautics and Space Administration under Contract NAS 8-30753. This is Columbia Astrophysics Laboratory Contribution No. 234.

4. REFERENCES

Charles, P. A., Culhane, J. L., Zarnecki, J. C. 1975, *Astrophys. J. (Letters)*, 196, p. L19.

Gorenstein, P., Harris, B., Gursky, H., Giacconi, R., Novick, R., and Van den Bout, P. 1971, *Science*, 172, p. 369.

Gronenschild, E.H.B.M., Mewe, R., Heise, J., Brinkman, A. C., den Boggende, A.J.F., and Schrijver, J. 1976, *Astron. Astrophys.* 49, p. 153.

Gull, T. R., Parker, R.A.R., and Kirschner, R.P. 1977, in *Supernovae*, ed. D. N. Schramm (Dordrecht: Reidel), p. 71.

Inoue, H., Koyama, K., Matsuoka, M., Ohashi, T., Tanaka, Y., and Tsunemi, H. 1980, *Astrophys. J.*, 238, p. 886.

Kahn, S.M., Charles, P.A., Bowyer, S., and Blissett, R.J. 1980, *Astrophys. J.*, 242, p. L19.

Kayat, M.A., Rolf, D.P., Smith, G.C., and Willingale, R. 1980, *Mon. Not. R. Astro. Soc.*, 191, 729-737

Keen, N.H., Wilson, W.E., Haslam, C.G.T., Graham, D.A., and Thomasson, P. 1973, *Astron. Astrophys.*, 28, p. 197.

Rappaport, S., Doxsey, R., Solinger, A., Borken, R. 1974, *Astrophys. J.*, 194, p. 329.

Rappaport, S., Petre, R., Kayat, M. A., Evans, K. D., Smith, G. C., and Levine, A. 1979, *Astrophys. J.*, 227, p. 285.

Raymond, J. C. and Smith, B. W. 1977, *Astrophys. J. Suppl.*, 35, 419.

Raymond, J. C., Davis, M., Gull, T. R., and Parker, R.A.R. 1980, *Astrophys. J.*, 238, p. L21.

Solinger, A., Rappaport, S., and Buff, J. 1975, *Astrophys. J.*, 201, p. 381.

Stark, J.P.W., and Culhane, J. L. 1978, *Mon. Not. R. Astr. Soc.*, 184, 509.

Stevens, J. C., and Garmire, G. P. 1973, *Astrophys. J. (Letters)*, 180, L19.

Weisskopf, M. C., Helava, H., and Wolff, R. S. 1974, *Astrophys. J. (Letters)*, 194, L71.

Woodgate, B. E., Stockman, H. S., Angel, J.R.P., and Kirschner, R. P. 1974, *Astrophys. J. (Letters)*, 188, L79.

A THEORETICAL MODEL OF THE CYGNUS LOOP AND ITS X-RAY EMISSION

S.A.E.G. Falle and A.R. Garlick
Department of Applied Mathematics
University of Leeds,
Leeds, England

ABSTRACT

In this paper we present a theoretical model of the Cygnus Loop which, with only a small number of free parameters, accounts for most of the gross features of the observations. We believe that none of the models described in the literature can explain the observations in such a simple way.

OBSERVATIONS

Bright Filaments

The Cygnus Loop consists of a number of bright filaments arranged in an incomplete ring on the sky. There also are some filaments both outside and inside the ring. These filaments have been studied by a number of authors, in particular:--

a) Hubble (1937) looked at the proper motion of the filaments. He found a proper motion of 0.03" yr^{-1}. If this is combined with the radius of the remnant, then we find a 'Hubble' age $t_H = \theta/\dot{\theta} = 150,000$ yr. Since the remnant has certainly been slowed down by interaction with interstellar material, the Hubble age represents an upper limit to the age.

b) Minkowski (1958) measured the radial velocity of some of the filaments. He found an expansion velocity ~116 km s^{-1}. If this is combined with a), then we find that the distance is 770 pc. This distance is somewhat uncertain since the radial velocity and proper motion do not refer to the same filaments. There is then the implicit assumption that the bright filaments constitute a roughly spherical expanding shell and that the proper motion measures the material velocity of the shell.

J. Danziger and P. Gorenstein (eds.), Supernova Remnants and their X-Ray Emission, 261–266.
© *1983 by the IAU.*

c) Miller (1974) made photoelectric measurements of the line and continuum intensities in the filaments. He found an electron density of ~200 cm^{-3} at a temperature of ~12000 K. This means that the pressure is ~6.6 x 10^{-10} c.g.s. If the shock velocity is 116 km s^{-1}, then the ambient density must be ~5 x 10^{-24} gm cm^{-3}. He also deduced from the H$_\beta$ intensity that the filaments are about 10 times deeper than their width, i.e. they are sheets seen edge on.

X-Ray Emission

a) Rappaport et al. (1974) found that although the X-ray emission correlates quite well with the optical, there is some X-ray emission outside the boundaries of the optical remnant.

The X-ray luminosity is 2 x 10^{36} erg s^{-1} in the band 0.15-1.5 keV. If the spectrum is fitted to a thermal spectrum, then the temperature is 2.9 $\pm$ 1.5 x 10^6 K.

b) Rappaport et al. (1979) made scans of the X-ray intensity in the North-South and East-West directions. They found that there is less limb brightening in the South than in other parts of the shell. Note that the optical shell also appears less complete in the South.

Fast H$_\alpha$ Emission

Kirschner and Taylor (1976) have used the H$_\alpha$ emission to measure the radial velocities in the remnant. They found radial velocities in the range +200 to -300 km s^{-1}, and that the high (S$_\alpha$ ~ 10^{-4}-10^{-5} erg cm^{-2} s^{-1} sr^{-1}) surface brightness features all have low velocities (<+56 km s^{-1}). The high velocity features have low surface brightness (S$_\alpha$ < 3 x 10^{-6} erg cm^{-2} s^{-1} sr^{-1}).

Radio Observations

a) Moffat (1971) has observed the Cygnus Loop at 1420 MHz with the Cambridge half mile telescope. The radio and optical emission correlate quite well except in the South, where there is a considerable amount of radio emission outside the boundary of the optical remnant. The spectrum is non-thermal with an index of ~0.45. Polarisation measurements show that the magnetic field is generally aligned with the optical filaments. This suggests that the radio emission is due to compression of the interstellar medium (van der Laan 1962).

b) De Noyer (1975) made 21 cm observations of the remnant with the NRAO telescope with a velocity resolution of 3.3 km s^{-1}. She was looking for the expanding HI shell that should exist if the remnant is in the radiative phase. She found that the surface density σ of such a shell must satisfy

$$\sigma \leq 2.7 \times 10^{18}\Delta V \text{ gm cm}^{-2}$$

where ΔV is the velocity dispersion in the shell (in km s^{-1}). If the remnant is in the radiative phase, we would expect $\sigma \sim 2 \times 10^{19}$ gm cm^{-2}. However, if the ambient density is slightly non-uniform, then there would not be a single shell, but a number of shells with velocities differing by ~ 10 km s^{-1}. The effective velocity dispersion would then be 10 km s^{-1}, and these observations would not preclude the remnant being in the radiative phase.

THEORETICAL MODEL

We shall now show that the observations just described, as well as other features of the remnant, can be explained if the interstellar medium has a steep density gradient. For simplicity we will consider a density discontinuity, but the results would not be greatly altered if the density changes more gradually. All that is required is that the density scale height be somewhat less than the present size of the remnant. The picture is as shown in figure 1.

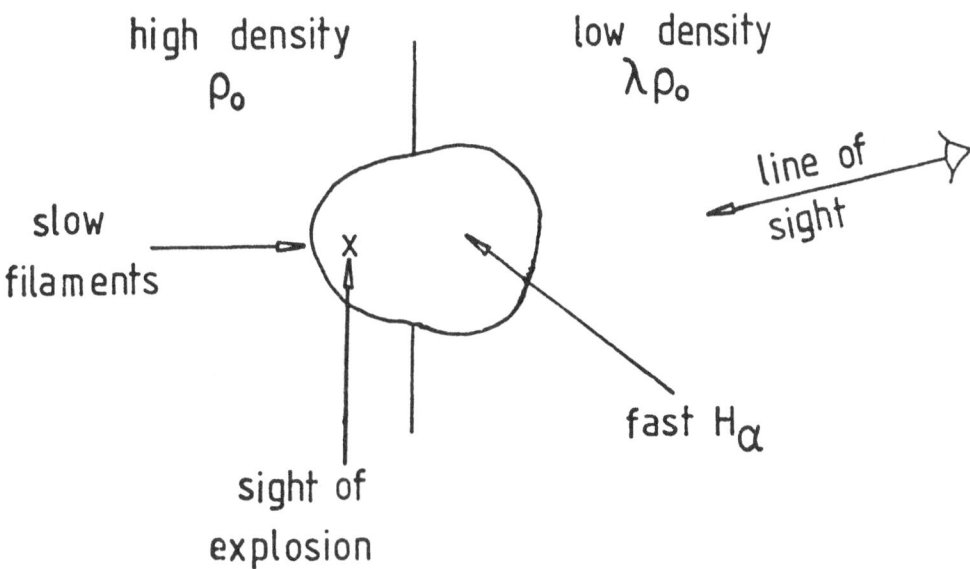

a) 'Breakout' leads to high velocities.

b) The fast H$_\alpha$ emission comes from originally dense gas which has been accelerated.

c) The X-ray emission from the shocked low density gas extends beyond the optical remnant.

d) The slow filaments and the soft X-rays are approximately as for a spherical remnant in the dense gas.

e) The orientation explains the incomplete optical shell and the X-ray limb brightening.

CALCULATION OF THE THEORETICAL PARAMETERS

We need to determine:—
  1) The energy of the explosion E_0
  2) High ambient density ρ_0
  3) The density ratio λ
  4) The distance of the explosion from
     the contact discontinuity d
  5) The age of the remnant t

From the observations we have:—

  1) Radius $R \sim 18$ pc
  2) Slow filament velocity $v_s \sim 100$ km s^{-1}
  3) Fast H_α velocity $v_f \sim 300$ km s^{-1}
  4) Pressure in the slow filaments $p_s \sim 6.6 \times 10^{-10}$ c.g.s.
  5) Fast H_α surface brightness $S_\alpha \sim 3 \times 10^{-6}$ erg cm^{-2}
                                                 sr^{-1} s^{-1}
  6) X-ray luminosity (0.15–1.5 keV) $L_x \sim 2 \times 10^{36}$ erg s^{-1}
  7) X-ray temperature $T_x \sim 2.9 \pm 1.5 \times 10^6$ K

We also know that:—

8) The radius of the remnant in the dense gas is about the same as that in the diffuse gas, since if it were smaller there would be no X-ray emission outside the optical remnant, while if it were much larger the X-ray emission would extend considerably beyond the boundaries of the optical remnant.

9) The X-ray emission from the diffuse gas is about the same as from the dense gas. This is because the X-ray intensity just outside the optical remnant, which comes from the diffuse gas, is about the same as that inside.

To determine E_0, ρ_0, λ, d and t we proceed as follows. From the pressure p_s in the bright filaments we find $\rho_0 = 5 \times 10^{-24}$ gm cm^{-3}. We then do some adiabatic axisymmetric numerical calculations with different values of λ, which is the only dimensionless parameter in such a calculation. Let t_b be the time at which the shock reaches the contact discontinuity, i.e. the breakout time. From the fact that the remnant must be the same size on both sides of the discontinuity, we find that $t/t_b \sim 9$ and from v_f/v_s we get $\lambda \sim 0.05$. From the observed radius we then get $d \sim 8$ pc.

It is not a bad approximation to use an adiabatic calculation for this as we shall find that cooling does not become important until after breakout. Cooling does then not have much of an effect on the dynamics of the gas which has broken out.

Up to the times which we consider, it can be shown that the behaviour of the remnant in the dense gas is the same as for a spherical remnant. We therefore carry out a spherically symmetric numerical calculation (Falle 1981) with cooling. This is done with different values of E_0 and with $\rho_0 = 5 \times 10^{-24}$ gm cm^{-2}. We then find the time at which the material velocity is equal to v_s. Then choose E_0 such that the radius and X-ray luminosity match the observed values. Note that since we have only one free parameter E_0 and two observables R and L_x, the fact that these two values can be matched provides a test of the validity of the model.

Finally we calculate the H_α surface brightness of the broken out material which is cooling now. It is this material which in our model produces the faint fast H_α emission.

The results of all this are:—

Parameter	Theoretical	Observed
E_0	7×10^{50} erg	
ρ_0	5×10^{-24} gm cm^{-3}	
λ	0.05	
t	7×10^4 yr	
t_b	7.4×10^3 yr	
t_c (cooling time)	1.4×10^4 yr	
v_s	100 km s^{-1}	100 km s^{-1}
v_f	300 km s^{-1}	300 km s^{-1}
L_x	2×10^{36} erg s^{-1}	2×10^{36} erg s^{-1}
p_s	6.6×10^{-10} c.g.s.	6.6×10^{-10} c.g.s.
R	20 pc	18 pc
T_x	Dense gas 10^6 K	
	Diffuse gas 2.4×10^6 K	$2.9 \pm 1.5 \times 10^6$ K
S_α	$\langle 5 \times 10^{-6}$ c.g.s.	$\langle 3 \times 10^{-6}$ c.g.s.

It can be seen that the observed and theoretical values agree very well. Furthermore there are more observed values than there are free parameters in the theoretical model, so that this agreement is a genuine test of the model. This is in contrast to other models which either have a large number of free parameters, or fail to account for all of the observations.

REFERENCES

De Noyer, L.K. 1975, Astrophys. J., 196, 479.
Falle, S.A.E.G. 1981, Mon. Not. R. astr. Soc., 195, 1011.
Hubble, E.P. 1937, Carnegie Yrb., No. 36, p. 189.
Kirschner, R.P. and Taylor, K. 1976, Astrophys. J., 208, L83.
Laan, H. van der, 1962, Mon. Not. R. astr. Soc., 124, 127.
Miller, J.S. 1974, Astrophys. J., 189, 239.
Minkowski, R. 1958, Rev. Mod. Phys., 30, 1048.
Moffat, P.H. 1971, Mon. Not. R. astr. Soc., 153, 401.

Rappaport, S., Doxsey, R., Solinger, A., and Borken, R.   1974,
    Astrophys. J., <u>194</u>, 329.
Rappaport, S., Petre, R., Kayat, M.A., Evans, K.D., Smith, G.C., and
    Levine, A. 1979, Astrophys. J., <u>227</u>, 285.

DISCUSSION

DICKEL: Geometrically just where is the discontinuity?

FALLE: Nearly perpendicular to the line of sight.

MCKEE: The gas producing the blueshifted emission lines was shocked at
a time of less than 10^4 yr and thus reached X-ray emitting temperatures.
Has there been time for this gas to cool off and produce the observed
optical emission?

FALLE: Yes. We used Kahn's approximation to estimate the rate at which
material was cooling.   This is how we calculated the fast H_α surface
brightness.

THE W28 SUPERNOVA AND THE SURROUNDING INTERSTELLAR MEDIUM

Zealey, W. J.   United Kingdom Infrared Telescope
                               of the Royal Observatory Edinburgh
MacGillivray, H. T.   Royal Observatory Edinburgh
Malin, D.   Anglo Australian Observatory
Hartl, H.   Astronomisches Institut der
                               Universitat Innsbruck

A large shell, 1.5 degrees in diameter, outlined by nebulosity and dust clouds has been found centred on the W28 supernova remnant. Obscuration has been studied over a 2 degree square area centred on W28. The W28 SNR is coincident with a region of reduced obscuration. The densest parts of the obscuration $A_B \geq 4^m$ is coincident with the molecular clouds associated with M20 and the OH maser to the south of M20.

The observed absorption is interpreted as being the result of the expansion of the W28 SNR into an older pre-existing SNR or stellar wind bubble.

1. INTRODUCTION

The young clusters NGC 6530, NGC 6514 and Bo 14 (10^6 yrs) the HII regions M8 and M20 and the W28 supernova remnant (6×10^4 yrs) all lie at distances of about 1.5kpc near $l = 6.0°$; $b = 0.0°$. The apparent proximity of these young objects has led to the suggestion that a physical link exists between them (Goudis, 1976; Stalibovskii and Shevchenko, 1981).

In 1979 a large, 1.5 degree diameter, low surface brightness, arc of nebulosity and dust clouds was found linking M8 and M20 (Figure 1) (Zealey et al. 1982, Hartl et al. 1981). This arc is centred on the W28 radio SNR and appears to pass across the face of M20. Here it is seen as the E-W dust lane of the Trifid. This extensive nebulosity is therefore foreground to M20, and possibly passes behind M8.

2. THE OBSCURATION

In order to further study the link between this large shell, W28, M8 and M20 we have carried out star counting procedures in their neighbourhood. A deep blue ESO/SERC Sky Survey plate of Field 521 was

J. Danziger and P. Gorenstein (eds.), Supernova Remnants and their X-Ray Emission, 267–272.

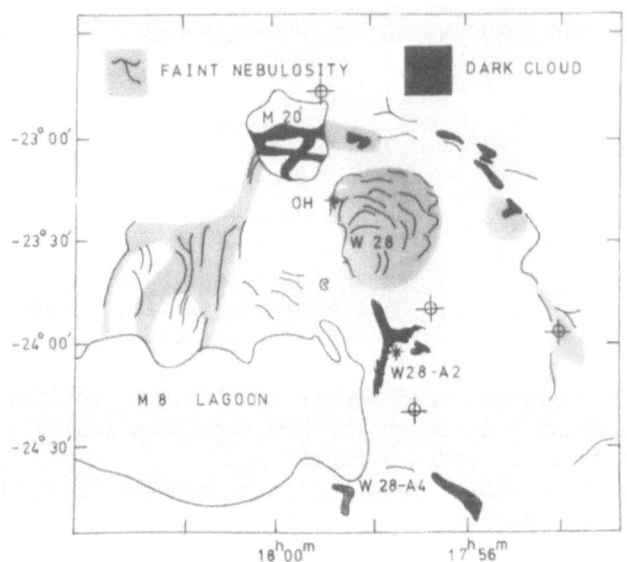

Fig. 1. A diagram of salient features in the M8, M20, W28 region.
        Indicated are the W28 SNR, the M8, M20 HII regions, the W28-A2
        compact HII region and the OH maser source.

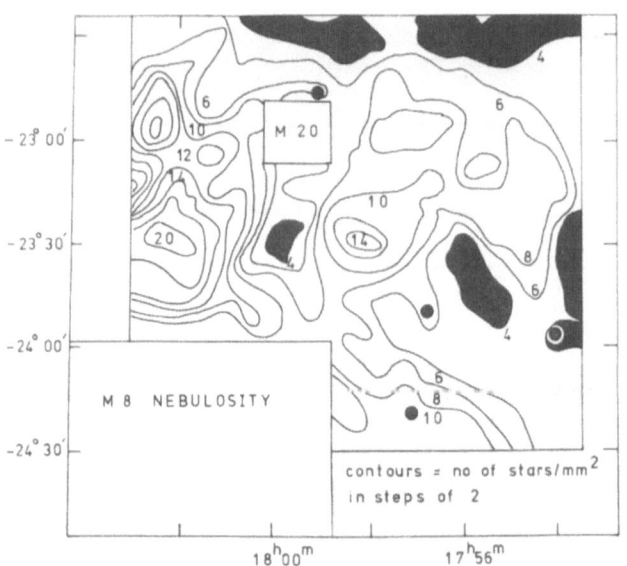

Fig. 2. A star density map of the M8, M20, W28 region.

measured using the fast, automated, measuring, machine COSMOS, operated
by the Royal Observatory Edinburgh. (Pratt et al. 1975; MacGillivray
1981). The data was analysed to provide star positions and magnitude.
The magnitudes were derived by using a standard photoelectric sequence

in NGC 6494 in a neighbouring field. The sequence was transferred to
the region of W28 using an iris photometer and working via intermediate
sequences in the overlap regions of the two fields. The NGC 6530
sequence (Walker 1957) was used to check the accuracy of the transfer
process. The data is displayed as a star density map (Figure 2). The
resolved elements are 3 millimetres by 3 millimetres.

Several regions were selected as being typical of heavily obscured
areas and lightly obscured or unobscured areas. A comparison of the
regions is shown in Figure 3, in the form of log N_m versus M

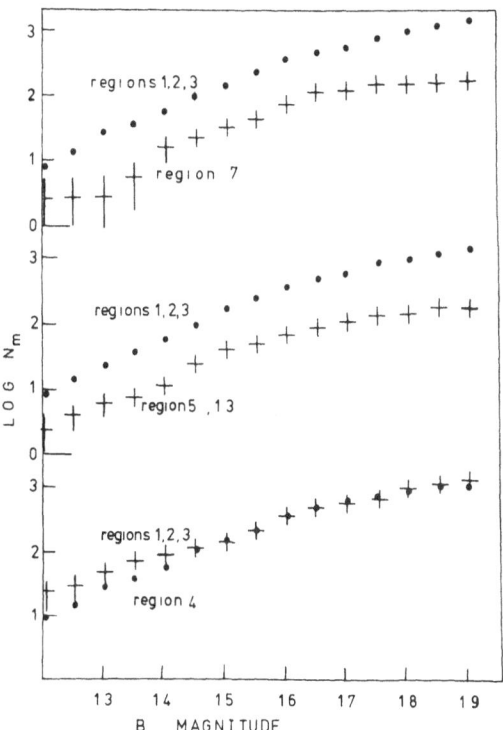

Fig. 3. Log N_m versus M diagrams for regions of heavy and intermediate
         obscuration. Crosses are for the obscured region, dots for the
         unobscured comparison region.
         Top diagram   W28-A2 compact HII region
         Middle diagram   CO/OH molecular cloud
         Bottom diagram   W28 + comparison

plots (where N_m is the number of stars per square millimetre brighter
than magnitude m).

The star density map shows three main features

(i)   A maximum in density coincident with the centre of the W28
          radio SNR (RA 17^H 57^M 35^S : DEC $-23°$ 28' 30"). Half density
          points coincide with the remnants edges.

(ii)   High densities are also found at the eastern edge of the area
          measured.

(iii)   Low densities exist between M8 and M20, near the molecular
          clouds found in CO and OH by Wootten (1981) and Pastichenke and
          Slysh (1974), and to the northern and western edge of the region.

3.   DISCUSSION

In view of the lack of known star clusters, coincident with the
observed positions of peak star density, and the similarity in peak
densities at well separated positions, it seems that the density map is
best interpreted as being due to varying absorption. The coincidence
between the position of maximum star density and the W28 SNR can be
interpreted as the SNR having made a cavity in an extensive sheet of
obscuration ($A_B \geq 4^m$). The similarity of the star density near W28 and
in other unobscured regions measured indicates that the W28 SNR has
almost totally penetrated the obscuring sheet. The sheet must
therefore be less than L = 20pc thick (dimension of W28 SNR at 1.5kpc).
The density of the sheet is therefore $N_H \geq 120$ cm^{-3} using $N_H = 630$ A_B/L
cm^{-3} (Jenkins and Savage 1974).

An upper limit to the distance of the sheet can be estimated by
assuming that it is totally opaque and all of the stars are foreground
objects (Armandroff and Herbst, 1981). Using d = 320 $N_{CT}^{0.57}$, where
N_{CT} is the number of stars in a 5' diameter area, they found an upper
limit of 2200 pc for the CO molecular cloud region. Star densities
show that the densest parts of the obscuring sheet have similar upper
limits to their distances.

4.   INTERPRETATION

If we discount the possibility of a chance alignment of two SNR we
conclude that the W28 SNR, the large 1.5 degree diameter shell and the
obscuration are coincident and lie in the immediate neighbourhood of
the M8, M20 HII regions. Two possible scenarios of the evolution of
this complex are:

(i) A single event

The large 1.5° diameter shell and the W28 SNR result from a single
event which occurred about 6×10^4 years ago. This event occurred at
the edge of a dense sheet of material, perhaps associated with the
Sagittarius spiral arm. Expansion occurred more rapidly in the less
dense material than into the dense sheet. The rapid expansion produced

the large shell. The slower expansion into the sheet produced the smaller, more intense W28 SNR. Star formation, as evidenced by the OH maser sources and W28 - A2 region has occurred at the edge of the W28 SNR. The HII regions M8, M20 and the young clusters are not a direct result of the W28 event's expansion.

(ii) Events of different periods

More than 10^6 years ago, star formation occurred in an expanding SNR or stellar wind shell. The M8 and M20 regions are a result of this epoch of star formation. More recently a high mass star, in or near this shell, evolved to become a supernova, producing the W28 SNR. This young, rapidly expanding remnant, has tunneled into the old fossil remnant, rejuvenating it. The old remnant is now seen as a 1.5 degree diameter shell. Star formation has been triggered in the old shell by W28.

In order to distinguish between the models, further molecular observations are required covering a large area. However, perhaps the existence of a pulsar PSR 1754-24 near the edge of the large shell, provides evidence for an old supernova event. The pulsar has a period of 0.234 seconds, indicating that the associated event is older than 10^6 years (Large 1970). It is possible for the pulsar to have travelled the projected distance of 29pc from the event centre in that time.

Acknowledgments

One of us (HH) wishes to thank the Austrian "Fonds zup Forderuny der Wissenschaftlichen Forschung (project 3487) for financial report.

REFERENCES

Armandroff, T. E. and Herbst, W. 1981 Astron J. 86, 1923.

Goudis C. 1976 Astrophys. & Space Science 40, 91.

Hartl, H., Malin, D. F., MacGillivray, H. T., Zealey, W. J., 1981, Proceedings Internationale Astronomishe Tagung Innsbruck 1981.

Jenkins, E. B. and Savage, B. D., 1974, Astrophys. J. 187, 243.

MacGillivray, H. T., 1981, SERC Measuring Machine Newsletter No. 1.

Pratt, N., Martin, R., Alexander, L. W. G., Walker, P. R., 1975, Image Processing Techniques in Astronomy Publ. Reidel, Dordrecht, Holland, edited de Jager and Nieuwenhuijzen, Page 217.

Pastichenko, M. I. and Slysh, V. I., 1974, Astron. Astrophys. 35, 153.

Stal'bov skii, O. I. and Shevchenko, V. S., 1981, Sov. Astron. 25, 1.

Zealey, W. J., MacGillivray, H. T., Hartl, H., Malin, D., 1982, submitted Astrophys. Letters.

EINSTEIN OBSERVATIONS OF THE SNRs IC443, W44 AND W49B

M.G.Watson, R.Willingale, J.P.Pye, D.P.Rolf, N.Wood, and
N.Thomas
X-Ray Astronomy Group, University of Leicester, England.

F.D.Seward
Center for Astrophysics, Cambridge, Ma., USA.

We present soft x-ray observations made with the Einstein Observatory
Imaging Proportional Counter (IPC) of IC443, W44 and W49B (for details
of the observatory and instruments see Giacconi et al.1979). The x-ray
emission from IC443 and W44 is clearly concentrated within the interior
of the remnant with little or no evidence for a limb-brightened shell.
Significant spectral differences are found across the x-ray images in
both remnants which are interpreted as being due to a combination of
differential absorption by molecular clouds and intrinsic spatial
temperature variations. The distant remnant W49B is only just resolved
in the IPC observations, but additional observations with the High
Resolution Imager (HRI) indicate a similar "infilled" morphology to
IC443 and W44.

IC443

The old supernova remnant IC443 shows a partial shell structure at
radio wavelengths, brightest in the NE, together with a complex system
of optical filaments which display a detailed correlation with the radio
emission (e.g. Duin and Van der Laan 1975). Studies at millimetre
wavelengths have revealed a dense CO cloud associated with IC443, which
lies across the centre of the remnant (Cornett et al.1977, Scoville at
al.1977). Early x-ray studies (e.g. Malina et al.1976) indicated an
age < 10^4 years, in disagreement with much larger estimates from the
observed low ($\sim$ 100 km s^{-1}) expansion velocity of the optical
filaments. The distance to IC443 is not well determined with estimates
in the range 0.5 to 2.5 kpc; we adopt d = 2 kpc.

In Fig.1 we show the soft x-ray image of IC443 overlaid on an
optical plate of the remnant. The x-ray image is a composite of three
overlapping observations with a total integration time of 5.1 ksec. The
results are summarised in Table 1. There is evident correlation with
the optical emission. Highest x-ray surface brightness occurs in the NE
quadrant, but the x-ray emission appears to originate predominantly from
the interior of the remnant. No significant emission is seen outside

J. Danziger and P. Gorenstein (eds.), Supernova Remnants and their X-Ray Emission, 273–280.
© *1983 by the IAU.*

Table 1   Summary of results

	IC443	W44	W49B
Assumed distance d [kpc]	2	3	10
Spectrum[1]: T_x $[10^7 K]$	2	2	2
(assumed) N_H $[10^{21}$ cm$^{-2}]$	4	5	30
Luminosity[2] $[10^{35}$ erg s$^{-1}]$	2.5	1.6	12
Maximum dimensions [arcmin]	50	25 x 35	6
[pc]	29	22 x 31	18
Electron density[3] $\langle n_e \rangle$ $[$cm$^{-3}]$	0.15	0.2	1
Total emitting mass[4] $[M_\odot]$	60	60	85
Total thermal energy[4] $[10^{50}$ ergs]	4	4	7

Notes

(1) Approximate spectral fit to the IPC data with an optically thin thermal spectrum following Raymond and Smith (1977).
(2) Intrinsic source luminosity [0.5 - 4.5 keV] corrected for absorbing column density.
(3) Derived electron density assuming emissivities for fitted spectrum. Note that a volume filling factor f = 1 has been assumed. For different filling factors and source distances $n_e \propto f^{-\frac{1}{2}} d^{-\frac{1}{2}}$.
(4) Total emitting mass and thermal energy content for f=1.   Both quantities scale as $f^{\frac{1}{2}} d^{\frac{5}{2}}$.

(The values quoted in Table 1 assume cosmic abundances and ionisation equilibrium. The inclusion of enhanced metal abundances and non-equilibrium ionisation (e.g. Gronenschild & Mewe 1982 and several contributors to this volume) can make substantial differences to the temperature and density values and would tend to lower the total mass and thermal energy estimates.)

the edge of the remnant as defined by the radio/optical boundary (except possibly in the NW), and profiles through the x-ray image confirm that there is no appreciable limb-brightening except in part of the NW quadrant.

The spectral parameters derived for IC443 (quoted in Table 1) are necessarily crude since the energy resolution of the IPC allows a wide range of temperature and column density values. Assuming $N_H = 4$ x 10^{21} cm^{-2} (cf. Malina et al.1976), the approximate x-ray temperature

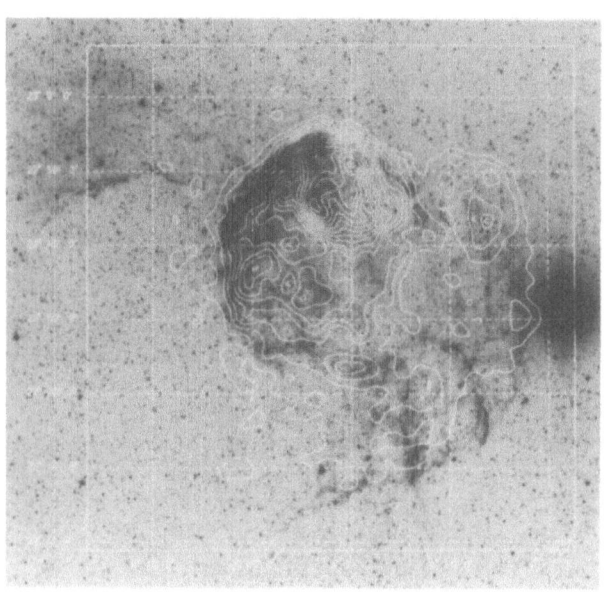

Fig.1 IPC x-ray image of IC443 (shown as contour map linear in x-ray surface brightness) overlaid on the PSS red plate. The x-ray image [~ 0.5 - 3 keV] has been Wiener filtered.

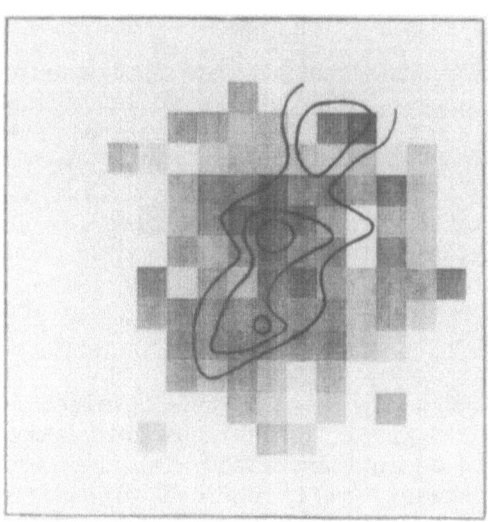

Fig.2 Spectral hardness map for IC443 (shown as linear greyscale) together with CO emission contours based on Cornett et al. (1977) shown to the same scale a Fig.1. Hardness ratio plotted is [counts in 1.3 - 3.1 keV band] / [counts in 0.2 - 1.3 keV band] for 256 arcsec. pixels.

2×10^7 K is somewhat higher than quoted by previous investigators (e.g. Galas et al.1981, Charles et al.1981). Nevertheless more detailed analysis of the IPC spectra for 5 regions inside IC443 confirms our approximate parameters.

In order to look for spectral variations across the x-ray image we have constructed a map of the hardness ratio (shown in Fig.2). (Note that changes in the hardness ratio can be due to either increased low energy absorption or increased x-ray temperatures, or both). Over the central region of the image there is clearly reasonable correlation of the hardness ratio with the CO emission contours (also shown in Fig.2), indicating increased low energy absorption in the dense molecular cloud. For $T_x \approx 2 \times 10^7$ K the hardness ratios observed in the region of the molecular cloud correspond to an additional column density of $\sim$ $(5 - 10) \times 10^{21}$ cm^{-2} through the cloud, consistent with density estimates from the CO measurements (Cornett et al.1977, Scoville et al.1977). Significant hardness variations are seen over other parts of the remnant which are probably not affected by localised absorption (on the basis of the CO map). The outer edges of the x-ray emission, in particular the brightest region to the N, have on average lower hardness ratios indicating lower temperatures ($\lesssim 5 \times 10^6$ K). The average x-ray spectra for IC443 quoted by Charles et al.(1981) and Galas et al.(1981) are consistent with this lower temperature (although it is clear from the IPC results that a single spectrum cannot be expected to be a good representation of the x-ray emission).

W44 (G34.6-0.5)

Radio maps of W44 show that it has a distorted shell structure approximately 20 x 30 arcmin. in extent with appreciable flattening in the NE (e.g. Clark et al.1975). No optical filaments have been detected, possibly as a result of reddening, since the (kinematic) distance to W44 is estimated to be 3 kpc (e.g. Radhakrishnan et al.1972). In the eastern part of W44 molecular line observations indicate the presence of a large dense molecular cloud associated with the remnant (e.g. Dickel et al.1976, Wooten 1977). Knapp and Kerr (1974) have reported the detection of a large, radially expanding shell of HI surrounding W44.

Fig.3 shows the IPC x-ray image of W44 obtained in an observation with integration time 2.3 ksec. Table 1 summarises the results. The x-ray emission appears to be anticorrelated with the radio emission in the sense that it is remarkably well confined within the boundary of the remnant as defined by the radio contours (see Fig.3). The x-ray emission is very clearly "centre-brightened" rather than limb-brightened, and significantly elongated in the N-S direction by about the same amount as the radio shell. Spectral parameters for W44 are quoted in Table 1. Note that in order to obtain a reasonable temperature estimate from the IPC spectrum, a column density of 5×10^{21} cm^{-2} appropriate for d = 3 kpc (e.g. Seward et al.1972) was assumed.

As for IC443 we have looked for spectral variations across the x-ray image by constructing a spectral hardness map. Significant hardness variations are seen, with the hardest region occurring in the NE near the peak of the radio emission, but there is no obvious correlation either with x-ray morphology, or the CO emission contours (e.g. Wooten 1977). If we assume that the emission from W44 is isothermal, the observed hardness variations correspond to the absorbing column density varying between 10^{20} to 10^{22} cm^{-2}. Alternatively, assuming a uniform column density of 5×10^{21} cm^{-2} would imply temperatures ranging from 10^7 to greater than 10^8 K. Because the implied range of column densities is so large, we suspect that most of the observed hardness variations are due to significant temperature differences in the emission region, although the increased hardness ratio in the NE could be due to increased absorption originating in the molecular cloud.

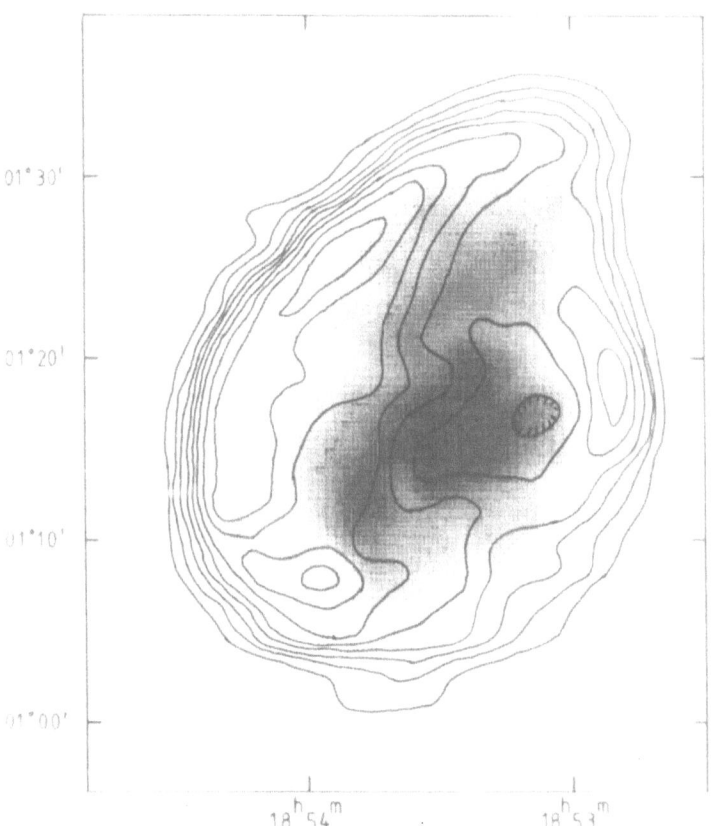

Fig.3 IPC x-ray image of W44 together with radio contours taken from the 408 MHz map of Clark et al.(1975) (HPBW ≈ 3 arcmin.). The x-ray image [∼ 0.5 - 4 keV] has been smoothed with a 30 arcsec. Gaussian. The greyscale is linear in x-ray surface brightness.

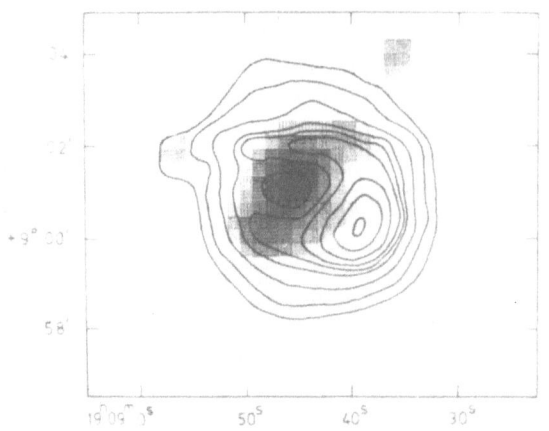

Fig.4 HRI x-ray image of W49B together with radio contours taken from the 10.7 GHz map of Downes & Wilson (1974) (HPBW = 1.3 arcmin.). The x-ray image [$\sim$ 0.2 - 4 keV] has been smoothed with a 30 arcsec. Gaussian. The greyscale is linear in x-ray surface brightness.

W49B (G43.3-0.2)

At an estimated distance of 10 kpc (e.g. Radhakrishnan et al.1972), and at low galactic latitude, W49B is a difficult object to observe at x-ray wavelengths. The 10.7 GHz radio map of Downes and Wilson (1974) with arcminute resolution shows a shell structure of diameter $\sim$ 3 - 4 arcmin. with enhanced emission in the SW (see Fig.4). (The shell morphology is confirmed by VLA mapping - D.Helfand, private communication.) Optical emission has not been observed from W49B, but this is hardly surprising given the distance and the optical absorption expected for a column density $N_H \approx 3 \times 10^{22}$ cm^{-2} (see below).

W49B was observed both with the IPC (integration time 1.6 ksec.) and HRI (integration time 11.1 ksec.). The results are summarised in Table 1. The IPC image shows a bright source coincident with W49B which is only marginally resolved. The HRI image (see Fig.4) shows very low surface brightness emission from W49B which is clearly extended. This is the first detection of W49B in x-rays. Within the limitations imposed by the low surface brightness of the x-ray image, it appears that the x-ray emission is slightly elongated in the N-S direction, and originates primarily from the region inside the radio shell, with no evidence for limb-brightening.

As with IC443 and W44, the IPC data can be used to obtain approximate spectral parameters for the remnant (see Table 1), although the angular resolution of the IPC is too low to allow examination of spectral variations over the image. The observed HRI flux is particularly sensitive to the absorbing column density and has been used as an additional constraint on the spectral parameters. The derived column density agrees well with the value estimated from HI 21 cm measurements (e.g. Akabane & Kerr 1965; Radhakrishnan et al.1972).

CONCLUSION : RADIO SHELLS AND X-RAY INTERIORS

The Einstein observations of IC443, W44 and W49B clearly show that in each case the x-ray emission is concentrated within the interior of the remnant with little or no evidence for limb-brightening. The lack of observed limb-brightening is easily understood if the shells are cool ($\sim 10^6$ K) since the attenuation by moderate column densities for low temperature thermal emission is large enough to hide the emission in the IPC. This explanation is particularly attractive for IC443 where there is some evidence (see above) for such a cool shell. Low x-ray shell temperatures are consistent with the estimated ages ($\gtrsim 10^4$ years) of both IC443 and W44, but may be problematic for W49B which is probably somewhat younger.

As there is no evidence for a central pulsar to power the interior x-ray emission by non-thermal processes, the simplest interpretation is that this emission originates in a hot ($\sim 10^7$ K) thermal plasma filling the remnants. This component of the x-ray emission is not expected for SNR evolution following the standard Sedov solution, in which most of the mass is swept-up in a thin shell which dominates the x-ray emission. Although SNR evolution in a "cloudy" medium (e.g. Cowie & McKee 1977) may allow a much larger fraction of the material to remain in the interior of the remnant, it is not clear how the interior can be heated. Because the cooling time is long, the emission seen could be interpreted as a fossil of an original hot plasma created early in the life of the remnant (e.g. by a reverse shock), but we note that x-ray observations of young remnants (e.g. SN1006, Pye et al.1981) show that the interior emission is dominated by a relatively cool plasma containing only a small fraction of the initial supernova energy release. In contrast the total thermal energy estimates (Table 1) indicate that appreciable fractions of the initial energy (i.e $\sim 10^{51}$ ergs) of the supernovae are locked up in this plasma. The total emitting masses estimated are also large, and imply that most of the material originates in the ISM, rather than from the supernova ejecta.

The morphology of the x-ray emission from IC443 and W44 is quite different to that exhibited by other remnants with similar sizes and ages (e.g. Cygnus Loop and Vela). Perhaps one clue to understanding the difference lies in the nature of the medium into which the remnant is expanding, rather than the properties of the initial explosion, since both IC443 and W44 are closely associated with (and may be interacting with) dense molecular clouds.

The Einstein Observatory is funded by NASA. MGW,RW,JPP,NW and NT acknowledge financial support from the SERC, and MGW thanks the Royal Society for a travel support.

REFERENCES

Akabane,K.,& Kerr,F.J.,1965. Aust.J.Phys.,18,91.
Charles,P.A.,Kahn,S.M.,Mason,K.O.,& Tuohy,I.R.,1981.
          Ap.J.(Lett.),246,L121.
Clark,D.H.,Green,A.J.,& Caswell,J.L.,1975.
          Aust.J.Phys.Ap.Suppl.,37,75.
Cornett,R.H.,Chin,G.,& Knapp,G.R.,1977. Astr.Astrophys.,54,889.
Cowie,L.L.,& McKee,C.F.,1977.   Ap.J.,215,213.
Dickel,J.R.,Dickel,H.R.,& Crutcher,M.,1976. P.A.S.P,88,840.
Downes,D.,& Wilson,T.L.,1974. Astr.Astrophys.,34,133.
Duin,R.M.,& Van der Laan,H.,1975. Astr.Astrophys.,40,111.
Galas, C.M.F., Venkatesan, D.,& Garmire,G.,1981. Ap.J.,250, 216.
Giaconni,R., et al. 1979. Ap.J.,230 540.
Gronenschild, E.H.B.M., & Mewe,R.,1982. Astr. Astrophys.,48, 305.
Knapp, G.R., & Kerr, F.J.,1974. Astr.Astrophys. 33, 463.
Malina,R.,Lampton,M., & Bowyer, S.,1976. Ap.J.,207, 894.
Pye, J.P., Pounds, K.A., Rolf, D.P., Seward, F.D., Smith,A.,
          & Willingale, R., 1981. M.N.R.A.S.194, 569.
Radhakrishnan,V., Goss,W.M., Murray,J.D., & Brooks, J.W., 1972.
          Ap.J.Suppl., 24,49.
Scoville, N.Z, Irvine, W.M., Wannier, P.G., & Predmore, C.R., 1977.
          Ap.J.,216, 320.
Seward, F.D., Burginyon, G.A., Grader, R.J., Hill, R.W., & Palmieri, T.M.
          1972. Ap.J, 178, 131.
Wooten, H.A., 1977. Ap.J.,216, 440.

THE X-RAY STRUCTURE OF THE SUPERNOVA REMNANT G78.2+2.1

L.A. Higgs and T.L. Landecker
Dominion Radio Astrophysical Observatory
Herzberg Institute of Astrophysics
Penticton, B.C., Canada

F.D. Seward
Harvard-Smithsonian Center for Astrophysics
Cambridge, Mass., U.S.A.

ABSTRACT

The south-eastern portion of the supernova remnant G78.2+2.1, in Cygnus, has been detected as a weak X-ray source by the Einstein Observatory. The X-ray structure is similar to that of the radio filaments in this region, and confirms that X-ray emission in this portion of the "Cygnus super-bubble" does originate in a known supernova remnant. Marginally significant variations in X-ray hardness across the mapped area have been detected and can be related to known radio and optical features of the remnant. In its X-ray properties, G78.2+2.1 resembles IC443.

1. INTRODUCTION

The supernova remnant (SNR) G78.2+2.1 lies in a heavily obscured region beyond the star γ Cygni, and is one of the brightest SNRs at radio wavelengths. Radio-continuum observations by Higgs et al. (1977) showed that it has a roughly circular structure, with two regions of enhanced radio emission: an area of intense filaments in the south-eastern quadrant and a weaker area of emission in the north-west quadrant. Embedded in the south-eastern filaments is the thermal radio source known as the γ-Cygni Nebula. Bychkov (1978) suggested that this emission nebula is ionized by photons from the supernova shock. Observations of 21-cm line emission from neutral hydrogen led Landecker et al. (1980) to propose a model in which the supernova blast occurred in a "slab" of interstellar material.

The first detection of X-ray emission from this SNR was made by Davidsen et al. (1977), who found a 0.5 to 2 keV source (Cyg X-7) coincident with it. Their observations indicated a temperature of 1.5 - 5 x 10^6 K and a column density of foreground hydrogen > 10^{22} cm^{-2}. A harder X-ray source just beyond the southern rim of the remnant was detected (Forman et al., 1979) by Uhuru (4U 2019+39). It may be

J. Danziger and P. Gorenstein (eds.), Supernova Remnants and their X-Ray Emission, 281–286.

282   L. A. HIGGS ET AL.

related to the supernova remnant.

Cash et al. (1980) detected a large ring of soft X-ray emission in the constellation of Cygnus, which they termed the "Cygnus super-bubble", and which includes the region of Cyg X-7. They claimed that, since the X-ray intensity from the direction of the SNR G78.2+2.1 is comparable to that from the rest of the large ring of emission, "it seems unreasonable to assign any of the flux to the supernova remnant". In order to clarify whether detectable X-ray emission from the SNR actually exists, we undertook a search of Einstein Observatory data.

2. OBSERVATIONS

An IPC observation of a field in the SNR was made on October 30, 1979, by the Center for Astrophysics. This field is indicated by the rectangle outlined in Figure 1. The contours define the 21-cm radio emission in this region (Higgs et al., 1977). The field for the 2800-s X-ray observation includes the region of most intense radio emission.

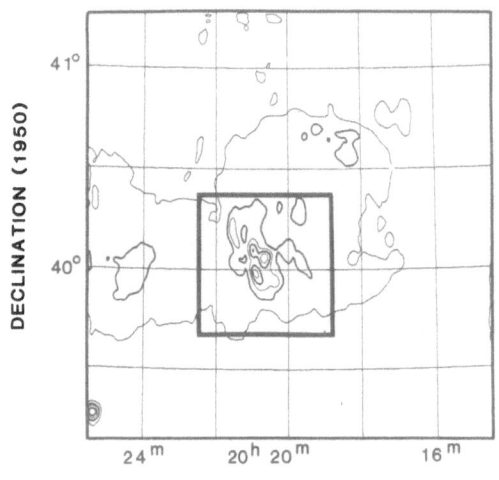

RIGHT ASCENSION (1950)

Figure 1. The X-ray field. The radio contours are 20, 40, 60, 80 and 100 K (T_B). The SNR is the approximately circular area of emission, 1° in diameter, in the centre. The emission to the left arises in the HII region IC 1318b.

Initial analysis of the X-ray data showed no conspicuous X-ray emission that could be identified with the SNR. A closer examination, however, indicated that smoothing of the data, combined with a subtraction of a blank comparison field, might reveal some weak X-ray structure. A suitable comparison image (a 6400-s observation of an effectively blank field, at about the same time as the γ-Cygni observation and with approximately the same instrumental parameters) was rotated to bring the IPC detector elements into alignment with those for the SNR image, and then both images were convolved to a resolution of 5'. The photon counts in the rotated and smoothed

comparison field (after scaling by the ratio of observing times) were
then subtracted from those in the smoothed γ-Cygni field.   The
resulting difference map is shown in contour form in Figure 2a.   The
peak photon flux in Figure 2a (at the lower right edge) is about 9 x
10^{-4} counts arcmin^{-2} s^{-1}, a 10-σ result.

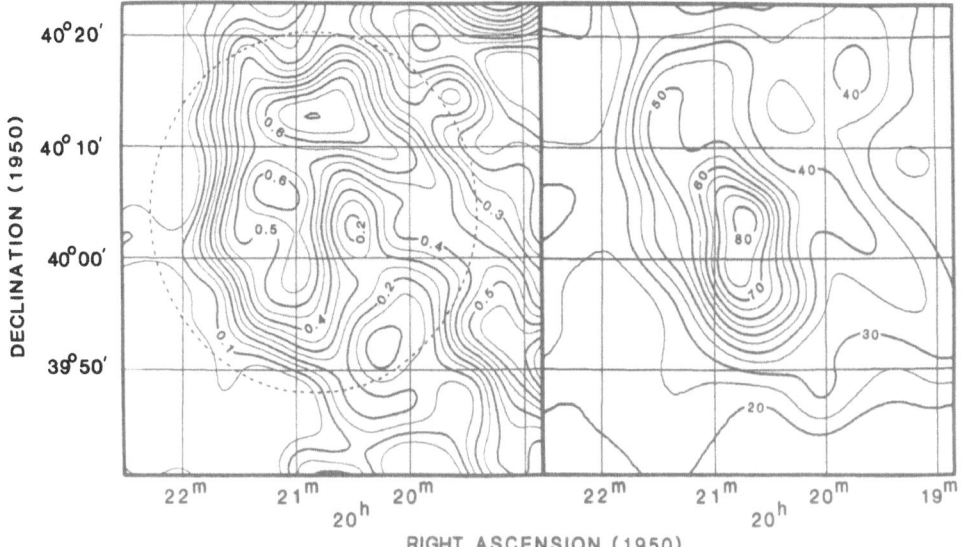

Figure 2. (a, <u>left</u>) X-ray emission (0.2 - 4.7 keV), smoothed
to 5'.   Contours are counts per 32" x 32" pixel for the
2800-s observation (step = 0.05).   The dashed circle
indicates a region chosen for detailed spectral analysis.
Estimated σ = 0.07.   (b, <u>right</u>) 21-cm radio emission,
smoothed to 5'.   The contours are brightness temperature
(step = 5K). The peak coincides with the γ-Cygni Nebula.

The 21-cm radio emission, convolved to the same resolution, is
shown in Figure 2b. Although a detailed correlation between the radio
emission and the detected X-ray emission does not exist, certain
similarities are apparent.   In both maps the emission is sharply
bounded to the south-east, and the bulk of the X-ray emission is near
the main radio filaments.   Moreover, the "bar" of X-ray emission
extending from top centre to the lower right edge of the map
corresponds to a radio feature seen in Figures 1 and 2b.   Clearly the
X-ray emission is related to the radio emission and can therefore be
attributed to the SNR.

3.   ANALYSIS AND DISCUSSION

To estimate the temperature of the X-ray emitting gas, a thermal
spectrum has been fitted to the photon counts (0.2 to 4.7 keV) detected

within the dashed circular area in Figure 2a. The observations are best fitted by temperatures 3.5×10^6 K $<$ T $<$ 4.5×10^7 K, and column densities of foreground hydrogen between 3×10^{22} cm^{-2} and 3×10^{21} cm^{-2}, respectively. For a column density of hydrogen of 10^{22} cm^{-2}, a reasonable estimate, the best-fitting thermal spectrum corresponds to T $\sim$ 1.5×10^7 K. (The HI observations of Landecker et al. (1980) yielded an estimated column density of <u>neutral</u> hydrogen of $\leqslant 6 \times 10^{21}$ cm^{-2} in front of the SNR, assuming the HI to be optically thin.) This temperature is somewhat higher than that deduced by Davidsen et al. (1977) for Cyg X-7.

On the assumption of a uniform T $\sim$ 1.5×10^7 K within the circular area of Figure 2a, the best-fitting thermal spectrum corresponds to a flux (corrected for absorption) of 6×10^{-11} erg cm^{-2} s^{-1} in the energy range 0.2 to 4 keV. Assuming the volume X-ray emissivity of Raymond and Smith (1977) and a distance of 1.5 kpc (Landecker et al., 1980), the r.m.s. density of the hot gas within the spherical volume corresponding to the dashed circle in Figure 2a is 0.1 to 0.4 cm^{-3} for 4.5×10^7 K $>$ T $>$ 3.5×10^6 K, respectively. Since X-ray-emitting filaments may fill less than 10% of the volume, the mean density is probably $\geqslant 0.7$ cm^{-3}. In a discussion of high-velocity HI cloudlets observed inside this SNR, Landecker et al. (1980) estimated that the density of the post-shock gas must exceed 2 cm^{-3}, assuming a blast velocity of 450 km/s. The higher X-ray temperature indicated by the current observations implies a higher shock velocity, approximately 1100 km/s. If this is the case, the velocities of the accelerated HI features would suggest that the hot post-shock gas has a density in excess of 0.4 cm^{-3}, in good agreement with the density deduced from the present observations.

The X-ray luminosity of the SNR can only be estimated roughly since the current observations cover only the south-east portion. The luminosity of the spherical region (33' in diameter) indicated in Figure 2a is between 5×10^{33} and 5×10^{34} erg s^{-1}, for a distance of 1.5 kpc. The radio SNR has a diameter of 62', so that the total X-ray luminosity probably lies between 3×10^{34} and 3×10^{35} erg s^{-1} (for the energy range 0.2 to 4 keV). This luminosity and a temperature of the order of 10^7 K indicate that the SNR is similar, in its X-ray properties, to IC443 (see, for example, Culhane, 1977).

Despite the poor photon statistics, we have attempted to detect any large temperature variations across the X-ray field. The same smoothing and subtraction of the comparison field were done separately for photons in the ranges 0.2 to 1.2 keV and 1.2 to 4.7 keV. The resulting maps are shown in Figure 3. There are three regions where marginally significant (2-σ) variations are seen (after first-order allowance for IPC gain variations across the image). These are indicated by the shaded areas in Figure 4.

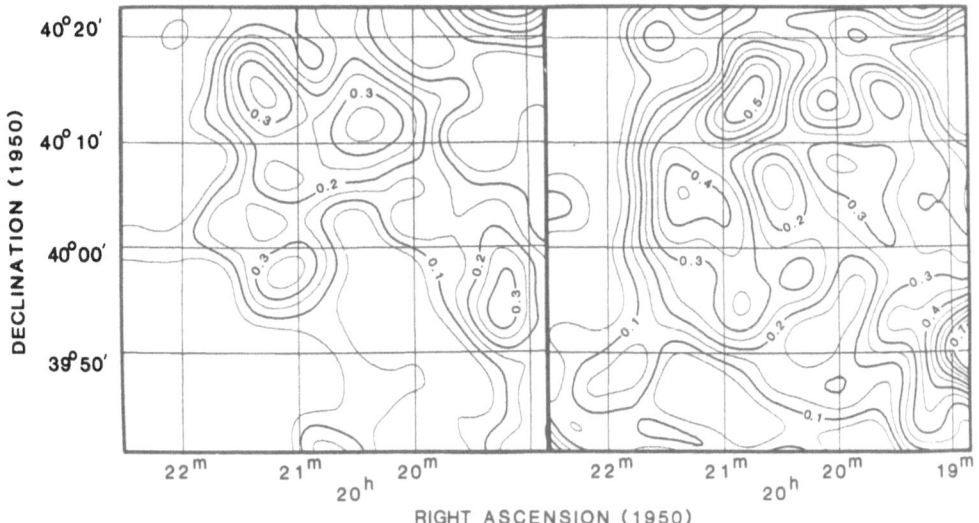

Figure 3. (a, <u>left</u>) X-ray emission (0.2 - 1.2 keV). Units are as in Figure 2. Estimated σ = 0.06. (b, <u>right</u>) X-ray emission (1.2 - 4.7 keV). Estimated σ = 0.05.

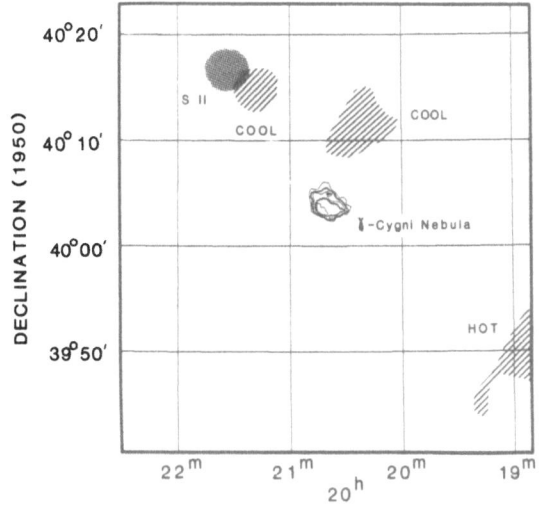

Figure 4. Regions (shaded) where X-ray temperature varies by > 2σ from average. The γ-Cygni Nebula is denoted by 6-cm radio contours (unpublished VLA data, Higgs et al. 1982). Contours are 0.5 to 2.5 mJy/beam (step = 0.5) where 11 mJy/beam ~ 3 K T_B.

Two "cool" regions and one "hot" region are shown. The cool region in the upper centre of the figure, situated above a "hole" in Figures 2a and 3b, extends in the direction of the γ-Cygni Nebula, shown by the radio contours. This thermal radio source, located in the midst of the non-thermal radio filaments, is somewhat of an enigma. In the model of the G78.2+2.1 SNR proposed by Landecker et al. (1980) (see their Fig. 9), the γ-Cygni Nebula is an interstellar cloud just encountering the blast, and ionized by UV from the shock. The present X-ray observations do not conflict with such an interpretation. The "hole" in the X-ray distribution, and the cooler temperature, may indicate that the shock has been slowed down owing to an interaction with a dense interstellar cloud. The γ-Cygni Nebula could then be a tenuous ionized fragment of the larger cloud. The other cool region that is seen corresponds closely with the area where van den Bergh (1978) detected SII emission from the SNR - perhaps another region of interaction with a dense cloud. This cool region is also close to the position of a high-velocity HI feature (+72 km/s) detected by Landecker et al. (1980).

The hot X-ray region at the lower-right edge of Figure 4 is also in agreement with the model of Landecker et al. (1980) where the blast is here propagating into a lower-density interstellar medium, with a correspondingly higher post-shock temperature.

ACKNOWLEDGMENTS

We would like to thank D. Harris of the CFA staff for assistance with the X-ray analysis.

REFERENCES

Bychkov, K.V: 1978, Astron. Zh. 55, pp. 1222-1227.
Cash, W., Charles, P., Bowyer, S., Walter, F., Garmire, G., and
    Riegler, G: 1980, Astrophys. J. Lett. 238, pp. L71-L76.
Culhane, J.L: 1977, in "Supernovae", ed. D.N. Schramm, D. Reidel
    (Dordrecht-Holland), pp. 29-51.
Davidsen, A.F., Henry, R.C., Snyder, W.A., Friedman, H., Fritz, G.,
    Naranan, S., Shulman, S., and Yentis, D: 1977, Astrophys. J.
    215, pp. 541-551.
Forman, W., Jones, C., Cominsky, L., Julien, P., Murray, S., Peters,
    G., Tananbaum, H., and Giacconi, R: 1978, Astrophys. J. Suppl.
    38, pp. 357-412.
Higgs, L.A., Landecker, T.L., and Roger, R.S: 1977, Astron. J. 82, pp.
    718-724.
Higgs, L.A., Landecker, T.L., and Roger, R.S: 1982, private
    communication.
Landecker, T.L., Roger, R.S., and Higgs, L.A: 1980, Astron. Astrophys.
    Suppl. 39, pp. 133-151.
Raymond, J.C., and Smith, B.W: 1977, Astrophys. J. Suppl. 35, pp. 419-
    439.
van den Bergh, S: 1978, Astrophys. J. Suppl. 38, pp. 119-128.

RADIO OBSERVATIONS OF SMALL DIAMETER SOURCES IN THE DIRECTION OF OLD
SUPERNOVA REMNANTS

E. Fürst, W. Reich
Max-Planck-Institut für Radioastronomie, Bonn, FRG

W. Hirth
Astronomisches Institut der Universität, Bonn, FRG

Radio observations with high sensitivity have shown that lots of
more or less compact structures can be found in the field of extended
and old supernova remnants (SNRs). These small diameter sources have
been subject to many recent observations. The aim of these studies is
to infer a possible physical association of these sources with the SNR
shell. The interest in this link is based on various aspects, instabilities of shocked interstellar matter, star formation, etc.

Unfortunately, it is difficult to ascertain any suggested association. Besides others, one possible criterium, sometimes used, is the
similarity of the radio spectral indices of both, the SNR shell and the
small diameter sources. This comparison seems to be suitable, if the
radio spectrum of the SNR is not straight but shows a break. Small
diameter sources in the field of those SNRs may be considered as candidates for an association, if they also have a bent radio spectrum similar to that of the SNR.

We have tested this method with the Effelsberg 100-m telescope for
the SNRs HB9 and S147, which are known to have bent radio spectra. In
case of HB9 only one source ($\alpha_{50} = 4^h55^m51^s$, $\delta_{50} = 45°48'12''$) with a
bent radio spectrum could be detected. However, at low frequencies the
spectral index of this source is probably smaller than that of HB9. Too
few flux values are known for this small diameter source, thus firm
conclusions cannot be drawn.

In case of S147 two neighboring sources have been identified, which
have the same spectral slope at high frequencies. The integrated spectrum of these sources is known between 179 MHz and 10.69 GHz and resembles that of S147 quite well (Figure 1).

The two sources have been described by Fürst et al. (1982). An
association is very probable. However, for a confirmation it is absolutely necessary to obtain the true sizes of the small diameter sources
by interferometric observations. In addition, 21 cm absorption line
measurements may reveal their galactic or extragalactic nature.

J. Danziger and P. Gorenstein (eds.), Supernova Remnants and their X-Ray Emission, 287–288.

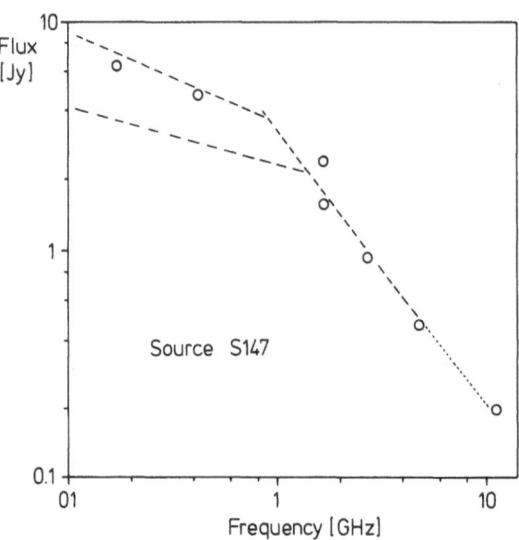

Fig. 1. The integrated flux (o) of two neighboring small
diameter sources in the field of S147. For comparison we
have plotted the slope of the radio spectrum of S147. At
low frequencies the two dashed lines reflect the uncertainty of the spectrum of S147.

If the association is true, some interesting conclusions can be
drawn. If equipartition magnetic field is assumed, the application of
the synchrotron theory on the high frequency part of the spectrum leads
to B $\approx$ 180 μGs. Assuming isotropic compression of interstellar matter
(B $\approx$ 3 μGs, n_0 = 0.5 cm^{-3}) as the origin of the sources, the number
density of the compressed small diameter objects is n $\approx$ 200 cm^{-3}.

There is still no clear explanation of the break in the radio spectra of SNRs. However, the frequency of the break depends certainly on
the compression rate. Because of the very similar radio spectra of S147
and the two small diameter sources, the compression of the SNR radio
shell should be similar. Therefore, the radio filaments of S147 are
very probably rather dense. It is interesting to examine the association of the sources also by applying the other methods mentioned above.

REFERENCE

Fürst, E., Reich, W., Beck, R., Hirth, W., Angerhofer, P.: 1982, Astron.
    Astrophys., in press

HIGH RESOLUTION X-RAY IMAGES OF PUPPIS A AND IC 443

R. Petre[1], C.R. Canizares[2], P.F. Winkler[3], F.D. Seward[4], R. Willingale[5], D. Rolf[5], and N. Woods[5]

[1]NAS/NRC Research Associate at Laboratory for High Energy Astrophysics, NASA/GSFC
[2]Center for Space Research, MIT; Alfred P. Sloan Foundation Fellow
[3]Middlebury College
[4]Harvard/Smithsonian Center for Astrophysics
[5]Leicester University

ABSTRACT

We present soft X-ray photomosaic images of two supernova remnants, Puppis A and IC 443, constructed from a series of exposures by the Einstein imaging instruments. The complex morphologies displayed in these images reflect the interaction between "middle-aged" supernova remnants and various components of the interstellar medium. Surface brightness variations across Puppis A suggest that inhomogeneities on scales from 0.2 to 30 pc are present in the interstellar medium, while the structure of IC 443 is apparently dominated by the interaction between the remnant and a giant molecular cloud.

In supernova remnants that have evolved well into their adiabatic expansion phase (t $\sim 10^4$ yr), the bulk of the soft X-ray emission arises from recently shock-heated interstellar matter. The soft X-ray morphology of such "middle-aged" remnants will be influenced by any large or small scale inhomogeneities in the interstellar medium (ISM) they encounter. Consequently, these supernova remnants serve as excellent remote probes of the structure of the ISM.

We present and briefly discuss here high-resolution soft X-ray images of two such "middle-aged" remnants, Puppis A and IC 443. Despite their similarity in age ($\sim 10^4$ yr), distance ($\sim$ 2 kpc) and diameter (25-30 pc), these two remnants display markedly different morphologies, presumably due largely to their different environments. A more extensive discussion of Puppis A may be found in Petre et al. (1982a); the IC 443 imaging results, along with spectral studies using the

J. Danziger and P. Gorenstein (eds.), Supernova Remnants and their X-Ray Emission, 289–293.

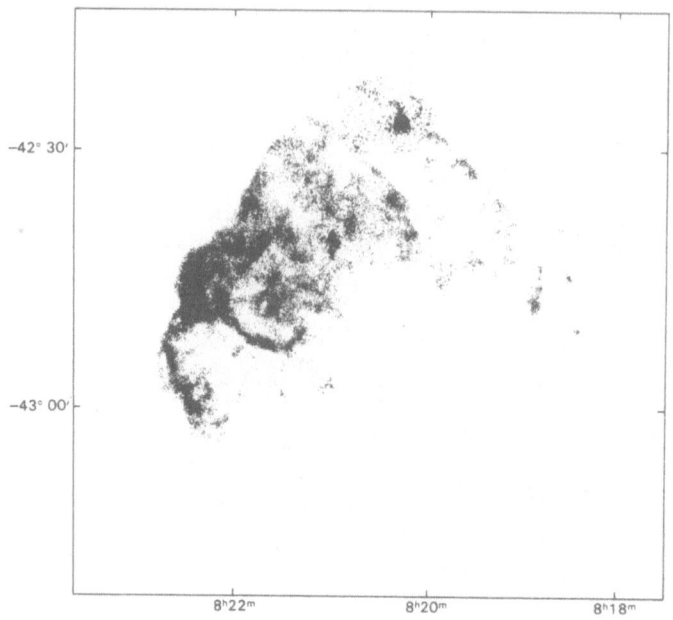

Figure 1: A high-resolution 0.1-4 keV photomosaic map of Puppis A.

appear in Petre et al. (1982b).

Puppis A

     Figure 1 shows a high-resolution X-ray image of Puppis A in the
energy range 0.1-4.0 keV. The map is an exposure-corrected photomosaic
of 11 Einstein High Resolution Imager (HRI) exposures, binned in 8"
pixels. The image reveals an ISM with structure on many scales. First,
there is a general decrease in X-ray surface brightness, by a factor of
at least 20, from northeast to southwest across the remnant,
perpendicular to the galactic plane. This effect is probably due to a
local density gradient in the ISM of at least a factor of 4 over a scale
of ~ 30 pc. Second, the surface brightness variations in the interior
and along the shell suggest a preshock ISM containing many small density
variations (factors of 2), about an average of ~ 1 cm^{-3}, over scales
from 0.2 to 10 pc. Despite the surface brightness variations along the
shell, the pronounced limb-brightening profiles are for the most part
consistent with that expected for an adiabatically expanding blast wave,
suggesting that the local clumpiness of the interstellar matter does not
prevent the adiabatic model from being locally valid. Finally, in
addition to the filamentary structure representing small density
perturbations, Puppis A contains two bright knots of emission along the
shell, one in the east and one in the northwest. These knots are
probably shocked interstellar clouds with preshock density of 10-20 cm^{-3}
and diameter 1-2 pc.

IC 443

Figure 2 depicts a 0.2-3.1 keV map of IC 443, overlayed on a Palomar Sky Survey red plate. The map is an exposure-corrected, Wiener-filtered photomosaic of three Imaging Proportional Counter (IPC) images,

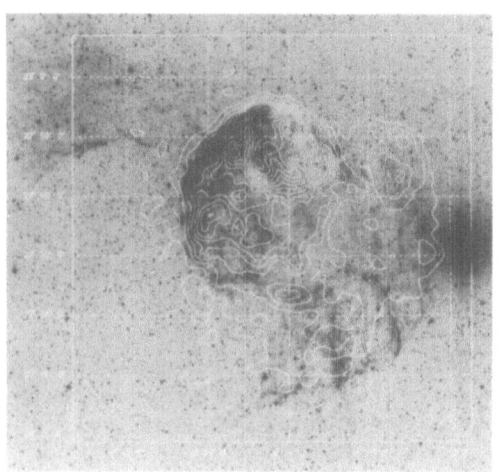 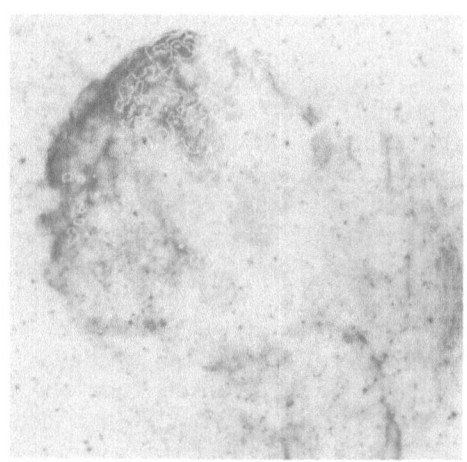

<u>Figure</u> <u>2</u> (left): A 0.2-3.1 keV image of IC 443, superposed on a Palomar Sky Survey red plate. This map is a photomosaic of three IPC images. <u>Figure</u> <u>3</u> (right): An HRI photomosaic of IC 443, superposed on the Palomar red plate.

binned in 32" pixels. The contours scale linearly with surface brightness. Although X-ray emission is detected from the entire SNR, the remnant does not appear to be limb-brightened. The average surface brightness in the vicinity of the bright northeastern optical filaments is a factor of ~ 4 higher than elsewhere. This network of filaments has apparently been created by the collision of the shock front with a giant molecular cloud, which is observed in CO to extend well beyond IC 443 to the northeast (Scoville <u>et</u> <u>al</u>. 1977; Cornett, Chin and Knapp 1977).

A sharper view of the northern portion of IC 443 is shown in Figure 3. This map is an exposure-corrected, Wiener-filtered photomosaic of three HRI images, with 16" binning, whose coverage of IC 443 is represented approximately by the area over which the grid appears. The contour levels were chosen to reveal the highest X-ray surface brightness features; X-rays were actually detected by the HRI from everywhere in the complex of optical filaments, with the notable exception of the finger of optical extinction extending northeastward into the filaments.

The HRI map emphasises the dominant X-ray feature, a ~ 15' diameter region which partially overlaps the optical filaments. The pressure of the X-ray emitting gas in this region, as inferred from spectral measurements by the Einstein SSS, is ~ 3 x 10^{-9} dyne cm^{-2}, placing it in approximate equilibrium with the cooler, denser, optically-emitting gas (Fesen and Kirshner 1980). Coupled with the general overlap between the regions of high X-ray surface brightness and the bright northern optical filaments, this suggests that the X-ray emitting gas may be interspersed among the sites of unstable radiative cooling which appear as optical filaments, and that both have arisen from the collision of the shock front with the molecular cloud.

The molecular cloud may affect the observed low-energy structure of IC 443 in a second way. As is visible in Figures 2 and 3, a lane of high optical extinction bisects IC 443 from northwest to southeast, extending a finger into the bright optical filaments. This lane demarcates another portion of the molecular cloud. The detection of shocked [H I] and a variety of molecular species with typical velocities around -30 km s^{-1} suggests that the front of the IC443 shell is colliding with the molecular cloud (see, e.g., Giovanelli and Haynes 1979; DeNoyer 1979). Although no general correlation between the X-ray surface brightness, as measured by the IPC, and the column density of the cloud, as inferred from the CO brightness temperature, is apparent, a correlation between column density and X-ray spectral hardness ratio

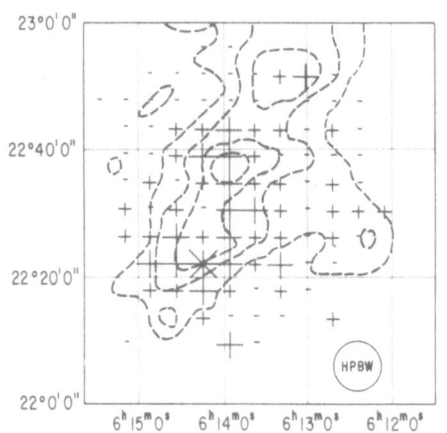

Figure 4: An IPC spectral hardness ratio map of IC 443, superposed on CO brightness temperature contours (from Cornett, Chin and Knapp 1977). Spectral hardness is defined as the ratio between 0.2-1.3 kev counts and 1.3-3.1 keV counts, with symbols of increasing prominence representing ratios of 0.6, 0.8, 1.0 and 1.2. Circle is HPBW of CO map.

(0.2-1.3 kev/1.3-3.1 kev) does exist. As seen in Figure 4, X-rays from regions coincident with the cloud are harder (average hardness ratio of ~ 0.8) than elsewhere in the remnant (average hardness ratio of ~ 0.6). To account for this increase of spectral hardness by absorption, a column density through the cloud of 2-4 x 10^{21} cm^{-2} is required (in addition to the column density to IC 443) if the characteristic temperature of IC 443 is uniform and within the range of 0.5-1.5 kev. This additional column density may be compared to the inferred average column density of H_2 through the cloud of 2 x 10^{21} cm^{-2} (i.e., 4 x 10^{21} cm^{-2} of [H I] – Cornett, Chin and Knapp 1977). It is thus probable that the cloud absorbs X-rays emitted by the interior of IC 443.

REFERENCES

Cornett, R.H., Chin, G., and Knapp, G.R. 1977, Astron. Ap., 54, pp. 889-894.

DeNoyer, L.K. 1979, Ap. J. (Letters), 228, pp. L41-L43.

Fesen, R.A., and Kirshner, R.P. 1980, Ap. J., 242, pp. 1023-1040.

Giovanelli, R., and Haynes, M.P. 1979, Ap. J., 230, pp. 404-414.

Petre, R., Canizares, C.R., Kriss, G.A., and Winkler, P.F. 1982a, Ap. J., 258, pp. 22-30.

Petre, R., Szymkowiak, A.E., Canizares, C.R., Seward, F.D., Willingale, R., Rolf, D., and Woods, N. 1982b, in preparation.

Scoville, N.Z., Irvine, W.M., Wannier, P.G., and Predmore, C.R. 1977, Ap. J., 216, pp. 320-328.

KINEMATICS OF SUPERNOVA REMNANTS IN THE GALAXY AND LMC

Peter Shull, Jr.
Max-Planck-Institut für Astronomie
and
Rice University

The optical emission lines of six SNRs have been observed at very high angular and kinematic resolutions. Kinematic ion temperatures were derived, and evidence was found in shocked regions for Maxwellian micro-turbulence on scales $\leq$ 0.01 pc, and for non-Maxwellian macroturbulence on scales > 0.1 pc. The widths of shocked regions in the Cygnus Loop and the existence of three types of spectral feature in the LMC remnants are discussed in terms of SNR evolution in cloudy interstellar media.

1. OBSERVATIONS AND RESULTS

The Cygnus Loop, IC 443, Puppis A, Vela X, N49, and N63A were observed with 4-m echelle spectrographs at Kitt Peak and Cerro Tololo with resolutions of 2" and 2.7 km s^{-1}. Widths, intensities, and radial velocities of bright spectral features (about 5" in size) visible in Hα, Hβ, and the doublets of [O III], [N II], and [S II] were measured. Thermal and non-thermal components of the Gaussians were determined from the differing widths of the H I and heavy-ion Gaussians. This report presents only portions of the results. Complete accounts may be found in P. Shull et al. (1982) and three in-press (1983) Astrophys.J. articles by P. Shull.

Bright knots in the smallest visible filaments of the galactic SNRs were observed. Mean values and standard deviations (not errors) of the line half-widths at half-maximum, v, corrected for the instrumental function, for knots in the eastern Cygnus Loop, are: <v> = 21 + 5, 14 + 5, and 18 $\pm$ 10 km s^{-1}, respectively, in H I, [N II], and [O III]. In the western part, the dispersions are 28 $\pm$ 4, 25 $\pm$ 8, and 29 $\pm$ 11 km s^{-1}.

The velocity dispersions for H I are larger than for [N II], as one would expect if thermal equilibrium prevailed. However, the [O III] dispersions are larger than the dispersions of H I and [N II]. On high-resolution spectra and photographs, the patterns of [O III] emission usually differ from the patterns of the other emission lines, indicating that the [O III] emission arises in physically distinct volumes. The mean ion temperatures and their standard deviations (not errors) for the

295

J. Danziger and P. Gorenstein (eds.), Supernova Remnants and their X-Ray Emission, 295–298.
© 1983 by the IAU.

galactic SNRs are in Table 1. The [O III] temperatures were derived by
assuming that the narrowest observed lines are purely thermal in origin.

TABLE 1. Mean Ion Temperatures of the Galactic SNRs

SNR	Region	<T([O III])> (1000 K)	<T([N II])>
Cygnus	E (SE2,3)	...	13 ± 4
	W	...	16 ± 9
	All 5 Regions	40-50	16 ± 9
IC 443	NE (A)	...	15 ± 9
	Both Regions	30	15 ± 9
Vela	SW (A)	...	27 ± 9
	SW (B)	...	22 ± 7
Puppis	NW (C)	...	19 ??

There is no apparent correlation between the nonthermal velocity
dispersions, u, within knots (microturbulence) and the standard
deviations of the radial velocities of the knots (macroturbulence). Both
types of turbulence are characterized by speeds of 10-30 km s^{-1}.

In LMC N49 and N63A, spectral features named spikes, narrow bands,
and broad bands may be distinguished on the basis of v and projected
expansion velocity (see Table 2). The bands expand relative to the
spikes, which are low-v, slit-length features with heliocentric radial
velocities typical of neighboring LMC H II regions. Kinematic ages for
N49 and N63A, based on maximum observed expansion velocities, are 16 000
and 10 000 years, respectively.

TABLE 2. Properties of Spikes and Bands in N49

Feature Type	v (km s^{-1}, HWHM)	v_{exp} (km s^{-1})	Surface Brightness
Spike	5-15	0	...
Narrow Band	30-60	100-140	Low
Broad Band	80-130	20-70	High

The mean values and standard deviations of v (corrected for
instrumental broadening) and of T in the eastern part of N49 are shown
in Table 3. For the spikes in N49 and N63A, the mean [O III], [N II],
and H I velocity dispersions become progressively larger. For the bands
of these remnants, however, the velocity dispersions of the heavier ion
species can exceed those of the lighter species. Therefore, the ion
temperatures derived for the bands will be useless (notice their large
standard deviations).

TABLE 3. Mean Velocity Dispersions and Ion Temperatures for N49 (East)

Feature Type	$\langle v \rangle$ (km s^{-1}, HWHM)			$\langle T \rangle$
	H I	[N II]	[O III]	(1000 K)
Spike	16 $\pm$ 5	12 $\pm$ 6	7 $\pm$ 1	16 $\pm$ 3
Narrow Band	53 $\pm$ 8	44 $\pm$ 9	47 $\pm$ 18	65 $\pm$ 69
Broad Band	88 $\pm$ 22	85 $\pm$ 15	109 $\pm$ 24	257 $\pm$ 336

2. INTERPRETATION

2.1. Cloud-Knot-Cell Hierarchy

The temperatures derived for the band features in N49 and N63A, too high for optically emitting regions, can be understood if the shocked medium consists of clouds (filaments about 1 pc in size), which contain knots (about 0.01 pc), which contain cells (unresolved). The knots' velocity distribution is non-Maxwellian, while that of the cells is Maxwellian. A knot and its cells are in thermal equilibrium.

With this model, the spectrograph slit (about 1" x 100") resolves the individual knots within clouds at 1 kpc (MWG SNRs); T and u can be derived because they are constant within a knot. At 50 kpc (LMC SNRs) the slit resolves only clouds; T and u cannot be derived because they vary from knot to knot within a cloud. Evidently, these variations become significant on length scales between 0.01 and 0.5 pc.

2.2. Knot Wandering

In high-resolution [OIII] and [S II] photographs of the eastern Cygnus Loop, the [O III] filaments are narrow (5"), and the knots are collinear; curiously, the [S II] filaments are broad (30"), and the knots are scattered. In contrast, the emission layer of a steady-state shock in the Cygnus Loop should be about 0".3 thick. However, the filament widths are comparable to the products of the knots' random velocities (10-30 km s^{-1}) with the emission lifetimes of postshock [O III] and [S II] gas, about 200 and 1000 years, respectively. Apparently, randomly wandering knots can broaden the filamentary structure of SNRs.

2.3 Cloud Accelerations and Velocity Dispersions

The overall symmetry in N49's spectra suggests that the SN produced one main blast wave. The acceleration of a cloud by a blast wave (Cox 1979) is proportional to (intercloud matrix density)/(cloud density). If the bands are produced by shocked clouds, the ratio of expansion speeds of the broad and narrow bands should give the relative densities of the clouds. Typical values of the narrow and broad bands' expansion speeds are 120 and 40 km s^{-1}, respectively, implying that broad bands are three times denser than narrow bands on the average. This is qualitatively confirmed by the greater surface brightness of the broad bands as compared to the narrow bands (Table 2). A 1000 km s^{-1} blast striking clouds having cloud/matrix preshock density ratios of 10 to 100 would accelerate them to 15-150 km s^{-1} (approximately what is observed).

The width of the band structure indicates that the blast produces shock wave ensembles as it propagates in the surrounding heterogeneous medium. The internal velocity dispersions, v, of clouds should be less than the velocities of the slow (optical) shocks within the clouds, postulated to be 70-120 km s^{-1}. Indeed, v is 30-130 km s^{-1}. If these shocks conserve kinetic energy, then v should be proportional to dρ $\rho^{-3/2}$ where dρ is the internal clumpiness of a cloud. Since v for the narrow bands is about 2-3 times smaller than for the broad bands, and broad bands have densities three times higher than narrow bands, the narrow-band clouds should be 15 times less clumpy than the broad-band clouds.

2.4. The N49 and N63A Spikes

The spike of N49 probably results from an H II regions ionized by UV photons from the cloud shocks. It has the spectral and kinematic proper-ties of nearby LMC H II regions, and exactly the same spatial extent as N49. Its brightness correlates with the SNR's band structure. J. Shull and McKee (1979) model a shock and its UV precursor flux for LMC-like abundances. The Stromgren layer and line intensities calculated from this model (corrected for the model's N overabundance) agree with observations although problems concerning the brightness of the spike remain.

For the same reasons as in N49, the N63A spike may be associated with an H II region. Actually, two superposed spikes are observed. The fainter, longer spike results from an extended region around the SNR. The shorter, brighter spike is produced by a round "blob" near the SNR's southwestern limb. One or both of these regions may be ionized by a star in the nearby cluster NGC 2030, although available data indicate that no star is hot enough ($T_{eff} \gtrsim$ 35000 K) to produce the high [O III]/Hα ratio observed in the extended region. Possibly the extended region, and maybe also the blob, are fossil regions, ionized by the SN progenitor or the UV/X-ray burst accompanying the SN explosion. Results based on the model burst spectrum by Falk (1978), especially involving the relative decreases of the [O III]/Hα ratios in both regions, support this idea.

It is a pleasure to acknowledge discussions with R. Dufour and R. Parker, NSF grant AST 80-07450 and NASA Contract 09-15940 to Rice University, and observing time grants from Cerro Tololo and Kitt Peak.

REFERENCES

Cox, D.P.: 1979, Astrophy.J., 234, p. 863.
Falk, S.W.: 1978, Astrophys.J.Lett., 225, p. L133.
Shull, J.M., and McKee, C.F.: 1979, Astrophys.J., 227, p. 131.
Shull, P., Jr.: 1982, Astrophys.J., 253, p. 682.

SUPERNOVA REMNANTS RESEMBLING THE CRAB NEBULA

K. W. Weiler
National Science Foundation, Washington, D.C. 20550, U.S.A.

While reviewing and systematizing the properties of the class of supernova remnants resembling the Crab Nebula it has been found that supernova remnants can be split into three morphological groups -- Class S (shells), Class P (plerions), and Class C (combinations) -- where the Class C objects appear to represent a new and especially interesting classification. In this overview, the identifying properties of all three classes are defined. Because the large Class S has been studied in detail many times previously, it is not discussed further here. For the smaller Classes P and C, the individual members and suspected members are presented and their properties reviewed. Finally, an origin and evolution for each class is suggested.

I. INTRODUCTION

In preparing a review of the members and radio properties of the galactic supernova remnants (SNR) which resemble the Crab Nebula (plerionic supernova remnants or plerions), it has become apparent that SNR exhibit a wider variety of forms than was previously recognized. Present information indicates that they can be divided into three morphological classes:

1) Class S (Shell supernova remnants)
2) Class P (Plerionic supernova remnants)
3) Class C (Combination supernova remnants).

Each of these will be discussed in turn.

II. CLASS S (SHELLS)

Of the 150-200 supernova remnants known in our Galaxy, it is apparent that the vast majority (>80%) are shells. Their identifying properties in each of the principal wavelength ranges are summarized in Table 1. Because they are the most common type of SNR, many members of the class have been studied in great detail and many reviews of their properties have been written. Thus, we will not concern ourselves

J. Danziger and P. Gorenstein (eds.), Supernova Remnants and their X-Ray Emission, 299-320.
© *1983 by the IAU.*

further with them here.

Table 1. Identifying Properties of Class S

       Radio
       1) Non-thermal emission
       2) Shell or partial shell form
       3) Spectral index $\alpha \sim -0.45$ ($S \propto \nu^{+\alpha}$)
       4) Weak integrated linear polarization
       5) Rough adherence to a Σ-D relation
       6) Circumferential or tangential magnetic field
       7) Shock wave generated

       Optical
       1) Thermal line-emitting filaments
       2) Shell or partial shell organization of filaments
       3) Shock wave generated

       X-Ray
       1) Thermal emission
       2) Both shell and filled-center forms seen
       3) Shock wave generated

III. CLASS P (PLERIONS)

The Class P remnants form a much smaller group of SNR. Only about 5% or
fewer of all known supernova remnants appear to fit the identifying
properties.

Table 2. Identifying Properties of Class P

       Radio
       1) Non-thermal emission
       2) Filled-center form extended emission
       3) Spectral index flat ($\alpha > -0.3; S \propto \nu^{+\alpha}$)
       4) Strong integrated linear polarization at high ν
       5) Well organized magnetic field
       6) General adherence to S_θ^2-d relation
       7) Central energy source

       Optical
       1) Non-thermal emission (?)
       2) Filled-center form extended emission
       3) Thermal line-emitting filaments
       4) Central energy source

       X-Ray
       1) Non-thermal emission

2) Filled-center form extended emission
3) Compact source present
4) Central energy source

γ-Ray
1) Emitter (?)

Even though the existence of this class resembling the Crab Nebula
was recognized over ten years ago (Weiler and Seielstad, 1971; Milne,
1971) and there have been many attempts to push any unusual object into
the class (see, e.g., a recent review by Weiler and Panagia, 1980), the
number of "true" plerions still appears to be very small. Only eight
reasonable examples are known at present and only four of these are well
established. The properties of the eight candidates are shown in Table
3 and discussed individually below.

G5.3-1.1 (Milne 56) - Table 3, Figure 1

After the detection of a number of x-ray sources near the galactic
center by Bradt et al. (1971), one, GX5-1, was suggested by Milne and
Dickel (1971) to be associated with a nearby supernova remnant G5.3-1.1.
They showed that the remnant was indeed linearly polarized, and thus
non-thermal, and had a filled-center form with the x-ray source included
within the outer contours. The spectral index appeared to be straight
from 0.6 to 5 GHz and unusually flat with an index of α=-0.2. They
suggested a distance of 2.4 kpc. Clark et al. (1975) confirmed the flat
radio spectrum down to 0.4 GHz and suggested that it might belong to the
class of "amorphous" supernova remnants. Angerhofer et al. (1977) also
confirmed the flat spectrum. Zealey et al. (1979) observed optical
filaments in the vicinity of the radio source and re-emphasized the
plerionic properties.

The properties of G5.3-1.1 are not well studied and the possible
association with GX5-1 is not proven. Even the detection of a weak 1.4
GHz radio source at the x-ray position (Braes et al., 1972) does not
necessarily establish a connection with the large radio supernova
remnant. However, until more detailed studies are carried out in the
radio and x-ray ranges, G5.3-1.1 must be considered as a possible
plerion.

G21.5-0.9 - Table 3, Figure 2

The object has the filled center form, linear polarized radio emission,
and straight and flat (α∼0) radio spectrum typical of the class. The
plerionic nature of the source, which has been discussed earlier by
Wilson and Weiler (1976b), Becker and Kundu (1976), and Weiler and
Panagia (1980), has been confirmed by Becker and Szymkowiak (1981) with

the detection of a filled-center, extended x-ray source coincident with the remnant. The properties listed in Table 3 are estimated for a distance of 15 kpc, the distance felt to be the most nearly correct (Weiler and Panagia, 1980; Becker and Szymkowiak, 1981). However, because HI absorption measurements (Caswell et al., 1975a) permit a smaller distance of $\sim$ 4.8 kpc, which is often used in the literature, calculated properties for the source at that distance are also included in the table.

G57.6-0.3 (4C21.53W) - Table 3, Figure 3

G57.6-0.3 is an unusual source which has been proposed by Erickson (1982) as a possible plerion. It has a filled center morphology and, at high frequencies (ν>150 MHz), a flat radio spectrum (α=-0.26). There is also a nearby (but not coincident) compact x-ray source (Erickson, 1982). The radio spectrum has a most unusual steepening at low frequencies (α=-2.4; ν<150 MHz) and there is no evidence for extended x-ray emission (Becker, private communication). No linear polarization is detected at either 6 or 20 cm (Becker, private communication). Direct application of the $S\theta^2$-d relation or a Σ-D relation yields the improbably large distance of $\sim$40 kpc. However, if a reasonable distance of $\sim$5 kpc is assumed for a galactic source, the object appears underluminous in the radio with respect to other plerions.

From the little known about it, the properties of G57.6-0.3 could also be consistent with those of an HII region or an extra-galactic source, but the unusual spectrum fits no single type of source well. Improved measurements are obviously needed at all wavelengths to determine the true nature of G57.6-0.3.

G74.9+1.2 (CTB87) - Table 3, Figure 4

An excellent example of the class, the radio properties of G74.9+1.2 have been thoroughly discussed by Weiler and Panagia (1980) and its plerionic properties by Weiler and Shaver (1978). The plerionic nature of the source has been confirmed by Wilson (1980) with the detection of a filled-center, extended x-ray source co-incident with the remnant.

G130.7+3.1 (3C58) - Table 3, Figure 5

The radio properties of G130.7+3.1 were first investigated by Weiler and Seielstad (1971) who suggested that it resembled the Crab Nebula. Wilson and Weiler (1976a) studied the source in more detail and Weiler and Panagia (1980) have discussed it in relation to other plerions. Van den Bergh (1978a, b) has found faint optical filaments at the position of the remnant and Weiler (1980) has discussed a correlation (like that known in the Crab Nebula) between the filaments and enhanced radio emission. Even though G130.7+3.1 has long been considered to be "the second Crab Nebula," strong confirmation by the detection of both compact and extended x-ray emission co-incident with the radio remnant has been gratifying (Becker et al., 1982).

The distance of >8 kpc given in Table 3 is estimated as a lower limit from HI absorption measurements (Hughes et al., 1971; Goss et al., 1973; Williams, 1973). A controversy has recently arisen due to new HI absorption measurements suggesting a smaller distance to the source (Green and Gull, 1983). However, with three independent older measurements in agreement and some question remaining concerning the sensitivity and spatial resolution of the new results, the best estimate of the distance remains >8 kpc.

G130.7+3.1 has been rather well connected to the supernova of AD1181 (Clark and Stephenson, 1977) and a Type II supernova origin (Panagia and Weiler, 1980). Thus, it is one of the small number of supernova remnants where the age and supernova type are reasonably well established.

G148+1 - Table 3

G148+1 is a small x-ray nebula $\sim$ 5' long trailing behind a pulsar with a period of ~ 0.15. No radio emission is yet known and there seems to be little information available except the description presented by Helfand (1981, 1983). Although more measurements are needed to determine its nature, since G148+1 does show the x-ray properties of a plerion, it is included here. V. Radhakrishnan (private communication) has suggested that the source might represent the ultimate state of a very old plerion.

G184.6-5.8 (Crab Nebula) - Table 3, Figure 6

The Crab Nebula is certainly one of the most studied objects in the Universe. Its plerionic properties have long been known and until other members of the class were identified, it was either considered unique or lumped together with the Class S supernova remnants. The radio properties were studied in detail in a series of papers in the early 1970's (Wilson, 1972 and following) and more recently by Swinbank and Pooley (1979 and following). The Crab is, of course, optically identified and shows both thermal optical filaments and non-thermal diffuse optical radiation (see e.g. van den Bergh et al., 1973). It is, in fact, the only galactic source proven to emit extended, non-thermal continuum radiation in the optical range, although Vela X possibly does so (Weiler and Panagia, 1980). The Crab is known to contain a compact x-ray source (the pulsar) surrounded by extended x-ray emission. Gamma rays have been detected from the Crab pulsar.

The Nebula is well connected with the supernova of AD1054 and the distance is generally accepted to be ~ 2 kpc. The supernova type of SN1054 has long been debated and was, for some time, considered to have been Type I. However, more recent results establish it rather firmly as having originated in a massive star (Hillebrandt, 1982; Davidson et al., 1982) and probably a Type II supernova (Chevalier, 1977).

An optical "halo" around the Crab has recently been reported (Murdin and

Clark, 1981) but Wilson and Weiler (1982) find that if the "halo" is evidence for a shell around the Crab, it is atypically weak in the radio range.

G328.4+0.2 (MSH15-57) - Table 3, Figure 7

G328.4+0.2 is often included in the lists of plerionic supernova remnants. It was studied by Goss and Shaver (1970) and Shaver and Goss (1970a) in their galactic survey work and included as a "possible" plerion by Weiler and Panagia (1980). It has a number of the radio properties of a plerion, including amorphous shape and flat spectrum (α =-0.24). Although it does not seem to have been studied for linear polarization, it has always been considered non-thermal in spite of its relatively flat spectrum. There appears to be no H109α emission (Shaver and Goss 1970b). No detection of x-ray emission from the source has yet been reported.

In spite of the relative lack of information and the fact that a thermal and/or extragalactic identification cannot be completely ruled out, G328.4+0.2 must be considered as a possible plerion due to its spectrum and radio morphology. Its known properties, however, are not entirely consistent. Caswell et al. (1975a) obtain a minimum distance of $\sim$20 kpc from HI absorption measurements. If this is correct, it implies from standard models that the object is half as old and twice as luminous as the Crab Nebula and expanding at the incredible average speed of 40,000 km s^{-1}. If the model of Weiler and Panagia (1980) is applied, the values (marked with superscript ^{d} in Table 3) are forced into a "reasonable" range but imply a much smaller distance. Thus, it appears that G328.4+0.2 is either mis-classified, much closer than 20 kpc, or a very extreme example of a plerion.

Thus, in spite of years of searching and many researchers trying to push new objects into the class, it appears that there are a maximum of 8 known Class P "true" plerions and two or three of these may be suspect. This implies that:

(a) plerions are short lived as Weiler and Panagia (1978) have suggested; or
(b) plerions are born only very infrequently as has been suggested by Srinivasan and Dwarakanath (1982); or
(c) both of the above.

However, the search for new objects has not been wasted. It has led to the development of what appears to be an entirely new class of very unusual objects -- the Class C plerion-shell combinations.

Table 3: CLASS P - Plerions

Name Galactic	Other	Position (1950) RA h m s	DEC ° '	Date SN	Age 10^3 Yrs	d kpc	$|z|$ pc	Size arcmin	pc	Ave. Expand Vel. km s^{-1}	Spectral Index $S_\alpha \nu^{+\alpha}$	S(1GHz) Jy	L_{Radio} 10^7-10^{11}Hz erg s	X-Ray Struct.	Comments	Fig.	References
G5.3-1.1	Milne 56	17 58 30	-24 50		12[a]	3[a]	60[a]	30x30	28x28[a]	1100[a]	-0.2	37	$1.9 * 10^{34}$	CMPT?	possible	1	2,3,4,5,6
G21.5-0.9		18 30 47	-10 36		[2.3[a]	[15, 4.8]	[230, 75]	1.3x1.3	[5.7x5.7, 1.8x1.8]	[1200[a]]	0.0	6.4	[$1.4 * 10^{35}$, $1.5 * 10^{34}$]	EXTD		2	1,7,20
G57.6-0.3	4C21.53W	19 37 30	+21 31		[1.7[a]	[40[a], 5]	[200[a], 26]	1.7x0.7	[19x8, 2.5x1.0]	[4000[a]]	-0.26	1.2	[$8.1 * 10^{34a}$, $1.3 * 10^{33}$]	CMPT?		3	8
G74.9+1.2	CTB87	20 14 10	+37 04		2.8[a]	≥12	250	9.4x5.9	33x21	4900[a]	-0.24	8.6	$5.4 * 10^{34}$	EXTD		4	1,9,10,21
G130.7+3.1	3C58	02 01 52	+64 35	1181	0.8	≥8	430	10x6	23x14	11000	-0.09	33	$1.5 * 10^{35}$	[CMPT EXTD]		5	1,11,12,13
G148+1		03 55	+54		No radio emission known at present									[CMPT EXTD]	[nature uncertain]		19
G184.6-5.8	Crab Neb.	05 31 31	+21 59	1054	0.9	2	200	7x5	4x3	1900	-0.26	1000	$1.6 * 10^{35}$	[CMPT EXTD]	[pulsar γ-rays]	6	1,14
G328.4+0.2	MSH15-57	15 51 44	-53 08		[2.6[a]	[≥20[a], 9]	[70[a], 30]	6x5	[35x30, 16x14[a]]	[2900[a]]	-0.24	15	[$2.6 * 10^{35}$, $5.7 * 10^{34a}$]		possible	7	[1,15,16,17,18]

[a]Estimated through the use of relations derived by Weiler and Panagia (1980)

References
1. Weiler and Panagia, 1980
2. Zealey et al., 1979
3. Milne and Dickel, 1971
4. Clark et al., 1975
5. Angerhofer et al., 1977
6. Bradt et al., 1971
7. Becker and Szymkowiak, 1981
8. Erickson, 1982
9. Wilson, 1980
10. Weiler and Shaver, 1978
11. Wilson and Weiler, 1976a
12. van den Bergh, 1978b
13. Becker et al., 1982
14. Wilson, 1972
15. Caswell et al., 1975a
16. Caswell et al., 1980
17. Goss and Shaver, 1970
18. Shaver and Goss, 1970a
19. Helfand, 1981, 1982
20. Wilson and Weiler, 1976b
21. Kazes and Caswell, 1977

IV. CLASS C (COMBINATIONS)

The Class C-Combination sources are poorly studied at present. Their morphology, however, appears to be describable as a simple combination of a Class S-Shell with a Class P-Plerion (see Tables 1 and 2). Oddly enough, the Class already contains more objects than there are "true" plerions and, because it is so poorly studied, essentially all members offer extremely interesting examples for further study. The properties of the proposed members of the class are shown in Table 4 and discussed individually below.

G6.5-0.1 (W28) - Table 4, Figure 8

G6.5-0.1 is a large, shell-shaped, non-thermal supernova remnant located in a complicated region of the galactic plane. Observers have mapped its radio emission at a number of frequencies and distinguish a partial shell approximately 40' in diameter with linear polarization and a normal spectral index of $\alpha \sim -0.4$ (see e.g. Goudis, 1976; Dickel and Milne, 1976; Milne and Wilson, 1971; Kundu and Velusamy, 1972; Altenhoff et al., 1978). However, it contains a compact region, known as G6.6-0.1, which has a flat radio spectrum ($\alpha \sim -0.2$) but no linear polarization (Becker, private communication) and a nearby compact x-ray source (Andrews et al., 1982). There is also extended x-ray emission from the area (Long, 1979). Although the positional accuracy is not good ($\sim +1^{\circ}$), there is known to be a γ-ray source (2CG006-00) near W28 (Swanenburg et al., 1981).

The general source properties are described in Table 4, with those of the plerion (P) separated as well as possible from those of the shell (S). Although more detailed studies are needed, W28 appears to be a possible member of Class C.

G27.3+0.0 (Milne 62, Kes 73) - Table 4, Figure 9

G27.3+0.0 is a difficult source to discuss. It is in a complicated region of the Galaxy and the source structure itself is complex. Although it has an apparent shell shape, it may consist of unrelated parts both thermal and non-thermal. Angerhofer et al. (1977) and the references in Table 4 provide a good discussion of what is known about the source.

The source is included here because it is reputed to contain a compact x-ray source and thus may have at least one of the properties of the Class C objects. However, more detailed information is necessary to establish its nature.

G29.7-0.3 (Kes 75) - Table 4, Figure 10

G29.7-0.3 was classified as a supernova remnant in older catalogues due to its being extended and having a relatively steep spectrum ($\alpha \sim -0.7$).

However, it was not until the source was well resolved by Becker and
Kundu (1976) that it was shown to consist of several components with
differing spectral indices. Recent work by van Gorkum, Shaver, and
Salter (private communication) has now shown that G29.7-0.3 consists of
a steep spectrum shell surrounding a flat spectrum filled-center
component (Fig. 10). Extended x-ray emission is associated with the
central component. The distance to G29.7-0.3 is estimated to be ~7 kpc
(Caswell et al., 1975a) from HI absorption measurements. From the
observations presently available, G29.7-0.3 appears to be an excellent
example of a plerion-shell combination.

G34.6-0.5 (W44) - Table 4, Figure 11

G34.6-0.5 is a large shell-type radio supernova remnant at a distance,
from HI absorption measurements by Caswell et al. (1975a), of ~3 kpc.
Except for a slightly, but not unusually, flat radio spectrum (α~-0.3)
its radio properties generally resemble a normal Class S SNR.

However, x-ray observations have been reported (Pounds, 1980) from the
Einstein IPC indicating that G34.6-0.5 does not show the usual thermal,
shell-shaped x-ray emission of normal Class S remnants. It has an
asymmetric, centrally peaked x-ray structure lying in the western half
of the radio map. The x-rays from the center of the shell are soft, but
the spectrum is harder from part of the region. This makes the remnant
somewhat unusual.

While this is rather tenuous evidence from which to determine that
G34.6-0.5 is anything but a normal, shell-type SNR, it is included here
as a possible Class C remnant until improved radio and x-ray information
becomes available.

G39.7-2.0 (W50) - Table 4, Figure 12

G39.7-2.0 is an object of currently great interest not so much for its
supernova remnant properties, although it is unusual in being a filled
shell, but because it contains, and is presumably powered by, the
unusual "star" SS433. The radio remnant has an elongated shape,
(Geldzahler et al., 1980; Downes et al., 1981) somewhat resembling the
elliptical form of G130.7+3.1 (3C58) but with the sharp outer edge
typical of a Class S-shell supernova remnant. The integrated spectral
index of the source is exactly that of an average Class S source
(α=-0.45) but the spectral index distribution is still not well known.
Associated with SS433 is both the compact and extended x-ray emission
(Seward et al., 1980) typical of plerions.

Weiler and Panagia (1980) and Panagia and Weiler (1981) have suggested
that G39.7-2.0 represents an extreme case of a shell-plerion combination
where the plerion completely fills the shell. SS433 then serves as the
energy source for the plerion in much the same way as the pulsar PSR0532
does in the Crab Nebula. This model predicts G39.7-2.0 will show a
definite spectral index change from the "steep" spectrum shell (α ~

-0.45) to the "flat" spectrum plerionic filling ($\alpha > -0.3$) and implies
that the remnant is quite old ($\sim$30,000 years) and expanding very slowly
($\sim$300 km s^{-1}).

G68.9+2.8 (CTB80) - Table 4, Figure 13

Although the Class C sources are often unusual, G68.9+2.8 has to qualify
as one of the most odd. Radio maps (Velusamy et al., 1976; Angerhofer
et al., 1981) show it to consist of a very complicated, steep spectrum
extended region with a smaller, flatter spectrum, possibly plerionic,
"plateau" containing a compact, very flat spectum radio core. The
source also contains both compact and extended x-ray emission from the
"core" (Becker et al., 1982). Despite the lack of a clear shell shape
in its large scale structure, measurements of optical line-emitting
filaments in the area confirm the identification as a supernova remnant
(Angerhofer et al., 1980). Angerhofer et al. (1980) estimate the
distance to be $\sim$3kpc from HI absorption measurements. The suggestion
has been made that G68.9+2.8 is the remnant of the supernova of AD1404
(Strom et al., 1980).

Although the compact "core" of the source exhibits definite plerionic
properies with resolved, filled-center radio structure at high
resolution, a flat radio spectrum, and x-ray emission, the question
remains whether the larger ($\sim$10'), somewhat steeper spectrum "plateau"
is also part of the plerion. Until better information is available, we
have included it as such. One must question further whether these are
associated with the large scale "shell" or "shells." Again, for lack of
other evidence, we have assumed so.

The possible association of the remnant with the "guest star" of AD1404
appears unlikely. Not only would such an age lead to an extremely high
average expansion velocity for the extended remnant, Clark and
Stephenson (1977) in their study of historical Chinese records accord
the "guest star" a low probability of having been a supernova.

G93.7-0.3 (DA551, CTB104A) - Table 4, Figure 14

G93.7-0.3 possesses an unusual filled-shell radio morphology which
reminds one of G39.7-2.0 (W50) (Mantovani, et al., 1982). Also, its
integrated spectral index of $\alpha \sim -0.3$ is somewhat flat for a normal Class
S supernova remnant, although it is certainly within the deviation for
normal shells. A weak, steep spectrum, compact radio source is known to
exist within the shell at α(1950) = 21^{h}27^{m}13^s.1, δ(1950) = +50^{0}23'37" but
its connection, if any, with the remnant is unknown. Although G93.7-0.3
is included here as a possible Class C source, its membership in the
class remains to be proven.

G263.9-3.3 (Vela XYZ) - Table 4, Figure 15

The Vela supernova remnant has recently been discussed in great detail
by Weiler and Panagia (1980) and needs little further discussion here.

It is probably the best prototype of the Class C sources combining a large, non-thermal, steep spectrum radio shell (Vela YZ), with associated thermal x-ray emission, and a non-thermal, flat spectrum, plerionic radio component (Vela X), with non-thermal compact and extended x-ray emission. Its pulsar is also a source of γ-rays.

The Vela remnant has extensive, thermal, optical filamentary emission and a relatively well established distance of $\sim$0.5 kpc. From the pulsar spin down rate, the age of the remnant is known to be $\sim$12,000 years. After the Crab Nebula, Vela X provides the best chance for detecting non-thermal optical emission from a galactic supernova remnant (see e.g. Weiler and Panagia, 1980).

G320.4-1.2 (MSH15-52) - Table 4, Figure 16

G320.4-1.2 appears at first glance to be a normal Class-S remnant (Caswell et al., 1981) with clear optical thermal filaments (van den Bergh, 1978b) and an HI absorption distance of $\sim$4kpc (Caswell et al., 1975a). However, Seward and Harnden (1982) have found the Class C properties of extended x-ray emission from the source and an x-ray pulsar within its confines. The pulsar has the fifth shortest period ($\sim$0.15 s) and the greatest rate of increase of period of any pulsar known. Radio pulses have also been detected (McCulloch et al., 1982; Manchester et al., 1982).

Although there is 1.4 GHz radio emission exceeding 30 mJy per beam area in the area of the pulsar, there is no evidence for a prominent plerion. More detailed radio studies and a spectral index distribution measurement will be necessary to determine if the emission is unusual. However, an obvious discrepancy exists. A shell-type supernova remnant as large as G320.4-1.2 would normally be considered quite old ($>10^4$ years) while the dynamic age of the pulsar is quite young ($\sim$1600 years). Although a recent model by Srinivasan et al. (1982) can explain the discrepancy, proving physical association between the two objects and determining the details of their radio and x-ray properties is still necessary.

G326.3-1.8 (MSH15-56, Kes 25) - Table 4, Figure 17

G326.3-1.8 was the first object proposed as a Class C, shell-plerion combination. It is a perfect example of a steep spectrum radio shell with a flat spectrum plerionic remnant within its confines. Zealey et al. (1979) have found thermal optical filaments associated with the shell confirming its supernova remnant nature. Caswell et al. (1975a) obtained two possible distances (1.5 kpc and 4.6 kpc) for the remnant from HI absorption measurements and Weiler and Panagia (1980) argue for the correctness of the greater distance.

In spite of being an excellent example of a Class C remnant in its radio properties, G326.3-1.8 shows only diffuse emission associated with its shell in the x-ray range. It has no x-ray enhancement in the vicinity of the radio plerion (R.H. Becker, private communication).

G327.4+0.4 (Kes 27) - Table 4, Figure 18

G327.4+0.4 has a partial shell shape (Caswell et al., 1975b) and strong
linear polarization confirming its non-thermal nature (Dickel and Milne,
1976). Its distance is estimated to be ~6 kpc (Lamb and Markert, 1981).
G327.4+0.4 shows both compact and extended x-ray emission near its
center (Lamb and Markert, 1981) which is near, but not co-incident with,
a radio peak. A comparison of the 0.4 GHz and 5 GHz maps of Caswell et
al. (1975b) suggests that this peak may have a flatter radio spectrum
than the larger shell. G327.4+0.4 is also located within the error
circle for the γ-ray source CG327-0 and may be associated with it (Lamb
and Markert, 1981). The source thus appears very similar in morphology
to G29.7-0.3 (Kes 75) and is a good candidate for being a Class C
remnant.

G332.4-0.4 (RCW103) - Table 4, Figure 19

G332.4-0.4 appears to be a normal Class S remnant with a well defined
radio shell shape, a spectral index of $\alpha \sim -0.5$ (Caswell et al., 1980),
and a bright filamentary shell (van den Bergh et. al., 1973).
Additionally, low resolution x-ray measurements show a circular region
of presumably thermal emission co-incident with the radio shell (Lamb
and Markert, 1981). There does remain some discrepancy over the
probable distance to the object, with ~3 kpc being obtained from HI
absorption measurements (Caswell et al., 1975a) and ~8 kpc from Σ-D
estimates (Caswell et al., 1980). For Table 4 we have taken an average
distance of ~6 kpc to estimate the intrinsic source properties.

G332.4-0.4 is unusual in that it contains a compact x-ray object
centered on the shell remnant (Tuohy and Garmire, 1980) and is possibly
associated with the γ-ray source CG333+0 (Lamb and Markert, 1981).
There is no apparent excess radio emission at the x-ray source position
and the radio shell does not appear unusually "filled." However, until
more detailed radio and x-ray information becomes available, G332.4-0.4
must be considered a possible Class C remnant.

Even from the small number of examples presently available and the
probability that some of the suggested members of Class C will prove to
be misclassified, it appears that Class C (shell-plerion combinations)
is larger than Class P ("true" plerions). Further, it is likely that
the former class will grow as more Class S (shell) SNR are studied in
detail. Class C is, in any case, unusual and provides an interesting
area for study.

Although it is risky to split Class C, containing at most a dozen
objects at present, into subclasses, there appears to be at least weak
evidence for doing so. A possible subgrouping is illustrated in Figure
20.

Table 4: CLASS C - Combinations

Name Galactic	Other	Position (1950) RA h m s	DEC ° '	Date SN	Age 10³ Yrs	d kpc	/z/ pc	Size arcmin	pc	Ave. Expand Vel. km s⁻¹	Spectral Index S α ν⁺ᵃ	S (1GHz) Jy	L Radio 10⁷-10¹¹ Hz erg s⁻¹	X-Ray Struct.	Comments	Fig.	References
G6.5-0.1	W28	17 57 47	-23 20			2	3.5 [P S]	1 / 40	0.6 / 23		-0.2 / -0.4	~3 / 330	6.0*10³² / 3.5*10³⁴	[CMPT / EXTD?]	γ-rays?	8	2,3,4,5,6
G27.3+0.0	[Milne 62 / Kes 73]	18 38 20	-05 06			Poorly defined remnant, structure uncertain								CMPT?	Class C?	9	7,8,9,10,11
G29.7-0.3	Kes 75	18 43 48	-03 02			7	37 [P S]	0.5 / 3	1 / 6		-0.15 / -0.65	8.5	6.6*10³³	[CMPT / EXTD]		10	12,13,14,15
G34.6-0.5	W44	18 53 45	+01 13			3	25	27	24		-0.3	230	7.4*10³⁴	EXTD	Class C?	11	3,5,12,29,33
G39.7-2.0	W50	19 09 21	+04 54			3	100	130x65	110x55		-0.45	86	1.8*10³⁴	[CMPT / EXTD]		12	[1,3,16,17, 18,19]
G68.9+2.8	CTB80	19 51 36	+32 49	1404?	0.6?	~3	150 [P S]	10x6 / 40	9x5 / 35	60000? / 300000?	0.0 / -0.6	2 / 100	1.8*10³³ / 1.4*10³⁴	[CMPT / EXTD]		13	20,21,22,23
G93.7-0.3	[DA551 / CTB104A]	21 28	+50 30			~2	10	90	50		-0.3	40	~6*10³³		Class C?	14	8,24
G263.9-3.3	Vela XYZ	08 33 05	-45 37		12	0.5	30	210x110 / 256	31x17 / 37	1000 / 1500	-0.1 / -0.6	1100 / 650	2.0*10³⁴ / 2.6*10³³	[CMPT / EXTD]	[pulsar / γ-rays]	15	1
G320.4-1.2	[Kes23 / MSH15-52 / RCW89]	15 10 30	-59 05			4	85	30	35		-0.3	70	3.6*10³⁴		pulsar	16	12,25,26,27
G326.3-1.8	[Kes25 / MSH15-56]	15 48 24	-56 04			5	160	15x8 / 36	20x12 / 30		-0.1 / -0.4	40 / 100	7.0*10³⁴ / 6.6*10³⁴		pulsar	17	1,12,28,29
G327.4+0.4	Kes27	15 45	-53 12			6	40	20	35		-0.6	33	1.9*10³⁴	[CMPT / EXTD]	γ-rays?	18	30,31
G332.4-0.4	RCW103	16 13 48	-50 55			6	40	10	18		-0.5	27	1.9*10³⁴	CMPT	γ-rays?	19	12,36,38

P = plerionic component
S = shell component

References for Table 4

 1. Weiler and Panagia, 1980
 2. Andrews et al., 1982
 3. Clark and Caswell, 1976
 4. Milne and Wilson, 1971
 5. Kundu and Velusamy, 1972
 6. Altenhoff et al., 1978
 7. Angerhofer et al., 1977
 8. Velusamy and Kundu, 1974
 9. Milne and Dickel, 1974
10. Dickel and Milne, 1976
11. Caswell and Clark, 1975
12. Caswell et al., 1975a
13. Shaver and Goss, 1970a
14. Becker and Kundu, 1976
15. Becker, 1982
16. Geldzahler et al., 1980
17. Downes et al., 1981

18. Seward et al., 1980
19. Panagia and Weiler, 1981
20. Angerhofer et al., 1980
21. Angerhofer et al., 1981
22. Velusamy et al., 1976
23. Becker et al., 1982
24. Mantovani et al., 1982
25. Caswell et al., 1981
26. van den Bergh, 1978b
27. Seward and Harnden, 1982
28. Zealey et al., 1979
29. Clark et al., 1975
30. Lamb and Markert, 1981
31. Caswell et al., 1975b
32. Tuohy and Garmire, 1980
33. Pounds, 1980

V.   SUMMARY

        The purpose of this review has mainly been to classify and briefly
discuss the known supernova remnants without an attempt to develop
models or explanations for their origin and evolution. However, from
what is already known about several members of the classes, it is
possible to assemble a somewhat speculative view of their origins. This
is presented in Table 5. Detailed discussion, however, is reserved for
future work.

Table 5.   Origin

          Class S (Shell)
          1)   White dwarf, low mass, old star
          2)   Type I optical supernova
          3)   No radio supernova
          4)   Complete stellar disruption
          5)   Strong shock wave formation
          6)   Shell-type remnant
          7)   Remnant long lived

          Class P (Plerion)
          1)   Massive, young star
          2)   Type II optical supernova
          3)   Radio supernova
          4)   Stellar remnant remains (pulsar?, SS433-type?)
          5)   No strong shock or shock dissipated early
          6)   Remnant short lived

Class C (Combination)
1) Massive, young star
2) Type II optical supernova
3) Radio supernova (?)
4) Stellar remnant remains (pulsar?, SS433-type?)
5) Shock not fully dissipated
6) Plerionic remnant + shell remnant

FIGURES

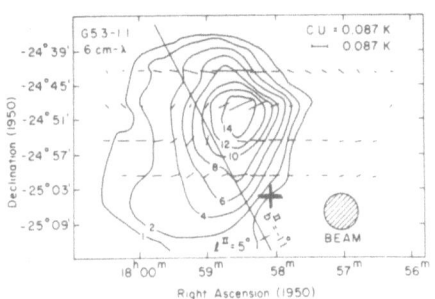 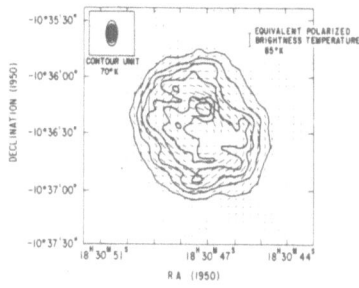

Figures 1 & 2: (Left) G5.3-1.1 (Milne 56) at 5 GHz (Angerhofer et al., 1977). The position of the x-ray source GX5-1 is indicated by a cross. (Right) G21.5-0.9 at 5 GHz (Becker and Szymkowiak, 1981).

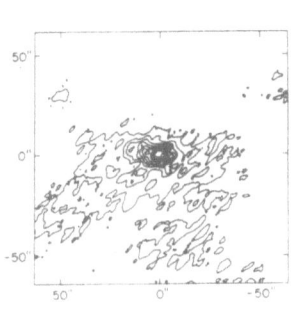 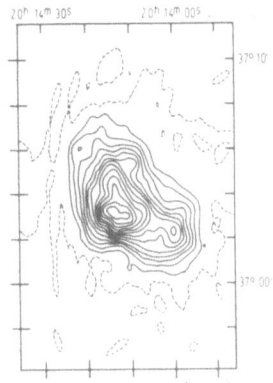

Figures 3 & 4: (Left) G57.6-0.3 (4C21.53W) at 1.5 GHz (Erickson, 1982). Field center coordinates are α (1950) = $19^h37^m29.61^s$, δ(1950) = $+21^\circ30'34''.0$. (Right) G74.9+1.2 (CTB87) at 1.4 GHz (Weiler and Shaver, 1978).

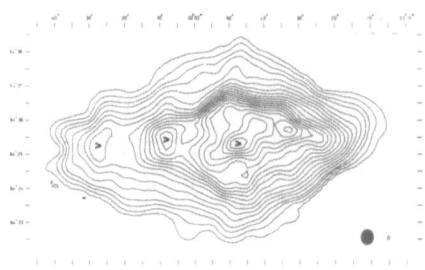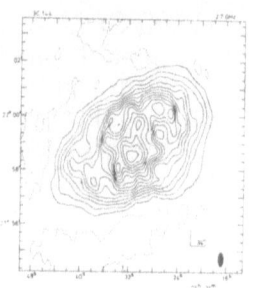

Figures 5 & 6: (Left) G130.7+3.1 (3C58) at 1.4 GHz (Wilson and Weiler, 1976a). (Right) G184.6-5.8 (Crab Nebula) at 2.7 GHz (Wilson, 1972).

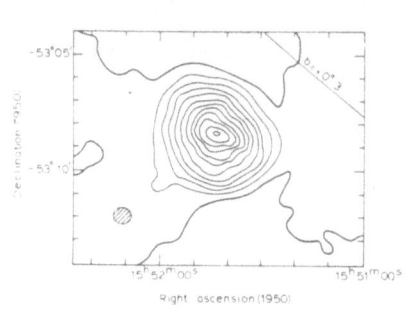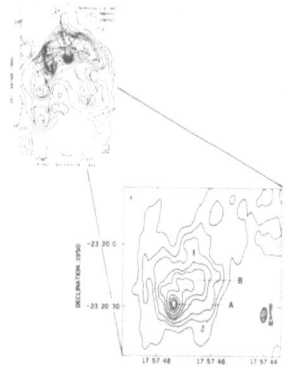

Figures 7 & 8: (Left) G328.4+0.2 (MSH15-57) at 1.4 GHz (Caswell et al., 1980). (Right, top) G6.5-0.1 (W28) at 5 GHz with a resolution of 4' (Milne and Wilson, 1971). (Right, bottom) Compact central region G6.6-0.1 at 1.4 GHz with a resolution of ∿5"x10" (Andrews et al., 1982).

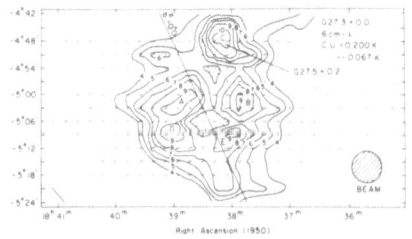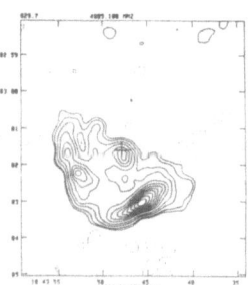

Figures 9 & 10: (Left) G27.3+0.0 (Kes73, Milne 62) at 5 GHz (Angerhofer et al., 1977). (Right) G29.7-0.3 (Kes 75) at 5 GHz (van Gorkum, Shaver, and Salter, private communication). The position of the x-ray emission is marked with a cross.

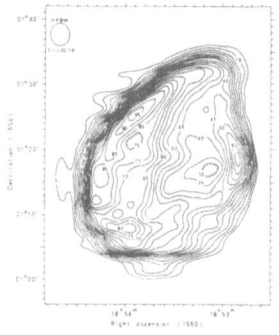 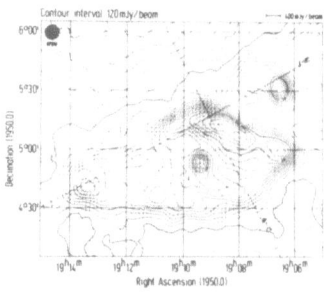

Figures 11 & 12: (Left) G34.6-0.5 (W44) at 0.4 GHz (Clark et al., 1975). (Right) G39.7-2.0 (W50) at 1.7 GHz (Downes et al., 1981). The compact source in the center is SS433.

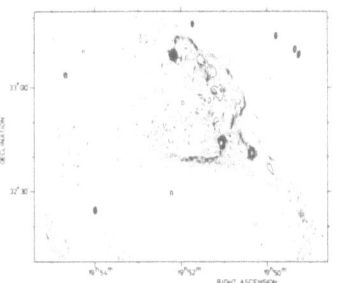 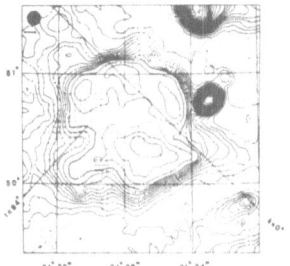

Figures 13 & 14: (Left) G68.9+2.8 (CTB80) at 0.6 GHz (Angerhofer et al., 1981). The compact feature near α =19^{h}51^m, δ =+32°45' has spectral index α=0 and coincides with the x-ray source. (Right) G93.7-0.3 (DA551, CTB104A) at 1.7 GHz (Mantovani et al., 1982).

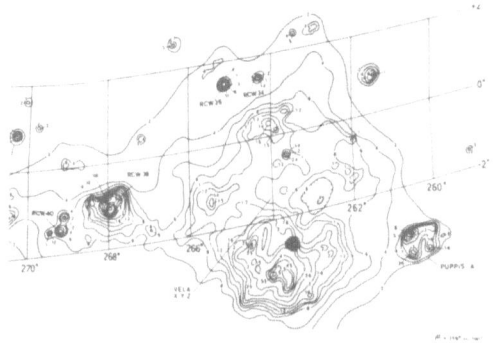 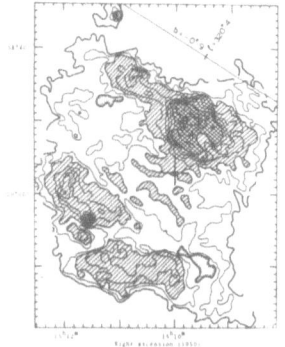

Figures 15 & 16: (Left) G263.9-3.3 (Vela XYZ) at 2.7 GHz (Day et al., 1972). The position of the Vela pulsar is marked with a spot. (Right) G320.4-1.2 (MSH15-52) at 1.4 GHz (Caswell et al., 1981). The position of the x-ray pulsar is marked with a cross.

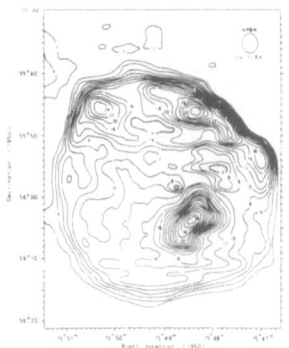
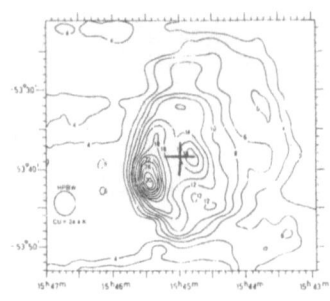

Figures 17 & 18: (Left) G326.3-1.8 (MSH15-56, Kes 25) at 0.4 GHz (Clark et al., 1975). (Right) G327.4+0.4 (Kes 27) at 0.4 GHz (Caswell et al., 1975b). Position of the x-ray source is marked with a cross.

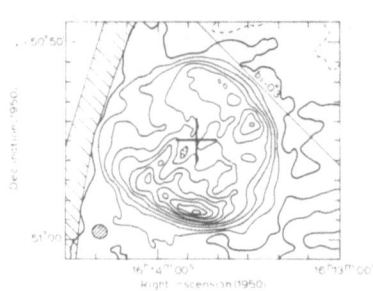
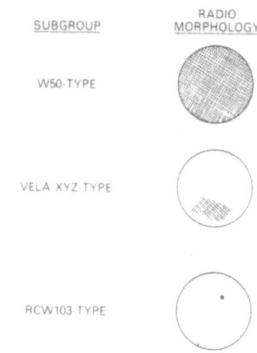

Figures 19 & 20: (Left) G332.4-0.4 (RCW103) at 1.4 GHz (Caswell et al., 1980). Position of the x-ray source is marked with a cross. (Right) Possible subgroupings of the Class C - Combination sources.

REFERENCES

Altenhoff, W. J., Downes, D., Pauls, T., Schraml, J. 1978, Astron.
   Astrophys. Suppl. 35, 23
Andrews, M. D., Basart, J. P., Lamb, R. C., Becker, R. H. 1982,
   Astrophys. J. (Lett.), in press
Angerhofer, P. E., Becker, R. H., Kundu, M. R. 1977, Astron. Astrophys.
   55, 11
Angerhofer, P. E., Wilson, A. S., Mould, J. R. 1980, Astron. Astrophys.
   236, 143
Angerhofer, P. C., Strom, R. G., Velusamy, T., Kundu, M. R. 1981,
   Astron. Astrophys. 94, 313
Becker, R. H. 1982, preprint
Becker, R. H., Kundu, M. R. 1976, Astrophys. J. 204, 427
Becker, R. H., Szymkowiak, A. E. 1981, Astrophys. J. (Lett.) 248, L23
Becker, R. H., Helfand, D. J., Szymkowiak, A. E. 1982, Astrophys. J.
   255, 557

Bradt, H., Burnett, B., Mayer, W., Rappaport, S., Schnopper, H. 1971,
    Nature 229, 96
Braes, L. L. E., Miley, G. K., Shoenmaker, A. A. 1972, Nature 236, 392
Caswell, J. L., Clark, D. H. 1975, Aust. J. Phys. Astrophys. Suppl. 37,
    57
Caswell, J. L., Murray, J. D., Rogers, R. S., Cole, D. J., Cooke, D. J.
    1975a, Astron. Astrophys. 45, 239
Caswell, J. L., Clark, D. H., Crawford, D. F. 1975b, Aust. J. Phys.
    Astrophys. Suppl. 37, 39
Caswell, J. L., Haynes, R. F., Milne, D. K., Wellington, K. J. 1980,
    Month. Not. Roy. Astron. Soc. 190, 881
Caswell, J. L., Milne, D. K., Wellington, K. J. 1981, Month. Not. Roy.
    Astron. Soc. 195, 89
Chevalier, R. A. 1977, "Supernovae," Ed. D. N. Schramm (Dordrecht:
    Reidel), p. 53
Clark, D. H., Caswell, J. L. 1976, Month. Not. Roy. Astron. Soc. 174,
    267
Clark, D. H., Green, A. J., Caswell, J. L. 1975, Aust. J. Phys.
    Astrophys. Suppl. 37, 75
Clark, D. H., Stephenson, F. R. 1977, "Historical Supernovae" (Oxford:
    Pergamon Press), pp 161 - 171
Day, G. A., Caswell, J. L., Cooke, D. J. 1972, Aust. J. Phys. Astrophys.
    Suppl. 25, 1
Davidson, K., Gull, T. R., Maran, S. P., Stecher, T. P., Fesen, R. A.,
    Parise, R. A., Harvel, C. A., Kafatos, M., Trimble, V. L. 1982,
    Astrophys. J. 253, 696
Dickel, J. R., Milne, D. K. 1976, Aust. J. Phys. 29, 435
Downes, A. J. B., Pauls, T., Salter, C. J. 1981 Astron. Astrophys. 103,
    277
Erickson, W. C. 1982, Astrophys. J. (Lett.), in press
Geldzahler, B. J., Pauls, T., Salter, C. J. 1980, Astron. Astrophys. 84,
    237
Goss, W. M., Shaver, P. A., 1970, Aust. J. Phys. Astrophys. Suppl. 14, 1
Goss, W. M., Schwarz, U. J., Wesselius, P. R. 1973, Astron. Astrophys.
    28, 305
Goudis, C. 1976, Astrophys. & Space Sci. 40, 91
Green, D. A., Gull, S. F. 1983, this volume, p. 329.
Helfand, D. J. 1981, NATO Advanced Study Institute, Cambridge, England
Helfand, D. J. 1983, this volume, p. 483.
Hillebrandt, W. 1982, Astron. Astrophys. 110, L3
Hughes, M. P., Thompson, A. R., Colvin, R. S. 1971, Astrophys. J. Suppl.
    23, 323
Kazes, I., Caswell, J. L. 1977, Astron. Astrophys. 58, 449
Kundu, M. R., Velusamy, T. 1972, Astron. Astrophys. 20, 237
Lamb, R. C., Markert, T. H. 1981, Astrophys, J. 244, 94
Long, K. S., 1979, HEAO Science Symposium, NASA CP-2113, p. 422
Manchester, R.N., Tuohy, I.R., D'Amico, N. 1982, Astrophys. J. (Lett.),
    in press
Mantovani, F., Nanni, M., Salter, C. J., Tomasi, P. 1982, Astron.
    Astrophys. 105, 176
McCulloch, P. M., Hamilton, P. A., Ables, J. G. 1982, I.A.U. Circular

No. 3704, June 15
Milne, D. K. 1971, "The Crab Nebula," IAU Symposium 46, eds. R. D.
    Davies and F. G. Smith, p. 248
Milne, D. K., Dickel, J. R. 1971, Nature Phys. Sci. 231, 33
Milne, D. K., Wilson, T. L. 1971, Astron. Astrophys. 10, 220
Milne, D. K., Dickel, J. R. 1974, Aust. J. Phys. 27, 549
Murdin, P., Clark, D. H. 1981, Nature 294, 543
Panagia, N., Weiler, K. W. 1980, Astron. Astrophys. 82, 389
Panagia, N., Weiler, K. W. 1981, Vistas in Astronomy 25, 87
Pounds, K. 1980, AAS-HEAD Meeting, Boston
Seward, F., Grindlay, J., Seaquist, E., Gilmore, W. 1980, Nature 287,
    806
Seward, F. D., Harnden, F. R. Jr. 1982, Astrophys. J. (Lett.), 256, L45
Shaver, P. A., Goss, W. M. 1970a Aust. J. Phys. Astrophys. Suppl. 14, 77
Shaver, P. A., Goss, W. M. 1970b, Aust. J. Phys. Astrophys. Suppl. 14, 1
Srinivasan, G., Dwarakanath, K. S., Radhakrishnan, V. 1982, Current
    Science 51, 596
Srinivasan, G., Dwarakanath, K. S. 1982, J. of Astron. Astrophys., in
    press
Strom, R. G., Angerhofer, P. E., Velusamy, T. 1980, Nature 284, 38
Swanenburg, B. N., et al. 1981, Astrophys. J. Lett. 243, L69
Swinbank, E., Pooley, G. 1979, Month. Not. Roy. Astron. Soc. 186, 775
Tuohy, I., Garmire, G. 1980 Astrophys. J. (Lett.) 239, L107
van den Bergh, S. 1978a, Astrophys. J. (Lett.) 220, L9
van den Bergh, S. 1978b, Astrophys. J. Suppl. 38, 119
van den Bergh, S., Marscher, A. P., Terzian, Y. 1973, Astrophys. J.
    Suppl. 26, 19
Velusamy, T., Kundu, M. R. 1974, Astron. Astrophys. 32, 375
Velusamy, T., Kundu, M. R., Becker, R. H. 1976, Astron. Astrophys. 51,
    21
Weiler, K. W. 1980, Astron. Astrophys. 84, 271
Weiler, K. W., Seielstad, G. A. 1971, Astrophys. J. 163, 455
Weiler, K. W., Shaver, P. A. 1978, Astron. Astrophys. 70, 389
Weiler, K. W., Panagia, N. 1978, Astron. Astrophys. 70, 419
Weiler, K. W., Panagia, N. 1980, Astron. Astrophys. 90, 269
Williams, D. R. W. 1973, Astron. Astrophys. 28, 309
Wilson, A. S. 1972, Month. Not. Roy. Astron. Soc. 157, 229
Wilson, A. S. 1980, Astrophys. J. (Lett.) 241, L19
Wilson, A. S., Weiler, K. W. 1976a, Astron. Astrophys. 49, 357
Wilson, A. S., Weiler, K. W. 1976b, Astron. Astrophys. 53, 89
Wilson, A. S., Weiler, K. W. 1982, Nature, in press
Zealey, W. J., Elliot, K. H., Malin, D. F. 1979, Astron. Astrophys.
    Suppl. 38, 39

DISCUSSION

WOLTJER: Before becoming too convinced that the Crab Nebula is the
remnant of a Type II supernova, it may be good to remember that at its
200 pc distance from the galactic plane, massive stars are extremely
rare. 3C58 may raise similar problems.

WEILER: Both historical and modern astrophysical arguments point toward a massive star - Type II supernova origin for the Crab. Although such stars are rare at that distance from the galactic plane, only one is needed to have given us the Crab Nebula. Even if all six relatively certain members of Class P are considered, the statistics are still too poor to estimate the scale height of their progenitors. If the pulsar in the Crab is typical of a young pulsar, then it is known that pulsars show a scale height distribution for their progenitors similar to that of Population I stars.

DICKEL: G21.5-0.9 showed tangential E vectors rather than a uniform field. Are there any other examples of that field orientation?

WEILER: No, not in any plerions where the polarization distribution is well known.

CHEVALIER: Although the Type II supernovae SN1979c and SN1980k were bright radio sources and SN1980k was a bright x-ray source, it is likely that the total radiated energy in the early phase was much smaller than the total kinetic energy of the explosion. Thus, this energy should still be available to drive a shock wave in the interstellar medium.

WEILER: I cannot disagree. One can speculate that very early dissipation of energy in the circumstellar material may weaken the shock enough that later interaction with the interstellar medium is insufficient to cause formation of a radio shell. However, there is no direct evidence for this. If this speculation is false, then the problem of what happened to the shells in the Class P ("true" plerionic) sources remains as before.

BISNOVATYI-KOGAN: When a plerion expands into the interstellar medium it may form a shell if the energy is sufficient. This was shown by T. Lozinskaya based on observations and on simple theoretical estimates. What can you say about this possibility and how many shell SNR could have been plerions?

WEILER: This is a difficult question. The Crab Nebula, for example, is certainly expanding supersonically into the interstellar medium, so that one might expect a shock wave to be present. However, it shows no evidence for limb brightening like historical SNR of similar age (Tycho, Kepler, etc.). Neither do any other of the Class P objects show limb brightening, although this is essentially by definition. If a plerion does eventually sweep up and accelerate relativistic particles to form a shell and if plerions are short lived with respect to shells as Weiler and Panagia have proposed, then the remaining shell would probably look quite normal leaving us little way of knowing which old shells were once plerions and which were always shells. One cannot rule out the possibility of an intermediate stage, however, where the plerion is fading but still visible. This might look something like Vela XYZ or the rest of our Class C-Combination remnants.

REGELMAN: Can you comment on the morphology of radio polarization in the plerions, particularly those which appear elongated?

WEILER: There are no obvious systematics. In the Crab and 3C58 we find in one that the intrinsic polarization direction is parallel to the direction of elongation and in the other that it is perpendicular. G21.5-0.9, which is not elongated in its radio morphology, has polarization vectors which appear to run around in a circular pattern.

X-RAY OBSERVATIONS OF CRAB-LIKE SUPERNOVA REMNANTS

R. H. Becker
Virginia Polytechnic Institute and State University

Abstract
    On the basis of extensive radio surveys of the galactic plane,
approximately 140 sources of diffuse radio emission have been classified
as supernova remnants (SNR). Using spectral index and spatial distribution as the primary selection criteria, these have been subdivided
into two groups, "shell" and "Crab-like". In each case, the radio
emission is assumed to be of non-thermal origin. The two distinct morphologies arise from two distinct energy sources. For shell remnants,
the energy is drawn from the reservoir of kinetic energy in the expanding shock front; in Crab-like remnants, the energy is drawn from the
rotational kinetic energy of a central stellar remnant.

    These two classes of remnants differ, significantly in their x-ray
emission. With few exceptions, radio shell remnants emit thermal x-rays
from shock heated gas which is itself distributed in a shell. Crab-like
sources (as defined by their radio properties) emit synchrotron x-rays
in a centrally-peaked spatial distribution. Presumably, the x-ray
emission from these objects is an extension of the radio spectrum. Crab-
like sources have a high probability of containing a compact (unresolved)
source of x-ray emission which in analogy to the Crab Nebula, is identified as the central stellar remnant.

    The general absence of either compact x-ray sources or Crab-like
diffuse nebulae within shell sources indicates that active pulsars, are not
usually formed in SN events which eventually form shell sources. However,
there are several examples of remnants which share both shell and Crab-
like characteristics so we cannot rule out an evolutionary connection
between these two classes of SNR.

Introduction
    The radio and optical morphology of the Crab Nebula are strikingly
distinct from those of most galactic supernova remnants (SNR). The distinguishing radio characteristics include a flat spectrum, a filled-
center brightness distribution, and a centrally-located pulsar. (Weiler
and Shaver 1978). In contrast, shell-type remnants exhibit relatively

J. Danziger and P. Gorenstein (eds.), Supernova Remnants and their X-Ray Emission, 321–328.
© *1983 by the IAU.*

steep spectrum and a shell structure. The radio emission from both the
Crab Nebula and shell remnants is linearly polarized indicating a synchrotron emission mechanism. Since the recognition that 3C58 shared
many of the radio characteristics of the Crab Nebula (Weiler and Seielstad
1971), a growing number of galactic radio sources have been suggested as
Crab-like SNR or plerions (for a review, see Weiler and Panagia 1980).

The study of Crab-like remnants as x-ray sources has developed much
more slowly. Prior to the launch of the Einstein Observatory, only two
Crab-like SNR had been detected, the Crab Nebula and Vela X. We might
note that these were the only two remnants known to contain radio pulsars
prior to Einstein. In fact, up till then, only one other SNR, W50, was
known to contain a stellar remnant of any kind. The improved sensitivity
of the Einstein Observatory over that of previous x-ray satellites
(Giacconi et al 1979) has allowed us to study many more SNR, both shell-
type and Crab-like, than previously possible. (for summary, see Table I).
In this paper, I hope to review the current status of x-ray observations
of Crab-like SNR.

TABLE I. - COMPACT OBJECTS IN CRAB-LIKE SNR

Name	%of X-ray Emission	Pulsed
Crab	$\sim$ 5	Yes
3C58	$\sim$ 5	No
Vela X	$\sim$ 33	No
MSH 15-52	$\sim$ 20	Yes
CTB 80	$\sim$ 25	No
G21.5-0.9	< 5	---
G29.7-0.3	< 15	---

Stereotypical Crab-like SNR

As discussed by Weiler and Panagia (1980) while some SNR share all
the morphological traits of the diffuse radio emission from the Crab
Nebula, others appear to be a conglomeration of both Crab-like and
shell attributes. There are only 5 unambiguous, Crab-like objects discovered to date. They are the Crab Nebula, Vela X, 3C58, (Weiler and
Seielstad 1971) G21.5-0.9, (Becker and Kundu, 1976) and G74.9+1.2 (Duin
et al 1975). All five have been studied with the Einstein Observatory.
The first three are the best studied, having been observed with the
IPC, the HRI, and the SSS, (see Becker, Helfand and Szymkowiak 1982 and
references therein). All three exhibit a filled-center x-ray brightness
distribution, all three contain unresolved x-ray sources (presumably
stellar remnants) and all have non-thermal x-ray spectra. Only the
stellar remnant in the Crab Nebula is observed to pulse.

The detection and study of the stellar remnant is hindered by the
filled center morphology of the diffuse emission. The emission from
the Crab pulsar is separable from the nebula by the very fact that it
pulses. For 3C58 and Vela X, the separation of compact and diffuse component is more difficult, requiring a modeling of the diffuse surface

brightness.

IPC and HRI observations of G21.5-0.9 reveal a filled-center brightness distribution coincident with the radio source but no point source
was detected (Becker and Szymkowiak 1981). IPC observations of G74.9+1.2
reveal diffuse emission consistent with a filled-center distribution
(Wilson 1980). The IPC, lacking sufficient resolution, would not have
been able to distinguish a point source within G74.9+1.2.

The x-ray data from the five detected remnants listed above were
discussed in detail by Becker et al (1982). They postulated that the
ratio of x-ray to radio luminosity (L_X to L_R) could be used as an indication of the age of the remnant, noting that as the break in the syn-
chrotran spectrum moved down in frequency, the ratio of L_X to L_R should
decrease. This led to the conclusion that 3C58 was much older than the
previously conjectured association with SN1181 would imply.

Ambiguous Crab-like Remnants
In recent years, the distinction between Crab-like and shell remnants has become blurred as some SNR have been found to share properties
of both groups. Typically, these SNR are composed of a Crab-like radio
component found inside a steep spectrum radio shell. Examples of this
morphology are CTB80, G326.3-1.8, W28, and G29.7-0.3. All four of these
SNR have been imaged by the Einstein Observatory.

The remnant CTB80 has a flat specrrum radio core with extensions to
the north, east, and west in which the spectra become progressively
steeper away from the core (Angerhofer et al 1981). The HRI image of
CTB80 reveals as unresolved x-ray source coincident with the radio core
surrounded by diffuse x-ray emission. (Becker et al 1982). The low
ratio of L_X to L_R suggested that this object is substantially older than
the Crab Nebula.

High resolution radio images of G29.7-0.3 have revealed that the
source is composed of several steep spectrum components arranged in a
shell, of about 2 arcmin diameter. An additional radio component within the source appear to have a flatter spectrum (Becker and Kundu 1976).
The IPC and HRI images revealed x-ray emission emanating only from the
flat Crab-like component (Becker, in preparation). The x-ray spectrum
was highly absorbed resulting in too low a count rate for the HRI to
detect the presence of a point source within the diffuse emission. The
ratio of L_X to L_R exceeds that of any other Crab-like SNR, suggesting
that G29.7-0.3 is a very recently formed remnant.

Two other remnants are known to be composed of steep spectrum
radio shells which contain nonthermal, flat spectrum components,
G326.3-1.8 (Milne et al (1979) and W28 (Andrews et al 1982). Both
objects have been observed with the IPC and HRI. Although G326.3-1.8
shows diffuse x-ray emission associated with its shell, there is no
enhancement in the vicinity of the Crab-like core (Becker, in preparation). The same is true for W28 (Long 1979). However W28 does

contain an unresolved x-ray source several arcmin outside of the Crab-like core but within the larger shell. (Andrews et al 1982).

Usually, remnants are catagorized by their radio attributes, primarily because remnants typically are studied best at radio wavelengths Therefore it is unusual to classify a remnant as Crab-like when it does not share the radio morphology of the Crab Nebula. If we expand our definition of Crab-like to include all SNR in which a compact stellar remnant appears to be injecting energy into the diffuse nebula, then we must include W50 and MSH 15-52 (G320.4-1.2) in the discussion.

The optical studies of SS433 which is situated within W50, certainly indicate that W50 meets this criteria. X-ray observations of SS433 with the IPC and HRI show it to be an unresolved x-ray source surrounded by a diffuse source of x-ray emission which is elongated in the east-west direction (Seward et al 1980). At present there is no evidence that W50 contains a flat spectrum radio component.

Similarly, radio maps of MSH 15-52 reveal a steep spectrum shell typical of shell remnants (Caswell, Milne, and Wellington 1981). However, Seward and Harnden (1982) have reported the presence of a compact x-ray source within the remnant which has a 0.150 s pulse period and is surrounded by diffuse x-ray emission. The x-ray data alone are convincing that we are observing a Crab-like object and one suspects that more sensitive radio observations will reveal a radio counterpart.

Discussion
        Although we generally use several specific radio characteristics to define the class of Crab-like SNR, these characteristics are really diagnostics used to infer the presence of a stellar remnant which is continually injecting energy into the remnant. To that extent, it is the existence of this stellar remnant which is the defining characteristic of a Crab-like or "pulsar-driven" remnant. Therefore, the observation of a compact object within such an object speaks most directly to its intrinsic nature. The discovery of compact x-ray sources in four additional Crab-like SNR (3C58, CTB80, W28 and MSH 15-52) is probably the Einstein Observatory's most important contribution in this area (see Table I for summary of compact source data).

In analogy to Crab Nebula, we assume that the diffuse emission at all wavelengths is the result of synchrotron emission from relativistic electrons accelerated by the central stellar remnant. The highest energy photons derive from the highest energy electrons which in turn have to shortest lifetimes. Therefore, the x-ray emission is related to the most recent pulsar energy losses, while the radio emission is an integrated measure to the pulsar's total energy loss. In effect, the x-ray emission should track the pulsar energetics more closely than longer wavelength emission. However, this conclusion relies heavily on the assumption that the x-ray emission is non-thermal synchrotron emission.

This has been shown conclusively for the Crab Nebula for which the
x-ray emission is known to be polarized and to obey a power-law spectrum.
Spectral data are also available for 3C58 (Becker et al 1982), Vela X
(Pravdo et al 1976) and MSH15-52 (Szymkowiak, private comm.), all of
which have nonthermal power-law spectra. X-ray spectral data do not
exist for the remaining Crab-like SNR and this is the area most in need
of additional observations. Until the x-ray emission from these other
remnants is demonstrated to be non-thermal, the interpretation of the
x-ray emission will remain uncertain.

If we accept the synchrotron nature of the x-ray emission, then it
follows that the x-ray emission will be the best indication of the
evolutionary state of a Crab-like remnant. Since the x-ray emitting
electrons are short-lived, the x-ray luminosity should decrease as the
energy injection rate decreases. In addition, as the relativistic
electron population evolves, the break in the emission spectrum will
move towards lower frequency, resulting in a decrease in the ratio of
L_X to L_R giving a distance independent measure of the age.

The time dependence of the x-ray luminosity of a Crab-like SNR will
be proportional to that of the pulsar itself if the efficiency for production of relativistic electrons remains constant. One functional form
for the time dependence of a pulsar, as suggested by Goldreich and Julian
(1969) and Pacini and Salvati (1973) is

$$L = L_0 / (1 + t/T)^{(n+1)/N-1}$$

where $2L_0$ is the initial rate of energy loss, T is the characteristic
spin down time, and n the braking index. For $t >> T$ the equation reduces
to

$$L = 2L_0 (T/t)^{(n+1)/(n-1)}.$$

Wilson used this relationship to compare the age of G74.9+1.2 to the
Crab Nebula with the implicit assumption that L_0, T, and n are the same
for both. If so, then the ratio of luminosities for the two SNR implies
an age of $6-9 \times 10^3$ years for G74.9+1.2. This procedure suffers from the
uncertainty in the remnant's initial conditions. Furthermore, the age
calculated for G74.9+1.2 is inversely proportional to its assumed distance,
a poorly determined quantity.

In an attempt to eliminate the dependence of age on distance, Becker
et al (1982) attempted to formulate an age dependent parameter based on
the evolution of the energy spectrum of relativistic electrons in a magnetic field. For two extreme cases, continuous injection of electrons vs
instantaneous injection of electrons, they found $L_X/L_R \propto B^{-1.5} t^{-1}$ or
$B^{-3} t^{-2}$ respectively, where B is the magnetic field strength and t the
remnant age. The ratio L_X/L_R would be a valid age indicator if we assume
the B is the same within all Crab-like remnants. However, high magnetic

fields would have the effect of "ageing" a remnant prematurely. For the Crab and Vela X, with ages of $\sim$ 900 years and $\sim$ 12,000 years respectively, the ratio of L_X/L_R is 140 and 0.1 respectively. This ratio suggests that either the injection of electrons approximate an instantaneous injection or that the fields in Vela X is higher than in the Crab Nebula. Values of L_X/L_R for the other Crab-like SNR are given in Table II. In the newly found x-ray object within MSH 15-52, no diffuse radio counterpart is observed (Caswell et al 1981), suggesting L_X/L_R is high, consistent with the young age implied by the spin down rate (Seward and Harnden 1982).

TABLE II. RADIO AND X-RAY LUMINOSITIES OF CRAB-LIKE SNR

NAME	DIST.(kpc)	DIAM.(pc)	L_R(ergs/s)	L_X(ergs/s)	RATIO
G29.7-0.3	19	3	5×10^{33}	3.5×10^{36}	500
Crab	2	3.5	1.8×10^{35}	2.5×10^{37}	100
G21.5-0.9	4.8	2.3	1.7×10^{34}	2.6×10^{35}	15
MSH 15-42	4.2	$\sim$2.0	2×10^{34}	2.6×10^{35}	10
CTB 80	3	$\sim$5	5.2×10^{32}	8×10^{32}	1.6
G74.9+1.2	12	27	6.1×10^{34}	8×10^{34}	1.3
3C58	8	8	1.7×10^{35}	1×10^{35}	.6
Vela X	.5	24	2.4×10^{34}	3×10^{33}	.1

The diffuse x-ray emission in a Crab-like remnant should serve as a guidepost to the location of the stellar remnant even when a stellar remnant is not observed. Since radio emitting electrons are long-lived, the centroid of the radio emission can be displaced from the pulsar location if the pulsar has significant space velocity, as is the case with Vela X. But short-lived x-ray emitting electrons do not have time to move away from the pulsar, so we expect the pulsar to be coincident with the diffuse x-ray centroid. Thus even when the pulsar is undetected, its location can be localized.

A related issue to the two contrasting types of SNR, ie Crab-like and shell, is the question of the origin of pulsars. Statistical studies of pulsars typically lead to the conclusion that the galactic pulsar birth rate is one every 10-25 years (for instance Vivehanand and Narayan 1981). With the near completion of surveys of SNR in radio and x-rays for Crab-like objects, we can examine results in a more quantitative way. What we have seen is that Crab-like SNR can be isolated objects or part of a shell remnant. Observationally, there is no discernible difference between Crab-like sources inside or outside of shells and similarly, the evolution of the shell may be unaffected by the presence of a Crab-like core. Therefore, these two aspects of SNR are decoupled, each following independent evolutionary tracks.

I'd also like to assert the likelihood that a large fraction of SN produce pulsars, at least to a 1st approximation. This is based on the pulsar space density, and the expected SN rate in our galaxy. Unless pulsars are formed in other ways, most SN must result in pulsars. But of the $\sim$ 100 shell remnants, only 4 contain pulsar-driven remnants, i.e., one in twenty-five. What of the other 96%? As has been suggested by others, perhaps most neutron stars are not formed as active pulsars, but only turn on much later at slower periods, long after the shell has dispersed and too weak to form observable Crab-like remnants. If so, then the data tells us the percentage of early pulsar turn-on, namely 4%.

But what if the pulsar-driven remnant development is truly independent of the shell development. Then for every shell-less Crab-like object there should be 24 shell-less remnants in which the pulsar does not turn on. Since the number of shell-less Crab sources equals the number inside of shells, then the number of shell-less pulsar-less remnants should equal the number of empty shell remnants. How does a shell-less, pulsar-less SNR appear? It doesn't, therefore 50% of all SN do not produce observable remnants.

In conclusion, I would stress the need for additional observations. Spectra are available for only three Crab-like SNR. Furthermore, in many cases where images are available, they are of very poor statistical quality. Hopefully these objects will be given high priority in any future x-ray mission.

Acknowledgements
This work was supported in part by Nasa Grant NSG 8-432. I want to thank Helen Hatcher for preparing the manuscript for publication.

References

Andrews, M. P., Basart, J. F., Lamb, R. C., and Becker, R. H. 1982. Ap. J. (Letters) in press.
Angerhofer, P. C., Strom, R. G., Velusamy. T., and Kundu, M. R. 1981. Astron. Astrophys. 94, 313
Becker, R. H. and Szymkowiak, A. E., 1981, Ap. J. (Letters) 248, L23.
Becker, R. H. and Kundu, M. R. 1976, Ap. J. 204 , 427.
Becker, R. H., Helfand, D. J., and Szymkowiak, A. E. 1932, Ap. J. J. 255,, 557.
Caswell, J. L. Milne, D. K., and Wellington, K. J. 1981 MNRAS, 195, 39.
Duin, R. M., Israel, F. P., Dickel, J. R., and Seaquist, E. R. 1975. Astron. Astrophys. 38, 461. [Pacini, F. and Salvati, M. 1973. Ap. J. 186, 249.]
Giacconi, R. et al. 1979, Ap. J. 230, 540.
Goldreich, P. and Julian, W. H. 1969. Ap. J. 197, 869.
Milne, D. K., Goss, W. M., Haynes, R. F., Wellington, K. J., Caswell, J. L., and Skellern, D. J. 1979. MNRS 188, 437.
Pravdo, S. H. et al. 1976, Ap. J. (Letters) 203, L67.
Seward, F., Grindlay, J., Seaquist, E., and Gilmore, W. 1980. Nature 287, 806.
Seward, F. D. and Harnden, F. R. 1982 preprint.
Weiler, K. W. and Seielstad, G. A. 1971. Ap. J. 163, 455.
Weiler, K. W. and Panagia, N. 1980. Astron. Astrophys. 90, 269.

Weiler, K. W. and Shaver, P. A. 1978. Astron. Astrophys. <u>70</u>, 389.
Wilson, A. S. 1980. Ap. J. (Letters) <u>241</u>, L19.

DISCUSSION:

HELFAND: To comment on Dr. Weiler's point, we have searched 30 of the
60 shell-like SNR within 5 Kpc with the Einstein Observatory. We could
have observed a Vela-like source in any of these and we have not seen
them. Thus, this limit mentioned by Dr. Becker is a meaningful one.

BECKER: Thank you, David. I'd also point out that Crab-like components are clearly absent from the very young SNR such as Cas A, Tycho,
Kepler, and SN1006. We can conclude that these objects do not contain
active pulsars.

TUOHY: I would like to point out that there is another Crab-like SNR
which has not yet been mentioned, namely N157B in the Large Magellanic
Cloud. This remnant has a filled centre radio structure and a flat
radio spectral index. Also, the recent solid state spectrometer
observations by Clark et al (1982) show that the x-ray spectrum
appears smooth, as expected for synchrotron emission.

BECKER: Yes, I agree that N157B should be included. A calculation of
the ratio of L_X to L_R gives a value of ~ 5. Neither at the high or low
ends of the range of observed ratios.

MCKEE: Since most of the Crab-like remnants you have discussed are
young, whereas most of the known shell-like SNR's in the Galaxy are
old, it appears to me that the Crab-like phase is short-lived and
that the fraction of SNR's with pulsars significantly exceeds the
fraction which are Crab-like. Would you comment?

BECKER: The Crab-like nebula surrounding the Vela pulsar is probably
10,000 years old, so since most observed SNR are no older than this,
the Crab-like components would still be present.

NEW HIGH RESOLUTION RADIO OBSERVATIONS OF 3C10, 3C58 AND PART OF THE CYGNUS LOOP.

D.A. Green and S.F. Gull
Mullard Radio Astronomy Observatory, Cavendish Laboratory,
Cambridge, United Kingdom.

Introduction

We present new high resolution observations of 3C10 (Tycho's SN) and 3C58, made at 2.7 GHz with the Cambridge 5 km telescope. These observations have a resolution of $\sim$ 4 arcsec, considerably better than previous observations, giving a detailed view of the radio morphology of these two remnants. These maps have also been processed by the Maximum Entropy Method (MEM).

We also present preliminary results from observations of part of the Cygnus Loop made at 408 MHz with the Cambridge One-Mile telescope (OMT).

Reduction of 3C10 and 3C58 Observations

Table 1 presents the relevant parameters of the 5 km telescope as used for the observations of 3C10 and 3C58. It should be noted that each remnant is smaller than the grating ring radius. In each case the contributions from several background sources were removed from the observed visibilities, as these sources had grating responses that fell across the face of the remnant.

These observations have been further processed by MEM (Gull & Daniell 1979) using an algorithm by Gull & Skilling (Skilling 1982). This processing increases the dynamic range of the maps and reduces the side lobe responses, 'Clean' being inappropriate for this type of extended source. The maps presented in figs. 1 and 2 are grey-scale representations made on the Cambridge SERC Starlink computer.

J. Danziger and P. Gorenstein (eds.), Supernova Remnants and their X-Ray Emission, 329–334.
© *1983 by the IAU.*

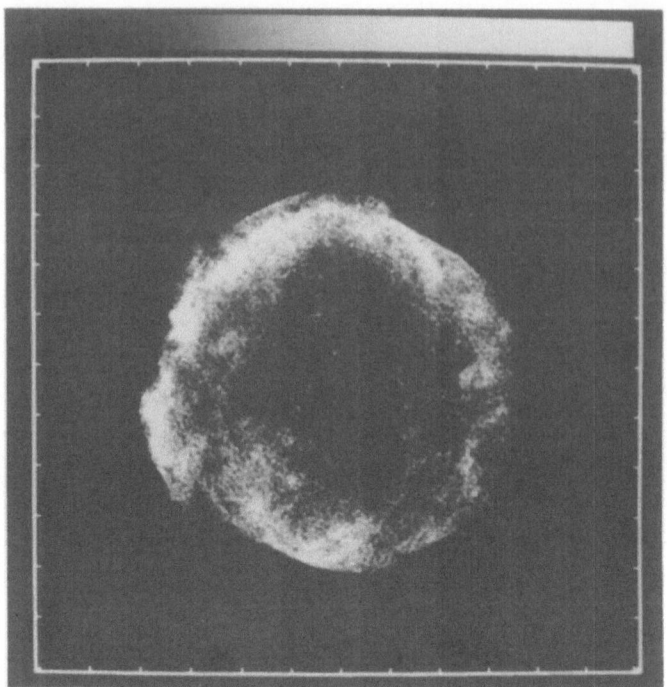

Figure 1. 3C10 at 2.7 GHz. Range, black to white, is 0.0 to 1.6 mJy/pixel, (pixel size is 2.0x2.0 arcsec). The pips on the frame are every 64 arcsec.

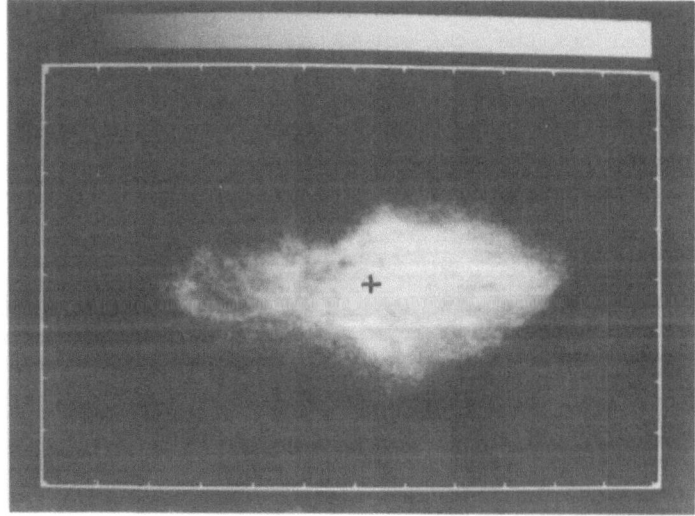

Figure 2. 3C58 at 2.7 GHz. Range, black to white, is 0.0 to 3.0 mJy/pixel, (pixel size is 2.0x2.0 arcsec). The pips on the frame are every 64 arcsec. The cross marks the position of the X-ray point source.

Table 1:   Details of the 5 km telescope as used for the observations
           of 3C10 and 3C58.

Frequency	2.695 MHz
Stokes parameter observed	I-Q
Number of spacings observed	128
Spacing increment	35.74 m
Grating ring radius (RA)	10.2 arcmin

	3C10	3C58
	h m s	h m s
Map centre: RA	0 22 30.0	2 1 55.0
(1950.0) Dec	63°52' 00".0	64°35' 30".0
Date of observations	Jun/Jul 1980	Oct/Nov 1981
Resolution (RA x Dec)	3.7 x 4.1 arcsec	3.7 x 4.1 arcsec
Angular size of source	8 x 8 arcmin	9 x 5 arcmin

Tycho - Discussion

The general morphology of the remnant is as seen previously (see,
for example, Duin & Strom 1975, Henbest 1980), but fig. 1 shows
several interesting features that have not been seen on lower
resolution radio maps. The western and southern edges of the remnant
are fitted very well (within 5 arcsec) by a circular arc of radius
226 arcsec, centred at $0^h22^m28^s.5$ (± $0^s.3$), 63°51'26" (± 2"). We shall
take this to be the 'geometrical' centre of the remnant, which is
presumably the direction to the site of SN 1572, as the arc fits the
perimeter of the remnant so well. The edge of the remnant is
extremely sharp over much of this arc, and shows considerable limb-
brightening, especially near the southern edge.

The eastern and northern edges of the remnant generally appear
more diffuse and brighter than the rest of the perimeter of the
remnant. In the NE the edge is very sharp, and strongly limb-
brightened (this is at a radius of 262 arcsec from the geometrical
centre of the remnant, larger than the radius for most of the remnant).
The bright emission in the N and NE appears to form a partial ring
of diffuse emission. It is in just these regions that the optical
filaments are seen (van den Bergh et al. 1973). The optical filaments
align extremely well with the bright radio emission.

Comparison with the Einstein HRI X-ray observations (Becker
et al. 1982) shows a very close correspondence between the X-ray
emitting clumps in the middle of the remnant and clumps of radio
emission. The brighter X-ray emission is in the NW rather than the
NE as for the radio. The radio emission being noticeably more limb -
brightened than the X-ray.

There is an unresolved source near the centre of the remnant.

This source was first reported by Gull & Pooley (1980) after the
original processing of these observations. This source is at
$0^h22^m31^s21$ ($\pm$ 0^s05), $63°52'16''8$ ($\pm$ $0''4$) 1950.0 epoch, (this is
0.4 arcsec from the originally reported position). Its 2.7 GHz
flux (1980.55 epoch) is 4.0 mJy, and it is 54 arcsec from the
geometrical centre of the remnant. Morbey & van den Bergh (1980)
report no optical stellar image at the position originally given by
Gull & Pooley.

       This source could be a background radio source (the chance of
finding a source of flux greater than 4 mJy within the boundary of
the remnant is 0.2), or it may be the radio remnant of the SN of 1572.
If this is the stellar remnant of the SN of 1572 and if the
'geometrical' centre of the remnant is the direction to the site
of the SN, then the proper motion of this source is 0.13 arcsec/year.
This puts a lower limit on its velocity of 1300 km s^{-1} (taking 3C10
to be at a distance of 2.3 kpc, Albinson & Gull 1982). High
resolution proper motion studies will be useful to clarify the
nature of this source.

3C58 - Discussion

       Fig. 2 shows that 3C58 has a large extent of low brightness radio
emission, with a distinct outer edge which is presumably the
position of the blast-wave shock. This feature has not been so
evident in previous observations, as this low brightness emission
would be lost in the noise or else not easily seen on contour
representations. Thus 3C58 appears larger, and relatively broader
than in previous observations, being 9 x 4.7 arcmin (RA x Dec),
(though Wilson & Weiler (1976) get a still larger size of 10 x 6 arcmin,
which we believe is due merely to their larger beam at lower frequencies).

       There is a bright region of radio emission at $2^h1^m46^s3$ ($\pm$ 0^s3),
$64°35'12''$ ($\pm$ 2''), which is 20 arcsec from the X-ray point source
position (Becker et al. 1982) which is marked on fig. 2. The
filamentary structure of the bright emission is complex and
reminiscent of that of the Crab Nebula.

Cygnus Loop

       Part of the Cygnus Loop has been observed (as a full '5C' survey:
5C8) with the OMT at 408 and 1407 MHz. Preliminary results at 408 MHz
are presented in Fig. 3. This map has a resolution of 80 x 160
arcsec (RA x Dec), it has been corrected for the primary response
of the OMT, and several of the brighter background sources in the field
have been removed from this map.

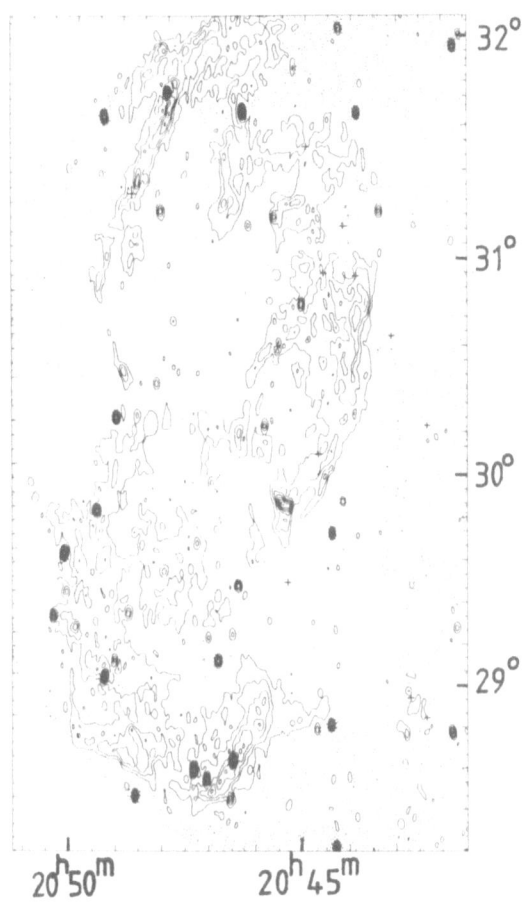

Figure 3. Part of the Cygnus Loop at 408 MHz.
The first solid contour is at ∿ 0.8 K, then every
∿ 1.6 K. Point sources have been removed from the
positions marked by crosses.

Acknowledgements

    We thank everybody who helped make and process these observations,
particularly Guy Pooley and staff of the 5 km, John Fielden and Martin
Brown for their programming. D.A.G. thanks the SERC for financial
support. We also thank Dave Griffiths for the preliminary map of the
Cygnus Loop.

References

Albinson, J.S. & Gull, S.F., 1982. In 'Regions of Recent Star
    Formation', (D.Reidel), eds. Roger, R.S. & Dewdney, P.E., p. 193.
Becker, R.H., Helfand, D.J. & Szymowiak, A.E.,1982. Astrophys. J.,
    255, 557.
Duin, R.M. & Strom, R.G., 1975. Astr. Asrtophys., 39, 33.
Gull, S.F. & Daniell, G.J., 1979. In 'Image Formation from
    Coherence Functions in Astronomy', (D.Reidel), ed. van
    Schooneveld, L., p. 219.
Gull, S.F. & Pooley, G.G., 1980. IAU Circular, No. 3502.
Henbest, S.N., 1980. Mon. Not. R. astr. Soc., 190, 833.
Morbey, C. & van den Bergh, S., 1980. IAU Circular, No. 3511.
Seward, F., Gorenstein, P. & Tucker, W., 1982. Preprint.
Skilling, J., at 'Maximum Entropy Data Analysis & Estimation Workshop',
    Laramie, (Wyoming, 1981).
Strom, R.G., Goss, W.M. & Shaver, P.A., 1982. Mon. Not. R. astr.
    Soc., 200, 473.
Wilson, A.S. & Weiler, K.W., 1976. Astr. Astrophys., 49, 357.
van den Bergh, S., Marscher, A.D. & Terzian, Y., 1973. Astophys. J.
    Suppl. Ser., 26, 19.

CONSEQUENCES OF A NEW DISTANCE DETERMINATION OF 3C58

D.A. Green and S.F. Gull
Mullard Radio Astronomy Observatory, Cavendish Laboratory,
Cambridge, United Kingdom

Introduction

In this paper we present new observations that show 3C58 is much closer than has been assumed in the past. This necessitates a reappraisal of many of the quantitative comparisons that have been made between 3C58 and the Crab Nebula or other 'filled-centre' remnants.

The observations also show that the structure of the neutral hydrogen in the ISM is, in practice, the limiting factor for HI absorption distance determinations, a factor that has not been taken into account in previous observations. This must raise doubts about the validity of many previous HI absorption distance determinations, which include the majority of the Galactic 'Σ-D' calibrators.

Observations

We have observed the supernova remnant 3C58 at 21 cm with the Cambridge Half-Mile Telescope (HMT). Two complete syntheses, overlapping in velocity coverage, were made in November 1981. Details of the telescope, receivers and the survey are given in table 1. In order to derive accurate HI absorption measurements towards the source it is important to include the contribution from HI emission on all scales. Data containing large-scale structure, unobtainable with the HMT, were derived from the Berkeley HI survey (Williams 1973a, Weaver & Williams 1973), and added to the synthesis maps. Continuum emission was subtracted from these 'composite' line maps to give the final channel maps, some of which are presented in fig. 1.

We derived optical depth and column density profiles for HI along the line of sight to 3C58 using (see, for example Spitzer 1978),

$$\tau = \mathrm{LOG}_e (T_{cont}/(T_{comp} \tau T_{em})), \tag{1}$$

$$\text{and} \quad N_{HI} = 1.823 \times 10^{23} \times V \times T_{em} \, \mathrm{m}^{-2} \text{ where } V = 2.0 \text{ km s}^{-1}. \tag{2}$$

J. Danziger and P. Gorenstein (eds.), Supernova Remnants and their X-Ray Emission, 335–340.
© 1983 by the IAU.

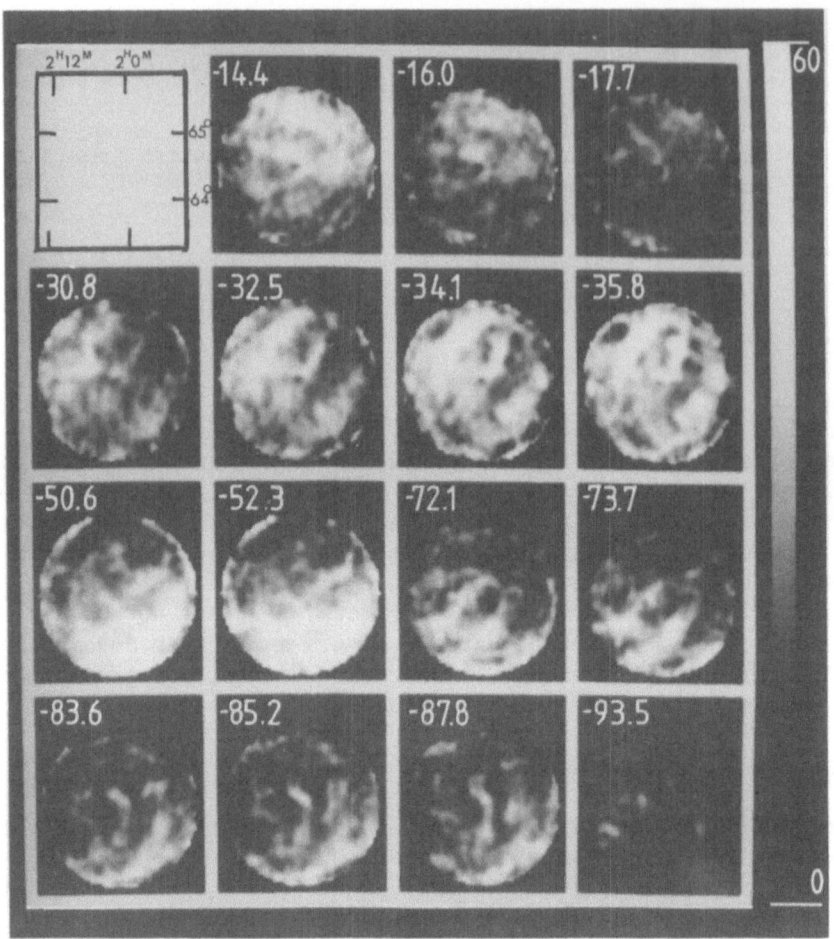

Figure 1. Photographic representations of some of
the channel maps. The range, black to white, is 0 to
60 K, as shown by the scale on the right. The velocity
of each channel (in km/s w.r.t. the Local Standard of
Rest) is marked above each map. The field of view is
sharply cut off at a radius of 80 arcmin. Notice the
dark 'spot' of absorption in the centre of the maps
with velocities below −36 km/s.

On-axis temperatures for the continuum (T_{cont}) and for each channel (T_{comp}) were taken directly from the continuum and composite maps. The emission temperature (T_{em}) in the direction of 3C58, for each channel, was estimated from the surrounding area on each channel map. The errors shown in fig. 2a, b are ± 2σ errors and are almost solely due to the complicated structure of the ISM

Table 1: Specification of the HMT as used in the observations in the direction of 3C58.

Primary beam	94 arcmin HPBW
Spatial resolution	7.1 x 7.9 arcmin (RA x Dec)
Continuum receiver	10 MHz
Line receiver	Digital cross-correlation spectrometer with 80 delay channels/spacing
	32 frequency channels for each synthesis channel separation 1.65 km s^{-1}
	(each of width 2.0 km s^{-1} to half power points)
Noise on synthesis maps	∿ 0.4 K
Field centre (1950.0)	RA = 2^h 2^m
	Dec = 64° 35'

Figure 2a shows only two distinct absorption features, one from −14.6 to −17.6 km s^{-1}, corresponding to the edge of the local arm, and another centred at −34.1 km s^{-1}, corresponding to the Perseus

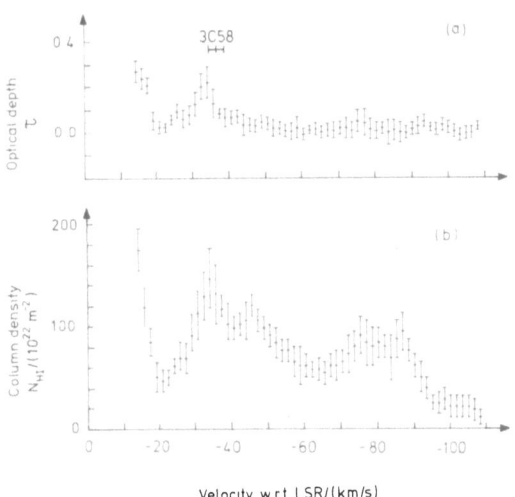

Velocity w rt LSR/(km/s)

Figure 2. a) Plot of optical depth (τ) against velocity. b) Plot of column density N_{HI} against velocity. (Error bars are 2σ on both plots).

arm. It is clear that there is no absorption past -39.1 km s^{-1}. Thus
we can place 3C58 as being at a distance equivalent to a velocity of
-36.6 ± 2.5 km s^{-1}. This corresponds to a distance of 2.6 ± 0.2 kpc
on the Schmidt model (Schmidt 1965), but any systematic difference
from the Schmidt model would give larger errors. For example the
work of Roberts (1972) would place 3C58 at only 2 kpc.

Consequences

With a distance of 2.6 kpc, 3C58 is comparable in size to the
Crab Nebula. Associating 3C58 with the SN of 1181 (Stephenson 1971),
the mean expansion velocity of the remnant is $\sim$ 3900 km s^{-1}. From
our data, and that of Williams (1973b) for velocities < -14.6 km s^{-1},
we estimate the integrated neutral hydrogen column density is to 3C58
of $2.55 \pm 0.3 \times 10^{25}$ m^{-2}. This is in good agreement with the value
of $2.0 \pm 0.5 \times 10^{25}$ m^{-2} derived by Becker et al. (1982) from Einstein
observations. Then, the optical absorption to 3C58 can be estimated
as 1.3 mag, giving an absolute magnitude for SN1181 at max of -14.4 mag.
This is subluminous compared to extragalactic SN, but these are in
any case biased towards the brighter SN.

It has been suggested (Becker et al. 1982) that 3C58 is not the
remnant of the SN of 1181. This was based on a comparison of several
'filled-centre' remnants for which X-ray observations are available.
However, it is not clear how the X-ray dependent properties of these
remnants indicate an evolutionary sequence, and indeed there may be
beaming effects. The simplest, and perhaps the safest comparison
is of linear sizes. With our revised distance 3C58 it is not much
larger than the Crab Nebula (AD 1054). Bearing in mind that we do not
know how typical the Crab or 3C58 are of 'filled-centre' remnants,
we conclude there is no compelling evidence for doubting the
association of 3C58 with the SN of 1181.

Acknowledgements

We thank all those who helped make the observations and reduce
the data. D.A.G. thanks the SERC for financial support.

References

Albinson, J.S. & Gull, S.F., 1982. 'Regions of Recent Star Formation',
    (eds. Roger, R.S. & Dewdney, P.E.), p. 193, (Reidel).
Becker, R.H., Helfand, D.J. & Szymkowiak, A.E., 1982. Astrophys. J.,
    255, 557.
Roberts, W.W., 1972. Astrophys. J., 173, 259.
Schmidt, M., 1955. In 'Stars and Stellar Systems Vol. 5', (eds.
    Blaaw, A. & Schmidt, M.), p. 513, (Univ. of Chicago Press).
Spitzer, L., 1978. 'Physical Processes in the Interstellar Medium',
    (Wiley).
Stephenson, F.R., 1971. Quart. J. R. astr. Soc., 12, 10.
Weaver, H. & Williams, D.R.W., 1973. Astr. Astrophys. Suppl. Ser., 8, 1.

Williams, D.R.W., 1973a. Astr. Astrophys. Suppl. Ser., 8, 505.
Williams, D.R.W., 1973b. Astr. Astrophys. 28, 309.

DISCUSSION

VAN DEN BERGH: You may have solved the pulsar problem but at the
cost of pushing the Hubble constant above 200 km s^{-1} Mpc^{-1}!

GREEN: I am afraid we have no control over where 3C58 actually is!
Our distance makes SN1181 severely subluminous; so much so that
supernovae of this absolute magnitude would scarcely be detectable
in the spiral arms of external galaxies.

WEILER 1) The "normal" HI absorption distance for 3C58 is based
partly on high velocity weak absorption at -60 to -100 km s^{-1}. These
are reported to have TAU 0.1 which is less than your error bars. How
can you rule these out? 2) You dismiss the utility of single-dish
HI absorption measurements, but by inserting a zero spacing in your
own observations you have, with great effort, synthesised a 100 m
single-dish. Why are your "single-dish" observations better than
previous ones? 3) If you feel you have maintained the advantage
of an interferometer, why do your results disagree with two previous
sets of interferometric measurements, at least one of which had super-
ior spatial resolution and sensitivity?

GREEN: 1) Previous measurements of weak high velocity absorption
features are spurious; they reflect the difficulties of obtaining
good spectral baselines. To do this it is necessary to take account
of the structure of the neutral hydrogen. Our measurements are in
fact very sensitive, but the thermal noise accounts for less than
20% of our quoted errors, the bulk being due to the uncertainty in
estimating the emission on the line of sight to 3C58. If there were
any large holes in the neutral hydrogen, we could detect a TAU of 0.005.
However, there are no holes of this type. 2) We believe that a
100 m single-dish would be just as good for work of this type,
provided that sufficiently large maps are made, to enable good
estimates of the emission on the line of sight. Averaging of a few
neighbouring points is simply not good enough! 3) The principal
previous interferometric measurement had a velocity resolution of
40 km s^{-1}. This is of the order of the velocity difference between
the LSR and the position of 3C58!

GOSS: In your abstract you appear to be dubious about the use of HI
absorption to obtain distances to SNRs. This criticism is based on a
sample of only one or two. A large number of Galactic SNRs have
been observed by the HI absorption technique. The distances derived
are, in general, lower limits; upper limits are more difficult.
(In fact the present determination for 3C58 is only a lower limit.)
The HI absorption technique in our galaxy has been tested using HII
regions. For these objects the intrinsic velocities are known from H
recombination lines and often spectroscopic parallaxes of the

exciting stars are known. These do NOT indicate that HI absorption
distances are, in general, unreliable.

GULL: If there are other factors that indicate a true distance, then
one knows which baseline wiggles in the absorption spectra are real,
and which to discount! The fact that our absorption baseline is
beautifully flat from -39 km s^{-1} tells us that this is a distance
measurement, not a lower limit. There are, for example, bright
(30 K) filaments which cross the line of sight at -45 km s^{-1},
and no absorption is seen. It is true that we have obtained, from
our sample of two (3C10 (Albinson & Gull 1982) & 3C58), distances
which are less than half of previous HI distance estimates. This
understandably makes us sceptical of many other similar distance
determinations. However, far from being dubious of the value of
neutral hydrogen absorption techniques, we believe that it is a
principal method for determining Galactic distances provided that
sufficient care is taken to estimate the emission on the line of sight.
This is painfully difficult; there is no cheap and cheerful way of
doing it.

THE RADIO MORPHOLOGY AND POSSIBLE NATURE OF 3C 58.

Ma Er[*]
Sterrewacht, Leiden, Netherlands

Richard G. Strom
Netherlands Foundation for Radio Astronomy,
Dwingeloo, Netherlands

Observations have been made of the supernova remnant 3C 58 with the Westerbork Synthesis Radio Telescope at 6 and 49 cm. These measurements provide us with greater resolution and sensitivity than that attained with previously published data. The 49 cm map has been used for comparison with an existing 21 cm one (Wilson and Weiler, 1976) to obtain information on the spectral index, rotation measure and depolarization. The 6 cm map is valuable both for its greater resolution, and for comparing with an observation made with the same instrument eight years previously. The total intensity distribution is shown in Figure 1.

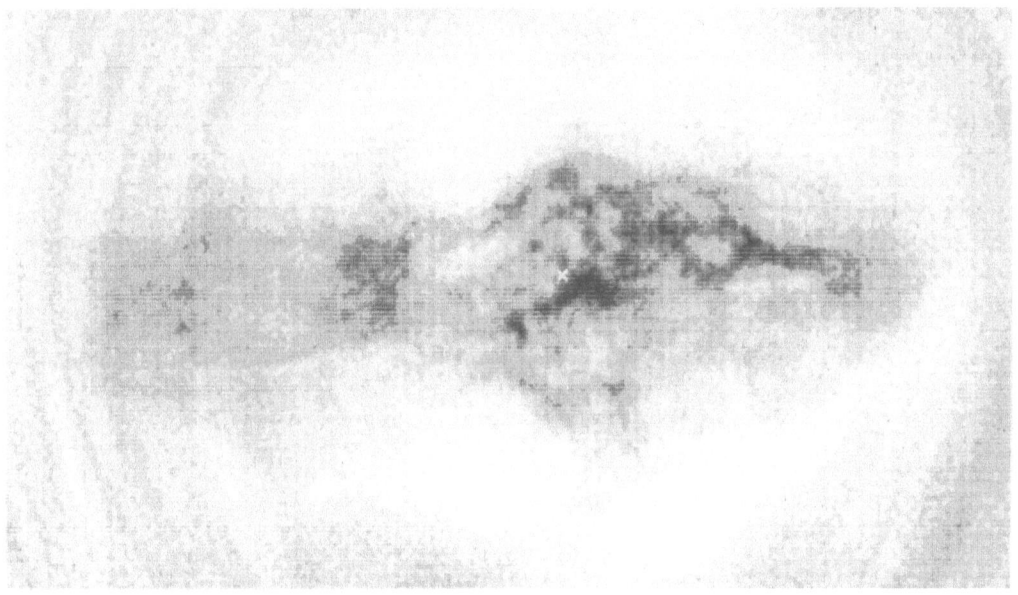

Figure 1. A 6 cm map of 3C 58 made with a resolution of 3" arc.

----
[*]Permanent address: Beijing Astronomical Observatory, Academia Sinica, Beijing, China.

J. Danziger and P. Gorenstein (eds.), Supernova Remnants and their X-Ray Emission, 341–342.

The possible association of 3C 58 with SN 1181 appears to be strengthened through the recent work of Liu Jin-yi (1982), who has discovered a 13th century star map with the 1181 guest star indicated. It is still too early to say whether the two-epoch 6 cm observations show evidence for changes, although there is a slight indication that one of the brightest features may have varied. The new 6 cm map shows that the X-ray point source (indicated by a cross in Figure 1) does not coincide with the radio peak. It lies, rather, on one of several ridges in the central region of the remnant. The two brightest peaks appear to be joined by such a bar-like structure, and are elongated roughly perpendicular to it. It is tempting to speculate that we are seeing a channel along which energy is flowing from a central compact object, and that the flow has been deflected near the bright peaks.

        The intercomparison of our three frequency maps leads to results similar to those obtained by Wilson and Weiler (1976). With our higher resolution we find that the gradient in rotation measure of some 60 rad m^{-2} is not smooth, but occurs rather in two fairly abrupt jumps. As these discontinuities do not appear to be associated with features in the brightness distribution or in the depolarization ratio we conclude that they arise in the interstellar medium between us and 3C 58 rather than in the source itself.

        A fuller account of this work is being prepared for publication elsewhere.

Acknowledgements.
M.E. was supported during his stay in Leiden by a grant provided by the Royal Netherlands Academy of Sciences (KNAW) from funds made available by the Ministry of Education and Science. The Westerbork Synthesis Radio Telescope is operated by the Netherlands Foundation for Radio Astronomy with the financial support of the Netherlands Organization for the Advancement of Pure Research (ZWO).

References.
Liu Jin-yi, 1982. Preprint (in Chinese).
Wilson, A.S., Weiler, K.W., 1976. Astron. Astrophys. **49**, 357.

CTB80: COSMIC COLLISION?

C.J. Salter
T.I.F.R.Centre
Bangalore
India.

F. Mantovani and P. Tomasi
Istituto di Radioastronomia
Bologna
Italy.

ABSTRACT. High resolution maps of the Galactic radio source CTB80 at three different frequencies are presented. A new interpretation in terms of a cosmic collision between two SNRs of different age is suggested.

CTB80 is one of the most mystifying Galactic Objects yet discovered and has recently attracted considerable attention from X-ray, optical and radio astronomers (see for example: Becker et al. (1981), Angerhofer et al. (1980, 1981), Velusamy et al. (1976), van den Bergh (1980).

The present observations of the radio continuum emission of the extended feature at 408 MHz, 1720 MHz and 4750 MHz and the linearly polarized emission at 1720 MHz and 4750 MHz (Figs 1,2,3) throw new light on the morphological, spectral and polarization properties of the whole source.

SPECTRAL INDEX

At present it is not possible to define the exact form of the continuum spectrum of CTB80. In Fig. 4 are reported the available data from the literature. On the basis of our data the most probable value for the spectral index seems to be $\alpha \simeq 0.45$ between 1 GHz and 5 GHz and $\alpha \simeq 0.0$ at frequencies below 1 GHz. $(S \alpha \nu^{-\alpha})$. The overall spectral index distribution between 408 MHz and 1720 MHz appears consistent with the total integrated flux density in the range of frequencies, being predominantly flat. The distribution of spectral index between 1720 MHz and 4750 MHz is also consistent with the total flux densities measured at these two frequencies. The distribution shows that: 1) there is a slope of 0.35 in the central component; 2) there is a steepening of the spectrum away from the central concentration along the eastern ridge (up to $\alpha = 0.7$) and the northern arc (up to $\alpha = 0.5$); 3) the spectrum flattens towards the south-western extremity.

J. Danziger and P. Gorenstein (eds.), Supernova Remnants and their X-Ray Emission, 343–346.

POLARIZATION

From the figures we can see that
considerable linear polarization
is found at both frequencies with
different distribution and polarization percentage.

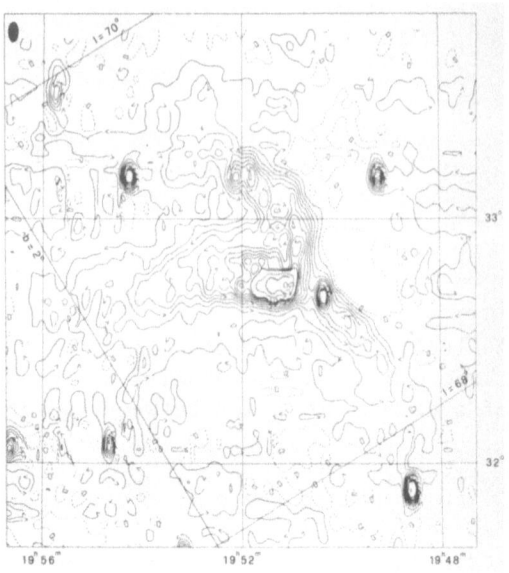

Figure 1. CTB80 at 408 MHz.

At 1720 MHz the source shows
linear polarization in the northern and eastern parts. The south-
western section of the southern
arc does not appear to possess
significant polarization at this
frequency. Also a minimum in the
polarization appears to exist on
the central source.

In the northern-arc the
polarization percentage rises to
around 10% in two regions, while
on the eastern end of the southern arc, percentage of about 15%
are present.

The results at 4750 MHz are
quite different. On the south-
western arc the linear polarized
radiation follows the total intensity distribution, and the
peak of polarized radiation
corresponds to the peak of the
source in the central plateau.
Everywhere in this region the
polarization percentage is
greater than 20%, increasing to
about 40% in the southern part
of the central plateau. On the
northern arc also the polarization intensity closely follows
the total intensity contours.

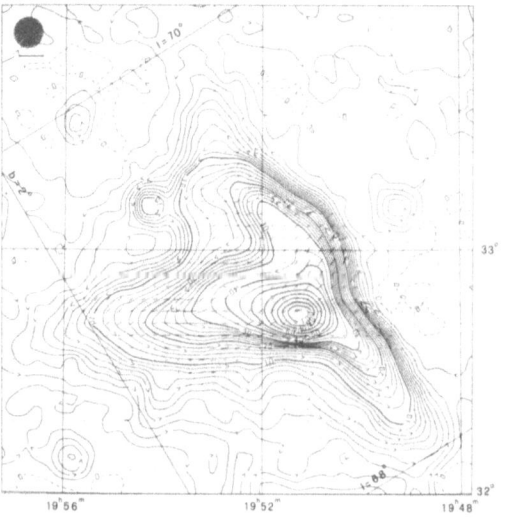

Figure 2. CTB80 at 1720 MHz.

CONCLUDING REMARKS

In view of the non-thermal origin
of the radio emission, we may
interpret the morphology of CTB80
as produced by a cosmic collision
between two SNRs of different age.
Their different age is suggested

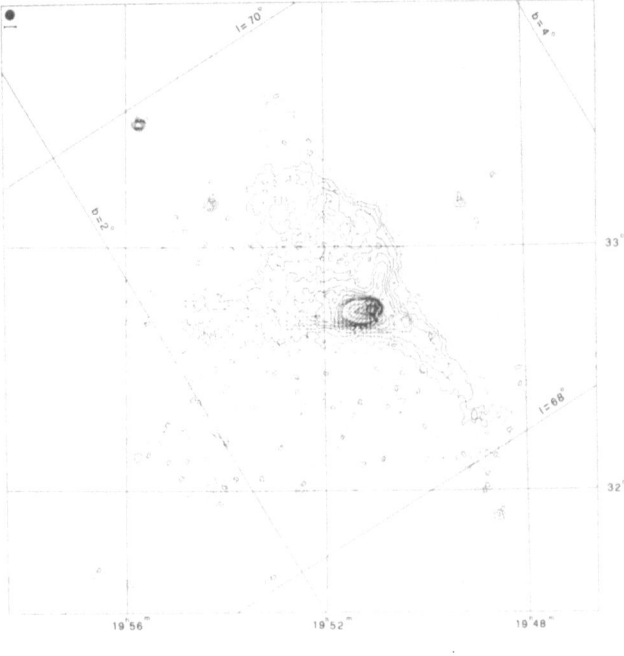

Figure 3. CTB80 at 4750 MHz.

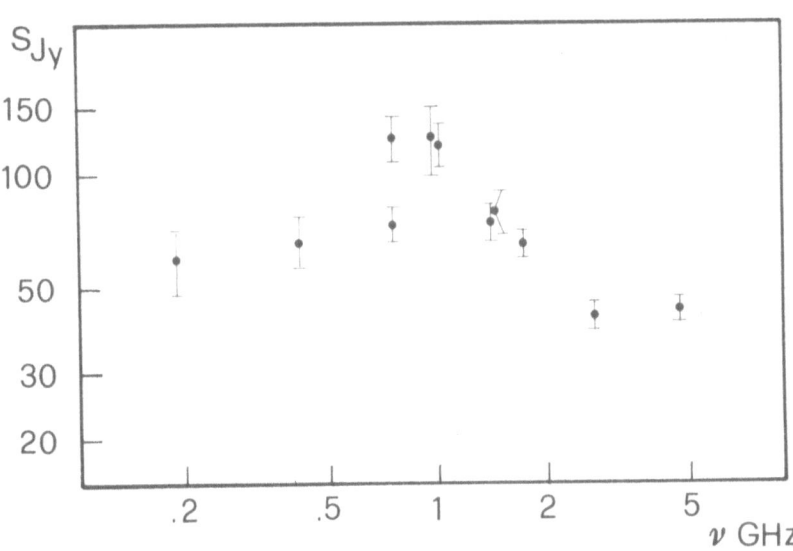

Figure 4. The continuum spectrum of CTB80.

by the different angular diameters of two expanding shells as clearly
visible in our maps. The northern arc might represent the youngest
SNR that is shocking the oldest one (southern arc). Only part of the
expanding shells are visible, but this is not unusual among SNRs. They
are expanding with respect to the two centres roughly at 50 pc of distance from each other (according with the estimated distance of 3+1 kpc,
Angerhofer et al.), perpendicularly to the line of sight.

We note on the outer western part, of the object, a steep gradient
in the radio brightness distribution, otherwise in the inner part of the
object the radio brightness increases slowly. This is not true for the
southern part of the central region where the brightness increases very
quickly and that might represent the colliding region.

From a general point of view the feature of CTB80 is not dissimilar
from the density distribution obtained by Jones et al. (1979) in a collisional simulation between two SNRs of different age expanding through
an uniformly dense medium after an explosion with a typical energy
emission of 10^{51} erg located in two points separated from each other by
40 pc. The most interesting feature of these calculations is the increase
of matter density of a factor 10 in a region, that might correspond, in
this framework, with the central part of CTB80. In a three dimensional
case, that enhanced density will be distributed in a "Mach Ring" that we
are seeing, because of our line of sight, tangentially. As most of the
emission from the relativistic particles is displaced in the western
part of the object, this means the western part of the "Mach Ring" will
be enhanced with respect to the western part. The collision between two
shock fronts might produce a reacceleration of the relativistic particles
changing the spectral shape of the continuum emission along the source.

REFERENCES

Angerhofer, P.E., Strom, R.G., Velusamy, T., Kundu, M.R.: 1981, Astron.
    Astrophys. 94, 313.
Angerhofer, P.E., Wilson, A.S., Mould, J.R.: 1980, Astrophys. J. 236,143.
Becker, R.H., Helfand, D.J., Szymkowiak, A.E.: 1981, Columbia Astrophys.
    Lab. Contr. No. 206.
van den Berg, S.: 1980, Public. of the Astron. Soc. of the Pac. 92, 768.
Jones, E.M., Smith, B.W., Straka, W.C., Kodis, J.W., Guitar, H.: 1979,
    Astrophys, J. 232, 129.

LARGE SCALE FEATURES OF THE HOT COMPONENT OF THE INTERSTELLAR MEDIUM

Gordon P. Garmire
The Pennsylvania State University
Department of Astronomy
University Park, PA 16802

The interstellar medium contains identifiable hot plasma clouds occupying up to about 35% of the volume of the local galactic disc. The temperature of these clouds is not uniform but ranges from 10^5 up to 4×10^{6}°K. Besides the high temperature which places the emission spectrum in the soft x-ray band, the implied pressure of the hot plasma compared to the cooler gas reveals the importance of this component in determining the motions and evolution of the cooler gas in the disc, as well as providing a source of hot gas which may extend above the galactic disc to form a corona. The following report presents data from the A-2 soft x-ray experiment on the HEAO-1 spacecraft concerning the large scale features of this gas. These features will be interpreted in terms of the late phases of supernovae expansion, multiple supernovae and the possible creation of a hot halo surrounding the region of the galactic nucleus.

INTRODUCTION

The low energy (0.1-1.4 keV) diffuse celestial X-ray emission is thought to be dominated by emission features originating mostly from hot plasma clouds in our galaxy. The very strong dependence of the photoelectric cross section upon energy implies that within this energy band we are sampling regions of the galaxy from a few tens of parsecs at the lowest energy up to several kiloparsecs at the upper end of the range in the galactic disk. At high galactic latitude, however, even at 0.2 keV it is possible to obtain information from sources outside of the galaxy with only moderate attenuation. The hot plasma clouds emitting the x-rays are most probably the remnants of old supernovae and interstellar gas heated by the expanding shock waves of these remnants. Results from the HEAO-1 A-2 Low Energy Detectors (LED's) on large scale features of the galactic diffuse x-ray emission are presented in the following sections.

THE OBSERVATIONS

Maps of the diffuse soft x-ray emission have been constructed using

J. Danziger and P. Gorenstein (eds.), Supernova Remnants and their X-Ray Emission, 347–355.

data from the 3° x 3° collimator of one LED on HEAO-1. The detectors have
been described by Rothschild et al 1979. By utilizing the pulse height
data which is telemetered every 10.24 seconds, maps in four energy bands
have been created: 0.18-0.54 keV, 0.56-1.0, 1.0-1.4 keV and 1.4-2.7 keV.
The data were all obtained from the second layer of the detector to avoid
electron contamination which was found to be present at all points in the
HEAO-1 orbit. The 10.24 second accumulation time adds an additional 2°
to the angular resolution along the direction of the scan path. The data
have been binned in approximately 3° x 3° cells on the sky. The resultant maps for the energy bands 0.18-2.7 keV, 0.18-0.56 keV, 0.56-1.0 and
1.0-1.4 keV are displayed in Figures 1 through 4. The 0.18-2.7 keV data
are obtained from a rate scaler which was read out every 1.28 seconds,
resulting in better angular resolution than the data obtained in the other energy bands. The highest energy data is omitted, since this map shows
no large features at the level of the noise in the map. The white areas
of the maps are regions not sampled during the survey.

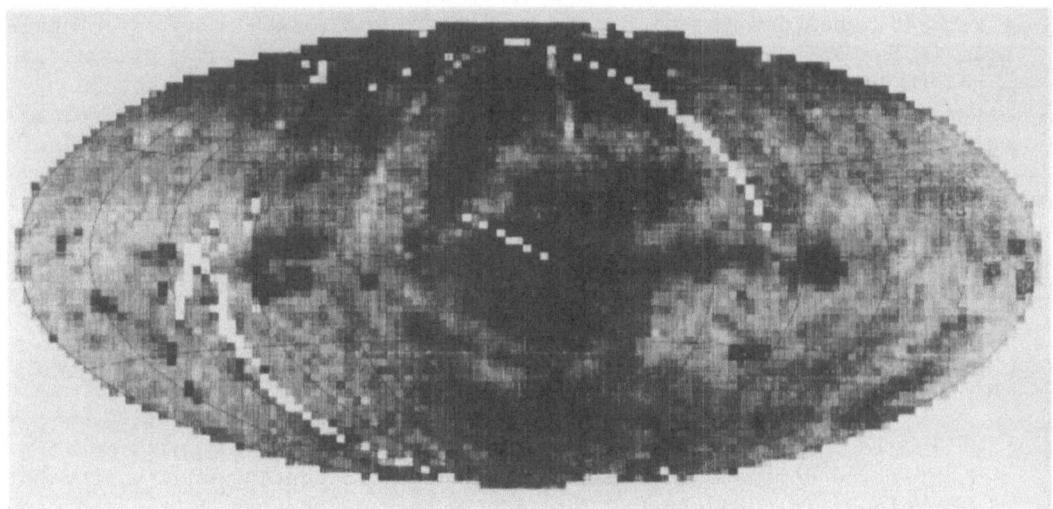

The 0.18-2.7 keV map of the diffuse x-ray background. The bin size is
approximately 3° x 3° varying somewhat from the center of the map toward
the Galactic Center to the edge of the Aitoff projection. White streaks
and bins are regions of no data. The North Galactic Pole is at the top.

THE LARGE SCALE FEATURES

    A number of extended regions of soft x-ray emission have been report-
ed in the literature and are presented here with somewhat better angular
resolution and sensitivity than displayed before, except in the cases of
the Gemini-Monoceros Ring and the Cygnus Superbubble which have been published on the basis of HEAO-1 data. In general the large angular diameter
regions are nearby (< 2 kpc), have temperatures between about two and four
million degrees, and are probably associated with one or more supernova
explosions which occurred at times greater than 10^4 years ago. The following discussion is organized in terms of the angular size of the feature,

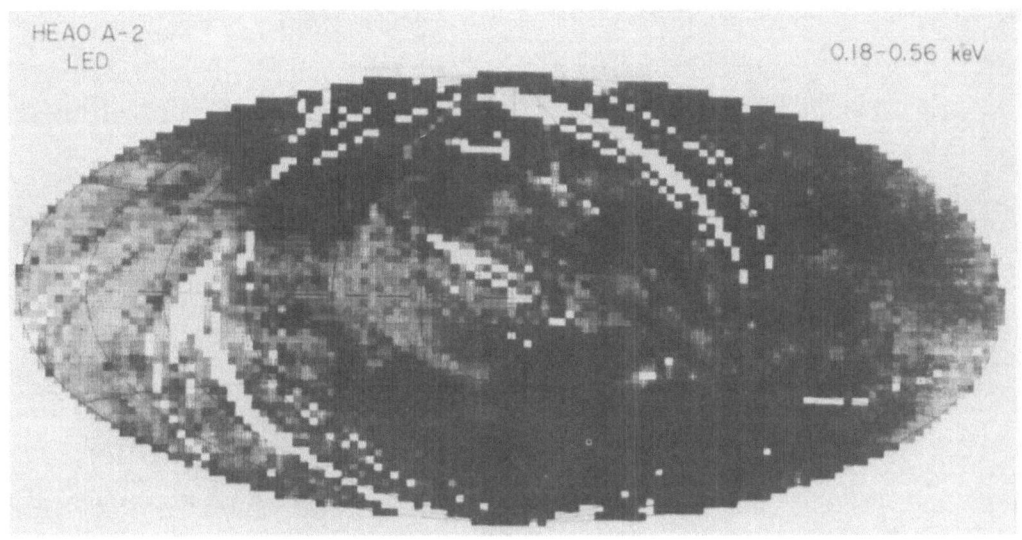

The 0.18-0.56 keV map of the diffuse x-ray background. Very dark narrow linear features (streaks) are spurious and should be ignored.

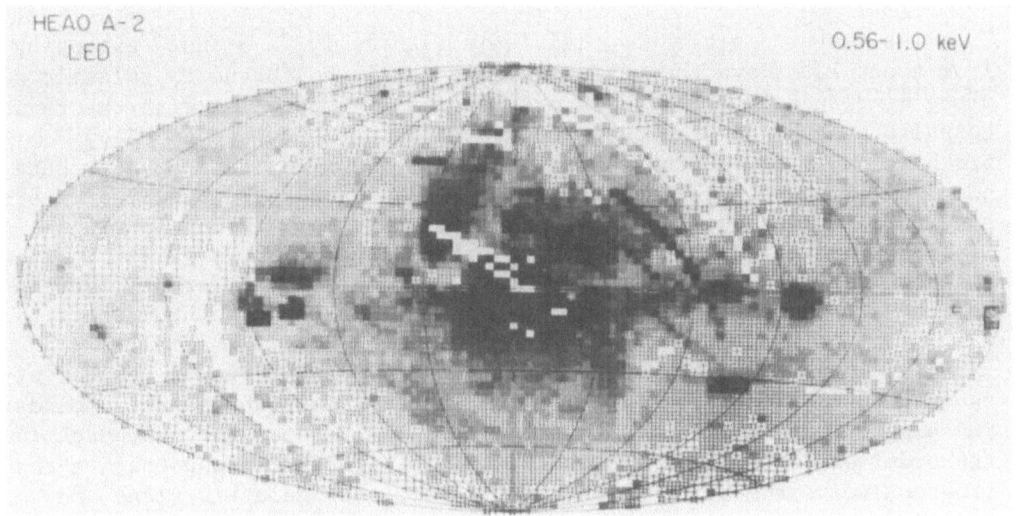

The 0.56-1.0 keV map of the diffuse x-ray background.

the largest diameter first.

NORTH POLAR SPUR

The association of the radio feature known as the North Polar Spur with an old supernova remnant was first suggested by Hanbury Brown et al., 1960. The X-ray emission has been studied by Bunner et al., 1972, Cruddace et al., 1976, Borken and Iwan 1977 and recently by Rocchia et

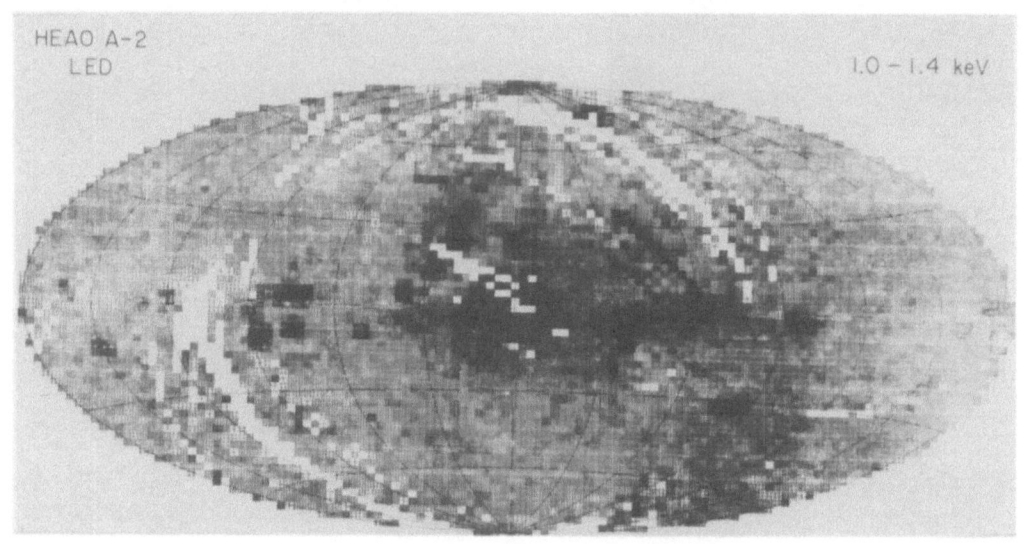

The 1.0-1.4 keV map of the diffuse x-ray background.

al., 1983.

        The feature is clearly visible in Figures 1, 2, 3 and 4 extending
from about 12° above the galactic plane nearly to the north galactic pole
along longitude 30°. The data are marginally consistent with the radio
Loop I contour along the outer edge away from the center noted by
Berkhuijsen 1972. There is a tendency for the x-ray data to bifrucate
near 30° latitude. The bifrucation could be either the result of a second
supernova remnant overlapping the NPS remnant or different portions of a
single expanding shell projected onto the plane of the sky. If it is from
a second supernova explosion the contour suggests a different location
for the explosion, closer to the galactic plane than the NPS center.

        The edge of the bright portion of the NPS away from the center of
the region is clearly resolved by the scanning data across this boundary.
The highest resolution data (1.5°) on this edge shows that the edge is of
the order of 2° to 3° at its sharpest. The spectrum along the arc changes
progressively from being the hardest towards the galactic plane and soft-
ening as the emission recedes from the plane, ranging from about 4×10^6°K
near the plane and decreasing to about 2×10^6°K at the highest latitudes.
There is a clear indication of absorption in the 1/4 keV channel at about
30° latitude, since presumably the plasma responsible for the radiation
provides photons in this channel with an increasing surface brightness
all the way down to the extreme tip of the spur at about 12° latitude
where the 0.5-1.0 keV flux is greatest. An absorbing column of at least
6×10^{20} hydrogen atoms/cm^2 using the cross section of Cruddace et al.
1974 is required to account for the attenuation of the 1/4 keV x-rays.
Such absorption is comparable to the total 21 centimeter column density
in this direction. There does not appear to be a sharp gradient in the
21 cm data at the point of disappearance of the 1/4 keV flux although a

triangular shaped cloud of HI in Heiles and Jenkins 1976 is associated
with a bifrucation of the tip of the 1/4 keV surface brightness of the
emission. At the inner edge of the NPS there appears to be an absorbing
cloud located at about 35 to 40 degrees latitude and 10 to 15 degrees
longitude associated with an HI wisp visible in Heiles and Jenkins 1976.
The column density in X-rays through the cloud must be about 2×10^{20}
atoms/cm^2, which implies a mean density of $4(\frac{d}{100} pc)$ atoms/cm^3 assuming
the wisp is as deep as it is wide in linear dimension and d is the wisp's
distance. A measure of the distance to the wisp would provide some constraints to the size of the local hot cavity in that direction,. since the
1/4 keV flux is not less than the mean toward this feature.

In Figure 1, a large extended region of emission is present interior
to the radio boundary of Loop I. By ascribing this radiation to the shell
of Loop I, an upper bound to the total luminosity can be obtained. It
is worth pointing out that the edge along longitude 30° is sufficiently
bright that the entire shell should be visible if emission were uniformly
distributed in the shell. Some of the interior region is as bright as
the edge of the shell, which implies that if this emission is truly in
the emitting volume the volume emissivity must be nearly an order of magnitude greater than in the observed edge, since the path length through the
center of the shell is so much less than the path through the edge of the
shell. An estimate to the luminosity in the 0.18-1.4 keV band is 2×10^{35}
ergs/sec based on the excess above background over an area of about 0.2
steradians and a distance of 240 pc to the center. The energy in the
remnant is much more difficult to estimate, since the density structure
strongly couples to the x-ray emission. A density contrast of a factor
of four changes the x-ray emission from the brightest region to being invisible against the general background. Based on a shell geometry and a
uniform filling of the observed shell, the energy content is roughly
9×10^{51} ergs. This is high for a single supernova explosion and lead
Borken and Iwan to suggest the region is the result of at least two explosions. The bifrucated shell mentioned earlier could be interpreted
as evidence supporting this conjecture.

THE GALACTIC BULGE

An alternate interpretation of the intense region of 0.56-1.4 keV
emission in the general direction of the galactic center is that the e-
mission is in fact originating from a cloud of hot plasma surrounding the
galactic center and extending out to some 2.5 kpc from the center. The
absorption by neutral gas at these energies does not preclude this interpretation for the emission a few degrees on either side of the galactic
plane. Near the plane, sources fill in any dip that might be expected
from the neutral gas absorption.

If we take this emission to originate at a mean distance of 10 kpc,
the luminosity is roughly 3×10^{38} ergs/s at a temperature of 3×10^{6}°K.
If the region is uniformly filled, the mean density is 0.003 electrons/cm^3
and the total energy content is 10^{55} ergs. Assuming that the gas in this

region is flowing outward roughly at the thermal velocity, the crossing
time is about 10^7 years. Since the cooling time scale is of the order of
10^9 years, evaporation is the dominant loss mechanism. The gravitational
escape velocity, assuming the material partakes of the local galactic rotation is of the same order as the thermal velocity, which means the evaporation time scale is probably greater than 10^7 years. In order to re-
place the 10^{55} ergs in a time greater than 10^7 years, about one SN of
energy must feed the bulge region every 10^3 years, a plausible rate for
the stellar content.

Clearly much more work is required to establish the distance to this
feature surrounding the direction of the galactic center. It is possibly
that the emission originates at intermediate distances between the galactic
center and the NPS in bubbles such as the one observed in Cygnus (Cash
et al 1979). However, in order for these bubbles to extend out of the
plane by nearly 20°, they must either be larger in radius than the Cygnus
superbubble or nearer. There is no known OB association in the direction
of the galactic center nearly the size of the Cygnus OB association. Consider further that the Orion association does not exhibit such a bubble
even though it is closer and contains more OB stars than the Cygnus association. Thus, the emission feature is quite possibly associated with
the galactic center.

THE ERIDANUS REGION

A bright region in Eridanus ($\ell \sim 205$, b ~ -50) has been noted by
Davidson et al., 1972, Williamson et al., 1974, and Long et al., 1977.
This region is revealed clearly on the 1/4 keV map as an arc of a circle
distorted by the Aitoff projection at the edge of the map. An HI feature
has been observed by Heiles 1976, which follows the x-ray arc quite precisely. It is possible that this feature represents an old SNR expanding
into the low density medium rather far off of the galactic plane. The
brightest portion of the remnant which is away from the plane is possibly
the result of an absorption effect in view of the HI distribution observed
by Heiles which surrounds the x-ray shell and shows low velocity gas obscuring the portion nearer to the plane than about -30° as well as obscuration for the portion at longitudes less than about 200° except near
the extreme edge of the loop near latitude -50°. As Heiles points out,
this is one of the best examples of a correlated x-ray and HI shell. A
detailed study of this feature is in preparation. The x-ray parameters
lead to the following estimates for the physical conditions in the remnant
for a distance of d_{150} of 150 pc: $n_e \sim 0.03/cm^3 \ d_{150}^{-1/2}$, $E_0 = 8 \times 10^{50}$ ergs,
$T \sim 2 \times 10^{6}°K$, $M_{hotshell} \sim 8M_\odot \ d_{150}^{5/2}$ and an adiabatic expansion age of
about 4×10^4 years. Heiles estimates the mass of the HI shell at
$7.3 \times 10^4 \ d_{150}^2 \ M_\odot$ which far exceeds the mass of hot gas. Such a gross
difference has not been observed in other supernova remnants, and further
discussion will be reserved for a future publication.

THE VELA REGION

Another region of extended emission which has not been noted by previous investigations is a low surface brightness region surrounding the Vela supernova remnant. The total emission is about 0.1 of the Vela SNR flux. Assuming that this emission originates from a uniform hot cloud at the distance of Vela which is taken to be 450 pc, the luminosity is 4×10^{34} ergs/sec if the temperature and column density of hydrogen are the same as the Vela remnant ($\sim 2.6 \times 10^{6}$°K and 2×10^{20}/cm^2, Moore and Garmire, 1976). The low surface brightness feature may be an earlier supernova in the same region. The outline fits roughly the angular extent of the Gum Nebula (Bok, 1971). Bok suggests that the Gum Nebula may be somewhat closer than the nebula associated with the Vela X and PSR0833-45 region.

THE LUPIS-CENTARUS REGION

There is an extended region of soft X-ray emission in the Lupis-Centarus constellations centered at about $\ell^6 = 320°$ and $b^4 = +20°$ extending over roughly 500 sq. deg. Several supernova remnants are bounded by this more extended feature, the Lupis Loop, the SN 1006 remnant and H1538-32 (Riegler et al 1980). The extended emission feature is comparable in size to the Gum Nebula in x-rays and of about the same surface brightness. Since the distance is undetermined at this time, estimates of the luminosity can not be made. It is possible that this feature extends all the way down to the galactic plane and that a lane of absorbing material prevents the observation of the true angular extent of the region based on the data shown in Figure 3. There is a clear displacement of the centroid of emission of this hot cloud with energy as can be noted between Figures 2 and 3, where the low energy data shows the bulk of the emission coming from about 315° longitude and the next higher energy channel shows the emission centered at about 340° longitude. In fact the data from the two energy bands barely overlap at 330°. A similar displacement of emission is present in the North Polar Spur only in latitude rather than longitude. It is not possible to rule out that these are two separate regions of emission at different distances that overlap in projection along 330° longitude

DISCUSSION

The Cygnus Superbubble (Cash et al 1980) and the Gemini-Monoceros Ring (Nousek et al 1981) have been discussed in details previously. Several new features detected by the HEAO-1 A-2 experiment have been presented here, together with all sky maps in several energy bands which are sensitive to emission from elderly supernova remnants. The fact that of order four old remnants (age $\stackrel{<}{\sim} 10^5$ years) are observed out to a distance of 500 pc (this excludes the Cygnus object which appears to be a different phenomenon and is more distant) is consistent with a galactic supernova remnant production rate of about one every fifty years.

If we assume that the five large features observed are within 400 pc, are roughly 70 pc in radius, and the disc is 300 pc thick, then the observed shells fill about 20% of the available volume of the galactic disc. If we are inside one such shell and we interpret several features in the 1/4 keV map as being additional shells that are less well defined with respect to the general background, the filling factor can possibly reach 35%. Only by going to even older remnants which have completely lost their identities can the filling factor reach 50% or higher. The gas pressure in these shells is of the order of 3 x 10^5 k dynes/cm^2 versus typical cool gas pressures of 300 k dynes/cm^2, where k is Boltzmann's constant. This difference serves to illustrate the potential for this gas to control the dynamics of cool gas clouds in the galactic disc.

I would like to thank J. Nugent and R. Truax for computational assistance in constructing the maps and J. Nousek for useful comments. This work was supported by NASA contract number NAS5-26809.

REFERENCES

Berkhuijsen, E.M.: 1973. Astron. Astrophys. 24, pp. 143-147.
Bok, B.J.: 1971. "Sky and Telescope" 42, pp. 64-69.
Borken, R.J. and Iwan, DeAnne: 1977. Astrophys. J. 218, pp. 511-520.
Brown, H., Davies, R.D., and Hazard, C.: 1960. Observatory. 80,
     pp. 191-198.
Bunner, A.N., Coleman, P.L., Kraushaar, W.L., and McCammon, D.: 1972.
     Astrophys. J. (Lett.) 172, pp. L67-L72.
Burrows, D.N., McCammon, D., Sanders, W.T. and Kraushaar, W.L.: 1982.
     Astrophys. J. (submitted).
Cash, W., Charles, P., Bowyer, S., Walter, F., Garmire, G.P. and Riegler,
     G. R.: 1980. Astrophys. J. (Lett.) 238, pp. L71-L76.
Cruddace, R.G., Friedman, H., Fritz, G. and Shulman, S.: 1976. Astrophys.
     J. 207, pp. 888-893.
Davidsen, A., Shulman, S., Fritz, G., Meekins, J.F., Henry, R.C. and
     Friedman, H.: 1972. Astrophys. J. 177, pp. 629-642.
Heiles, C.: 1976. Astrophys. J. (Lett.) 208, pp. L137-L139.
Heiles, C. and Jenkins, E.B.: 1976. Astron. and Astrophys. 46, pp. 333-360.
Long, K.S., Patterson, J.R., Moore, W.E. and Garmire, G.P.: 1977.
     Astrophys. J. 212, pp. 427-437.
Moore, W.E. and Garmire, G.P.: 1976. Astrophys. J. 206, pp. 247-253.
Nousek, J.A., Cowie, L.L., Hu, E., Lindblad, C.F. and Garmire, G.P.:
     1981. Astrophys. J. 248, pp. 152-160.
Riegler, G.R., Agrawal, P.C., and Gull, S.F.: 1980. Astrophys. J. (Lett.)
     235, pp. L71-L75.
Rocchia, R., Arnand, M., Blondel, C., Cheron, C., Christy, J.C., Koch,
     L., Rothenflug, R., Schnopper, H.W. and Delvaille, J.P.: 1983.
     this volume, p. 357.
Rothschild, R., Boldt, E., Holt, S., Serlemitsos, P., Garmire, G.,
     Agrawal, P., Riegler, G., Bowyer, S. and Lampton, M.: 1979. Space
     Science Inst. 4, pp. 269-301.
Williamson, F.O., Sanders, W.T., Kraushaar, W.L., McCammon, D., Borken,
     R. and Bunner, A.N.: 1974. Astrophys. J. (Lett.) 193, pp. L133-L137.

DISCUSSION

SANDERS: To answer the question of how far away is the 0.56-1.0 keV enhancement towards the galactic center, couldn't you use the column density of neutral hydrogen that you fit to your data to see if it's consistent with the extinction measured to the galactic center by other means?

GARMIRE: The problem here is made very difficult by the variations in the local emission which makes the subtraction of this emission from the more distant features uncertain and somewhat arbitrary. For this reason, I'm hesitant to quote a figure for the amount of X-ray absorption toward the direction of the galactic bulge region.

DAVELAAR: 1) Is there evidence for Loop IV emission in the HEAO-A2 maps? 2) Do you see indication for any connection between Loop I and possibly Loop IV?

GARMIRE: 1) It is not very clear, if present. 2) The only emission in the Loop IV region essentially overlaps Loop I. I don't see an obvious way to disentangle the projection effects.

DANZIGER: Does the extended region around Vela correspond to the optical extent of the Gum Nebula?

GARMIRE: Approximately, yes.

MCKEE: Garmire's model for Loop I involves a high energy for one SNR, whereas Cox described a one SNR model for the Loop. Could you reconcile this apparent discrepancy?

GARMIRE: The total energy content of the NPS (Loop I) is quite uncertain. The bifurcation of the shock front near $\ell \sim 30°$ and $b \sim 30°$ could be interpreted as evidence for a second SNR.

BLEEKER: What is the observational basis for the statement in your abstract that the interstellar medium is occupied with a dominant hot gas component of at least 50% (50%-90%)?

GARMIRE: The fraction of the local galaxy filled by identifiable features is up to 35%. Older features which have lost their identifiable structure undoubtedly exist and could double the filled volume with hot plasma over that obtained from identifiable features.

SPECTRAL OBSERVATION OF THE SOFT X-RAY BACKGROUND AND OF THE NORTH POLAR SPUR WITH SOLID STATE SPECTROMETERS

R. Rocchia, M. Arnaud, C. Blondel, C. Cheron, J.C. Christy
L. Koch and R. Rothenflug
Section d'Astrophysique, Centre d'Etudes Nucléaires de
Saclay, France

H.W. Schnopper and J.P. Delvaille
Harvard/Smithsonian Center for Astrophysics
Cambridge, Massachusetts, USA

In this paper, we present preliminary results of soft X-ray diffuse background observations. We observed two particular regions of the sky in the 0.3-1.5 keV range. The detection system consisted of three independent, 1 cm diameter, cooled solid state detectors. Nearly overlapping fields of view subtended a solid angle of approximately 1/4 sr. Except for the field of view, the whole set was similar to that described in Schnopper et al. (1982) (hereafter referred to as paper 1). This system was flown on board a three-axis stalibized rocket. The flight took place at White Sands Missile Range on 1981 May 4 at 0755 UT.

The first target ($\ell^{II} \sim 120°$, $b^{II} \sim 70°$) is a region of the sky close to the North galactic pole. We consider its spectrum as representative of the soft X-ray foreground produced by the local hot bubble. 155 seconds of useful data were obtained. The second target ($\ell^{II} \sim 25°$, $b^{II} \sim 25°$) is the brightest region linked with the North Polar Spur (Iwan 1980) and was observed for 170 seconds. Only data above 400 eV are presented in this preliminary report, a careful analysis of the noise made sure that no contamination occurs above this energy.

1. THE SOFT X-RAY FOREGROUND SPECTRUM

The chosen region of the sky lies more than 30° away from the radio ridge of the North Polar Spur (NPS) (Berkhuijsen et al. 1971) and it is among the brightest regions of the sky in the B-band and the C-band maps of the Wisconsin group (see for instance Mc Cammon 1979).

The contribution of the high energy X-ray background was subtracted using the spectrum $11.E^{-1.4} \exp(-\sigma(E).N_H)$. The column density $N_H = 3.10^{20}H$ atoms/cm^2 includes neutral atomic hydrogen on the line of sight (Daltabuit and Meyer 1972) and a probable contribution of molecular hydrogen (see paper 1). Our results are, however, not very sensitive to this parameter.

After this subtraction, we obtained the residual spectrum plotted on figure 1. It is now largely proved that this emission has a thermal origin (Inoue et al. 1979, Schnopper et al. 1982). The feature around 530 eV is attributed to the emission of O VII ions. We compared these data with a model of plasma X-ray emission developped at Saclay which includes the atomic data of Kato (1976). (See

357

J. Danziger and P. Gorenstein (eds.), Supernova Remnants and their X-Ray Emission, 357–360.

paper 1). Runs were performed with local galactic abundances (Meyer 1979) and with an absorption of 6.10^{19}H atoms/cm of intervening material (Inoue et al. 1979).

The first test was made with a single-temperature model with line emission. We get a chi-square of 37 for 22 degrees of freedom, but the fit is not good above 800 eV. But by adding a second component we succeed to lower the chi-square to about 18. However, the poor statistics of this experiment does not allow a definite conclusion about the spectrum shape of this second component (line spectrum or pure bremsstrahlung). We displayed on figure 1, as an example, the curve obtained from two line emission spectra. The absorbing material for the second component was arbitrarily fixed to the same value used for the first. The temperature ranges are, at the 90% confidence level :

$$T_1 = 8.7 \,{}^{+2.1}_{-2} \cdot 10^{50} \text{ K} \qquad\qquad T_2 = 5.6 \,{}^{+6}_{-3} \cdot 10^{6} \text{ °K}$$

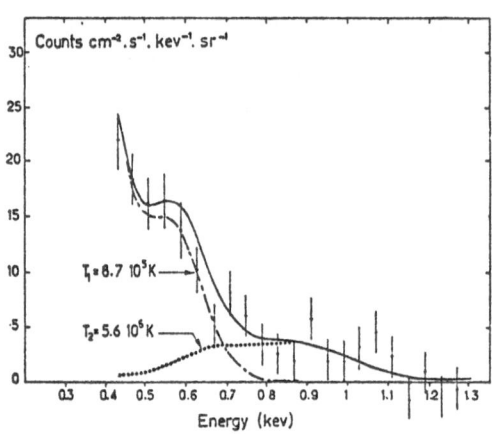

Fig.1. Spectrum of the foreground emission. The experimental points are compared with the convolution of the best fit model and the detector response. The vertical error bars represent $\pm 1\,\sigma$ on each experimental point.
The contribution of each component is shown in the case of the two-temperature model (with line emission).

These results confirm the temperature value of the soft X-ray foreground measured during the first flight (Rocchia et al. 1981). The existence of a second component in the soft X-ray background was extensively discussed in Nousek et al. (1982) : our results seem to show that it does exist, at least in this particular direction. We checked that the contribution of point X-ray sources is negligible, and we are left with the possibility that this component is due to the Galactic Halo.

2. THE NORTH POLAR SPUR SPECTRUM

The NPS is a prominent feature in surveys of the soft X-ray background in the energy range 0.5-1 keV (see for instance, Hayakawa et al. (1977) and Burstein et al. (1977). The region centered on ($\ell^{II} \sim 25°$, $b^{II} \sim 25°$) appears as the brightest region associated with this old supernova remnant (Iwan 1980, Davelaar et al. 1980).

The contribution of the extragalactic background was subtracted with the same spectrum as above, but with a total column density of 7.10^{20}H atoms/cm for the intervening material. The residual spectrum is plotted on figure 2. This emission includes contributions from the NPS itself and of course the soft X-ray foreground. It seems more complex that the previous spectrum, with Fe XVII, Ne IX and O VIII line contributions. A slight enhancement can be seen around 1350 eV at the energy of the Mg XI lines.

As a first test, a two-temperature model was tried with the following parameters :

a) local galactic abundances
b) an intervening material of 6.10^{19} H atoms/cm^2 for the lower temperature as above
c) a column density of 7.10^{20} H atoms/cm^2 for the higher temperature assuming all the absorbing material lies between the local hot bubble and the NPS.

On figure 2, the best fit curve is compared with the experimental points. The contribution of the two temperatures, with the following best fit values, are also plotted :

$$T_1 = 1.10^6 \text{ K} \qquad\qquad \text{E.M.} = 6.7 \; 10^{-2} \text{ cm}^{-6} \text{ pc}$$

$$T_2 = 4.7 \; 10^6 \text{ K} \qquad\qquad \text{E.M.} = 1.25 \; 10^{-2} \text{ cm}^{-6} \text{ pc}$$

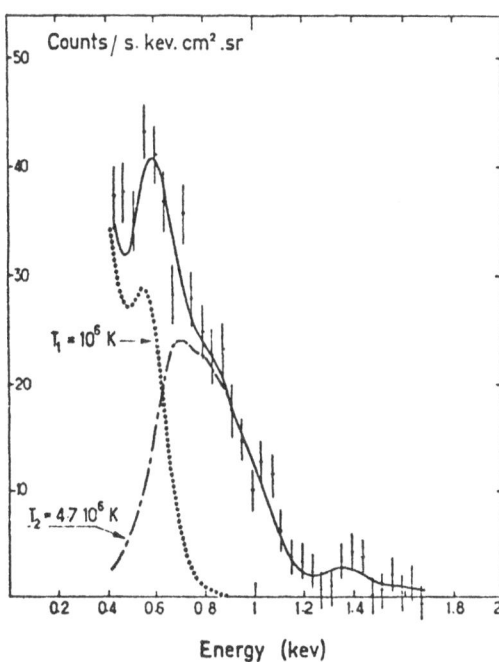

Fig.2. Spectrum obtained in the direction of the brightest region in the North Polar Spur (ℓ^{II}~25°, b^{II}~25°). In addition to the total best fit spectrum, the contributions of the two components are plotted separately. Local galactic abundances are assumed for both components.

We get a chi-square of 33 for 32 d.o.f. in the case of this best fit. The two contours at 90% and 68% confidence levels are given on figure 3 for this two-temperature model.

This sharing in two temperatures is somewhat arbitrary, since part of the low temperature emission is probably due to the NPS (Iwan 1980). The high emission measure we obtained for the low temperature can be an indication of the existence of this contribution.

In a future paper, we will try to compare these data with previous results on the NPS.

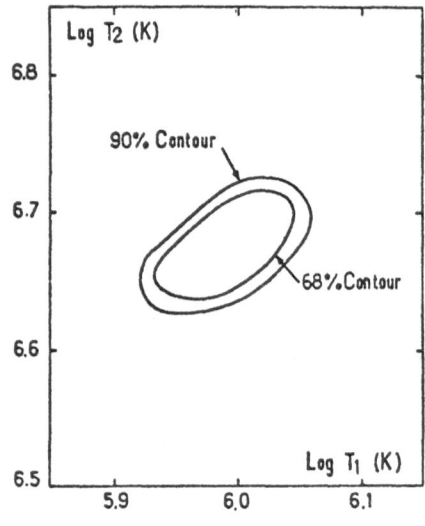

Fig. 3. Contours at 90% and 68% confidence levels obtained by fitting the data depicted on fig.2 with a two-temperature model.

Acknowledgements. This program is supported in part by NASA (USA) Grant 5304 and by CNES and CEA (France). We thank J. Gomes and J. Jenkins of SAO and the GERES staff of Saclay for their engineering and technical assistance. We thank O. Testard for the cryogenics. We also thank the Sounding Rocket Division staff at GSFC and WSMR and the staff at LURE, Orsay, for their assistance and hospitality. We thank Mr Friant for the detectors.

REFERENCES

Berkhuijsen, E.M., Haslam, C.G.T., Salter, C.J. 1971, A. and A. 14, 252.

Davelaar, J., Bleeker, J.A.M. 1980, A. and A. 92, 231.

Inoue, H., Koyama, K., Matsuoka, M., Ohashi, T., Tanaka, Y., Tsunemi 1979, Ap.J.Lett., 227, L85.

Inoue, H., Koyama, K., Matsuoka, M., Ohashi, T., Tanaka, Y., and Tsunemi, H. 1980, Ap.J. 238, 886.

Iwan, D. 1980, Ap.J. 239, 316.

Kato, T. 1976, Ap.J. suppl. 30, 397.

Meyer, J.P. 1979, in "Les Éléments et leurs Isotopes dans l'Univers", 22nd Liège International Astrophysical Symposium (Univ. of Liège Press), p. 153, 465, 477, 489.

Nousek, J.A., Fried, P.M., Sanders, W.T., Kraushaar, W.L. 1982, Ap.J. 258, 83.

Rocchia, R., Arnaud, M., Blondel, C., Cheron, C., Christy, J.C., Ducros, R., Koch, L., Rothenflug,R., Schnopper, H.W., and Delvaille, J.P. 1981, Space Sci. Rev. 30, 253.

Schnopper, H.W., Delvaille, J.P., Rocchia, R., Blondel, C., Cheron, C., Christy, J.C., Ducros, R., Koch, L., Rothenflug, R. 1982, Ap.J. 253, 131.

POSSIBLE CONTRIBUTIONS OF SUPERNOVA REMNANTS TO THE SOFT X-RAY DIFFUSE BACKGROUND (0.1 - 1 keV)

W. T. Sanders, D. N. Burrows, D. McCammon, and W. L. Kraushaar
Department of Physics, University of Wisconsin-Madison
Madison, Wis 53706 U.S.A.

ABSTRACT

Almost all of the B band (0.10-0.19 keV) and C band (0.15-0.28 keV) X-rays probably originate in a hot region surrounding the Sun, which Cox and Anderson have modeled as a supernova remnant. This same region may account for a significant fraction of the M band (0.5-1 keV) X-rays if the nonequilibrium models of Cox and Anderson are applicable. A population of distant SNR similar to the local region, with center-to-center spacing of about 300 pc, could provide enough galactic M band emission to fill in the dip in the count rate in the galactic plane that would otherwise be present due to absorption of both the extra-galactic power law flux and any large-scale-height stellar (or galactic halo) emission.

INTRODUCTION

The data were obtained in a series of ten sounding rocket flights. The payloads had two proportional counters collimated to $\sim 6°5$, one with a boron-coated window and one with a carbon-coated window. Our low energy data (0.1-1 keV) were binned into the three broad energy bands defined above. Figure 1 shows the response curves for these bands.

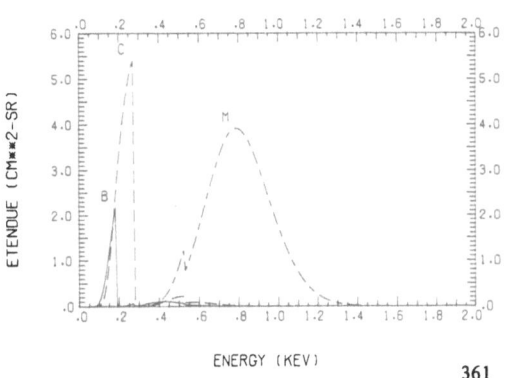

Figure 1. Effective area-solid angle product for the B, C, and M bands.

361

J. Danziger and P. Gorenstein (eds.), Supernova Remnants and their X-Ray Emission, 361–365.
© 1983 by the IAU.

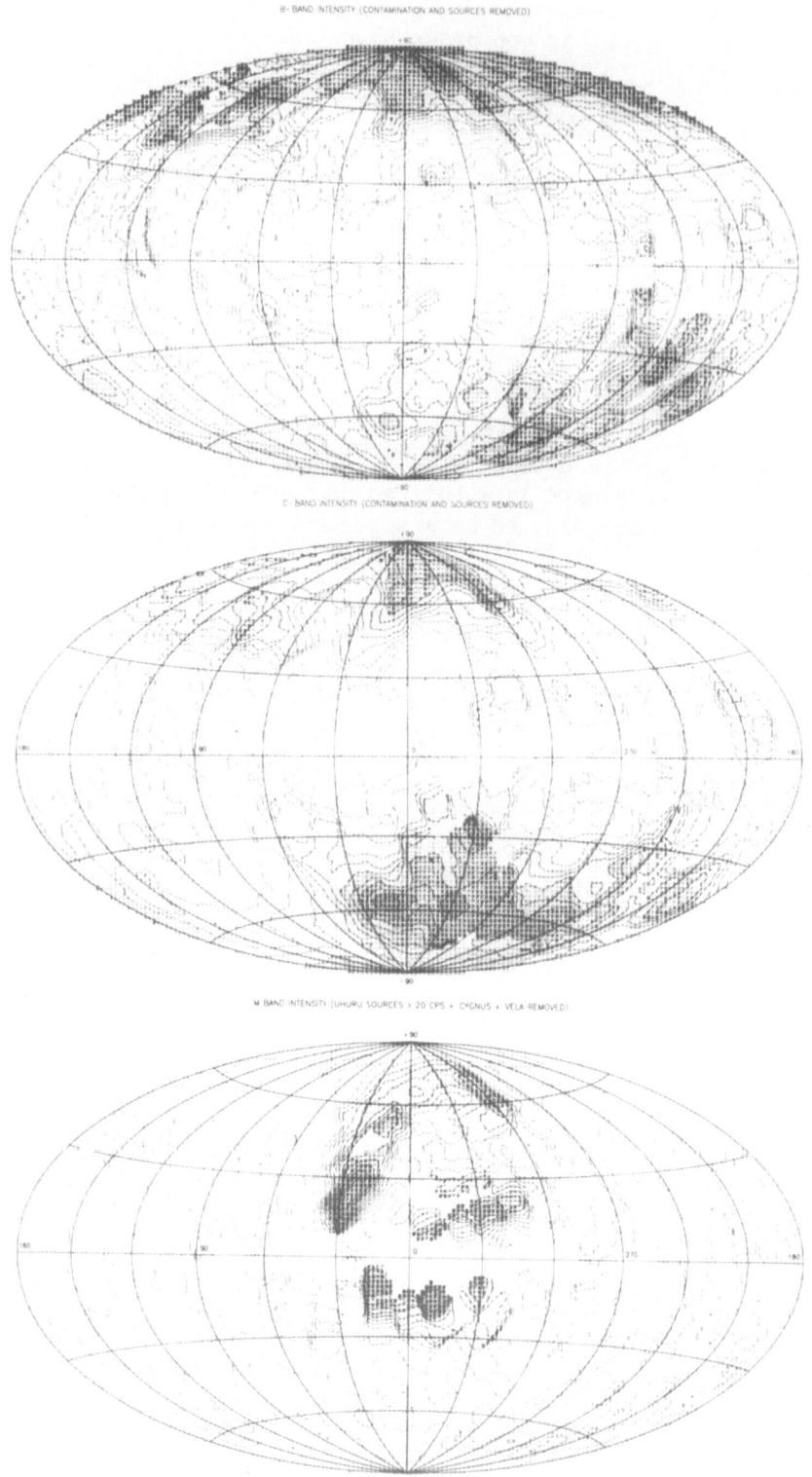

Figure 2.   B, C, and M band all-sky maps.

Maps of the B, C, and M band data are presented in Figure 2. Two very bright supernova remnants, Vela and the Cygnus Loop, have been removed from these maps. Fainter SNR (SN1006, Lupis Loop) or probable SNR (North Polar Spur, Eridanus, and Gemini-Monocerous enhancements) are still visible, but are excluded from any further analysis here. The experiments and data reduction are described in detail by McCammon et al. (1983).

THE LOCAL CAVITY

The intensity of the low energy diffuse background is everywhere greater than that predicted by an extrapolation to lower energies of the power law spectrum observed at energies > 2 keV. At least two additional sources of X-rays seem to be required: one primarily for the quite anisotropic B and C bands and a different one primarily for the rather isotropic M band. Sanders et al. (1977) proposed that the source of most of the B and C band X-rays is a hot ($\sim$ million degree) region surrounding the Sun that has uniform temperature but varying emission measure in different directions due to the varying extent of this region. We interpret this region as a local supernova cavity surrounding the Sun.

Theoretical support for such a picture comes from the work of Cox and Anderson (1982). They have calculated the dynamical, thermal, ionization, and spectral structures for blast waves of energy 5×10^{50} ergs in a hot low-density interstellar environment. Their nonequilibrium model shows that the B and C band intensities are reproduced by such an explosion if the ambient density is about 0.004 cm^{-3}, the blast radius is roughly 100 pc, and the solar system is located inside the shocked region. However, the M band count rate produced by this model is only 20-30% of the observed rate, consistent with the suggestions that most of the M band counts have a different origin.

THE M BAND DATA

Nousek et al. (1982) found that the M band data from latitudes <-15° were consistent with an absorbed power law plus an absorbed large-scale-height disk-shaped distribution of galactic emission. We compare the latitude dependence of the data to a model consisting of these two M band sources plus the local cavity contribution. The M band data, averaged over ten degree latitude intervals, are shown in Figure 3. The modeled contribution of the absorbed extragalactic power law is indicated by the line with alternating long and short dashes. We assume that the large-scale-height disk-shaped emission component of Nousek et al. is the integrated contribution of a population of dM stars (Rosner et al. 1981). Thus, the short-dash line shows the predicted contribution from an exponential distribution of stars (scale height = 350 pc, $n_s(0) = 0.065$ pc^{-3}, $L_x = 2 \times 10^{28}$ ergs s^{-1}) absorbed by an exponential distribution of gas (scale height = 110 pc, $n_H(0) = 1$ cm^{-3}). The long-dash line gives the contribution from the local SN cavity, assuming that the M band

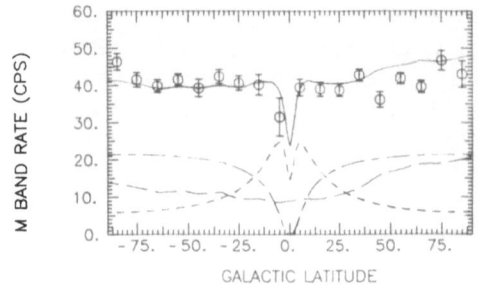

Figure 3. M band count rate as a function of galactic latitude. The modeled contribution of the extragalactic power law (— - —), stars (---), and local cavity (— — —) are shown, along with their total (solid line).

count rate from the cavity is 10% of the C band rate from the cavity (Cox and Anderson).

Figure 3 shows reasonable agreement between the data and the model, except perhaps within 10° of the galactic plane where the model shows a dip. A more detailed examination of the data near the plane reveals that in the longitude interval 90°<ℓ<180°, there is a broad dip at low latitudes. In this interval, the above model, with slightly different parameter values, is a reasonable fit to the data. But in the longitude interval 180°<ℓ<250°, the data show no dip in the plane. Since the dip in the model is an unavoidable feature of both the large-scale-height and extragalactic components, some additional X-ray emission component is needed which can fill in at low latitudes.

MORE DISTANT CAVITIES

The data from several scans in the interval 200°<ℓ<250° were averaged in longitude and plotted in Figure 4a along with the model lines of Figure 3 (convolved with our collimator response). We have calculated the flux from a population of old SN cavities like our own to see if

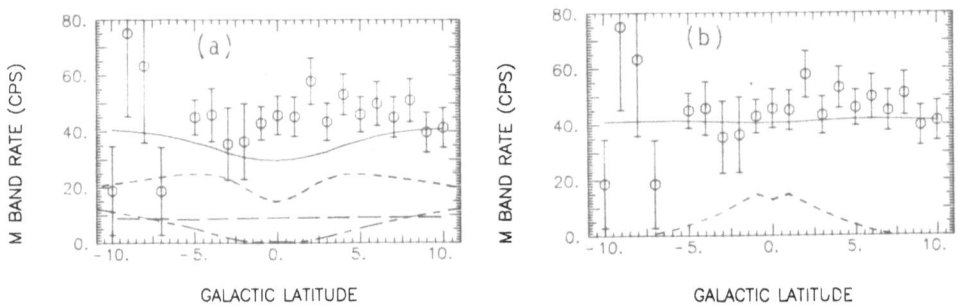

Figure 4. The data are from scans across the plane in the interval 200°<ℓ<250°. (a) The model lines are as in Figure 3, except they have been convolved with our collimator response function. (b) The dashed line shows the calculated contribution from a population of old SN cavities, and the solid line gives the total model with this component added in.

this is a reasonable candidate for the X-ray emission required to fill in the dip in the plane. These cavities are all assumed to be spherical with the same radius and surface brightness as our cavity and were randomly placed with their centers in the galactic plane with a mean distance between centers of 300 pc. The dashed line of Figure 4b shows the contribution from such cavities. The solid line shows that the addition of this component (again smoothed with our collimator response) provides a nearly latitude-independent M band flux. A question that must be answered, however, is why two adjacent longitude intervals along the galactic plane show such a different M band latitude dependence.

This research was supported by the National Aeronautics and Space Administration under grant NGL 50-002-044. We thank Don Cox for many helpful discussions.

REFERENCES

Cox, D. P. and Anderson, P. R.: 1982, Ap. J., 253, pp. 268-289.
McCammon, D., Burrows, D. N., Sanders, W. T., and Kraushaar, W. L.: 1983, Ap. J., submitted.
Sanders, W. T., Kraushaar, W. L., Nousek, J. A., and Fried, P. M.: 1977, Ap. J. (Letters), 217, pp. L87-L91.
Nousek, J. A., Fried, P. M., Sanders, W. T., and Kraushaar, W. L.: 1982, Ap. J., 258, pp. 83-95.
Rosner, R., Avni, Y., Bookbinder, J., Giacconi, R., Colub, L., Harnden, F. R., Jr., Maxson, C. W., Topka, K., and Vaiana, G. S.: 1981, Ap. J. (Letters), 249, pp. L5-L9.

HI SHELLS AND SUPERSHELLS

Carl Heiles
Astronomy Department
University of California, Berkeley

I. HI SURVEY DATA

A. Existing Survey Data

The entire sky has now been surveyed in the 21-cm line with an angular resolution of about 0.5 degree. In the north, above declination -20° or so, the "galactic plane" $|b|<10°$ has been completely sampled Weaver and Williams (1973; 1974). Above declination -30° or so, the sky outside the galactic plane has been almost completely sampled (Heiles and Habing, 1974; Heiles, 1975).

In the south Kerr, Kerr, and Bowers have surveyed the galactic plane; the data are in an advanced state of reduction and will soon be available. Outside the galactic plane, both Cleary, Haslam, and Heiles (1979) and Colomb, Poppel, and Heiles (1980) have surveyed the sky outside of the galactic plane. There is reason for two surveys, not just one, in the southern sky: the Magellanic Stream (Mathewson, Cleary, and Murray, 1974). HI in the Magellanic Stream is, in some regions, brighter than the ordinary gas in our galaxy, and widely separated in velocity. The Cleary et al. survey, with its wide velocity coverage, allows one to determine accurate column densities--not just within the local gas, but within the Magellanic Stream as well (Heiles and Cleary, 1979). The other survey has better velocity resolution and is necessary for mapping the kinematics of the gas.

B. The Stray Radiation Problem

There is a significant systematic error in some of the high-latitude data. This arises from "stray radiation," which is the response of a telescope to radiation coming from directions other than that where it is pointed. It makes observed column densities larger than they should be. The effect is particularly serious at high latitudes, where the 21-cm line is weaker than it is everywhere else in the sky. The derived column densities can be too high by factors of

J. Danziger and P. Gorenstein (eds.), Supernova Remnants and their X-Ray Emission, 367–372.
© *1983 by the IAU.*

order two (Heiles, Stark, and Kulkarni, 1981). This makes a very significant difference for low-energy X-ray observations, and particularly for extragalactic studies.

Stark, Heiles, and others have used the Bell Labs horn reflector at Crawford Hill, Holmdel, New Jersey, to make a new survey of the northern sky. This telescope is the same one used for the original discovery of the cosmic background radiation and is essentially free of stray radiation. This survey will be available on magnetic tape from me in a few months.

II. MORPHOLOGICAL AND PHYSICAL PROPERTIES OF HI SHELLS AND SUPERSHELLS

A. Morphological Properties

When HI data are presented as photographic maps of HI column density in small velocity ranges, a spherical expanding shell appears as circles of different diameter at different velocities. Easily visible shells include one around Radio Loop I (the North Polar Spur), visible in the photographs of Colomb et al. (1980) and the Eridanus shell, discussed by Heiles (1976); the interiors of these shells are the classical high-latitude sources of diffuse X-ray emission. The only other HI shell that exhibits X-ray emission is the Monogem ring (Nousek et al., 1981); this HI shell is discernable only with great difficulty. A number of large shells and supershells are easily visible in the galactic plane; an excellent example is GS096+04-113 in Figure 2 of Heiles et al. (1979). Larger structures are visible in the HI photographs that combine HI data from both the galactic plane and high latitude Hat Creek surveys. These photographs are presented in Heiles (1982b), who also gives a new list of shells and filaments.

There are many curved HI filaments that do not change size with velocity. None have significant X-ray emission. The proper interpretation of such filaments is not certain. They may be portions of shells that have evolved to the point where they are no longer expanding; in a homogeneous interstellar medium, such a shell would be complete and the filament would be that part of the shell seen tangentially. On the other hand, they may be filamentary condensations within a larger shell--or might be simply condensations in the ambient medium, and not parts of formerly expanding shells at all. Detailed analysis of HI data, not yet performed, should enable these questions to be resolved.

Some shells are huge. Diameters range above 2 kpc, and observed kinetic energies of the largest shells range above 10^{53} erg. These "supershells" tend to be located at fairly large galactic radii, outside the solar circle. Although this preferential location might be an observational bias, reflecting the fact that shells are easier to recognize outside the solar circle, it is likely to be real. Large HI shells have been observed in external galaxies--in M101 by Allan, Goss, and van Woerden (1973), and Allan and Goss (1979), and in M31 by Brinks (1982)--and the same tendency is observed.

Expanding shells tend to show only one hemisphere. That is, only the approaching or the receding hemisphere is seen, and in those few cases in which both are seen one of the two hemispheres is very much weaker and more difficult to recognize. The implication is that most shells are not fully complete, a point which also emerges from the simple fact that the circular filaments on the sky are usually only partial circles and not equally discernable over their full circumferences.

B. Magnetic Fields in Shells

A puzzling feature about expanding HI shells is the relatively large thicknesses. Consider the gas pressure in the shell, given by the product of density and temperature, nT. Temperatures in shell structures are generally low, less than about 200K; the shells are in fact the low temperature "cloud" component of the interstellar medium seen in the 21-cm line (Heiles, 1982). Volume densities in shell structures are much lower than those we have come to accept as characteristic of the cloud component; the shell geometry forces a change in the derived densities. For the Eridanus shell, for example, the HI volume density is less than 3 cm^{-3} at the distance of 500 pc (see Reynolds and Ogden, 1979 and Heiles, 1976).

Therefore, the gas pressure in the shell is only about 300 cm^{-3} K. This is much lower than the ram pressure of the shell expansion, and also much lower than generally-accepted values for the pressure of the interstellar medium (see Cox's paper in this volume). How can the shell maintain such a low gas pressure in the presence of such high external pressures?

The answer is the shell magnetic field, measured using Zeeman splitting of the 21-cm line to be 7 µGauss in two cases by Troland and Heiles (1982). A measurement of Zeeman splitting yields only the longitudinal component of the field, i.e. that component oriented along the line of sight to the observer. From geometrical considerations alone, with a probability of 1/2, the field is at least twice as large as the measured value. Suppose the field strength is really 10 µGauss; then the equivalent magnetic pressure is 3 x 10^4 cm^{-3} K, 100 times larger than the gas pressure. Clearly the magnetic field plays a crucial role in the gas dynamics of the shell.

C. Unusual and Peculiar Morphology in Apparent Shell Structures

There are some cases in which the velocity structure does not mimic that expected for an expanding shell. Two spectacular examples are discussed in more detail in Heiles (1982b). One shell, about 30 degrees

in diameter and located in the galactic anticenter, can be discerned
over a velocity range of 100 km/sec or more. There is no systematic
change of filament diameter with velocity. There is velocity structure,
but it exhibits no systematic trends. The other consists of
large-diameter HI filaments that overlap radio loops II and III. In the
vicinities of these loops are curved HI filaments with comparable
angular scales. These filaments exist in velocities ranging from +20 to
-170 km/sec. Some filaments with widely different velocities are
superimposed or lie parallel to each other. In one case, the end of a
long filament abuts the end of another filament whose velocity is 70
km/sec different. A final example is the peculiar structure discovered
by Lockman and Genzel (1982).

The juxtaposition of HI filaments at widely differing velocities is
not expected if the filaments are parts of expanding shells. It seems
to me that these structures are intrinsically different from the usual
shells. They have velocity structure, but it is not simple expansion
and they are visible over very wide velocity ranges. Therefore, even
though some of the filaments appear roughly to be circular on the sky,
they might not be parts of shells.

III. ORIGIN OF SHELLS AND SUPERSHELLS

Many shells are of moderate size and energy, and some are clearly
associated with star clusters. In these cases it is most reasonable to
assume that the source of energy is stellar winds and supernovae from
the stars. Bruhweiler et al. (1980) have developed a theory that nicely
accounts for the properties of many HI shells using this idea.

It is natural to try to extend the application of this idea to all
shells and supershells. However, for the largest supershells this meets
with difficulty because of the energy requirements. Five supershells in
Heiles (1979) are estimated to have kinetic energies of more than
10^{53} erg, with radii ranging up to 1.3 kpc and expansion velocities up
to 24 km/sec. Such shells are much too energetic to have been produced
by a "typical" OB associated with 28 BO stars and earlier (equivalent to
the Sco OB1 association); a much, much larger association would be
required. However, there is no independent evidence that such large
star clusters exist in our galaxy, particularly outside the solar circle
where most supershells reside and where star formation activity is
presumably quite moderate. Similarly, the structures with peculiar
kinematic morphology are quite possibly not expanding shells at all.
And if they are, the very large velocity spreads and peculiar
juxtapositions of filaments at widely different velocities imply that
peculiar, unknown physical processes are at work.

There seems to be little reason to assume that such structures are
produced by the "conventional" energy source, stellar winds and
supernovae in star clusters. The most prominent alternative is

collisions of high-velocity clouds with the galactic disk, a possibility
explored theoretically by Tenorio-Tagle (1980, 1981). An infalling
cloud transfers a large fraction of its kinetic energy to the ambient
gas in the disk; energy requirements are met fairly easily for clouds
similar to those observed. The fact that only one hemisphere of the
expanding supershells is observed is a prediction of the theory. The
juxtaposition of gas filaments at widely different velocities can be
modelled in terms of colliding gas clouds (see Cohen, 1981). However,
the presence of adjacent gas filaments at different, widely-spaced
velocities does not seem to be a natural consequence of this model.
More detailed theoretical and observational studies--perhaps using
X-rays--may clarify the explanation of these interstellar gas
structures.

REFERENCES

Allen, R.J., Goss , W.M., and van Woerden, H.: 1973, Astron. Ap. 29,
     447.
Allen, R.J. and Goss, W.M.: 1979, Astron. Ap. Suppl. 36, 135.
Brinks, E.: 1983, "Leiden Workshop on Southern Galactic Surveys."
     W.B. Burton and F. Israel, (eds.), D. Reidel, p. 239.
Bruhweiler, F.C., Gull, T.R., Kafatos, M., and Safia, S.: 1980, Ap. J.
     238, L27.
Cleary, M., Heiles, C. and Haslam, G.: 1979, Astron. Ap. Suppl. 36, 95.
Cohen, R.J.: 1981, MNRAS 196, 835.
Colomb, F.R., Poppel, W.G.L., and Heiles, C.: 1980, Astron. Ap.
     Suppl. 40, 47.
Heiles, C.: 1975, Astron. Ap. Suppl. 20, 37.
Heiles, C.: 1976, Ap. J. 208, L137.
Heiles C.: 1979, Ap. J. 229, 533.
Heiles, C.: 1982a, Ap. J., in press.
Heiles, C.: 1982b, Ap, J., in preparation.
Heiles, C. and Cleary, M.: 1979, Aust. J. Phys. Ap. Suppl. No. 47.
Heiles, C. and Habing, H.J.: 1974, Astron. Ap. Suppl. 14, 1.
Heiles, C., Stark, A.A., and Kulkarni, S.: 1981, Ap. J. 247, L73.
Lockman, F.J., and Genzel, B.L.: 1982, Ap. J., submitted.
Mathewson, D.S., Cleary, M.N. and Murray, J.D.: 1974, Ap. J. 190, 291.
Nousek, J.A., Cowie, L.L., Lindblad, C.J., and Garmire, G.P.: 1981, Ap.
     J. 248, 152.
Reynolds, R.J. and Ogden, P.M.: 1979, Ap. J. 229, 942.
Tenorio-Tagle, G.: 1980, Astron. Ap. 88, 61.
Tenorio-Tagle, G.: 1981, Astron. Ap 94, 338
Troland, T.H., and Heiles, C.: 1982, Ap. J. 260, L19.
Weaver, H.F., and Williams, D.R.W.: 1973, Astron. Ap. Suppl. 8, 1.
Weaver, H.F., and Williams, D.R.W.: 1974, Astron. Ap. Suppl. 17, 1.

DISCUSSION

GOSS: In the Westerbork HI observations of M101 (an ScI galaxy), we
observe one or two large HI holes. The diameters are 2 to 4 kpc. Can
these be related to your large galactic shells?

HEILES: I would think that they are the same type of object. It is
comforting to see them in other galaxies, because their appearance in our
own galaxy is severely influenced by the presence of unrelated HI along
the line of sight. Do you see expansion motion around the HI holes in
other galaxies?

GOSS: We observe no large systematic velocities in the M101 holes, but
they would be difficult to detect because this galaxy is very face-on.

GULL: Partial shells are notable in other regions of the spectrum. About
half a dozen SNR's show complementary structure in the form of
a [OIII] half ring and the remainder in a Hα, [NII], [SII] half ring.
Other SNR's show only half a shell. Around OB associations it is rare
that more than half to two-thirds of the shell is visible. This may be
explained by differences in shock velocities, densities, and especially
nonuniformities in the interstellar medium.

RADIO SYNTHESIS MAPS OF LARGE SUPERNOVA REMNANTS

R. Braun, H. van der Laan
Sterrewacht, Leiden, Netherlands

R.G. Strom
Netherlands Foundation for Radio Astronomy,
Dwingeloo, Netherlands

Several large (at least $0.^{\circ}5$ diameter) supernova remnants (SNR) located at $2.^{\circ}5$ or more from the galactic plane have been mapped with the Westerbork Synthesis Radio Telescope (WSRT) at 49 cm. The sample, which includes IC443, DA530, VRO42.05.01, CTA1 and OA184, is particularly suitable for complementary studies in other spectral regimes. By choosing objects at relatively high galactic latitudes we have consciously selected SNR which are likely to suffer less than average extinction and are probably nearer to the sun than most. This makes them particularly attractive for optical and X-ray studies which, along with IR and further radio observations, are either in progress or being planned. These are summarized in Table 1.

Table 1

SNR	Observation	Date
IC443	WSRT 49 cm continuum	Spring 1982
	WSRT 21 cm HI line	Summer 1982
	IRAS 4 band IR mapping	proposed
	optical velocity field mapping	planned
DA530	WSRT 49 cm continuum	Spring 1982
	WSRT 21 cm continuum	planned
	EXOSAT soft X-ray mapping	planned
VRO 42.05.01	WSRT 49 cm continuum	Spring 1982
	WSRT 21 cm continuum	planned
	IRAS 4 band IR mapping	proposed
	optical velocity field mapping	planned
	EXOSAT soft X-ray mapping	planned
CTA1	WSRT 49 cm continuum	Spring 1982
OA184	WSRT 49 cm continuum	Spring 1982.

J. Danziger and P. Gorenstein (eds.), Supernova Remnants and their X-Ray Emission, 373–376.

One aspect of the project is to search for small scale radio spectral index and polarization variations to investigate the particle acceleration mechanism and magnetic field structure. Another aim is to study how SNR interact with their environment. Neutral hydrogen cloudlets associated with IC443 which were discovered by De Noyer (1978) are being investigated in greater detail. We also hope to study how the SNR shock affects an adjacent molecular cloud by observing the expected IR emission. In several intermediate age remnants it should be possible to examine the dynamics using high resolution optical spectra of their filaments. Finally, we plan to map the hot thermal gas through the soft X-ray emission from several remnants to further clarify their evolutionary state and the prevailing physical processes.

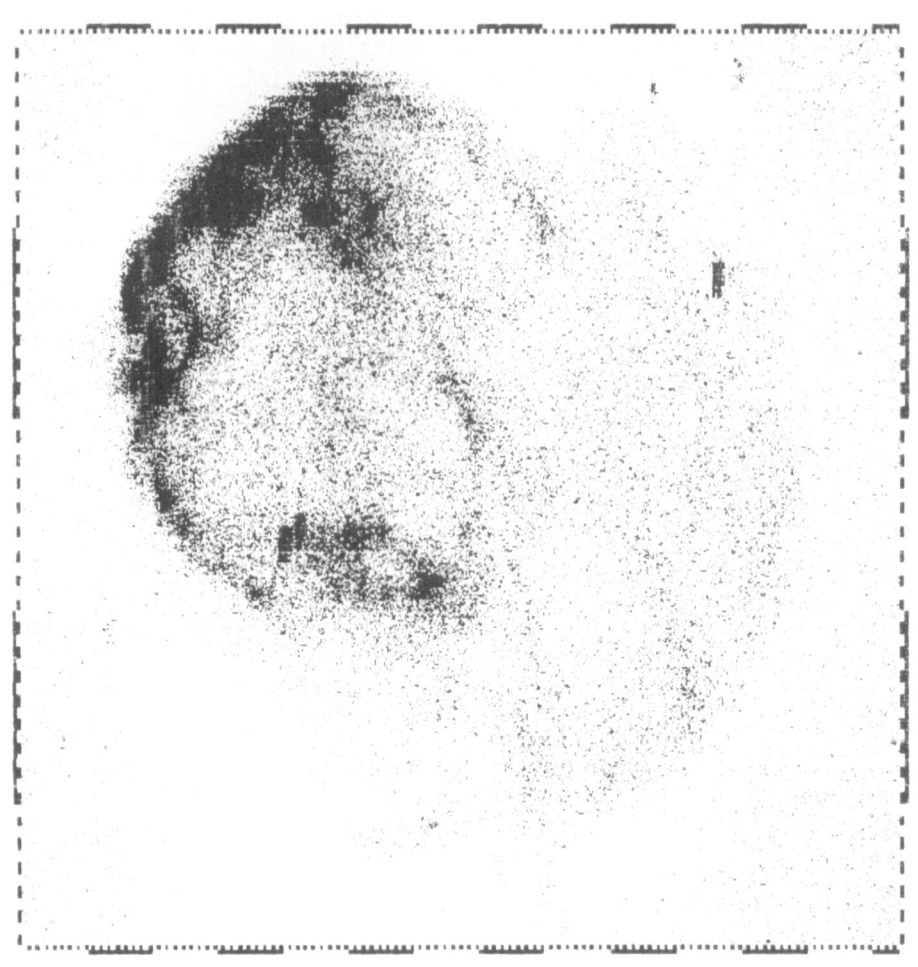

Figure 1.   The 49 cm map of IC443.

In addition, there are a number of general questions regarding SNR structure which call for clarification. As was clear from earlier work (e.g. Duin and Van der Laan, 1975), there is often an excellent correlation between the (thermal) optical nebulosity and (nonthermal) radio brightness distribution. IC443 (Figure 1) is a fine example of this. Another feature seen in several remnants including IC443 is for the radio emission on one side of the rim to be dominant. In a few objects such as DA530 (Figure 2), by way of contrast, the shell appears to be bracketed by two symmetric arcs of roughly equal brightness.

The tentative and rather general nature of this report reflects the very preliminary stage of much of the data reduction. Results on the various remnants will be published as the work progresses.

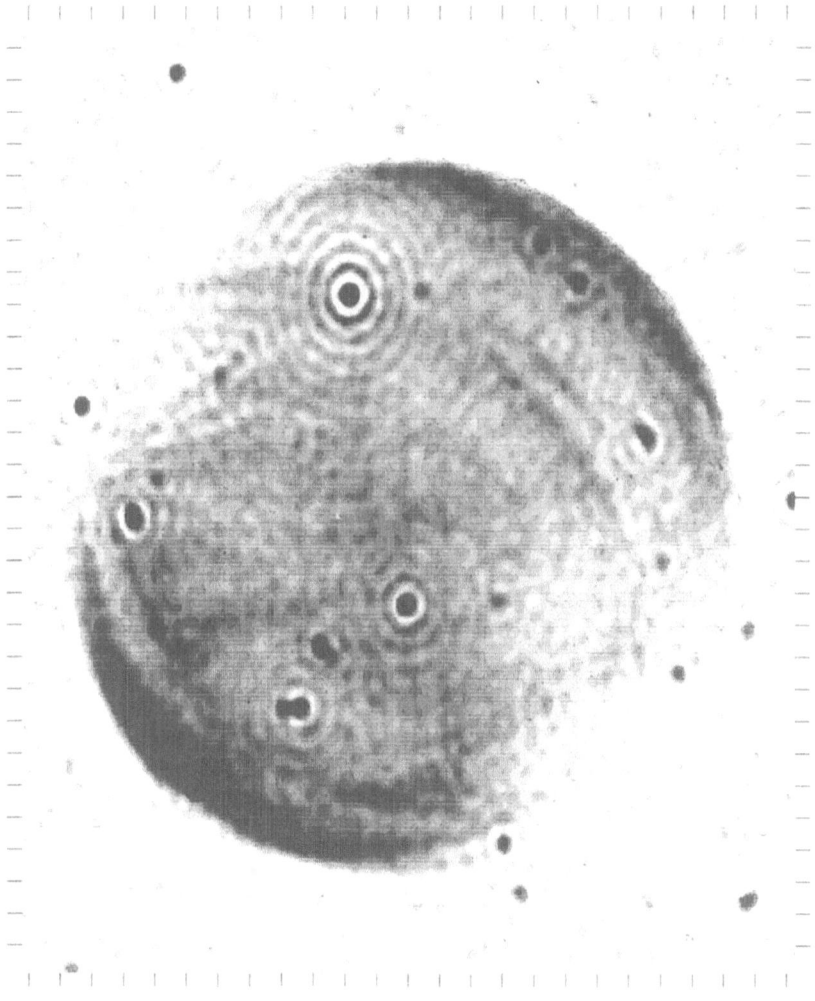

Figure 2.   The 49 cm map of DA530.

ACKNOWLEDGEMENTS

R.B. was supported during the course of this work by the Dutch Ministry of Education and Science through a Netherlands Government Scholarship. The Westerbork Radio Observatory is operated by the Netherlands Foundation for Radio Astronomy with the financial support of the Netherlands Organization for the Advancement of Pure Research (Z.W.O.).

REFERENCES

De Noyer, L., 1978. Mon. Not. Roy. Astron. Soc. 183, 187.
Duin, R.M., Van der Laan, H., 1975. Astron. Astrophys. 40, 111.

RADIO CONTINUUM OBSERVATIONS OF LARGE SUPERNOVA REMNANTS

W. Reich, E. Fürst, W. Sieber
Max-Planck-Institut für Radioastronomie, Bonn, FRG

Radio observations of large supernova remnants (SNRs) with high angular resolution have been provided by modern synthesis instruments preferentially at frequencies below 2 GHz. Since these instruments are sensitive mainly to unresolved emission spots, weak extended SNRs usually remain undetected. Besides this, there are numerous physical parameters, which can be studied more properly at higher frequencies. In particular, the polarization characteristics can be more easily analyzed and reduced to the intrinsic magnetic field orientation. In some cases foreground effects substantially disturb the SNR's field structure at low frequencies.

The determination of the intrinsic magnetic field vectors serves as an essential prerequisite for a detailed examination of the interaction between interstellar matter/magnetic field and the supernova remnant. This interaction should be reflected in the morphology of the remnant.

Polarization and morphology of large (old) supernova remnants beyond 1 GHz can be easily observed with the 100-m telescope at Effelsberg. As a single dish the 100-m telescope is well suited to map extended areas of low surface brightness with high reliability. The natural advantage of single dish instruments lies in the fields of high dynamic range, high sensitivity, low spurious polarization, and a nearly unlimited field of view. Instrumental characteristics of the new 2.7 and 4.75 GHz receiver systems at Effelsberg are given in the table. Two

Frequency	4.75 GHz	2.7 GHz
Beamwidth	2.5 arcmin	4.5 arcmin
Bandwidth	500 MHz	50 MHz
System	3-channel paramp	3-channel FET
Mode	dual/single horn	single horn
Polarization	I,Q,U	I,Q,U
System noise temperature	65 K	60 K
Aperture efficiency	49%	55%
$T_A(K) / S(Jy)$	1.4	1.56

J. Danziger and P. Gorenstein (eds.), Supernova Remnants and their X-Ray Emission, 377 379.

examples of recently observed sources can be found in the figures which represent a small sample from a still running extensive observing program.

Besides the determination of morphology and polarization structure we intend to obtain high quality spectral index maps, which in connection with magnetic field estimations may serve as a test for acceleration models for energetic electrons as, for instance, the model proposed by Bell (1978).

Finally, we hope to find further large (old) remnants with bent radio spectra. In combination with small scale structures this might provide information on the compression rate of the radio emitting regions.

Bell, A.R.: 1978, Monthly Notices Roy. Astron. Soc. 182, pp. 147-156

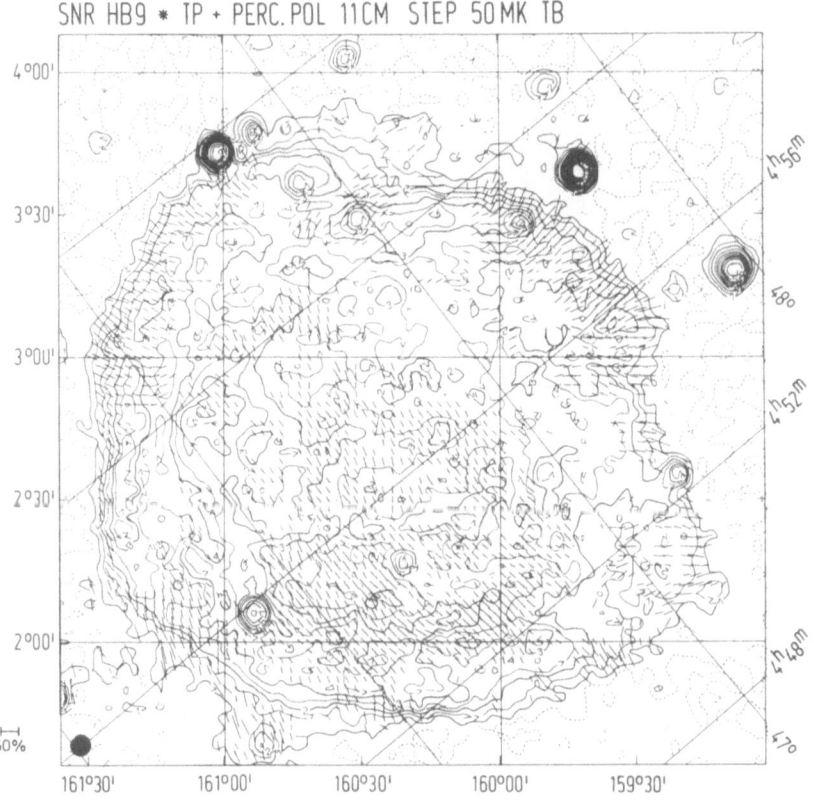

Fig. 1: 2.7 GHz observations of HB9 with the Effelsberg 100-m telescope with an angular resolution of 4.'5. Contours run in steps of 50 mK T_B. Percentage polarization is represented by bars in the direction of the electric vector. Percentage polarizations are plotted above 15%.

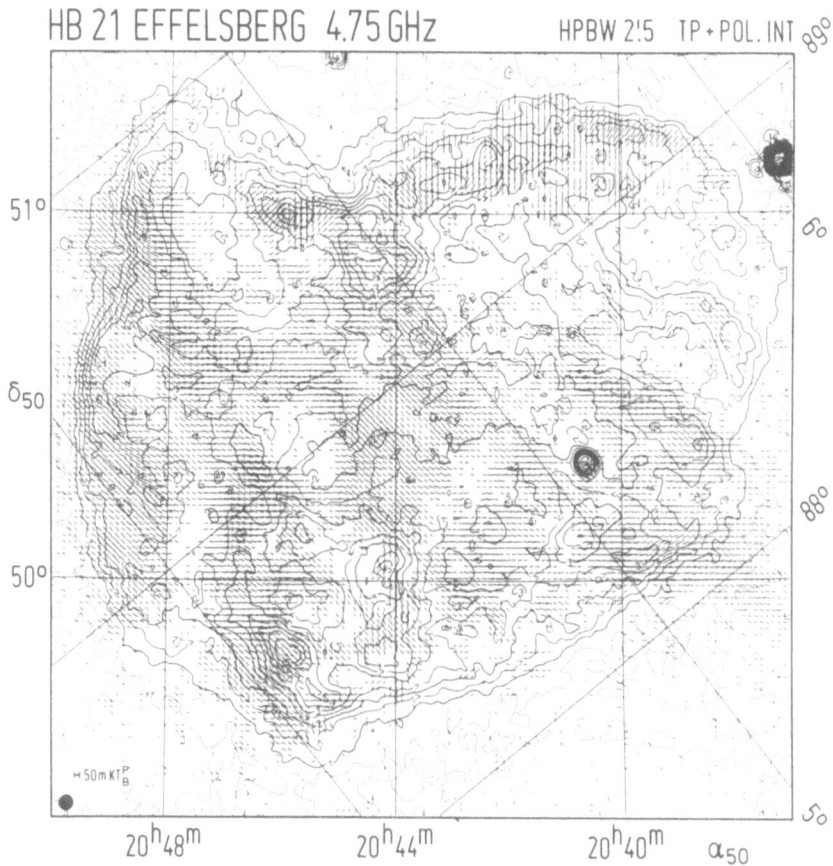

Fig. 2: 4.75 GHz observations of HB21 with the Effelsberg 100-m telescope with an angular resolution of 2!5. Contours run in steps of 50 mK
T_B. The polarized intensity is represented by bars in the direction of
the electric vector.

ARE FILAMENTS CYLINDRICAL OR SHEET-LIKE?

M. D. Smith
Astronomy Program, University of Maryland
and
J. R. Dickel
Department of Astronomy, University of Illinois.

ABSTRACT. Rapid cooling and collapse in a thermally unstable medium can lead to filamentary structures that are comparable to those observed in old supernova remnants. Here, simple formulae are derived for the shape of a filament forming in this manner. The cross-sectional shape of a filament will flatten and the ratio of major to minor axes (aspect ratio) increases in proportion to the density as it becomes sheet-like. Thus the aspect ratio of observed filaments is expected to be of the same order as the temperature change during the collapse, T_1/T_0. In very old remnants the filaments may appear to be tubes or sheets depending on perspective with aspect ratios less than 10. This model predicts a correlation between optical morphology and the temperature of the shock-heated gas. For T_1 between 10^5K and 10^6K, the thin sheet that forms is likely to split up through asymmetric warping disturbances, leading to several parallel striations, consistent with observations in the Cygnus Loop. Above 10^6K, the instability may cause the sheet to degenerate before it has cooled to 10^4K. Hence, an absence of well-defined filaments is expected near X-ray emitting regions.

1. THE COLLAPSE

As a region of gas cools behind a shock front, it can collapse and fragment. These collapsed regions are thought to correspond to observed optical filaments in old supernova remnants. Thermal instabilities cause the fragmentation. That is, regions of greater density cool faster than regions of lower density, and thus, to keep in pressure balance, become even denser. They will apply to regions in which the sound crossing time, t_s, is shorter than the cooling time, t_c. Using a mathematical technique involving a small parameter called the slenderness ratio, $\varepsilon = t_s/t_c$, it is possible to accurately follow the collapse in two dimensions (Smith and Dickel 1983).

The results obtained by the above method are easily generalized. This is best seen by taking an ellipsoid with axes a, b and c, and external pressure P_e and central pressure P_c. As a result

J. Danziger and P. Gorenstein (eds.), Supernova Remnants and their X-Ray Emission, 381–383.
© *1983 by the IAU.*

of radiative cooling within the ellipsoid the central pressure is
slightly less than the external pressure. It is this small pressure
difference which drives the collapse. The rates of change of the
lengths of the axes depends on the pressure gradients $(P_e-P_c)/a$, $(P_e-P_c)/b$ and $(P_e-P_c)/c$. Since the pressure gradient is greatest along
the shortest axis, the collapse is mainly in this direction. Thus,
the ellipsoid flattens and once, say, a $\ll$ b, c the collapse is
essentially only along the a-axis.

With the collapse, the magnetic field becomes more ordered.
Thermal conduction inhibits temperature variations in the field
direction and we can then consider a two dimensional collapse. The
cross-sectional area of a filament is inversely proportional to the
mass density (mass conservation) and, further, is proportonal to the
temperature provided the external pressure remains unchanged. It
follows that the axial ratio is given by

$$a/b \sim O(T_i/T_f) \quad \text{for } a > b, \text{ and} \tag{1}$$

$$a/b \propto T_i/T_f \quad \text{for } a \gg b, \tag{2}$$

where T_i and T_f are the initial and final temperatures of the
collapsing region.

Through the slenderness method we find that a sheet is unstable
to warping. Initial irregularities in the shape of a region are preserved in size while the region collapses. Thus irregularities are
amplified relative to the sheet thickness. Given that the initial
shape is far from elliptical, the instability will split up the filament. We do not expect a thin sheet to survive intact for very long.

2. OBSERVATIONAL IMPLICATIONS

Putting $T_f \sim 10^4$K, we can find the structure at the optically
observable stage. With $T_i \lesssim 10^5$K we do not expect a great collapse
or change in shape. For 10^6K $> T_i > 2.10^5$K we expect several
parallel striations in the form of sheets wiht axial ratios of
10:1. For $T_i > 10^6$K, highly unstable thin sheets are predicted,
resulting in more diffuse filamentary structures. Hence, even though
there is no observed correlation between X-ray and optical brightness
distributions, we predict a correlation between X-ray emission and
the optical structure. A further test of the model would be to
determine the depth of filaments as done by Miller (1974), who
concluded that certain filaments in the Cygnus Loop are sheets.

As J. Dickel has remarked in this Symposium, much work still
needs to be done in this area. The full story must include thermal
conduction, inertial collapse and magnetic field effects. We
encourage observations of individual filaments in both the radio and
optical. This will lead to tighter constraints on physical
parameters and give the theoretician a more precise task.

REFERENCES

Miller, J.S. 1974, Ap. J., 189, 239.
Smith, M. D., and Dickel, J. R. 1983, Ap. J. February 1.

DISCUSSION

DANZIGER: I performed a similar excercise as Miller (on Cygnus) for
a filament in the Vela supernova remnant, and obtained a similar
result, viz. a length in the line of sight about an order of
magnitude greater than the observed width.

BENFORD: How much does the presence of a significant embedded
magnetic field - which will be compressed and perhaps amplified -
affect your results?

SMITH: One must first determine the strength of the magnetic field
in the interstellar medium. In cool clouds the field and thermal gas
may be in approximate pressure balance, whereas the thermal pressure
may dominate in the hot intercloud gas. In either case, the
compression and heating by a blast wave leaves the magnetic pressure
relatively small. In the subsequent collapse, the magnetic field
within the region rapidly increases. For high T_1 the magnetic
pressure can balance the external pressure before the filament has
cooled to 10^4K. At this stage, however, the filament is likely to be
collapsing inertially i.e. ahead of pressure equilibrium, and hence,
the collapse towards a sheet continues. Finally it should be noted
that it is difficult to make dense filaments if the field pressure or
its turbulent amplification is of greater significance.

HEATING OF THE INTERSTELLAR MEDIUM BY SUPERNOVA REMNANTS

Donald P. Cox
Department of Physics, University of Wisconsin, Madison

We observe the heating of interstellar material in young supernova remnants (SNR). In addition, when analyzing the soft X-ray background we find evidence for large isolated regions of apparently hot, low density material. These, we infer, may have been heated by supernovae. One such region seems to surround the Sun. This has been modeled as a supernova remnant viewed from within. The most reasonable parameters are ambient density $n_0 \sim 0.004$ cm^{-3}, radius of about 100 pc, age just over 10^5 years (Cox and Anderson 1982).

Besides what is and what may be heated by SNR, there is also that category of what ought to be so heated. The story is by now familiar and not particularly sensitive to who is telling it. The underlying assumptions seem to be few: If supernovae as energetic as several times 10^{50} ergs occur in the Galaxy more frequently than one per 100 years, if the spatial and temporal positioning of the explosions (or a signif- icant subset of them) is fairly random, and if most of the interstellar hydrogen is in clouds, then the picture should be appropriate.

The interstellar volume should be dominated by a low density high temperature matrix. The hot gas should not be gravitationally bound to the disk so should rush outward to form a fountain, a quiescent corona, or a wind (or some combination of the three). Pressure vari- ations in the hot component beat on cloud boundaries and contribute significantly to the heating of low density HI regions. This heating alone is sufficient to sustain a neutral "intercloud" phase with $T \sim 10^4$ K. Finally, the far-reaching SN shockwaves in the hot matrix can potentially accelerate the cosmic rays by mechanisms which seem able to provide both the required power and the spectrum.

The basic idea of this model is that supernovae mechanically heat not only their surrounding matrix material, but also some material which was not previously part of the hot phase. One does not at the outset assume just how much material is heated in this way but there are limits. For example, with an average density of 1 cm^{-3}, there are 1000 $M_\odot$ in a sphere of radius 20 pc. But calculations for SNR evolving

J. Danziger and P. Gorenstein (eds.), Supernova Remnants and their X-Ray Emission, 385–392.
© 1983 by the IAU.

in a homogeneous medium (so that all mass enclosed is heated) reveal
that remnants radiatively cool by the time they heat about 1000 M_{Θ}
$(1 \text{ cm}^{-3}/n_0)^{2/7}$. The insensitivity to density implies that this result
will be approximately valid for the inhomogeneous case as well. In
order for remnants to reach immense sizes (R > 100 pc) the fraction of
enclosed mass which is heated and transferred to the hot phase (the
mass acquisition) must be quite small.

Fountain or wind models rely on three basic equations and are inher-
ently two parameter models. One parameter is the supernova power,
E_0/τ_{SN}, the other is the mass acquisition rate, $\dot{M}$. The three equations
are

$$\frac{5kT}{m} \sim \frac{E_0/\tau_{SN}}{\dot{M}} \tag{1}$$

(the temperature of the hot phase is just the ratio of energy to mass
inputs),

$$\left[\frac{3}{2} p\right] \left[2\pi R_g^2\right] c \sim E_0/\tau_{SN} \tag{2}$$

(the thermal energy is convected out of the plane at its sound speed,
$c = (10 \ kT/3m)^{1/2}$ where p = 2nkT, and finally

$$L(T)n^2 (2\pi R_g^2 \ H_0) = F \ E_0/\tau_{SN} \tag{3}$$

where F is the fraction of the energy which is radiated by matrix
material while still in the plane. In the equations above, E_0 is the
explosion energy, τ_{SN}^{-1} the supernova rate, R_g is the galactic radius,
H_0 the plane thickness, p, T, n are the hot matrix pressure, temperature,
and number density, respectively, m is the mean nuclear mass, and L(T)
is the cooling coefficient which I take to be the usual $L(T) \sim 1.3\text{x}10^{-22}$
$(T/10^6 \ K)^{-1/2}$ erg cm^3 s^{-1} approximation (for coronal temperatures) to
the collisional equilibrium calculations of Raymond, Cox, and Smith
(1976).

In preparation for this presentation I calculated the detailed
structure of a one dimensional gas flow approximation for the behavior
of this material. I wanted to see just how sensitive the above
equations were to detailed choices. The results for the model which I
made were that if the equations were to be used to represent conditions
in the plane, the right-hand side of equation (1) should be multiplied
by 6/5, while the right-hand side of equation (2) should be multiplied
by 0.97. In short, the equations are very insensitive to model details.

The flow away from the plane had a sonic point just above the
region of SN input. The temperature there was 3/4 of the central plane
value, while the pressure was down by a factor of 3. The magnitude of
the pressure drop is an interesting find; it is clearly required to
accelerate the material to supersonic velocities.

It is useful now to substitute some numbers into the two parameter model to see just how many things fall out. Any two parameters will do and I will use the supernova power E_0/τ_{SN} = 5×10^{50} ergs/50 years and the midplane temperature T = 10^6 K. (I also assumed R_g = 15 kpc and H_0 = 100 pc.) The derived properties of the system are then c = 152 km s^{-1}, $\dot{M}$ = 17 $M_\odot$ yr^{-1} (corresponding to 850 $M_\odot$ per SN), p = 1.0×10^{-12} dyn cm^{-2} (p/k = 7250 cm^{-3} K), n = 3.6×10^{-3} cm^{-3}, and F = 0.02. That is, the choice of supernova power and any one of the other parameters provides all of the other familiar sounding results above.

Two of the best known models (McKee and Ostriker 1977, hereafter MO, and Cox 1981, hereafter paper II) attempt to provide an understanding of mass acquisition which will reduce the above model to one parameter. The model is then the system response to the supernova rate. In fact, paper II goes on to look for ways in which the system decides its own supernova rate, removing even the last free parameter.

MO use thermal evaporation of clouds as their mass acquisition process. This is extremely temperature sensitive, and operating on the interstellar clouds it offers an additional relationship between T and $\dot{M}$. They went beyond that, however, arguing that in effect, F $\approx$ 1. By making this choice, they already eliminate the second parameter, fixing T, p, n, $\dot{M}$, etc. for a given supernova rate. Their use of thermal evaporation then serves to determine the parameters of clouds required to provide $\dot{M}$ in equilibrium. It also gives a concrete mechanism for maintaining that equilibrium. (Insufficient evaporation raises T which very much raises the evaporation rate.)

One could generalize the MO model by abandoning F $\approx$ 1 in favor of F $\lesssim$ 1 in which case it would no longer be overdetermined. It would allow fountains and winds for sufficiently high supernova rates or when there were too few clouds available for evaporation.

It is well-known (to those who know me well) that I have long been suspicious of thermal evaporation. I could visualize far too many processes which would provide tangential magnetic fields along cloud boundaries, among them the prior evaporation of those clouds which did not have such fields. Nevertheless, I came to realize that mass acquisition in some form was necessary for fountain models, because the cooling of material in the fountain removes material from the hot phase at a terrific rate.

In the model in paper II I reasoned that suitable mass acquisition could be achieved mechanically if there were, in addition to the hot matrix, also a warm neutral intercloud medium (something like MO's WNM but heated differently and not necessarily bounding denser cores) which occupied something like half the volume. This intercloud component was fed both by returning fountain material and by mass loss from stars, and it needed to be close to the point of instability for condensing into clouds, so that it could occasionally rid itself of excess material.

This is all the more important if there is a substantial infall of new
material into the Galaxy. The mechanical heating rate per unit volume
of this material was estimated to equal the supernova power per unit
volume. A study of the phase diagram indicated that this material
needed to have $T \sim 10^4$ K and that on average the heating had to balance
a cooling coefficient $L_d \sim 5 \times 10^{-25}$ erg cm^3 s^{-1}. (The subscript d refers
to destabilization.) Owing to partial magnetic support, the thermal
pressure was taken as half the matrix pressure, p. This combination of
results implies

$$h \sim E_0/(2\pi R_g^2 H_0 \tau_{SN}) \sim L_d \left[\frac{p/2}{k \cdot 10^4 K}\right]^2. \tag{4}$$

Also, because it assumes that half the supernova power goes into
maintaining intercloud material, and only half the plane volume is in
matrix form, there are factor of two changes in parts of each of the
first three equations.

Since equation (4) provides one more relationship, the model is
reduced to one parameter. Oddly enough, the result is an $\dot{M}$ which is
independent of the supernova rate. The matrix temperature is then
proportional to the supernova power, while the pressure increases only
as the square root, and the density inversely to the square root of
supernova power. In any case, a one parameter system is achieved,
different from the $F \simeq 1$ model of MO or even the $F \lesssim 1$ generalization
of their model.

One or the other of these models would have been dismissed if they
weren't each providing about the same results for the supernova rate
the Galaxy has. Since they do, it falls to a good hearted squabble
among us modelers to resolve the issue.

Meanwhile the Galaxy is telling us some things we should be
listening to. One way to picture one of these is to go back to the two
parameter model (skirting the issue of how to determine $\dot{M}$) and simply
write the general expression for F (the fraction of the SN power
radiated in the plane) in terms of combinations of two of the other
parameters. The result is

$$F \sim 0.02 \ (10^6 \ K/T)^3 \ (p/10^{-12} \ dyn \ cm^{-2})$$
$$\propto (E_0/\tau_{SN})/T^{7/2} \quad \propto (\dot{M})^{7/2}/(E_0/\tau_{SN})^{5/2}. \tag{5}$$

What we have is a parameter which is extremely sensitive, and
which also has strong observational restrictions. From the soft X-ray
background, we know that the entire background would derive from
material with $p = 10^{-12}$ dyn cm^{-2}, $T = 10^6$ K, and path length ~ 100 pc,
that is, from material in the plane, while something like 50 times as
much emission (for F = 1/50) will derive from the fountain. This
factor of 50 is the ratio of what we believe to be the supernova power
to the fraction of that power which the soft X-ray background implies
is being radiated at 10^6 K.

In short, the bulk of the supernova power is not being radiated by a plasma at 10^6 k, not in a fountain, not in a corona, not in the plane, nowhere. We could see it and we don't. There are only two ways around this within the context of this model. The first, espoused both in MO and in paper II, is that the energy is being radiated at a lower temperature, perhaps 3×10^5 K where it could hide from the X-ray telescopes. From equation (5) it is clear that this option implies large F and both papers essentially espouse F $\sim$ 1 with most of the radiation coming from material in or near the plane. The second option is that the energy is not radiated at all. It is instead used to drive material out of the galactic plane. The required value of T is at least 2×10^6 K to get the material out. A higher temperature yet, perhaps 4×10^6 K, is required to prevent greater than observed X-ray emission. This higher temperature is also desirable because it lowers the mass outflow in the wind ($\sim 17 \times 10^6/T$ $M_\odot$ yr^{-1}) toward more acceptable values.

I personally do not accept the possibility of such a hot wind bearing most of the supernova power. It is certainly not allowed if thermal evaporation is taking place at close to the MO value. It is not allowed if $\dot{M}$ is at the paper II value (although the need for destabilization is also not so pressing in this case). What troubles me more is universality.

Let me restate the situation. If the average supernova heats less than $\sim$ 100 $M_\odot$, there will be a strong hot wind, probably not too bright in X-rays. If it heats $\gtrsim$ 1000 $M_\odot$, the energy will be radiated in or near the plane as hard UV and will also not appear too bright in soft X-rays. The intermediate regime is forbidden by observations. It is forbidden for our Galaxy on average, or even as a significant fluctuation. It is forbidden in other galaxies as well, as more and more are found not to show strong X-ray emitting coronae.

That wouldn't be so bad, if we knew that supernovae managed never to heat even 100 $M_\odot$. But they seem to. The Cygnus Loop already has. So we require some mechanism that shuts off acquisition after 100 $M_\odot$, to assure a sufficiently hot wind. Or we need one which assures that it continues beyond 1000 $M_\odot$ to assure dissipation as UV. My lack of acceptance of the hot wind possibility thus derives from my inability to visualize a mechanism which restricts the energy to 100 $M_\odot$ and then liberates that heated gas intact. Despite my prejudice, I would urge caution on this point, however, because the remnants in the LMC seem to show a disturbing tendency to disappear after heating about 100 $M_\odot$. This disappearing act evidences itself as an apparent constant velocity expansion law, since the oldest and most common remnants have about the same expansion rate after interacting with the same mass.

The high mass acquisition end is, however, self-regulatory, particularly in the MO model. Mass acquisition simply remains high until the temperature gets too low. Their model is even stable as regards fluctuations in the supernova rate. Consider equation (5) with F = 1. We find that $T^{7/2} \propto (E_0/\tau_{SN})$. But in the MO model, $\dot{M} \propto T^{5/2}$ so $\dot{M}T$ is

D. P. COX

proportional to the supernova power as in equation (1), with no change required in the required cloud population. The temperature of the hot component is a weak function of supernova power, and hot emissive coronae do not have access to a major part of the SN energy.

This kind of rigidity is not as clear in the paper II model for which Ṁ is constant, and therefore depends on the additional regulation of the massive star formation and supernova rates.

Returning finally to my original point, for the supernova rate which the Galaxy is thought to have, the hot gas parameters, system pressure, and mass exchange rate are all essentially the same for either the MO model (as herein simplified) or the paper II model, because that information all follows from equations (1) through (3), subject to $F \sim 1$ in the MO case, or equation (4) in the paper II case. Both satisfy the soft X-ray background constraint. The difference between the two (at the given SN rate) is the manner in which Ṁ is achieved. The MO picture requires remnants to show strong evaporative effects. The paper II picture requires the presence of a considerable amount of warm intercloud material ($n \sim 0.3$ cm^{-3}).

Before closing, there are three other points to which I would like to call attention:

(1) A blast wave model now exists which includes shock heating of electrons and thermal conduction (Edgar and Cox 1983) and will soon be available for nonequilibrium ionization modeling of SNRs.

(2) If cosmic ray acceleration takes place efficiently in shock waves in time scales of 10^5 years, and if the soft X-ray background truly derives from our being inside an explosion remnant with that age, then the locally measured cosmic rays could contain a substantial component of essentially zero age.

(3) If the cosmic ray pressure tracks the variations expected in the thermal pressure in the matrix, either because of efficient acceleration plus localization by scattering, or perhaps even because of localization alone, the expected variations in cosmic ray pressure will be markedly at odds with the apparent near constancy of the cosmic ray flux at Earth over the last few million years (Szentgyorgyi and Cox 1983).

REFERENCES

Cox, D. P.: 1981, Ap. J., 245, p. 534 (paper II).
Cox, D. P., and Anderson, P. R.: 1982, Ap. J., 253, p. 268.
Edgar, R. J., and Cox, D. P.: 1983, in preparation.
McKee, C. F., and Ostriker, J. P.: 1977, Ap. J., 218, p. 148 (MO).
Szentgyorgyi, A., and Cox, D. P.: 1983, in preparation.

DISCUSSION

MCKEE: In your model of the ISM you find a mass exchange rate from clouds to the hot matrix of 17 $M_\odot$ yr^{-1} or 850 $M_\odot$ per SNR, which is comparable to the value Ostriker and I obtained. A key question is how this mass exchange affects the evolution of individual remnants; Ostriker and I found rather dramatic effects. What effects do you think the mass exchange has on SNR evolution?

COX: Ach! You're a good man, McKee. You know I'm reluctant to answer that question but maybe you don't know why. I think of the ISM as a very chaotic place about which it is dangerous to be too specific. I think remnants interact with a variety of environments and that getting mass into the low density phase may well be a two step or collective process. I am reasonably certain that we are seeing acquisition taking place as the Cygnus Loop shocks material with ambient density around 0.2 cm^{-3} to something like 3×10^6 K. We see roughly 200 $M_\odot$ which has been so acquired. In the case of the Cygnus Loop, the problem is how to stop acquisition before it goes too far. And for that we need for it to find the matrix phase and to begin propagating in that lower density environment. I'm not sure it has, as yet, and that troubles me.

Roughly speaking, however, my point of view is that when a remnant evolves in the hot phase, it engulfs, pressurizes, and therefore heats intercloud material substantially, as long as R $\leqslant$ 25 pc. You have made a calculation of the radiative lifetime of this material and concluded that, if I remember correctly, not enough of it would stay hot until another blast wave happened by. I haven't yet checked that calculation but, given that it is correct, I still am not convinced that acquisition has failed. What is needed is not to keep the material hot (which reverberations might anyway) but to keep it diffuse. It competes for the available volume only with the previous generation of matrix material, which has by then gotten so diffuse that I think we have to rethink the notion that it can be treated as a fluid.

DOPITA: Have your Galactic fountain/corona/wind models taken into account the resultant radial zone mixing that will occur? In view of the relatively large mass transfer occurring now (which may have been still larger in the past) such an effect could be very important in determining the present Galactic abundance gradient.

COX: No. Joel Bregman may have done some work in this area, but I think that by and large one should at present keep this mixing as a free parameter in the chemical evolution models with an eye on the fountain as a possible mechanism. Because of the extreme sensitivity of F to the parameters, the mass flux above the plane could be anything from zero to perhaps 20 $M_\odot$ per year.

FEDORENKO: You have argued that the Sun is situated within a SNR of radius 100 pc. But there are Soviet observations of Lyα from the environment of the Sun of order 20 a.u. They indicate that we are placed in a much denser medium with $n \sim 0.1$ cm^{-3}. How can you resolve this contradiction?

COX: There are a number of observations bearing on this question and the picture which emerges is that the Sun is in a small piece of inter-cloud material, probably partly ionized, with a temperature $\sim 10^4$ K, density ~ 0.1 cm^{-3}, and radial extent of at most 10^{19} cm. Beyond that, the average neutral hydrogen density is extremely low for a long way. I know that sounds implausible, but one supposes that our little cloud is but one of many. If so, however, their total volume occupation must be quite low, much less than the 1/2 that I have suggested for this component.

This research was supported in part by NASA grant NGL 50-002-044 at the University of Wisconsin. I am grateful to Wilt Sanders for a critical reading of the manuscript, and to the symposium organizers, the IAU, and the Graduate School of the University of Wisconsin for welcome funds.

THEORETICAL THERMAL X-RAY MAPS OF SUPERNOVA REMNANTS IN NON-UNIFORM MEDIA

H.W. Yorke [1], G. Tenorio-Tagle [2], P. Bodenheimer [3]
1 Universitäts-Sternwarte, Göttingen, FRG
2 Max-Planck-Institut für Astrophysik, Garching b.München, FRG
3 Lick Observatory, U.C. Santa Cruz 95064

Abstract.

The evolution of a supernova remnant resulting from an explosion near the edge of a molecular cloud is calculated with a 2-D hydrodynamical code (axial symmetry assumed). Cooling effects have been included. 1 kev X-ray maps of the remnant at different evolutionary times and different viewing angles are presented.

1. INTRODUCTION

A number of possible physical effects could cause supernova remnants to be non-spherical, including (1) an asymmetric stellar explosion due to rotational effects (Bodenheimer and Woosley 1982, see also Bisnovatyi-Kogan 1983), (2) small scale inhomogeneities in the interstellar medium (Chevalier and Theys 1975), (3) the interaction of two remnants (Ikeuchi 1978, Cox and Smith 1974, Jones et al. 1979, Bodenheimer et al. 1982), (4) an explosion inside of and near the edge of a molecular cloud (Tenorio-Tagle et al. 1982), (5) the interaction of a supernova remnant with an exterior cloud (Sgro 1975, McKee and Cowie 1975, Bychko and Pickel'ner 1975), and (6) the effect of a density gradient perpendicular to and out of the galactic plane (Chevalier and Gardner 1974, Sanders 1976, Möllenhoff 1976, Bodenheimer et al. 1983, Tenorio-Tagle et al. 1983).

In the following we shall consider mechanism (4) using a 2-D hydrodynamical code (see Bodenheimer et al. 1982,1983 and references therewithin). We assume that a 10^{51} erg explosion occurs 1 pc inside a cloud of molecular hydrogen of density $n = 10^3$ cm^{-3} and temperature $T = 10$ K. The partially ionized intercloud medium was assumed to have a density of $n = 1$ cm^{-3} and to be in pressure equilibrium with the cloud material. Rotation and self-gravity were not included, and cooling within the remnant was calculated according to Cox and Daltabuit (1971).

From the resulting density and temperature structure of the remnant we were able to solve the equation for radiation transfer in the optically

393

J. Danziger and P. Gorenstein (eds.), Supernova Remnants and their X-Ray Emission, 393–397.
© 1983 by the IAU.

thin case for various evolutionary times and different "viewing" angles. We present some of the results of these calculations in a series of diagrams over the next few pages. A more complete description including other cases considered is forthcoming (Tenorio-Tagle et al. 1982). Similar X-ray maps resulting from mechanism(3) are also in preparation (Bodenheimer et al. 1982).

2. RESULTS

The numerical results are shown in Fig. 1 to 5 for selected evolutionary times. In Fig. 1a (1b) we show the density (temperature) contour lines in a meridional cut 1.58 million years after the supernova explosion. The corresponding 1 kev X-ray map for an "observer" located in a direction 90 degrees from the symmetry axis (edge-on view) is shown in Fig. 1c. A similar set of contour diagrams are shown in Fig. 2abc for the supernova remnant 3.97 million years after the explosion (note the change in distance scale). In Fig. 3 (6.95 million years) the temperature structure is not shown. Instead, we display the 1 kev contour maps for three viewing angles: 90° (3b), 60° (3c) and 30° (3d). At this time the total 1 kev X-ray flux had decreased to 10^{14} erg s^{-1} Hz^{-1}. In Fig. 4 (11.3 million years) and Fig. 5 (33.9 million years) the density, velocity and temperature structure is shown for still later evolutionary times. Note that even at these late stages the part of the remnant moving into the molecular cloud evolves as a spherical explosion would.

Fig. 1a Density contours and velocity structure of remnant 1.58 million years after a supernova explosion inside a molecular cloud (bottom part of diagram). The explosion site is indicated with a cross. The maximum (minimum) contour level is indicated by an M (m) corresponding to the density given in the lower left-hand corner. The arrows give the direction and magnitude of the velocity at the arrows' tips. The magnitude for the standard length arrow is given in the lower right-hand corner.

Fig. 1b Same as Fig. 1a for the temperature contours. The maximum (minimum) contour level is indicated by a T (t) corresponding to the temperature given in the lower left-hand corner.

Fig. 1c X-ray (1 kev) contour map for the supernova remnant displayed in Fig. 1ab at a viewing angle of 90°. The total 1 kev flux of the source is given in the lower right-hand corner. The contour levels are spaced logrithmically in intervals of one order of magnitude.

Fig. 2 Same as Fig. 1 for an evolutionary age of 3.97 million years.

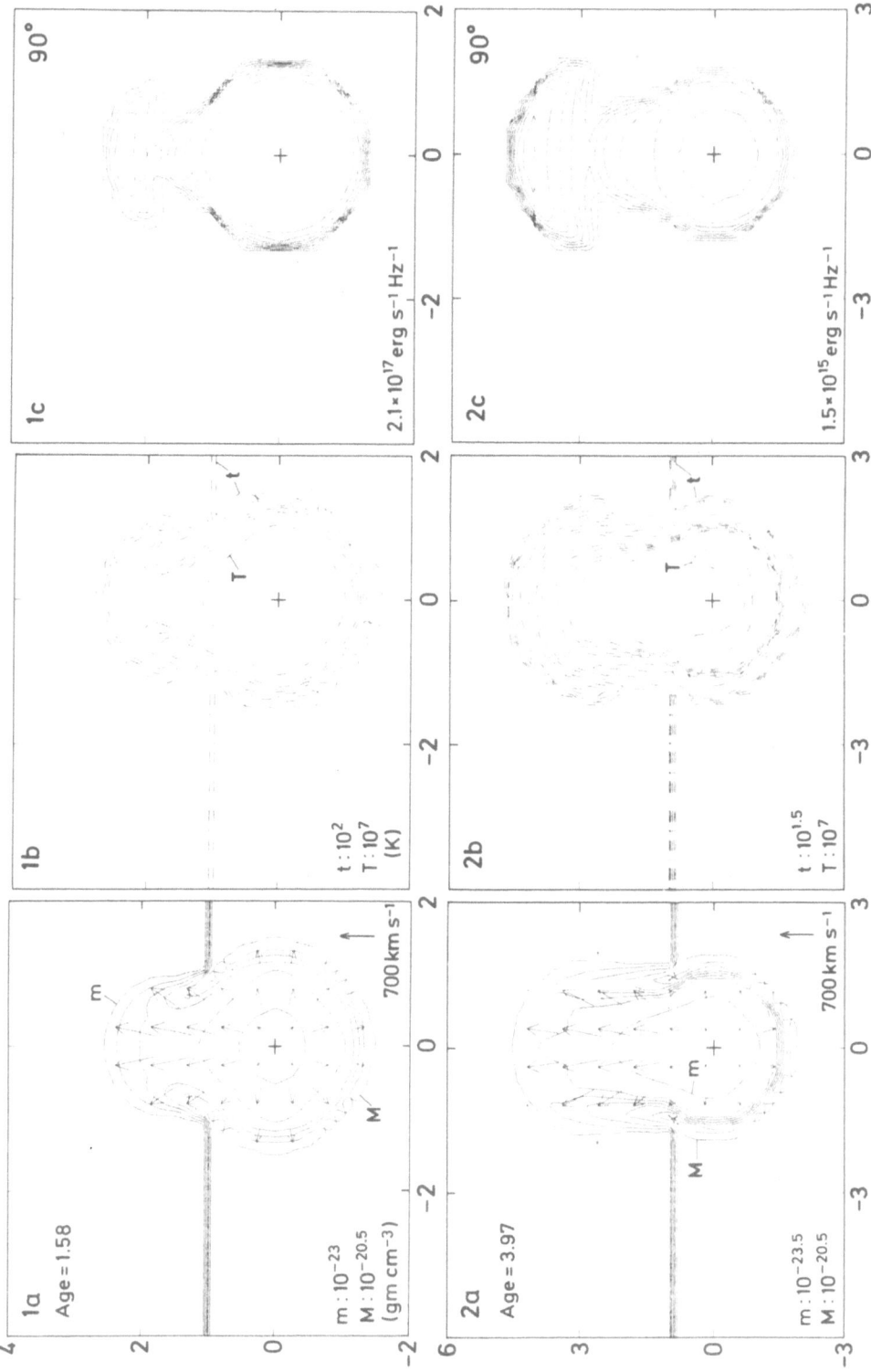

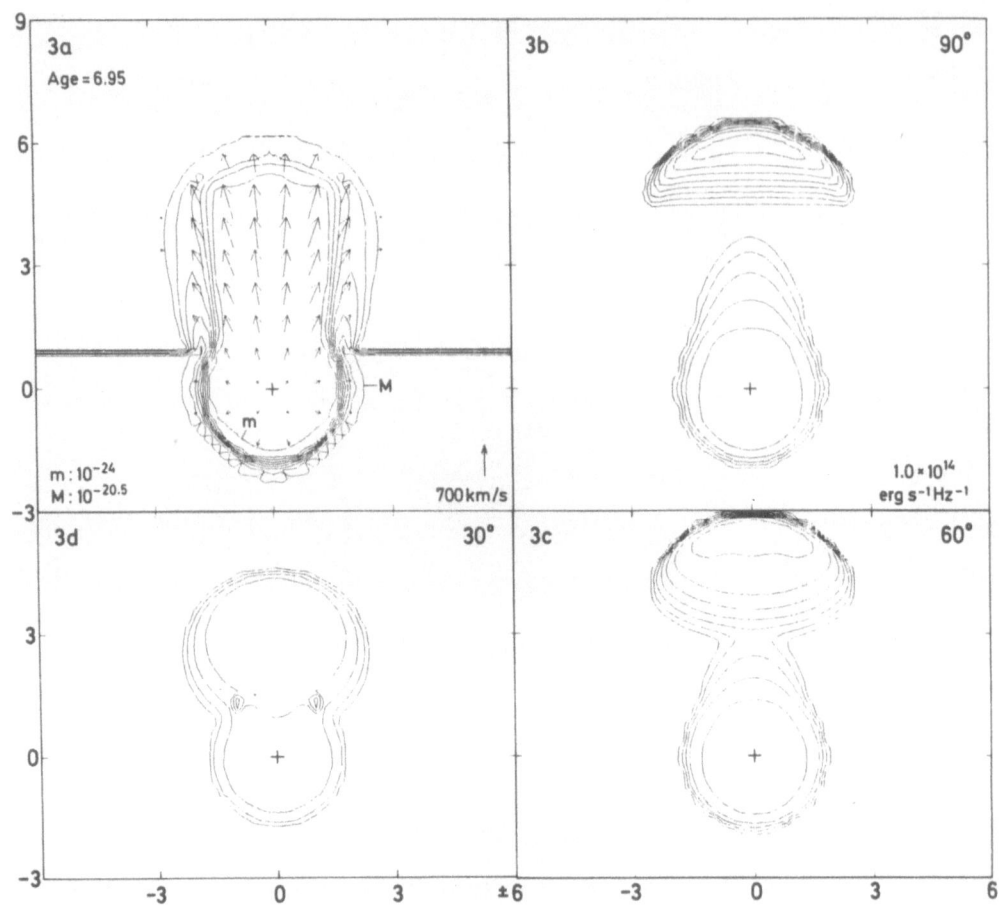

Figure
3a)   *Same as Fig. 1a for an evolutionary age 6.95 million years.*
3b)   *Same as Fig. 1c for the remnant shown in Fig. 3a (viewing angle: 90°)*
3c)   *Same as Fig. 1c for the remnant shown in Fig. 3a (viewing angle: 60°)*
3d)   *Same as Fig. 1c for the remnant shown in Fig. 3a (viewing angle: 30°)*

REFERENCES

Bisnovatyi-Kogan, G.: 1983, this volume, p. 125.
Bodenheimer, P., Woosley, S.E.: 1982, in preparation
Bodenheimer, P., Yorke, H.W., Tenorio-Tagle, G.: 1982, in preparation
Bodenheimer, P., Yorke, H.W., Tenorio-Tagle, G., Beltrametti, M.: 1983,
     this volume, p. 411.
Chevalier, R.A., Gardner J.: 1974, Astrophys. J. 192, 457.
Chevalier, R.A., Theys, J.C.: 1975, Astrophys. J. 195, 53
Cox, D.P., Daltabuit, E. 1971, Astrophys. J. 167, 113
Cox, D.P., Smith, B.W.: 1974, Astrophys. J. (Letters) 189, L105.
Ikeuchi, S.: 1978, Publ. Astron. Soc. Japan 30, 563.

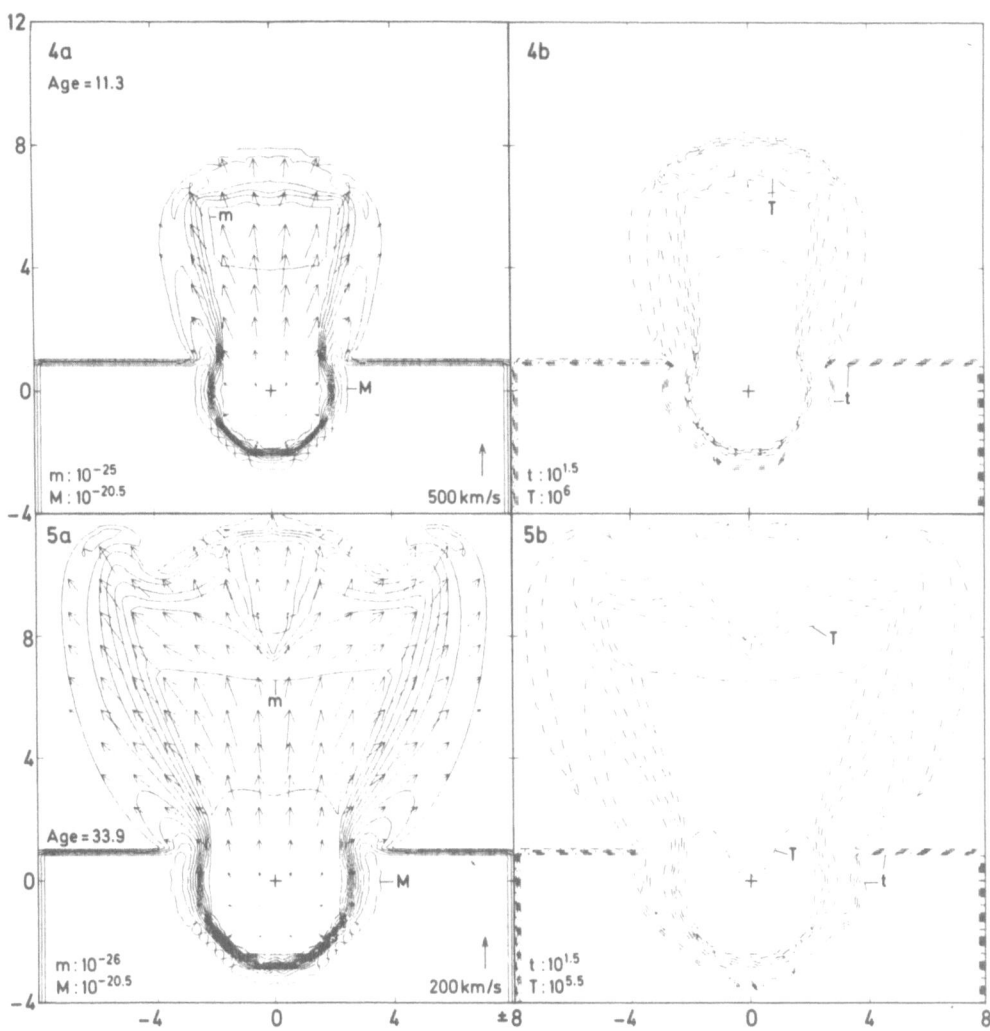

Fig. 4 Same as Fig. 1ab for an evolutionary age: 11.3 million years
Fig. 5 Same as Fig. 1ab for an evolutionary age: 33.9 million years

Jones, E.M., Smith, B.W., Straka, W.C., Kodis, J.W., Guitar, H.: 1979,
    Astrophys. J. 232, 129.
McKee, C.F., Cowie, L.L.: 1975, Astrophys. J. 195, 715.
Möllenhoff, C.: 1976, Astron. Astrophys. 50, 105.
Sanders, R.H.: 1976, Astrophys. J. 205, 335.
Sgro, A.: 1975, Astrophys. J. 197, 621.
Tenorio-Tagle, G., Bodenheimer P., Yorke, H.W.: 1982, in preparation
Tenorio-Tagle, G., Bodenheimer, P., Yorke, H.W.: 1983,
    this volume, p. 399.

THE EVOLUTION OF SUPERNOVA REMNANTS IN A NON-UNIFORM MEDIUM:
THE FATE OF AN EVOLVING OB ASSOCIATION

P. Bodenheimer[1,3], H.W. Yorke[2],
G. Tenorio-Tagle[3], M. Beltrametti[3]
[1] Lick Observatory, U.C. Santa Cruz 95064
[2] Universitäts-Sternwarte, Göttingen, FRG
[3] Max-Planck-Institut für Astrophysik, Garching b. München, FRG

Abstract: We study the effect of multiple supernova explosions in a non-uniform interstellar medium using a two dimensional hydrodynamical code (axial symmetry assumed). Cooling effects were included. In a uniform medium two or more supernovae exploding at the same place but at different times result in a remnant with less energy than the sum of the individual explosion energies - when cooling effects are important. As a case of special interest the evolution of an OB association with many massive stars experiencing supernova explosions is presented. We show that it is difficult to produce large supershells with multiple supernovae alone, even if the sum of their explosion energies appear to be sufficient. The existence of a giant HII region preceding the explosions does not alter this conclusion. Other mechanisms should be considered to produce supershells.

1. INTRODUCTION

Several mechanisms for the production of HI shells and supershells, as observed and discussed by Heiles (1979, 1983), have been considered by a number of investigators. Tomisaka et al. (1981), Bruhweiler et al. (1980) and Kafatos et al. (1980) for example maintain that sequential (multiple) supernova explosions arising from an OB association can explain their existence. In the latter two papers the effect of stellar winds preceding the explosion are also considered. Tenorio-Tagle (1981) and Tenorio-Tagle et al. (1981) show that high velocity, high latitude HI clouds colliding with the galactic disk can produce such supershells with the observed characteristics. Originally, Heiles (1979) thought the supershells could be produced by a single explosive event (a Type III supernova?).

In the following consider the effect of supernova explosions within an OB association. We have assumed a stratified interstellar medium with a scale height of 150 pc and a density $n = 1$ cm^{-3} at the symmetry plane ($z = 0$). These are values typical for our Galaxy at a distance of 10 kpc from the center (Paul, et al. 1976). We first allow the UV photons from the OB association to ionize the surroundings and we follow the evolution of the resulting HII region. The time dependent UV flux was calculated

J. Danziger and P. Gorenstein (eds.), Supernova Remnants and their X-Ray Emission, 399–404.
© *1983 by the IAU.*

assuming 2 10^4 stars distributed in mass according to Salpeter's initial
mass function (slope: 2.45), described in more detail in Beltrametti et
al. (1982). The association has about 20 stars more massive than 17 M$_\odot$ and
180 stars in the range 8 M$_\odot$ < M < 17 M$_\odot$ (see Ostriker et al. 1974). After
7.27 million years we allowed the 20 massive stars to explode simultaneously, each with 10^{51}erg. At 9.47 million years the remaining 180 OB-stars
exploded, again each with 10^{51} erg.

2. THE NUMERICAL MODEL AND SOME TESTS

We use the 2-D hydrodynamic code (with some modifications) discussed
by Bodenheimer et al. (1979). The version of the code which includes supernova explosions is also used by Tenorio-Tagle et al. (1983) and Yorke et
al. (1982). Numerical tests performed with the code are to be discussed
in a forthcoming paper (Bodenheimer et al. 1982). Here we shall summarize
some of the test results.
We place a bubble of hot gas with 10^{51} erg thermal energy and mass
1 M$_\odot$ in a medium of uniform density. During the subsequent expansion the
remnant remains spherically symmetric; the increase of radius with time
agrees very well with the solutions presented by McKee and Hollenbach (1980)
for the expansion of a supernova remnant with and without cooling. We have
performed tests in which we allow multiple explosions in a uniform medium.
When two explosions occur within 10^4 years, the shock front of the second
explosion merges with that of the first while the original remnant is still
in its adiabatic phase. The expansion of the resulting remnant approaches
the solution for a single 2 10^{51} erg explosion. When three explosions
occur separated by 10^4 years, the solution for a 3 10^{51} erg explosion is
attained. When, however, the two explosions occur within 10^5 years, cooling
within the first remnant is important. The expansion falls short of the
2 10^{51} erg solution (see Fig. 1).

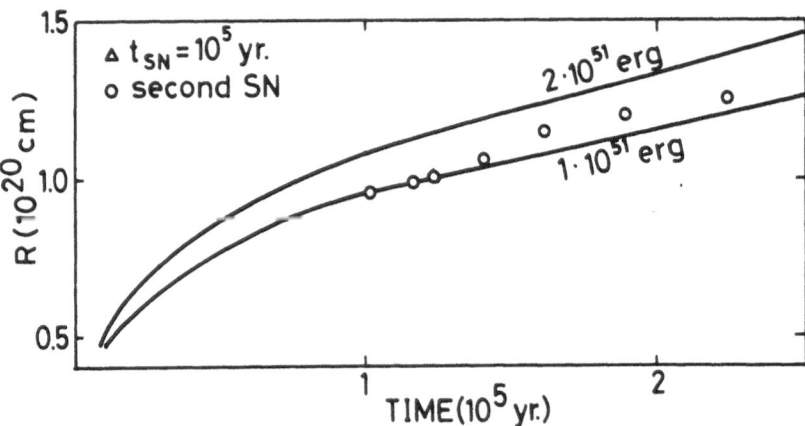

*Fig. 1. The radius of supernova remnants as a function of time. The solid
curves show the solutions of McKee and Hollenbach for explosions of the
indicated energy in a constant density (n = 1 cm^{-3}) medium including the
effects of cooling. The open circles show the solution for two explosions
of 10^{51} erg spaced 10^5 years apart.*

3. THE RESULTS

Figure 2a shows a cross-sectional view of a stratified galactic disk with parameters (scale height and density) typical of a 10 kpc galacto-centric radius, shortly after an OB association is born at the galactic plane. The density contours (solid lines) mark the decrease in density ($\Delta/\log\rho = 0.5$) and the boundary between the disk and a constant density halo ($\sim 1.6 \times 10^{-28}$ gm cm^{-3}). The ionizing photon output from the cluster of 2×10^{51} photons s^{-1} produces a giant HII region which, due to the stratification, expands mainly in the direction normal to the plane. As a result the disk becomes distorted as shown in figure 2b. At t = 7.3×10^6 yr, the most massive stars in the cluster become supernovae. Note that the decrease in ionizing radiation, a consequence of stellar evolution, has led to the recession of the ionization front and to a large expanding ring of neutral gas with a radius of 400 pc and a thickness of 50-100 pc. (see figure 2b)

The first explosion with a total energy of 2×10^{52} ergs was followed after 2.5×10^6 yr by a second explosion of 1.8×10^{53} ergs caused by 180 stars in the cluster with masses between 8 M$_\odot$ and 17 M$_\odot$ - typical of an OB association. Figure 2c shows the resultant flow, i.e. a single, almost spherical, shell expanding with a velocity $\lesssim 100$ km s^{-1}. Note that the dimensions of the original HII region are much larger than the present dimensions of the supernova remnant. Furthermore, even if one assumes that the velocity of the shell remains constant, it would require further 10^7 yr to catch up with the undisturbed galactic disk and acquire the dimensions typical of supershells (see Heiles 1979). However, after such a long time differential galactic rotation would have distorted them, making it hard to explain their usual round appearance.

One can therefore conclude that in order to construct a supershell the required energy should be deposited all at once. This, however, seems to rule out supernova explosions as it would imply a very unusual initial mass function for the stellar cluster. Therefore other mechanism(s) should be considered to explain the formation of super/rings and supershells.

ACKNOWLEDGEMENT

M.B. greatfully acknowledges support from the Thyssen Stiftung.

REFERENCES

Beltrametti, M., Tenorio-Tagle, G., Yorke, H.W. 1982, Astron. Astrophys. 112, 1.
Bodenheimer, P., Tenorio-Tagle, G., Yorke, H.W. 1979, Astrophys. J. 233,85.
Bodenheimer, P., Yorke, H.W., Tenorio-Tagle, G. 1982, in preparation
Bruhweiler, F.O., Gull, T.R., Kafatos, M., Sofia, S. 1980, Astrophys. J. (Letters) 238, L27.

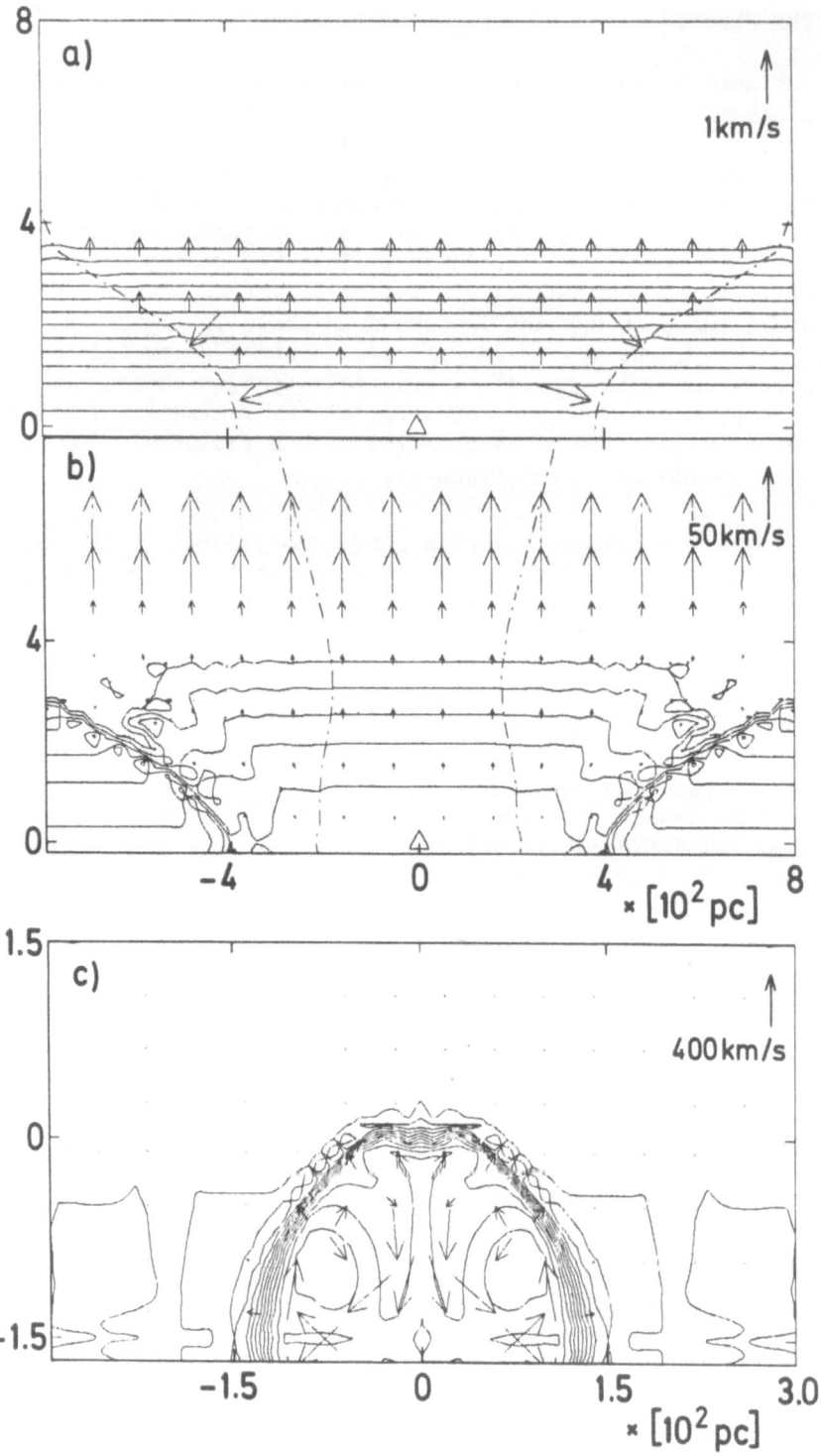

Figure 2. The Gas Dynamics around an OB Association.

a)   Cross-sectional view of the galactic disk shortly after the birth of the OB-association (triangle). The interstellar medium is stratified with a scale height of 150 pc and has a density of 1 cm⁻³ at the symmetry plane. The density contours are shown. The UV photons of the association have ionized the medium inside the dotted line. The length of the arrows are proportional to the gas velocity.

b)   Same as in figure 2a at t = 7.27 million years. The ionization front has receded since the most massive stars have moved away from the main sequence and soon will become supernovae.

c)   Enlargement of the region around the OB-association at t ∿ 10 million years.

Heiles, C. 1979, Astrophys. J. 229, 533
            1983, this volume, p. 367.
Kafatos, M., Sofia, S., Bruhweiler, F., Gull, T. 1980, Astrophys. J.
    242, 294.
McKee, C.F., Hollenbach, D.J. 1980, Ann. Rev. Astron. Astrophys. 18, 219.
Ostriker, J.P., Richstone, D.D., and Thuan, T.X., 1974, Astrophys. Letters,
    188, L87
Paul, J., Cassé, M., Cesarsky, C.J. 1976, Astrophys. J. 207, 62.
Tenorio-Tagle, G. 1981, Astron. Astrophys. 93, 338.
Tenorio-Tagle, G., Yorke, H.W., Bodenheimer, P. 1981, in "Mechanisme de
    Production d'Energie dans le Milieu Interstellaire", ed. P. Sivan,
    Marseille: Laboratoire d'Astronomie Spatiale du CNRS, p. 181.
Tenorio-Tagle, G., Bodenheimer, P., Yorke, H.W. 1983, IAU Symp. No. 101,
    this volume, p. 399.

DISCUSSION

MONTMERLE: 1) What kind of IMF did you use, and how sensitive are your results to the adopted spectral index? 2) With so many massive stars, you would expect SN explosions to take place in low-density cavities created by their stellar winds. Did you take this effect into account?

YORKE: 1) We adopted the Salpeter IMF with slope $\alpha=2.45$. We also computed the UV flux of an OB association with an IMF of $\alpha = 2$ and 3. We find that the rate at which the UV flux decreases as the association ages is almost identical, so that we expect the same behaviour of the HII region prior to the SN explosions. An association with $\alpha=2$ has ∿5 times more massive stars than an association with $\alpha=2.45$ provided that the total number of stars stays constant. Five times more energy would not lead to the formation of superrings.

2) The energy deposited into the gas by the ionizing flux of the OB association is greater than $6 \cdot 10^{54}$ ergs, while the energy released by stellar winds produced by 28 BO stars with -10^{-6} M_o/yr each lasting $3 \cdot 10^6$ yrs and wind velocities of $2 \cdot 10^3$ km/sec (Bruhweiler et al., 1980) is $\sim 7 \cdot 10^{50}$ ergs. Therefore, the evolution of the H II region prior to the SN explosions has greater influence on the environment of the association. Furthermore, there is still no proof that such strong winds operate during the whole lifetime of the stars. Therefore we believe that stellar winds produce only a negligible effect compared to supernovae explosions and the H II region evolution.

GULL (T.R.): 1) Did you include stellar winds in your model? 2) Why do you wait so long before your model allows SNs to occur? 3) Did you consider closer and more distant radii from the Galactic Center?

YORKE: 1) We did not include stellar winds and as stated above we do not believe to neglect an important effect.
2) According to recent stellar evolution calculations (Weaver et al., 1978, Ap.J. 225, 1021) massive stars (15 M_o < M < 25 M_o) become SN after $7 \cdot 10^6$ years.
3) Calculations already exist and will be presented in a future paper.

SUPERNOVA REMNANTS WITH CENTRAL PULSARS

Frederick D. Seward
Harvard-Smithsonian Center for Astrophysics, Cambridge, Mass.
and
Institute of Astronomy, Cambridge, England

ABSTRACT

The recent discovery of a central pulsing X-ray source makes MSH 15-52 the third SNR to contain a radio pulsar surrounded by diffuse X-ray emission. The pulsar periods are all increasing with time and the consequent loss of rotational kinetic energy is enough, in each remnant, to power a synchrotron nebula with the observed luminosity and volume.

After a review of the properties of the Crab Nebula it will be shown that both Vela X and MSH 15-52 have the same relationship between central pulsar and diffuse emission. Using empirical rules derived from these SNR, it is demonstrated that other plerionic remnants have similar characteristics. Two accretion-powered central sources can be distinguished from radio pulsars in SNR by the relatively high X-ray luminosity of the central source compared to that of possible synchrotron diffuse emission.

I. THE CRAB NEBULA

The Crab contains the fastest pulsar and the brightest synchrotron nebula. Although unique in appearance it is not unique in mechanisms of energy transfer from pulsar to synchrotron nebula. Both Vela X and MSH 15-52 contain pulsars and diffuse nebulae which are probably synchrotron nebulae powered by their respective pulsars. A good summary of the Crab nebula is given by Manchester and Taylor (1977), and IAU 46 (Davies & Smith 1971) contains much observational detail.

The distance to the Crab, 2.0 kpc is well determined for a SNR. The nebula is contained within bright optical filaments which in projection form a 5' x 7' ellipse. The enclosed volume of 4×10^{57} cm^3 is filled by diffuse radio and optical emission. The X-ray nebula, however, is considerably smaller with angular dimension $\sim 1'$ and volume $\sim 1 \times 10^{55}$ cm^3. There is no doubt this diffuse emission is synchrotron radiation since it is strongly polarized at all wavelengths.

405

J. Danziger and P. Gorenstein (eds.), Supernova Remnants and their X-Ray Emission, 405–416.
© 1983 by the IAU.

We consider only the X-ray data since this paper is a comparison of X-ray characteristics of different SNR. Knowing the luminosity L, and the volume, V_x; the magnetic field, H, and energy contained in relativistic electrons, E_p, can be calculated. We assume: 1) the nebula contains a uniform field, 2) the particles all have energy such that maximum synchrotron emission is at 1 keV, the center of the Einstein HRI energy range, 3) energy in the nebula is equally divided between relativistic electrons and magnetic field.

Then: $H = 25\ L_x^{2/7}\ V_x^{-2/7}$ Gauss, $E_p = E_H = \dfrac{H^2}{8\pi}\ V_x$, and the radiation lifetime of the electrons is, $\tau = 3000\ H^{-3/2}$ sec.

Measured and calculated properties of the Crab Nebula are listed in Table 2. The X-ray luminosity in the Einstein band (0.2 - 4 keV) is 2×10^{37} erg/s divided 95% from the nebula and 5% from the pulsar. The above formulae give: $H = 2 \times 10^{-4}$ Gauss, $E_H = 2 \times 10^{46}$ ergs and $\tau = 30$ years.

Since the Crab radiates strongly at all wavelengths ranging from radio to high energy γ-rays, there is a large population of electrons ignored by the above restriction to the X-ray band. The calculated H and E_H are lower limits. When the entire electromagnetic spectrum is taken into account, $L_{tot} = 2 \times 10^{38}$ erg/s, $H \approx 5 \times 10^{-4}$ Gauss and $E_H \approx 1 \times 10^{49}$ ergs. (The volume of the radio and optical nebula is ~ 100 x greater than that of the X-ray nebula and most of the particle energy is contained in the electrons emitting from IR - soft X-ray band. The high field is only required in the central region and H may well vary throughout the nebula.)

Thus if X-rays are considered to be the only appreciable radiation the derived magnetic field is only a factor of 2.5 less than the actual field but the total energy in particles and fields is grossly underestimated.

Now consider the central pulsar which has a strong internal magnetic dipole moment, m, and is slowing down. It radiates strong low frequency EM waves and the radiation torque exerted by this magnetic dipole radiation causes the neutron star to lose angular momentum.

If the star has rotational moment of inertia I (taken as 10^{45} g cm^2) and the period, p, and period derivative $\dot{p}$, are known; the magnetic moment is, $m = \left(\dfrac{3c^2}{8\pi}\ I\ p\dot{p}\right)^{\frac{1}{2}}$; the rate of rotational energy loss is,

$\dot{E} = -4\pi^2\ I\ \dfrac{\dot{p}}{p^3}$; and the "characteristic age" of the pulsar is, $A = \dfrac{1}{2}\dfrac{p}{\dot{p}}$

For the Crab Pulsar p is the smallest of any of the ~ 300 known radio pulsars and $\dot{p}$ is the second largest, so $\dot{E}$ is high and A is small. Table 1 lists measured and derived characteristics of the Crab and other pulsars.

E ≈ 5 x 10^{38} erg/s so the Crab pulsar can supply the energy
necessary to maintain the luminosity of the synchrotron nebula and the
efficiency of transferring rotational energy to radiation is ∿ 0.4.   If
soft X-rays were the only radiation observed this efficiency would
appear to be only .06.

The characteristic age of the pulsar, 1230 years, is slightly
larger than the known age of 920 years. The magnetic moment corresponds
to a surface field of ∿ 10^{13} Gauss.

II. MSH 15-52

The X-ray and radio pulsar in MSH 15-52 is surrounded by diffuse
X-ray emission and shows promise of being truly Crab-like: a synchrotron
nebula powered by spindown of a central pulsar. Although the diffuse
emission surrounding PSR 1509-58 has not yet been proved to be synchrotron emission, the high energy X-ray spectrum (Chaippetti & Bell-Burnell
1982) which is hard like that of the Crab, indicates that this is likely.

The pulsar period of 150 ms is not unusually rapid - it is the 6th
fastest radio pulsar. The period derivative, however, is 3 times
higher than that of the Crab pulsar (Table 1) and the highest of any
radio pulsar. Ė = 2 x 10^{37} erg/s; m = 3 x 10^{31} cm^3 and A = 1660 years.
Thus the pulsar is only slightly older than the Crab pulsar, has a magnetic moment 3 x higher, and is losing energy at ≈ 1/30 the rate.

The X-ray nebula observed by Seward *et al.*(1982) is much larger
than that surrounding PSR 0531+21 but not as luminous. The nebula is
extended N-S and fills a volume of 5 x 10^{57} cm^3. The observed X-ray
luminosity of 1.5 x 10^{35} erg/s is only ∿ .01 of Ė so the pulsar easily
supplies the necessary energy. The calculated nebular magnetic field,
∿ 10^{-5} Gauss, is at least a factor of 20 below that in the Crab and
characterizes the great difference in X-ray surface brightness.

The much larger envelope of MSH 15-52 allows the pulsar-supplied
magnetic field to expand more freely resulting in a larger, less luminous nebula. This is probably the usual situation and a search for Crab-
like objects should find many like MSH 15-52 and few as bright as the
Crab. This is true at least for our galaxy.

A radio map of MSH 15-52 shows only weak emission from the region
surrounding the pulsar (Caswell *et al.* 1981) and probably this emission
is from the expanding shell. If the synchrotron spectrum of MSH 15-52
were the same as that of the Crab except with the break in the spectrum
at the X-ray band where electron lifetime equals the age of the remnant,
optical brightness can be predicted to be V ≈ 28 mag/$arcmin^2$ and the
total radio signal expected to be only 10^{-3} Jy. Thus the synchrotron
nebula of MSH 15-52 could easily be much too faint to be seen at
radio and optical wavelengths.

III. VELA X

The SNR Vela X, containing bright optical filaments and several
regions bright in soft (presumably thermal) X-rays, is only $\sim$ 400 pc
distant. The characteristics of PSR 0833-45 and the diameter of the
shell both indicate an age of $\sim$ 1 x 10^4 years. The pulsar is surrounded
by X-ray nebular emission (Harnden 1983) which we will treat as showing
2 distinct phenomena.

First there is the Einstein HRI observation which shows a small,
1' diameter (0.1 pc), nebula surrounding a point-like source coincident
with the position of PSR 0833-45. The unresolved source is probably
the pulsar itself although the X-rays are not pulsed. The X-ray luminosity of the small surrounding nebula (herein called Vela A) is only
5 x 10^{32} erg/s. 3 x 10^{-5} that of PSR 0531+21 and 2 x 10^{-2} that of
PSR 1509-58. If either of these more luminous pulsars were surrounded
by an identical nebula, the nebular contribution would be too small to
be observed. Vela A then, might indicate a process present, but not
observed, in the vicinity of both the Crab and MSH 15-52 pulsars.

On the other hand, if emission with extent 0.1 pc and strength
equal to that of the emission from the PSR itself, were present, it
would be easily seen as an apparent steady flux from the PSR even if
spatially not resolved at a distance of 2 or 4 kpc. Since both the
stronger pulsars appear 100% pulsed in the HRI, strong phenomena analagous to Vela A have not yet been distinguished in the Crab Nebula or in
MSH 15-52.

The second phenomenon in Vela X is nebular emission $\sim$ 1° (7 pc)
in extent and surrounding the PSR. We suspect that much of this
emission is powered by the pulsar, based solely on the evidence of
spatial arrangement, although an accidental superposition of pulsar and
a region of bright thermal emission cannot at present be excluded.
Spectra from the Einstein detectors should provide an answer. We note
that Smith (1978) found a high energy nebula of extent 2° around the
pulsar and that the 2-10 keV flux from this region measured by Pravdo
et al. (1978) is higher than that expected from the pulsar alone. For
present purposes we will assume that *all* X-rays from this larger nebula,
Vela B, are synchrotron in origin and as before derive the magnetic
field. Results are listed in table 2. The field is only 4 x 10^{-6} Gauss,
close to the average interstellar field of 3 x 10^{-6} Gauss. The X-ray
luminosity, $\approx$ 10^{34} erg/s, is easily maintained by the pulsar rotational
energy loss, $\approx$ 7 x 10^{36} erg/s, and we conclude that Vela B is analagous
to the Crab synchrotron nebula.

As for MSH 15-52 the radio and optical brightness of Vela A and
Vela B can be predicted. Predicted radio and optical emission from
Vela B are too weak for detection. Vela A, however, is perhaps detect-
able optically with predicted signal V = 15 mag/arcmin^2.

Diffuse X-ray emission has been found around solitary radio pulsars
and results are summarized by Helfand (1983). The extent and luminosity
(10^{30} - 10^{33} erg/s) are close to the properties of Vela A. These small
nebulae are closely coupled with the pulsar itself. The relatively
strong magnetic fields, small size, and short particle lifetimes make
it unlikely that emission from these nebulae is influenced by the existence or non-existence of a SNR shell.

IV. THE CENTRAL PULSARS

The X-ray emission from the three central pulsars themselves is a
topic of great interest and is discussed by Bignani & Hermsen (1982),
Seward & Harnden. (1982) and by Manchester *et al.* (1982). PSR 0531+21
radiates essentially the same double pulse at all frequencies from radio
to γ-ray. PSR 0833-45 shows a double pulse at γ-ray energies, a single
radio pulse, and in the X-ray band the emission is apparently not pulsed.
X-rays from PSR 1509-58 are ∿ 100% pulsed with one pulse/cycle and there
is also one radio-pulse cycle. The variety of pulse shapes from these
three pulsars is difficult to interpret although the similarity of the
γ-ray pulses from 0833-45 to those of 0531+21 is significant since
0833-45 radiates most strongly at γ-ray energies.

Pulsar properties are listed in table 1 and Fig. 1 shows both
pulsar and diffuse luminosity as a function of $\dot{E}$. Both soft X-ray

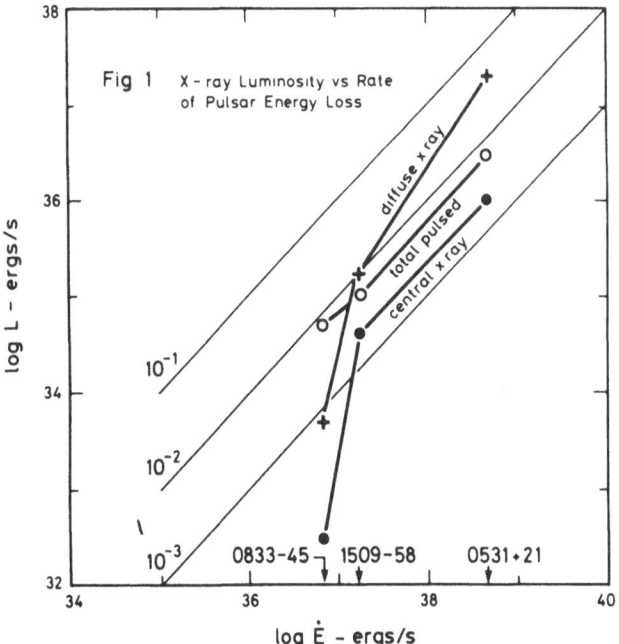

Fig 1   X - ray Luminosity vs Rate of Pulsar Energy Loss

and total (radio to γ-ray) luminosity are given when known. Considering
only X-rays we note that L_x (pulsar) is always less than L_x (diffuse)
by a factor of 4 - 20, and that the efficiency for generating diffuse
synchrotron X-rays decreases rapidly from ∿ 5% in the Crab nebula to
∿ 10^{-3} in Vela X. Including radio and γ-ray luminosity, the pulsed
emission in all three cases approaches 1% of $\dot{E}$ and the diffuse synchrotron luminosity of the Crab rises to ∿ 50% of $\dot{E}$.

The dependence of pulsed luminosity on $\dot{E}$ points to the magnetosphere of the pulsar as the origin of the radiation. The other possible
mechanism, radiation from high emissivity regions on the surface of the
neutron star, is less likely to be connected with $\dot{E}$.

A pulsar appearing as a point-like source of unpulsed X-rays either
has a high surface temperature, is surrounded by a small unresolved
diffuse nebula, or is oriented so emission which would appear pulsed
from another direction appears unchanging with time. The weak point-
like X-ray emission from PSR 0833-45 could be any of these. Again, as
for Vela A, if unpulsed emission this weak were present in 0531+21 or
1509-58 Einstein would not have been capable of detecting it.

V. FILLED-SHELL (PLERIONIC) SNR AND OTHER SNR WITH CENTRAL OBJECTS

In the radio band a number of SNR have characteristics similar to
the Crab Nebula: There is no shell-like structure, emission is polarized
and maximum at the center, and the radio spectrum is ∿ flat. These have
been listed and their radio properties studied by Lockhart *et al.* (1977)
Weiler & Shaver (1978), and Caswell (1979). Four of these were observed
by Einstein and X-ray results are given by Wilson (1980), Becker &
Szymkowiak(1981), and Becker, Helfand, & Szymkowiak(1982). Diffuse
X-rays were seen from all four and, like the Crab Nebula, X-ray emission
was strongest at the center, the diffuse X-ray source was smaller than
the radio SNR, and in the one case where a good spectrum was obtained
(3C58) there was no evidence for line emission. The X-ray morphology
of the central regions of 3C58 and of CTB80 is sharply peaked indicating
the existence of a central source, and accurate locations for these
sources were obtained. No point-like X-ray sources could be separated
from the diffuse emission observed from G21.5-0.9 or from G74.9+1.2 and
upper limits are given. Pulsations were not observed from any of these
central regions because the flux from all the sources is below the
threshold for detection of pulsations. The HRI counting rate of central
objects in SNR is listed in Tables 1 and 2. The 4 plerionic radio
remnants are all below .003 c/s, a factor of > 3 below PSR 1509-58, the
weakest pulsar detected. There were just not enough X-rays collected
to make a reasonable search for regular pulsations. 15,000 sec at
.002 c/s yields only 30 events. These SNR are prime targets for a more
sensitive high resolution instrument which should be able to resolve
central objects from surrounding diffuse emission and to find the
period if emission is pulsed as in PSR 1509-58.

Table 2 lists the observed and calculated properties of these SNR. As before, we have used the luminosity and volume of the X-ray sources to derive a magnetic field strength.

The remnants: W28, Kes 27, and W44 have also been observed by Einstein. They appear as filled-shell X-ray sources with no definite indication of a central source and are not included in this review because the data are not published or because the radio remnant is not filled-shell (Lamb & Markert 1981).

The remnant RCW 103 has a central point-like X-ray source and no obvious surrounding synchrotron nebula. The central source is too weak to reasonably search existing data for regular pulsations, and the apparent absence of a synchrotron nebula would favor radiation from the surface of a hot neutron star or a background object not connected with the remnant. Since the optical candidates for the central source are all very faint, Tuohy & Garmire (1980) favor the neutron star explanation.

Emission from the 8' diameter shell, however, produces 300 times more counts in the Einstein HRI than does the point-like source, and this strong emission from the shell could easily obscure a faint synchrotron nebula. Fig. 1 and the measured $L_X = 1.5 \times 10^{34}$ erg/s of the point source can be used to preduct a pulsar $\dot{E} \approx 1.5 \times 10^{37}$ erg/s and a diffuse nebula with $L_X \approx 1 \times 10^{35}$ erg/s, only 1/50 the luminosity of the shell. Thus radiation from the magnetosphere of a pulsar with small $\dot{E}$ is still a definite possibility for the X-ray emission mechanism of this central source.

VI. A SIMPLE MODEL

How do the magnetic fields calculated for the various synchrotron nebulae compare with expectations? Consider the following simple model. A spherical SNR of radius R, contains a central pulsar with rate of rotational energy loss $\dot{E}$. This energy input is equally divided between magnetic field and relativistic charged particles. The dynamics of the shell are completely determined by the initial explosion and the ISM rather than by internal energy deposited by the pulsar. Assume that the field is not particularly ordered so magnetic pressure is $\frac{1}{3}\frac{B^2}{8\pi}$ and that expansion of the remnant is adiabatic. Conservation of energy leads to the expression $3\dot{E} = 2BR^3\dot{B} + 4B^2R^2\dot{R}$.

Now assume $\dot{E}$ is constant, and that radiation energy losses are negligible. Neither is correct but the two effects do go in opposite directions and, at this point, we need to understand the gross properties.

The solution, written to explicitly contain the age of the remnant, t, is $B = \left(\frac{f\dot{E}t}{R^3}\right)^{\frac{1}{2}}$ where f is a constant dependent on the variation of $\dot{R}$ with time.

If the shell is stationary, $\dot{R} = 0$, f = 3, and internal energy

$= \dot{E}t = \dfrac{B^2 R^3}{3} = 2 \left(\dfrac{B^2}{8\pi}\right) \left(\dfrac{4\pi}{3} R^3\right)$ which is exactly correct since energy is distributed equally in field and particles. If the shell motion is a free expansion, f = 3/2, and the field is smaller since some energy is lost doing work. If the remnant is in the Sedov phase, $\dot{R} = \dfrac{2}{5} \dfrac{R}{t}$, and $f = \dfrac{15}{7}$ Since the expansion slows with time not as much energy goes into work.

These examples are plotted as straight lines in fig. 2 and compared with equipartition fields derived from observations. The points are of varying quality. The Crab, MSH 15-52, and Vela SNR points are good since E and t are known or calculated from the pulsar properties. The uncertainty shown for MSH 15-52 is due to the irregular nature of the shell, which in the E and S is 18 pc distant from the pulsar but in the NW the dense region RCW 89 is only ∿ 8 pc distant.

The other points are considerably more uncertain: We have assumed that 3C58 is the remnant of SN 1181 and have set the age at 800 years. $\dot{E}$ was estimated from the X-ray luminosity of the central object under the assumption that Fig. 1, based on the 3 Crab-like remnants, shows a general relationship between $\dot{E}$ and L_x. The ages of G21.5-0.9 and of G74.9+1.2 are estimated from the size of the remnant and as

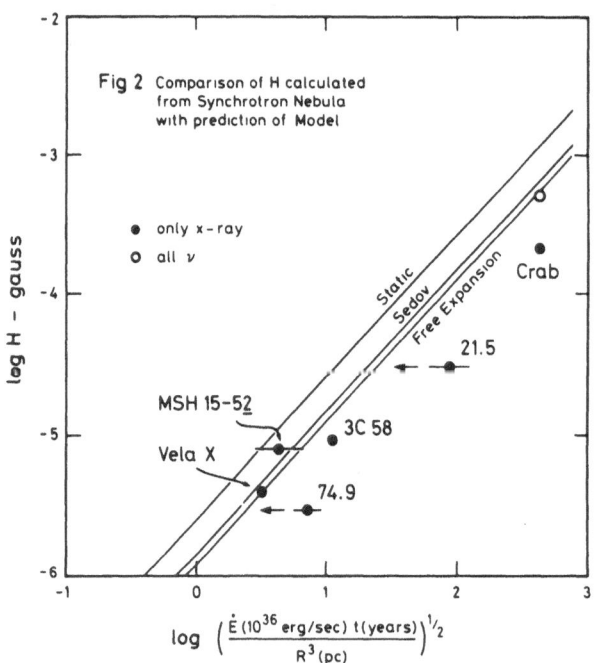

Fig 2 Comparison of H calculated from Synchrotron Nebula with prediction of Model

0.5-1.5 x 10^3 years and as 0.5-1.5 x 10^4 years respectively. Since no central sources were detected $\dot{E}$ estimated for these 2 remnants are upper limits. CTB 80 is not plotted. The shell of this object is so irregular that it is difficult to determine a size, or even to identify it as an SNR with confidence.

Here is a way of organizing these data. Note that: the X-ray derived fields are not far from and generally below the model prediction. The properties of these X-ray synchrotron nebulae are determined both by the energy input from the pulsar and by the expansion of the SNR shell. The high field in the Crab is a consequence of both the high energy input of the pulsar and of the low expansion velocity of the filaments which confine the radiated pulsar energy to a small volume, resulting in a uniquely bright synchrotron nebula.

VII. TWO ACCRETION-POWERED CENTRAL SOURCES

Both W50 and CTB109 contain central compact X-ray sources which are members of binary systems. The 3.5 sec X-ray pulsar within CTB109 shows frequency changes indicating orbital motion with a period of 1.9 hours (Gregory 1983). SS433 in the center of W50, is a bright optical source showing a period of $\approx$ 13 days (Crampton *et al*. 1980).

W50 contains 2 lobes of diffuse X-ray emission which because of their orientation along the long axis of the radio nebula are almost certainly connected with the high velocity beams of SS433 (Watson *et al*. 1983). The emission mechanism has not been determined but, assuming synchrotron radiation, the magnetic field and energy in particles can be determined as before. Results are in table 2. X-ray luminosity of each lobe is 5 x 10^{34} erg/s, H = 3 x 10^{-6}G, and energy in field + particles in each lobe is 5 x 10^{46} ergs.

CTB109, a SNR with X-ray bright shell on the eastern side only and with an X-ray bright central source, contains a diffuse feature interior to the shell and pointing at the pulsar L_x = 8 x 10^{24} erg/s. Assuming this "jet-like" feature is associated with the pulsar and is synchrotron emission we get H = 8 x 10^{-6} G and energy in field + particles = 2 x 10^{46} ergs.

The fields and energy associated with these X-ray diffuse features are similar to those of the MSH 15-52 and Vela X synchrotron nebulae, and we can again argue that corresponding optical and radio emission will have surface brightness too low to be detectable.

In both remnants, the relatively high luminosity and the binary nature of the central system point to accretion as the most likely source of X-ray emission. The jet-like morphology of the diffuse nebulae is also quite different from that observed in the other remnants discussed.

Fig. 3 illustrates that only in these two SNR is the X-ray luminosity of the central source greater than that of the diffuse features.

This empirical rule, at least in the X-ray band, might be useful in distinguishing accretion sources from radio pulsars. Morphology depends on orientation. W50 observed from a point on the axis of the jets would appear as a point source centered in a small diffuse nebula, and the spatial distribution of nebular emission as an indicator of type of source would be misleading.

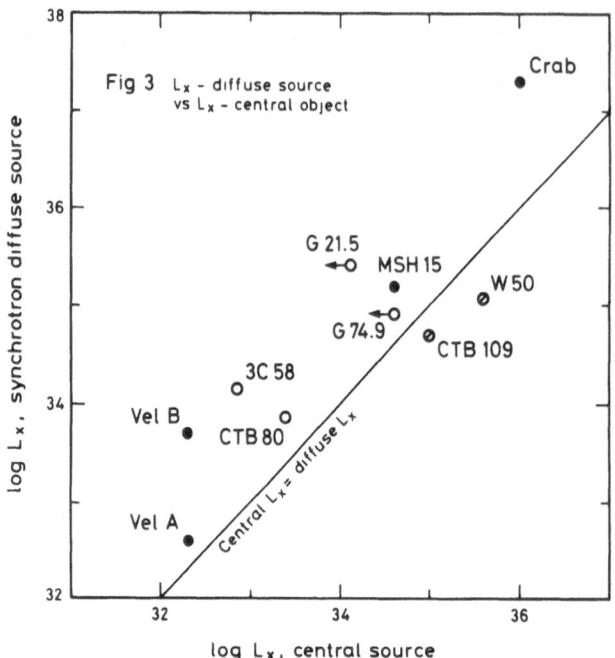

If the diffuse emission is synchrotron radiation the energy must come from a different source than electromagnetic radiation from the central pulsar. The high velocity particle beams of SS433 are an obvious choice for W50. Indeed the geometry requires this association.

Finally, it is interesting to speculate that diffuse X-ray "jets" might be a feature of most binary X-ray sources. The Radio lobes observed $\sim 2'$ on either side of Sco X-1 (Hjelming 1971) have never been explained. Jets similar to those in W50 with $L_x \sim 6 \times 10^{34}$ erg/s would probably be obscured by a bright central source, and $L_x \sim 10^{38}$ erg/s for many binaries. Long observations of a source in eclipse or in an off-state would be capable of detecting diffuse features.

IX.   SUMMARY

The discovery of a pulsar and synchrotron nebula within MSH 15-52 gives a second example of a system similar to the Crab Nebula and its pulsar. Using the observed properties of MSH 15-52 we identify a likely synchrotron nebula around the Vela PSR. These three remnants show an

Table 1 - Properties of Pulsars

Pulsar	0531+21	0833-45	1509-58
SNR	Crab	Vela X	MSH 15-52
p (sec)	.033 a	.089 a	.150 d
$\dot{p}$ (10^{-15} s/s)	423 a	125 a	1490 d
$\dot{E}$ (erg/s)	4.6(38)	6.9(36)	1.7(37)
m (gauss.cm^3)	6.7(30)	6.0(30)	2.6(31)
age (years)	1.23(3)	1.12(4)	1.66(3)
L_r ($10^7 - 10^8$ Hz)(erg/s)	2(30) b	1(29) b	-
L_x (0.2-4 KeV)(erg/s)	1(36) c	3(32) c	4(34) e
L_γ (.05-10 GeV)(erg/s)	2(35) b	4(34) b	-
L_{tot} (erg/s)	3(36)	5(34)	1(35)
C (HRI cts/s)	5. d	0.25 c	0.01 e

Table 2 - Properties of Diffuse Synchrotron Nebulae

SNR	Crab	MSH15-52	Vela A	Vela B	G 21.5-0.9
distance (Kpc)	2.0	4.2	0.4	0.4	5
size (arcmin)	1.4 x .8	7 x 4	1 x .8	30 x 60	1 x 1
V_x (cm^3)	9(54)	6(57)	4(52)	4(57)	9(55)
L_x (erg/s)	2(37)f	1.6(35)e	4(32)	5(33)	2.6(35) i
H (gauss)	2(-4)	9(-6)	5(-5)	4(-6)	3(-5)
E_H (ergs)	2(46)	2(46)	4(42)	2(45)	4(45)
τ (yrs)	30	4(3)	3(2)	1.5(4)	5(2)
C (HRI c/s)	109	.08 g	0.25	4 h	.023 i
C central source					< .001
L_x central source					< 1.3(34)i

3C 58	G 74.9+1.2	CTB 80	RCW 103	W 50	CTB 109
3	12	3	3.3	5	4
3 x 3	4 x 4	1.5 x 1.5	-	17 x 6	6 x 4
5(56)	8(58)	6(55)	-	5(58)	4(57)
1.4(34)	8(34) k	8(33) ℓ	-	6(34) n	5(34)
9(-6)	3(-6)	1.4(-5)	-	4(-6)	7(-6)
1.5(45)	4(46)	5(44)	-	3(46)	8(45)
4(3)	1.6(4)	2(3)	-	1.5(4)	5(3)
.03 g	.0024 g	.009	-	.02 g	.05 g
.0016 j	< .001	.0025 j	.0038 m	.09 g	.09 g
7(32)	< 4(34)	2(33) ℓ	1.5(34) m	4(35) n	1(35)

a Manchester & Taylor (1977)
b Bignami & Hermsen (1982)
c Harnden (1983)
d Seward & Harnden (1982)
e Seward et al. (1982)
f Toor & Seward (1974)
g IPC c/s x 0.1

h IPC c/s x 0.2
i Becker & Szymkowiak (1981)
j Becker et al. (1982)
k Wilson (1980)
ℓ unpublished IPC observation
m Tuohy & Garmire (1980)
n Watson et al.(1982), one lobe only

empirical connection between L_X of the synchrotron nebula, L_X of the central pulsar, and the pulsar $\dot{E}$. Properties of the diffuse nebula are determined by $\dot{E}$ and the size of the SNR. (The Crab is bright because $\dot{E}$ is high and because R is small.) The magnetic fields derived from the observed X-ray nebulae are not far from those predicted by a simple model in which the pulsar provides the energy. The Vela pulsar is also surrounded by a smaller nebula of high surface brightness but low luminosity, analogous to the X-ray nebulae found around isolated radio pulsars. X-rays observed from 4 other filled shell radio SNR and from RCW 103 show that a central pulsar is a definite possibility. The central objects in W50 and in CTB 109 are accretion powered and more luminous than diffuse interior emission which might be synchrotron radiation. The relative strength of X-rays from the central source is a possible way of distinguishing true pulsars from accretion powered sources.

This work was completed while a Guest Research Fellow of the Royal Society. I would like to thank the Institute of Astronomy for their hospitality and Linda Sparke for several enlightening discussions.

REFERENCES

Becker, R. & Szymkowiak, A. (1981) Ap.J.Letters, 248, L23.
Becker, R., Helfand, D. & Szymkowiak, A. (1982) Ap.J., 255, 557.
Bignami, G.F. & Hermsen, W. (1982) Ann.Rev.Ast.Ap. 21 in press.
Caswell, J.L. (1979) MN 187, 431.
Caswell, J.L., Milne, D.K. & Wellington, K.J. (1981) MN 195, 89.
Chiappetti, L. & Bell-Burnell, S.J. (1982) MN in press.
Crampton, D., Cowley, A. & Hutchings, J. (1980) Ap.J.Letters 235, L131.
Davies, R.D. & Smith, F.G. ed. (1971) "The Crab Nebula"IAU Symposium 46.
Gregory, P. (1983), this volume, p. 441.
Harnden, F.R. Jr. (1983), this volume, p. 131.
Helfand, D. (1981) IAU Symposium 95, 343.
Helfand, D. (1983), this volume, p. 483.
Hjelming, R.M. & Wade, C.M. (1971) Ap.J.Letters 164, L1.
Lamb, R.C. & Markert, T.H. (1981) Ap.J. 244, 94.
Lockhart,J.A., Goss,W.M., Caswell,J.L. & McAdam,W.B. (1977) MN 179, 147.
Manchester, R.N. & Taylor, J.H. (1977) "Pulsars" W.H. Freeman & Co.
Manchester, R.N., Tuohy, I. & D'Amico, N. (1982) Ap.J.Letters in press.
Pravdo, S.H., Becker, R.H., Boldt, E.A., Holt, S.S., Serlemitsos, P.J.
     & Swank, J.H. (1978) Ap.J. 223, 537.
Seward, F.D. & Harnden, F.R. Jr. (1982) Ap.J.Letters 256, L45.
Seward, F.D., Harnden, F.R. Jr., Murden, P. & Clark, D.H. (1982)
     submitted to Ap.J.
Smith, A. (1978) MN 182, 39P.
Toor, A. & Seward, F.D. (1974) Ast.J. 79, 995.
Tuohy, I. & Garmire, G. (1980) Ap.J.Letters 239, L107.
Watson, M.G., Willingale, R., Grindlay, J.E., Seward, F.D. (1982) in
     preparation.
Weiler, K.W. & Shaver, P.A. (1978) Astron. & Ap. 70, 389.
Wilson, A.S. (1980) Ap.J.Letters 241, L19.

SUPERNOVA REMNANTS WITH COMPACT X-RAY SOURCES

A, MSH 15-52

F.D.Seward, F.R.Harnden, Jr (Harvard-Smithsonian Center
                                for Astrophysics)
P.G.Murdin (Royal Greenwich Observatory)
D.H.Clark (Rutherford-Appleton Laboratory)

B, The Crab Nebula

D.H.Clark (Rutherford-Appleton Laboratory)
P.G.Murdin, R.Wood,R.Gilmozzi* (Royal Greenwich
                                    Observatory)
I.J.Danziger, (European Southern Observatory)
A.W.Furr (Astronomy Centre, University of Sussex)

        *Present address: Istituto di Astrofisica Spaziale,
Frascati,

A, MSH 15-52 : A SUPERNOVA REMNANT CONTAINING TWO COMPACT X-RAY SOURCES,

Einstein observations show two small-diameter, bright X-ray sources within the shell of the Galactic radio supernova remnant (SNR) MSH 15-52. X-ray spectra and optical extinction indicate that both compact sources are at least as distant, 4.2 kpc, as the diffuse emission from the shell. There is also enhanced, diffuse emission in the vicinity of both compact sources, as well as coincident with the radio shell. The source closest to the middle of the remnant is hard, unresolved by the HRI, was previously detected as 4U1510-59, and shows strong regular X-ray pulsations with a period of 0.150 s. The second source is associated with a collisionally-excited, high-density, optical knot within the Hα nebula RCW 89 close to the northwestern rim of the SNR.

CCD observations with the Anglo-Australian Telescope yield possible reddened candidates for 4U1510-59 and map the optical knot. Infrared spectral lines at 1.6μ of [FeII] appear in the knot. This is the first time these lines have been seen in an SNR; they are at an intensity comparable to the Balmer emission. UK Schmidt

417

J. Danziger and P. Gorenstein (eds.), Supernova Remnants and their X-Ray Emission, 417–419.

Telescope photographs of RCW 89 show new filaments which extend into
MSH 15-52 and support the hypothesis that the Hα nebula is part of
the whole SNR.

It is difficult to incorporate the apparently young pulsar
(characteristic age ~1600 years), the apparently old SNR (> 10,000
yr), and the bright knot into a single consistent system. We
discuss the following possibilities: the interaction of the
supernova ejecta with the interstellar medium producing the X-ray,
optical, and radio appearance of the nebula; the ages of the pulsar
and the SNR being less discrepant than they first appear; and a
second pulsar powering the knot and RCW 89.

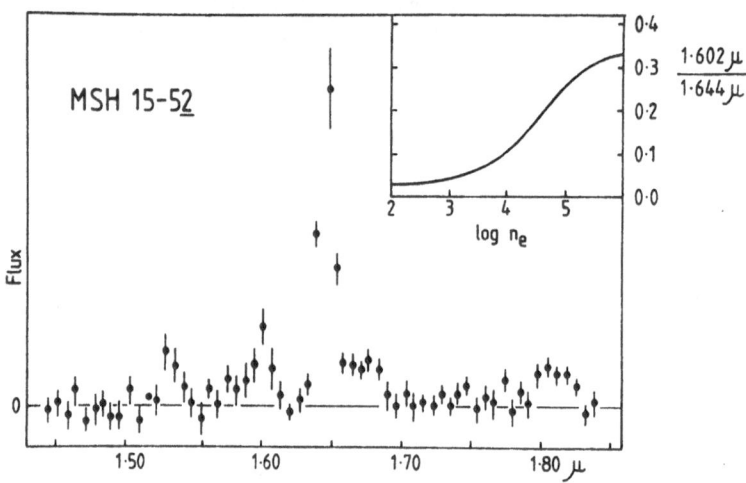

Figure 1: Infrared spectrum around wavelength 1.6 microns of the
northern bright knot in MSH 15-52, showing emission lines of [FeII].
This is the first time such lines have been reported in a supernova
remnant; in total they are of intensity comparable to the Balmer
emission. Inset is the density dependence of the line ratio 1.602μ
/1.644μ. In this object a high density, $\log n_e$ ~ 4.7, is indicated.

B. THE THREE DIMENSIONAL STRUCTURE OF THE CRAB NEBULA

More than 3000 radial velocity observations across the face of
the Crab Nebula are used .to investigate its 3-dimensional
properties. In the standard model it consists of a thick hollow
shell with synchrotron emission from within. We show that the thick
shell is composed of bright inner and faint outer components

surrounded by a higher velocity halo. Filaments are generally circumferential, but radial "spokes" link the inner and outer shells. There are spectral differences between the two extremities of the thick shell. The Crab Nebula's synchroton emission is confined within the shell system with a sharp discontinuity in brightness at the bright inner shell.

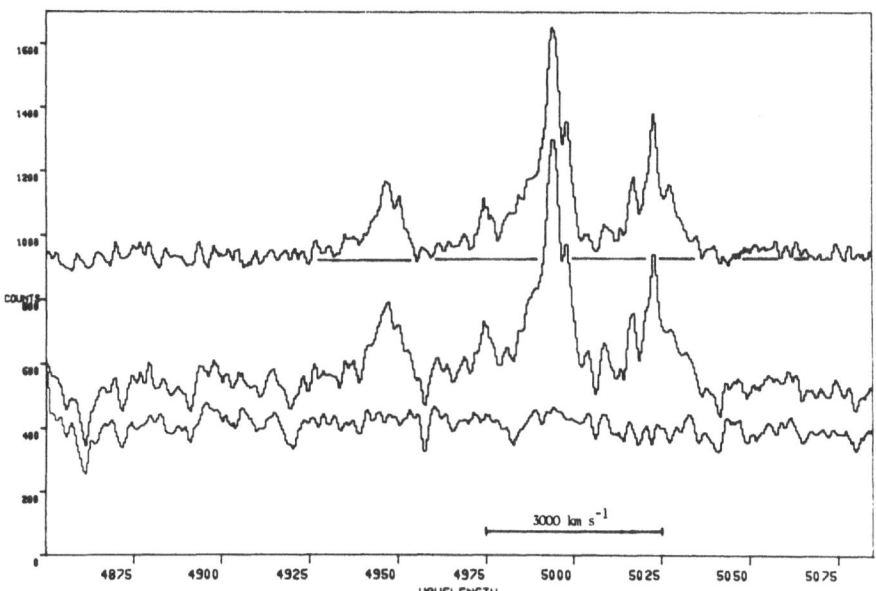

Figure 2: The middle spectrum represents the sum of many increments of a long slit spectrum through the Crab Nebula and centred on the pulsar. The lower spectrum is formed from an equal area of sky. The difference between the two (offset by 800 counts) is displayed at the top and represents filamentary emission on the synchrotron continuum. Interpolating from assumed continuum positions at about 4910 Å and 5075 Å gives the base level shown. Filamentary emission extends to 5065 Å (+3600 km s^{-1} if identified with λ5007) and 4920 Å (−2400 km s^{-1} for λ4959).

The full texts of these papers will appear in Astrophysical Journal and Monthly Notices respectively.

OBSERVATIONS OF G320.4-1.2 AND SPECULATIONS ON SUPERNOVA
REMNANT MORPHOLOGY

R.N. Manchester
Division of Radiophysics, CSIRO, Sydney, Australia
J.M. Durdin
Molonglo Radio Observatory, Hoskinstown, N.S.W., Australia

The galactic radio source G320.4-1.2 (MSH15-5̲2) consists of
several components, the most prominent of which is situated in the
north-west quadrant and is associated with the Hα nebula RCW89.
Caswell et al. (1981) mapped the source at 1.4 GHz with a resolution
of 50" arc and concluded that it was a single supernova remnant with
all components having spectral index $\alpha \approx -0.34$. This SNR has become
more significant with the recent discovery (Seward and Harnden, 1982)
of an X-ray pulsar of period 150 ms at the position (1950) R.A.
$15^h09^m59^s.5$, Dec. -58°56'57" near the centre of the remnant and the
detection of this pulsar at radio frequencies (Manchester et al., 1982).
The pulsar has some similarities to the Crab pulsar in that its period
derivative is extremely high and hence its characteristic age low,
~1570 years, comparable to that of the Crab pulsar. Timing observations
(Manchester and Durdin, unpublished) indicate that the pulsar is not
a member of a binary system and hence that the pulsed X-ray emission is
powered by rotational energy, as in the Crab pulsar.

We have mapped G320.4-1.2 with improved sensitivity using the
Molonglo Observatory synthesis telescope (MOST), which operates at a
frequency of 843 MHz (Mills, 1981). At this declination the beamwidth
is 43"×50" arc (R.A. ×Dec.). Two maps, shown in Figures 1 and 2, were
made, on 1982 June 15 and 17, the first of the whole field, 35'×41' arc,
and the second of the central portion, 11'×13' arc, both centred on
the pulsar. The two maps are independent, each requiring 12 h of
observation. Neither map has been cleaned; there are some negative
sidelobe responses around strong components in the south-east of the
map but generally their effect is believed to be small.

The overall appearance of the remnant in Figure 1 is similar to
that in the 1.4 GHz map of Caswell et al. (1981), which has comparable
resolution. This is consistent with their observation that all portions
of the remnant have approximately the same spectral index. A significant feature of the maps is the extension from the north-west component
towards the pulsar, evident in Figure 1 but shown much more clearly in
Figure 2, which has higher sensitivity (noise ≲1 mJy/beam). The pulsar
421

J. Danziger and P. Gorenstein (eds.), Supernova Remnants and their X-Ray Emission, 421–427.
© 1983 by the IAU.

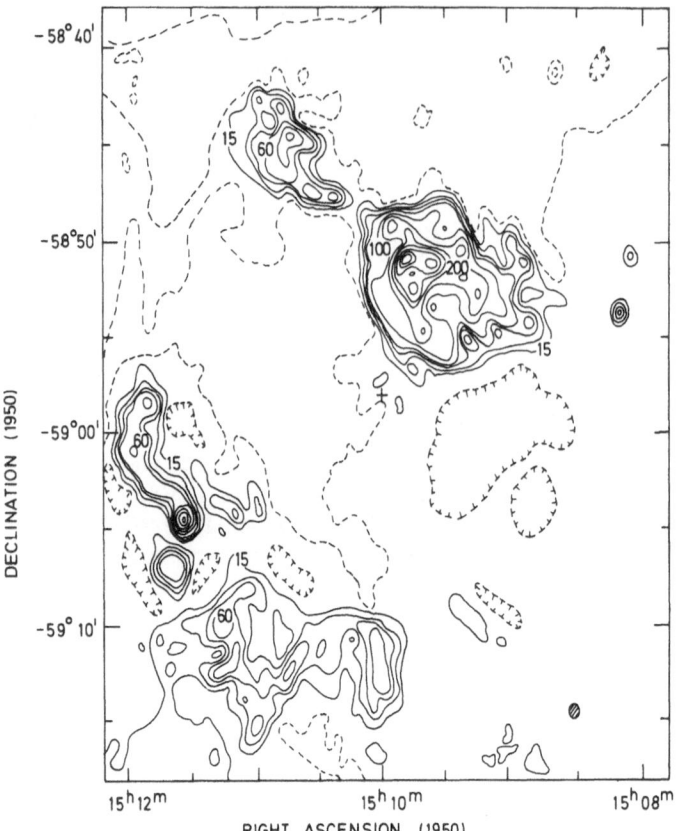

Figure 1. SNR G320.4-1.2 at 843 MHz. The half-power beamwidth is 43" × 50" arc (filled ellipse). Contour intervals are 0 (dashed), 15, 30, 45, 60, 100, 150, 200, 250, 300, 350 and 400 mJy/beam with respect to a locally defined zero. The position of PSR 1509-58 is indicated by the cross and the shaded ellipse represents the half-power beam.

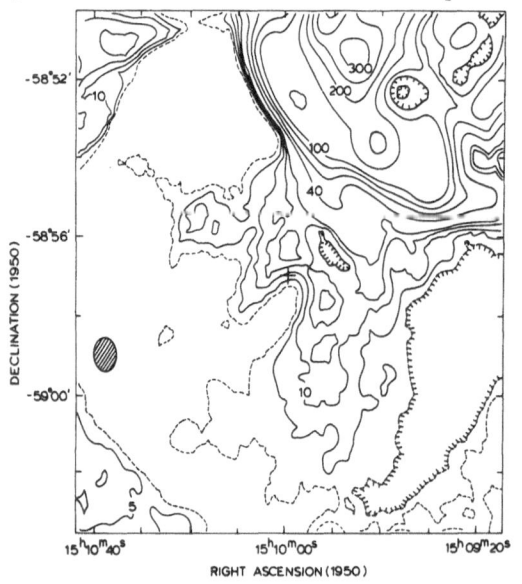

Figure 2. Region surrounding PSR 1509-58 at 843 MHz. The half-power beamwidth is 43" × 50" arc (filled ellipse). Contour intervals are 0 (dashed), 5, 10, 15, 20, 40, 60, 80, 100, 150, 200, 250 and 300 mJy/beam with respect to a locally defined zero. The pulsar position is indicated by the cross and the shaded ellipse represents the half-power beam.

itself lies on a steep gradient and is not detected, although its
expected mean flux density is $\gtrsim$2 mJy. Taken together with the central
location of the pulsar in the SNR and the agreement in estimated
distances (Manchester et al., 1982), this extension provides convincing
evidence for the association of PSR 1509-58 and G320.4-1.2. This is
only the third such association known (after Crab and Vela) and hence
of considerable significance. Of particular importance is the fact
that this remnant clearly does not fall into the filled centre or
plerion class. Several authors (e.g. Weiler and Panagia, 1980) have
suggested that only plerions contain active pulsars.

The north-west (RCW89) component is elliptical in outline and has
been interpreted as a separate SNR (e.g. Milne, 1970). However, it is
unlikely that the pulsar is associated with this component alone as the
implied pulsar space velocity is $\gtrsim$5000 km s^{-1}. Much more plausible is
the interpretation of Caswell et al. (1981) that the entire source is
a single remnant. We are then left with the problem of explaining the
morphology of the remnant and the relative location of the pulsar.

A possible solution to this problem, which has wide implications,
can be obtained by noting the existence of a mapping of the elliptical
north-west component on to the south-east component with all lines con-
necting corresponding points passing through the pulsar position. This
correspondence, illustrated in Figure 3a, immediately suggests that the
SNR morphology is dominated by the conical loci of two diametrically
opposed beams emitted by the rotating pulsar. We postulate synchrotron
electrons which are (or were) generated by locations where these beams
impacted on to regions of relatively high density, possibly the outer
boundary of the expanding SNR cavity. The proposal is in this sense
similar to beam models for extragalactic double radio sources (e.g.
Blandford and Rees, 1974).

The beam loci shown in Figure 3a pass relatively close to the
pulsar, implying that at the point of closest approach the beams are
(or were) essentially directed toward us. Since we detect this pulsar
in the radio and X-ray bands, the radio and X-ray beams must also be
directed toward us at some rotational phase. If one accepts the oblique
rotator magnetic-pole models for pulsar emission (e.g. Ruderman and
Sutherland, 1975), which have strong observational support, then the
radio and presumably the X-ray beams are emitted in opposite directions
along the magnetic axis. Clearly the most economical hypothesis is to
assume that the energetic beams are (or were) collimated along the same
axis.

There are a number of obvious tests of this proposal. Firstly, is
it energetically feasible? This question was answered in the affirm-
ative by Ostriker and Gunn (1971), who proposed that the entire supernova event was driven by the pulsar. The present proposal is more
limited: only the nebular synchrotron emission is driven by the pulsar,
although one could not rule out the possibility that impact of the beams
significantly affects the nebular expansion.

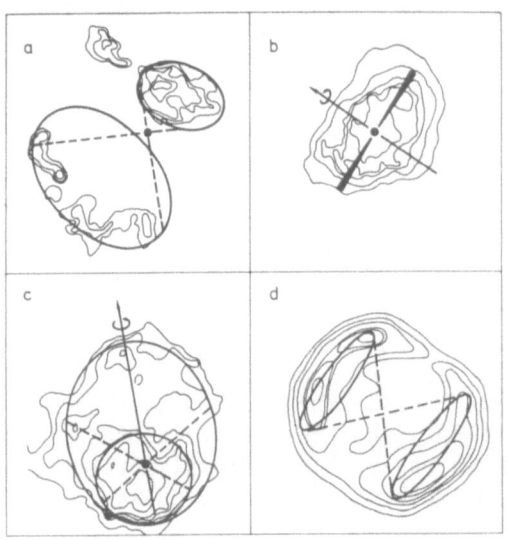

Figure 3. Schematic diagrams illustrating the loci of beams emitted by the central pulsar of (a) G320.4-1.2, (b) Crab nebula, (c) Vela SNR and (d) SN1006. A selection of contours is used to indicate the radio morphology of the SNR and the heavy ellipses represent the loci of the point of impact of the outer edge of the beam. Where known the orientation of the pulsar rotation axis is indicated. In the case of the Crab the configuration is degenerate (see text). The radio contours were obtained from the present paper (G320.4-1.2), Wilson (1972) (Crab), Day et al. (1972) (Vela), and Stephenson et al. (1977) (SN1006).

Secondly, are the other known examples of pulsar-SNR associations compatible with the proposal? Briefly, the answer is yes. In the case of the Crab pulsar, optical polarization observations (Kristian et al., 1970) indicate that the pulsar rotation axis is approximately perpendicular to the major axis of the nebula. Furthermore, the presence of a strong interpulse suggests that the rotation and magnetic axes are approximately perpendicular and that the rotation axis is approximately in the plane of the sky. This configuration, illustrated in Figure 3b, suggests that the Crab nebula is an oblate spheroid which we see edge-on. Because of the degeneracy involved this case cannot be claimed as strong evidence in support of the model but at least it is not inconsistent with it. The Vela association is more interesting. Weiler and Panagia (1980) suggest that the pulsar is associated with the more intense southern component, Vela X, but other authors (e.g. Milne, 1968) interpret the whole Vela XYZ complex as a single SNR. As for G320.4-1.2, a striking correspondence exists in the Vela SNR; as illustrated in Figure 3c, Vela X maps through the pulsar on to the outer boundary of the entire source with an approximately 1:2 ratio of radii. The orientation of the pulsar rotation axis, based on the intrinsic position angle of the radio pulse polarization (Hamilton et al., 1977) is consistent with the interpretation of the two ellipses shown in Figure 3c as the loci of beams emitted from the pulsar. In this case also a beam passes relatively close to the line of sight.

One can ask a third question: how do SNRs which have no detectable pulsar fit into the proposal? Let us first assume that all SNRs contain active pulsars. According to the model proposed here the radio morphology should consist of two elliptical components of similar ellipticity. The ellipticity, separation and relative size of the two components will be determined by the pulsar geometry and variations in the density of the interstellar medium. Most recognized SNRs have a shell morphology with a central minimum. It has long been known that many of these shell-type remnants have a bimodal structure - a good example is SN1006, shown in Figure 3d. Shaver (1969) suggested that this bimodal structure resulted from compression of the interstellar magnetic field. However, the orientation of the structure in many SNRs (e.g. Caswell, 1977) is not aligned with the galactic plane, as would be predicted by this mechanism. We propose that this bimodal structure results from the beam modulation described above when the pulsar rotation axis is roughly in the plane of the sky. Provided the rotation and magnetic axes are not perpendicular (as in the case of the Crab pulsar) one will then see the two ellipses approximately edge-on, as indicated in Figure 3d. An important point is that in this situation the pulsar beams are *not* directed in the line of sight and hence no pulses will be detectable. This will also be true when the rotation axis is close to the line of sight and the angle between the magnetic and rotation axes is not small. The SNR will then consist of two ellipses with a large degree of overlap. Such remnants are also classified as shell-type. The model therefore provides a natural explanation for the non-detection of pulsars in shell-type SNRs - only when the beams are directed toward the observer will pulsars be detectable and only in this case will the SNR have significant central emission and be classified as a plerion.

On the basis of this model one can predict which SNRs have potentially detectable pulsars and furthermore the exact location of the pulsar or, at least, its exact location when the SNR was very young. Clearly it will be of interest to search for these pulsars. It will also be of considerable interest to test the applicability of the model to a wider range of SNRs - for this we require high-resolution, high-sensitivity radio maps.

REFERENCES

Blandford, R.D., and Rees, M.J.: 1974, *Mon. Not. R. Astron. Soc.* 169, p. 395.

Caswell, J.L.: 1977, *Proc. Astron. Soc. Aust.* 2, p. 130.

Caswell, J.L., Milne, D.K., and Wellington, K.J.: 1981, *Mon. Not. R. Astron. Soc.* 195, p. 89.

Day, G.A., Caswell, J.L., and Cooke, D.J.: 1972, *Aust. J. Phys. Astrophys. Suppl.* No. 25, p. 1.

Hamilton, P.A., McCulloch, P.M., Manchester, R.N., Ables, J.G., and Komesaroff, M.M.: 1977, *Nature* 265, p. 224.

Kristian, J., Visvanathan, N., Westphal, J.A., and Snellen, G.H.: 1970, *Astrophys. J.* 162, p. 475.

Manchester, R.N., Tuohy, I.R., and D'Amico, N.: 1982, *Astrophys. J. (Lett.)* 262 (in press).

Mills, B.Y.: 1981, *Proc. Astron. Soc. Aust.* 4, p. 156.

Milne, D.K.: 1968, *Aust. J. Phys.* 21, p. 201.

Milne, D.K.: 1970, *Aust. J. Phys.* 23, p. 325.

Ostriker, J.P., and Gunn, J.E.: 1971, *Astrophys. J. (Lett.)* 164, p. L95.

Ruderman, M.A., and Sutherland, P.G.: 1975, *Astrophys. J.* 196, p. 51.
Seward, F.D., and Harnden, F.R.: 1982, *Astrophys. J. (Lett.)* 256, p. L45.
Shaver, P.A.: 1969, *Observatory* 89, p. 227.
Stephenson, F.R., Clark, D.H., and Crawford, D.F.: 1977, *Mon. Not. R. Astron. Soc.* 180, p. 567.
Weiler, K.W., and Panagia, N.: 1980, *Astron. Astrophys.* 90, p. 269.
Wilson, A.S.: 1972, *Mon. Not. R. Astron. Soc.* 157, p. 229.

DISCUSSION

SALVATI: a) Given your geometry for MSA 15-52, we should be outside the precession cone of the beams. How can we detect the radio pulses? b) In order to perturb appreciably the structure of the radio remnant, the beams must be energetic and the pulsar quite fast. How is this reconciled with the evidence that pulsars are born slow?

MANCHESTER: a) Firstly, I should point out that in this model the beam motion is assumed to result from rotation of the pulsar, not precession as in the case of SS433. In answer to your question, small differences in orientation of the radio and energetic beams might be expected if they are collimated at, for example, different radial distances from the star in the pulsar magnetosphere. Also there is some evidence (Narayan and Vivekanand, 1982, preprint) that pulsar radio beams are elongated in the latitude direction. b) It is true that pulsars must be born with relatively short period, say <10 ms, to be sufficiently powerful. In my view the evidence that pulsars are born slow is not strong. The number of missing short-period pulsars is quite small and these could easily have been missed in the large scale surveys (see Manchester,et al., this volume).

DANZIGER: A third epoch direct plate of the Vela Pulsar has been obtained at La Silla in order to measure the proper motion of this pulsar. The epoch and positions are summarised below. Within the errors there is no detectable proper motion. This means that if the pulsar is at a distance of 500 psc its maximum motion in the plane of the sky is 70 - 100 km/sec. It could not have originated near the centre of the radio shell if the age is $\sim 10^4$ years, but must have been born near its present position in the Vela X complex.

Epoch	Position
1975.2	$08^h33^m39^s.22 \pm 0^s03$ $- 45°00'10''1 \pm 0''33$
1977.3	$08^h33^m39^s.30 \pm 0^s02$ $- 45°00'10''3 \pm 0.3$
1981.9	$08^h33^m39^s.23$ $- 45°00'10''2$

MANCHESTER: This result is very interesting and is completely consistent with the present proposal which requires the pulsar still to be near its position at birth. In a few more years it should be possible to determine whether or not the pulsar was born at the centre of Vela X.

GIOVANNELLI: If all the SNRs (at least Type II) contain active pulsars; i) which is the birth rate of SN you can compute? ii) is this value in agreement with those given in different papers during this meeting ($\sim$ 50 - 70 yr)?; it seems to me, at the first look, higher than the mentioned ones.

MANCHESTER: Recent calculations of pulsar birthrates (Lyne, Manchester and Taylor, in preparation), give mean intervals of 20 - 60 years between births. This is fully consistent with the value of 30 years quoted by Mills and the range of 25 - 80 years quoted by Dickel at this meeting for the mean interval between SNR births in our Galaxy.

PRECESSING JET MODEL FOR THE SUPERNOVA REMNANT G109.1-1.0

P.C. Gregory, Department of Physics
G.G. Fahlman, Department of Geophysics and Astronomy
University of British Columbia
Vancouver, British Columbia, V6T 2A6

We propose that the central X-ray pulsar in G109.1-1.0, designated 1E 2259+586, ejects two oppositely directed precessing jets or beams, which give rise to the observed radio structure. The radio emission is interpreted as synchrotron emission from electrons accelerated at the interface of the jets with the walls of the SNR. Thus the observed intersecting arcs of radio emission represent the trace of the precessing jets on the supernova remnant walls. The precession axis is inclined at 37 degrees to the line of sight and the precession cone half angle is 55 degrees. The observed large scale X-ray jet in G109.1-1.0 is found to coincide in position with the precession axis as was found for the X-ray jets from SS 433.

INTRODUCTION

The extraordinary supernova remnant G109.1-1.0 has been under intensive study since its discovery in 1980 (Gregory & Fahlman, 1980; Hughes, Harten & Van den Bergh, 1981). It is remarkable because of the presence of an X-ray pulsar (Fahlman & Gregory, 1981 & 1983) at the center of curvature of the semi-circular shaped remnant and because of a jet like feature emerging from the pulsar on the eastern side. A detailed comparison of the X-ray and radio morphology of the supernova remnant (SNR) and the relationship of the remnant to the neighbouring molecular cloud has been presented by Gregory et al. (1983). Although the overall extent of the radio and X-ray emission is similar, in detail the distributions are very different (see figures 3 & 4) with the radio emission resembling two intersecting arcs. There is also clear evidence to suggest that the supernova is interacting with a molecular cloud on the western side (Gregory et al., 1983; Heydari-Malayeri et al., 1981). In overall morphology the SNR resembles a hemisphere. In this paper, we interpret the radio structure in terms of a simple kinematic precessing jet model as first proposed by Gregory and Fahlman in 1980. In this model, the radio emission is synchroton radiation from relativistic electrons accelerated in the interaction of the twin precessing jets with

429

J. Danziger and P. Gorenstein (eds.), Supernova Remnants and their X-Ray Emission, 429–436.

the walls of the SNR. We do not consider here the details of this interaction or the nature and generation of the jets themselves.

PRECESSING JET MODEL

    Figure 1(a) shows a sketch of the proposed jet model for a perfectly
spherical SNR cavity. The precessing jets lie along the surfaces of two

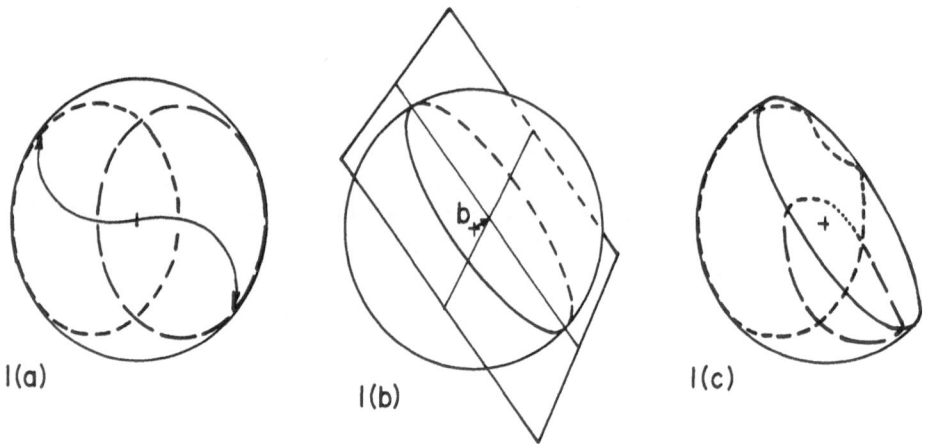

Figure 1.   The precessing jet model for a perfectly spherical SNR is
            shown in 1(a). The dashed ellipses are the traces of the two
            jets on the walls of the SNR. The interaction with the mole-
            cular cloud is represented by the intersecting plane in 1(b).
            The traces of the two jets on the resulting SNR cavity appear
            in 1(c).

cones which intersect the spherical cavity in two circles. For a precession cone half angle of 55 degrees and precession axis inclined at
37 degrees to the line of sight the two circular traces appear as intersecting ellipses as shown by the dashed lines. Because of the interaction
with the neighbouring molecular cloud, the SNR cavity resembles a hemisphere. For the purpose of this analysis, we have modelled the interaction with the molecular cloud by a plane which cuts off a portion of
the spherical cavity as shown in figure 1(b). This interface causes the
trace of the jets to depart from perfect ellipses (figure 1(c)).

    The coordinate system employed for this kinematic model is shown in
figure 2. The pulsar which is assumed to be the source of the two precessing jets is at rest at the origin of the right-handed coordinate
system x', y', z'. The y' axis is in the sky plane, and z' is the axis
about which the jets precess with an angular velocity Ω and at an angle
ψ. Consider a second right-handed coordinate system x, y, z which is
formed by rotating the x', y', z' axes about the y' axis until the z axis

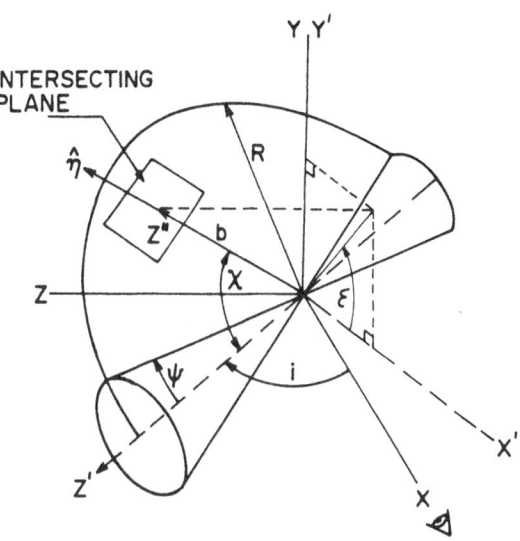

Figure 2.   The coordinate system used to discuss the geometry of the
            kinematic model.

is in the sky plane and the x axis coincides with the line of sight to
the observer. The precession axis z' lies in the x-z plane and is inclined to the line of sight by the angle i. We use a third coordinate
system x", y", z" to describe the orientation of the intersecting plane
representing the molecular cloud interface. The z" axis is parallel to
the $\hat{n}$ unit vector which denotes the normal to the plane and b is the distance of the plane from the pulsar in units of R, the radius of the
spherical portion of the SNR cavity. The x", y", z" axes are formed by
rotating the x', y', z' axes about the z' axis through an angle ε followed
by a rotation of χ about the new y" axis. Thus the interaction plane is
specified with respect to the x', y', z' axes by the coordinates b, ε, χ.

    Let $\underset{\sim}{r}$ represent the instantaneous point of intersection of either
jet with the walls of the SNR. The vector $\underset{\sim}{r}$ can be specified in terms of
spherical polar coordinates with ψ as the polar angle with respect to the
z' axis, θ, the azimuthal angle measured in the x'-y' plane and r, the
distance to the wall of the SNR in that direction. The projection of $\underset{\sim}{r}$
on to the sky plane (y, z plane) is given by the following components.

$$\underset{\sim}{r} \cdot \hat{y} = s_{jet} \, r \sin \psi \sin \theta \qquad (1)$$

and

$$\underset{\sim}{r} \cdot \hat{z} = s_{jet} \, r \{ \sin i \cos \psi - \cos i \sin \psi \cos \theta \} \qquad (2)$$

where s_{jet} is a sign parameter equal to +1 for the jet moving mostly
towards the observer and -1 for the counter jet. For a spherical SNR

cavity, the parameters i, ψ and R are given by:

$$2 \cos i = \frac{(\underline{r} \cdot \hat{z})\max - (\underline{r} \cdot \hat{z})\min}{(\underline{r} \cdot \hat{y})\max} \qquad (3)$$

$$2 \sin i \cot \psi = \frac{(\underline{r} \cdot \hat{z})\max + (\underline{r} \cdot \hat{z})\min}{(\underline{r} \cdot \hat{y})\max} \qquad (4)$$

$$R \sin \psi = (\underline{r} \cdot \hat{y})\max \qquad (5)$$

In the case of a spherical cavity intersected by a plane, the value of r will vary with the beam direction and depend on the angle ϕ between $\hat{\eta}$, the normal to the plane, and $\underline{r}$. The angle ϕ is given by

$$\cos \phi = \hat{\eta} \cdot \underline{r} = \sin \chi \cos \varepsilon \sin \psi \cos \theta + \sin \chi \sin \varepsilon \sin \psi \sin \theta$$

$$+ \cos \chi \cos \psi \qquad (6)$$

If $\phi \geq \cos^{-1}$ (b) then r = R.

If $\phi < \cos^{-1}$ (b) then b R sec ϕ.

For a particular choice of parameters i, ψ, b, χ and ε, equations 1, 2 & 6 describe the trace of the two beams on the walls of the SNR cavity as projected onto the sky plane.

To compare the model with observations, in particular CO observations of the molecular cloud, we also require equations describing the locus of intersection of the plane and spherical cavity projected on to the sky plane. If we let q be the locus of this intersection, then the y and z components of q are given by:

$$\underline{q} \cdot \hat{y} = R[\sin \phi_o \cos A \sin \varepsilon \cos \chi + \sin \phi_o \sin A \cos \varepsilon + \cos \phi_o \sin \varepsilon \sin \chi]$$

$$(7)$$

$$\underline{q} \cdot \hat{z} = R[\sin \phi_o \{\cos A(-\sin i \sin \chi - \cos \varepsilon \cos i \cos \chi) + \sin \varepsilon \sin A$$

$$\cos i\} + \cos \phi_o (\sin i \cos \chi - \cos i \cos \varepsilon \sin \chi)] \qquad (8)$$

where $\phi_o = \cos^{-1}$ (b). The locus is then obtained by varying the parameter A from $0 \rightarrow 2\pi$.

The model described above has been fit to the 20 cm VLA radio observations of Gregory et al. (1983). Equations 3, 4, & 5 were used to determine i, ψ and R and the remaining parameters b, χ and ε found from trial model calculations. Figure 1 (c) shows the best fitting model and figure 3 shows this model superposed on the 20 cm radio contours. The jet traces are shown as dashed lines and the SNR walls are the solid

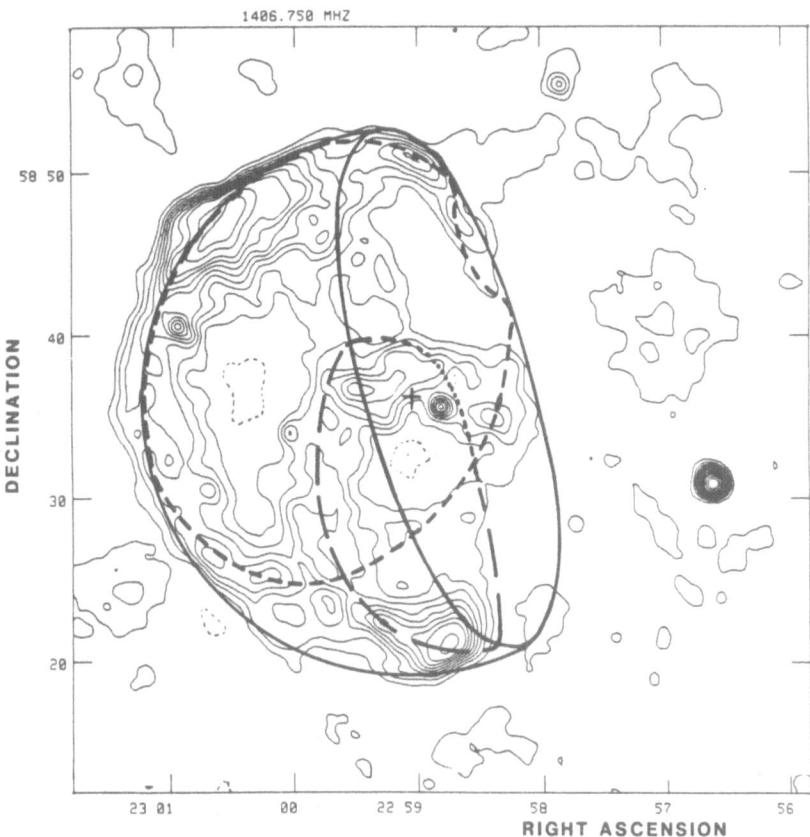

Figure 3.   The precessing jet model superposed on our 20 cm VLA radio
             map.   The dashed curves are the traces of theprecessing jets on
             the walls of the SNR.   The solid curves indicate the boundaries
             of the SNR cavity.   The cross indicates the location of the
             pulsar.

lines.   The same model fit is obtained for two sets of model parameters
depending on which jet is assumed to be towards the observer.

$$ i = \begin{pmatrix} -37 \\ +37 \end{pmatrix} \pm 3, \quad \psi = 55° \pm 5, \quad b = 0.15 \pm .05, $$

$$ \chi = \begin{pmatrix} 45 \\ 135 \end{pmatrix} \pm 5, \quad \varepsilon = 55° \pm 5 \text{ and } R = 16.6 \pm 0.5 \text{ arcmin.} $$

Where two values are quoted the upper value corresponds to the western
jet towards the observer.   For the western jet toward the observer, the

orientation of the molecular cloud interaction plane is such that the
eastern most extension of the cloud would give rise to a wedge like
absorption screen over the western portion of the SNR.

From the model parameters derived from the fit to the radio data,
we can deduce the orientation of the precession axis of the jets. The
jet model and precession axis, indicated by the arrow, are superposed on
the X-ray contours in figure 4. We note that the X-ray jet lies along
the direction of the precession axis derived from the radio structure.

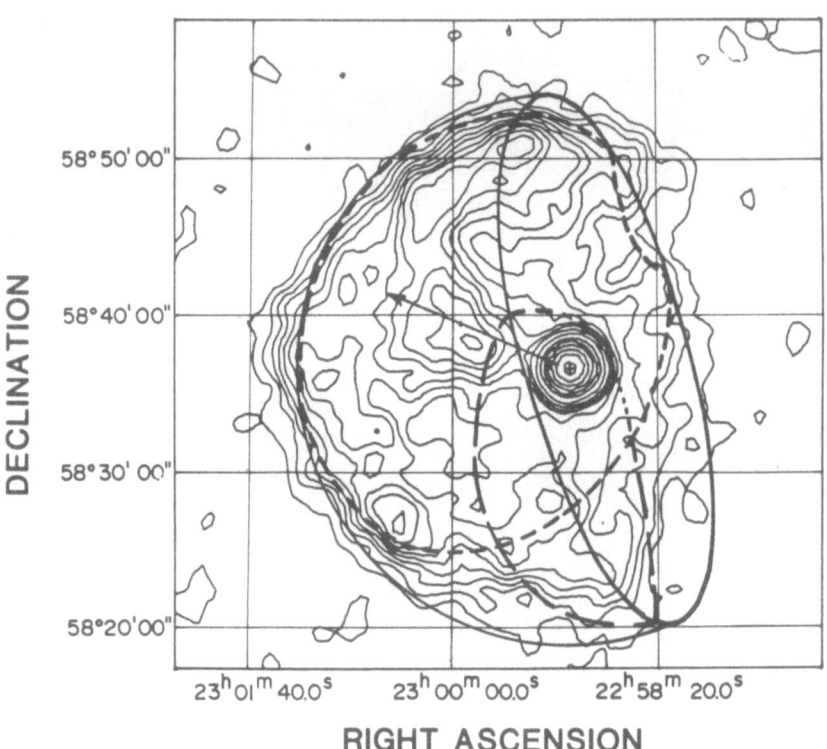

Figure 4.   The precessing jet model superposed on the X-ray image ob-
            tained with the IPC detector of the Einstein Observatory. The
            precession axis of the jet model is indicated by the arrow.

DISCUSSION

It is clear from figure 3 that the precessing jet model provides a
remarkably good qualitative fit to the radio structure in view of the
simple plane geometry assumed for the interaction of the supernova with
the molecular cloud. This interaction surface is unlikely to be a smooth
fat plane and the departures of the radio coutours from the fit close to

the pulsar undoubtedly reflect this fact. In principle, it is possible, knowing the jet parameters, to extract a three dimensional cross section of this interaction surface close to the pulsar, from the radio observations.

One check of the model is to compare the orientation of the inter-acting molecular plane derived from the radio continuum structure with the CO contours of Israel (1980) and Heydari-Malayeri et al. (1981). The position angle of the derived plane agrees within 10° to the orientation of the CO contours along the western edge of the SNR.

We note that portions of the two radio loops predicted by the model are missing or very weak in the SW quadrant. Since there is considerable variation in the emission strength about the loops, these portions may simply be much fainter. The two brightest radio features are almost diametrically opposite each other which suggests they might reflect the present locations where the two jets are exciting the cavity walls. Also the tail like structure exhibited by the SW feature suggests a sense of rotation consistent with the counter-clockwise rotation implied by the X-ray emission from the eastern jet. The lifetime of the synchrotron emitting electrons in years is $\sim -0.85 \; \nu^{-1/2} \; B^{-3/2}$, where ν is the ob-serving frequency in GHz and B is expressed in gauss. For there to be a noticeable decay in the emission as a function of position from the pre-sent location of the jets the time scale of this decay must be comparable to or less than the precession period of the jets. For a precession period similar to that of SS 433, this would require large field strengths in the interaction region of ~ 1 gauss. Our existing radio observations taken over a time span of one year set an upper limit on possible rota-tion of the hot spots of < 5 degrees per year.

At X-ray wavelengths, it is apparent that we may be seeing emission from one of the jets or from gas heated by the jet along its path. We note that similar to SS 433 (Seward et al., 1980; Grindlay et al., 1983) the X-ray jet in G109.1-1.0 is most intense along the direction of the precession axis as derived from the radio observations (see figure 4). The model described in this paper requires that a second oppositely directed jet emerge from the pulsar on the western side. Why is no X-ray emission detected from this jet? A close examination of the X-ray emis-sion indicates that there is a gap in the X-ray emission in the eastern jet extending from the pulsar out for an angular distance of ~ 6 arcmin. We note that a similar gap is observed in the X-ray emission from the SS 433 jets (Grindlay et al., 1983). On the western side of G109.1-1.0, the X-ray and radio emission declines rapidly and disappears completely at an angular distance of 8.5 arcmin from the pulsar. Thus at the point where we might expect to see X-ray emission from a western jet, the X-ray emission has all but disappeared from the nebula as a whole. The dis-appearance of the radio emission as well argues that this is not just X-ray absorption by the cloud although that may be partly the case. Thus the present X-ray observations do not rule out the existence of a western jet.

In spite of the parallels that have been drawn between the jet structure of SS 433/W50 and that proposed for 1E 2259+586/G109.1-1.0 it is important to note the significant differences that exist between the two central objects. SS 433 does not exhibit X-ray pulses and while the X-ray luminosities of the two objects are similar the ratio of optical to X-ray luminosity is more than 1000 times greater for SS 433. In addition the upper limit on the radio flux density of 1E 2259+582 of $\leq$ 200 μJy (Gregory et al., 1982) implies a radio luminosity more than 3000 times greater for SS 433. The X-ray, radio and optical properties of 1E 2259+586 are in fact much more like those of low mass close binary X-ray sources which are not known to exhibit jets. If 1E 2259+586 is a member of this class it is the first one to be found in association with a SNR.

At X-ray wavelengths both G109.1-1.0 and the SNR W50 show evidence for a large scale jet like feature emerging from the central source. In the case of SS 433/W50 there is direct radio and optical evidence for the existence of twin precessing jets which interact with the SNR. In this paper, we have shown how the existence of twin precessing jets from 1E 2259 + 586 could account for the observed radio structure of G109.1-1.0. A search for high velocity emission lines from the faint m_B = 22, optical counterpart of the pulsar reported by Fahlman et al. (1982) may help to confirm this picture.

The precessing jet model presented here may account for the radio structures of a number of other supernova remnants. In particular the SNR DA 530 is an obvious candidate and we recommend X-ray observations of this remnant to search for a central compact source and jet. This research was supported by grants from the Canadian Natural Science and Engineering Research Council. The NRAO is operated by Associated Universities Inc. under contract with the NSF.

REFERENCES

1.   Fahlman, G.G. & Gregory, P.C.: 1981, Nature 293, p. 202.
2.   Fahlman, G.G., Hickson, P., Richer, H.B., Gregory, P.C. & Middleditch, J.: 1982, Astrophys. J. Lett., in press.
3.   Gregory, P.C. & Fahlman, G.G.: 1980, Nature 287, p. 805.
4.   Gregory, P.C., Braun, R., Fahlman, G.G. & Gull, S.F.: 1983, IAU Symposium No. 101, this volume, p. 449.
5.   Gregory, P.C. & Fahlman, G.G.: 1981, Vistas in Astronomy, 25, p. 119.
6.   Grindlay, J., Band, D., Seward, F. & Stella, L.: 1983, IAU Symposium Symposium No. 101, this volume, p. 471.
7.   Heydari-Malayeri, M., Kahane, C., Lucas, R.: 1981, Nature 293, pp. 549-50.
8.   Hughes, V.A., Harten, R.H., Van den Bergh, S.: 1981, Astrophys. J. Lett. 246, L127.
9.   Israel, F.P.: 1980, Astron. J. 85, p. 1612.
10.   Seward, F., Grindlay, J., Seaquist, E.R. & Gilmore, W.: 1980, Nature 287, p. 806.
11.   Fahlman, G.G. & Gregory, P.C.: 1983, this volume, p. 457.

COMPARISON OF RADIO AND X-RAY OBSERVATIONS OF SNR G109.1-1.0

P.C. Gregory & R. Braun*, Department of Physics
G.G. Fahlman, Department of Geophysics and Astronomy
University of British Columbia, Vancouver, Canada

S.F. Gull, Mullard Radio Astronomy Observatory
Cambridge University, Cambridge, U.K.

In this paper, we present a comparison of the radio and X-ray morphology of the supernova remnant G109.1-1.0, based on recent radio observations at 6 and 20 cm and investigate the relationship of the SNR to a neighbouring molecular cloud.

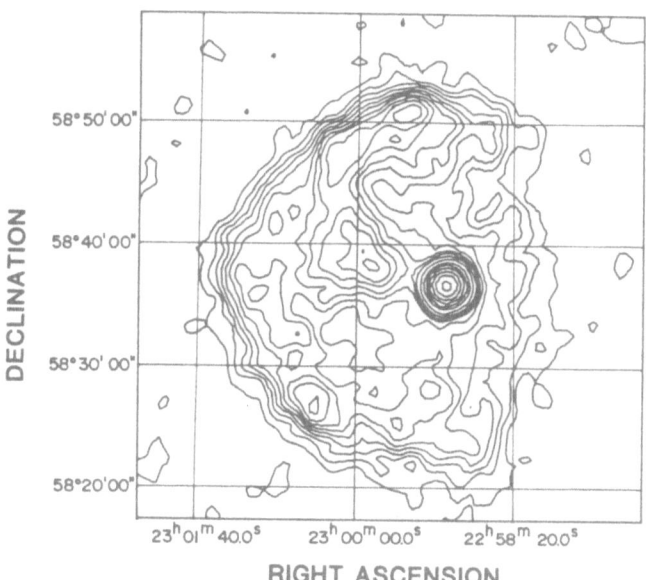

Figure 1. X-ray image of G109.1-1.0 obtained with the Einstein IPC detector, July 1980. Relative contour levels are 20, 30, 40, 52, 67, 83, 100, 118, 142, 200, 332, 470, 664, 940, 1328, 1878 and the map peak corresponds to a level of 2134.

---

* Present address: Sterrenwacht Leiden

J. Danziger and P. Gorenstein (eds.), Supernova Remnants and their X-Ray Emission, 437–443.

Figure 1 shows a contour plot of the X-ray emission (Gregory and Fahlman, 1980) with a resolution of 2.5 arcmin, obtained with the Einstein IPC detector in July 1980. At the center of curvature of the semi-circular shaped remnant is an X-ray pulsar (Fahlman and Gregory, 1981 & 1983). Also visible is a jet like feature emerging from the X-ray pulsar on the eastern side.

20 CM VLA OBSERVATIONS

VLA observations of G109.1-1.0 were obtained using C configuration in Sept. 1980 and D configuration in Oct. 1981 and the results combined to provide a map of the SNR at intermediate resolution. With a minimum baseline of 40 m, the array only provides reliable information on angular scales less than 15 arcmin. With an angular diameter of 33 arcmin, the SNR is slightly larger than the HPBW of 30 arcmin for the individual array antennas.

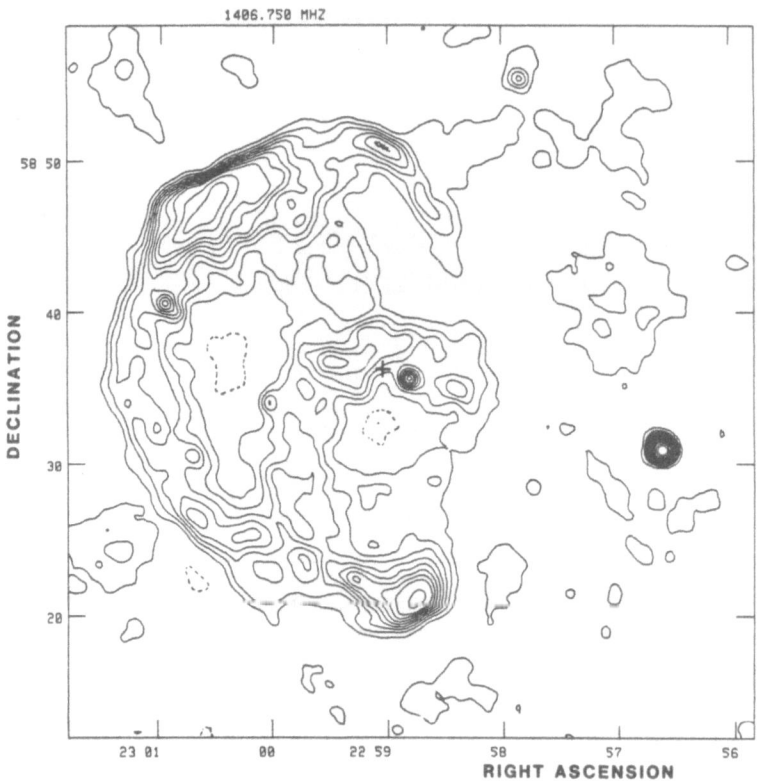

Figure 2. VLA 20 cm map of G109.1-1.0 at a resolution of 1 arcmin, not corrected for the primary beam of the array antennas.

Early maps made from the data showed the shell structure of the SNR within a negative bowl shaped region, characteristic of a source with undersampled low spatial frequencies. Experiments were carried out to remove the bowl by inserting a zero spacing flux density and through the use of the clean algorithm. Although the negative bowl was largely removed by this procedure, the maps obtained only provided valid information about angular scales < 15 arcmin.

The resolution of the VLA data is limited by the maximum baseline in C configuration to 12 arcsec, however, the map shown in figure 2 has a resolution of 1 arcmin as a result of tapering in the uv plane. Figure 3 shows the same result after correction for the primary beam response of the individual antennas.

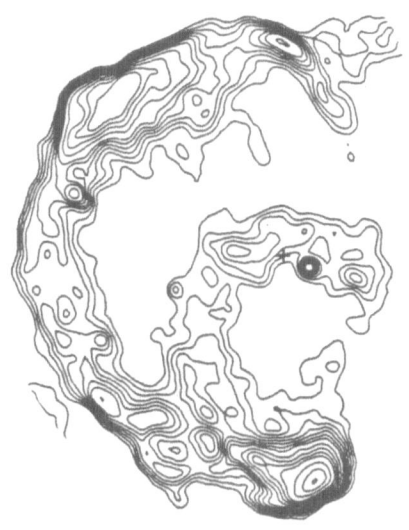

Figure 3. The same as figure 2 but corrected for the primary beam of the array antennas. Peak flux = 0.60 Jy/Beam. Contour levels = 0.30E-02 x (-2, 2, 3, 4, 5, 6, 7, 8, 10, 12, 14, 16, 20, 24, 28, 32, 36, 40).

6 CM SINGLE DISH OBSERVATIONS

The 6 cm radio observations were obtained in August 1980 with the NRAO 91 m transit telescope, as part of an on going survey of the Northern Milky Way for variable radio sources (Gregory & Taylor, 1981).The observations consists of driving the telescope in declination back and forth across the galactic plane, with earth rotation providing motion in the right ascension coordinate. Full sampling of the region was obtained by interleaving scans on successive days,to provide a series of parallel north bound scans ,separated by 1.5 arcmin and a series of south bound scans with the same separation. The telescope HPBW is 2.9 arcmin in RA and 3.2 arcmin in declination. A dual channel beam switching receiver system was used with an effective system temperature of 46 K and a band-width of 580 MHz. The dish gain at the source declination is 0.8 K/Jy. Both telescope feeds are sensitive to right hand circulation polarization and are separated by 7.2 arcmin. Because of beam switching, the observations yield directly a differential map. The differential observations were converted to a total intensity map using two different methods. The first of these is the analytic reconstruction method developed by Emerson, Klein and Haslam (1980). The second approach involved the use of a maximum entropy deconvolution program developed by S.F. Gull, G.J. Daniels and J. Skilling at Cambridge University.

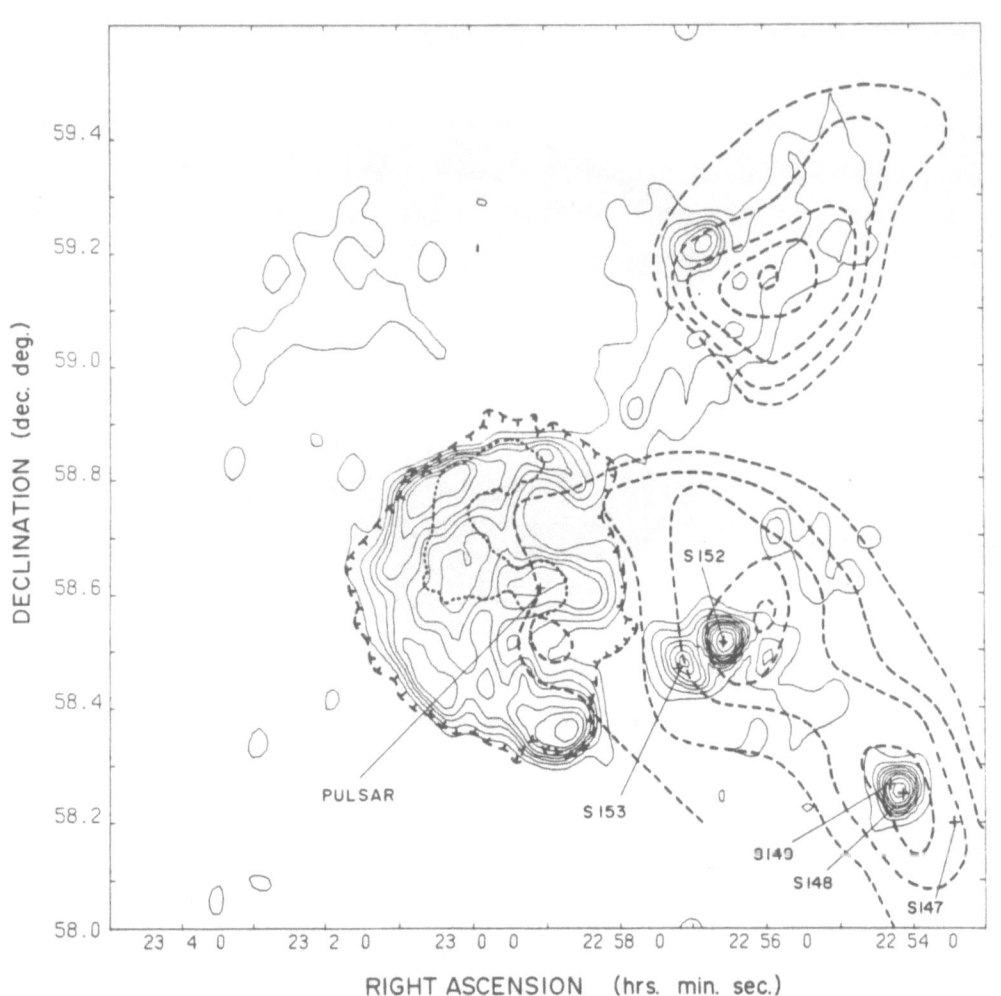

Figure 4. Radio map of G109.1-1.0 obtained with the NRAO 91 m telescope at a wavelength of 6 cm. Superposed are X-ray and CO emission contours shown as dashed lines.

RESULTS

Figure 4 shows the maximum entropy map derived from the 6 cm, 91 m telescope observations. The map is of a 1°6 field and shows clearly the SNR and its environment. Superposed on the radio map are contours of $C^{12}O$ emission (thick dashed line), a hatched "T" contour representing the outer boundary of the X-ray emission and the X-ray contour which best defines the jet (thin dashed line). The $C^{12}O$ emission is taken from the results of F. Israel (1980). At least 4 neighbouring sharpless HII regions are indicated on the map, S152, S153, S148 and S149. The maximum entropy method yields an image with a resolution dependent on the signal to noise ratio. Cross-sections through the peak of S152 exhibit a FWHM of 1.9 x 1.7 arcmin. The intrinsic size of S152, as determined by VLA observations (Gregory and Fahlman, 1981), is approximately 40 arcsec.

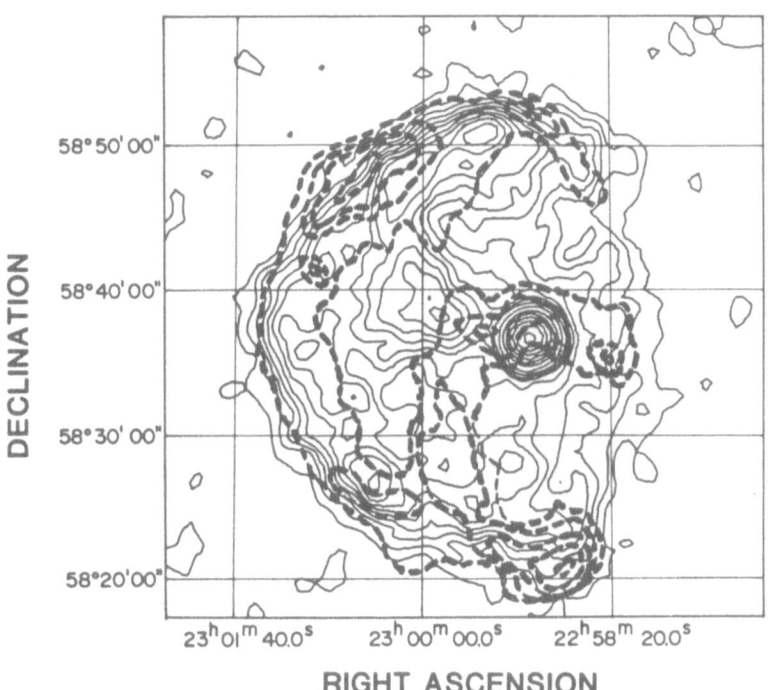

RIGHT ASCENSION

Figure 5. X-ray contours of figure 1 with 20 cm VLA contours superposed as dashed lines.

There is a very good agreement between the 6 and 20 cm radio maps. The emission in both is strongest in the south west with a second bright region in the north east. In both cases, the emission resembles two intersecting arcs. The fainter of the two arcs starts in the south west and passes through the central region of the remnant, finally curving westward just north of the pulsar. Several compact sources are also evident in the VLA map, including one, only 2 arcmin west of the X-ray pulsar. These have been studied at higher resolution and will be discussed in a subsequent paper. Also, a very sensitive search for a radio counterpart to the X-ray pulsar was carried out, but none was found (3σ upper limit = 200 μJy).

The 6 cm map shows radio emission extending to the north west of the SNR. There also appears to be a second molecular cloud overlapping some of this emission, at a considerable distance from the SNR. This radio emission is probably unrelated to the SNR and additional VLA observations of this region show that two of the components, between the SNR and the second molecular cloud, have a double structure, characteristic of extragalactic sources.

The X-ray and radio emission are compared in detail in figure 5, which shows 20 cm VLA contours superposed on the X-ray contours of figure 1.

CONCLUSIONS

Comparison of the radio, X-ray and CO contours leads to the following main conclusions.

1) The overall extent of the radio and X-ray emission is very similar, however, there are marked differences in the distribution of X-ray and radio emission within the remnant.

2) The jet like structure, apparent at X-ray wavelengths, doesn't have a radio counterpart.

3) The radio emission resembles two intersecting arcs.

4) There is no compact radio source associated with the X-ray pulsar. The 3σ upper limit on the continuous radio emission from the X-ray pulsar, at 20 cm, is 200 μJy.

5) Radio and X-ray peaks do not coincide in position. The radio peaks occur on the outer sloping edges of the X-ray emission.

6) The location of the $C^{12}O$ emission strongly suggests a physical interaction between the SNR and the neighbouring molecular cloud.

This interaction would provide a natural explanation for the semi-circular appearance of the SNR. A similar conclusion was reached by Heydari-Mayleri et al. (1981) based on $C^{13}O$ observations.

This research was supported by grants from the Canadian Natural Science and Engineering Research Council. The NRAO is operated by Associated Universities, Inc., under contract with the National Science Foundation.

REFERENCES

Emerson, D.T., Klein, U., Haslam, C.G.T.: 1979, Astron. Astrophys. 76, 92.

Fahlman, G.G. & Gregory, P.C.: 1981, Nature 293, 202.

Fahlman, G.G. & Gregory, P.C.: 1983, this volume, p. 457.

Gregory, P.C. & Fahlman, G.G.: 1980, Nature 287, 805.

Gregory, P.C. & Taylor, A.R.: 1981, Astrophys. J. 248, 596.

Gregory, P.C. & Fahlman, G.G.: 1981, Vistas in Astronomy 25, 119.

Heydari-Malayeri, M., Kahane, C., Lucas, R.: 1981 Nature 293, 549.

Israel, F.P.: 1980, Astron. J. 85, 1612.

THE PULSATION PERIOD AND POSSIBLE ORBIT OF 1E2259+586

G.G. Fahlman and P.C. Gregory
University of British Columbia
Vancouver, B.C. V6T 1W5 Canada

The currently known characteristics of the unique x-ray pulsar 1E2259+586 are briefly reviewed. The results from a recently completed analysis of the x-ray photon arrival times recorded by the Einstein Observatory are presented. The pulsar light curve has changed shape between two epochs separated by six months. The true pulse period is 6.978632 sec, double the previously reported value. In addition, there is evidence for orbital motion with a period close to 2300 sec. Some implications of this orbit are briefly discussed.

1. INTRODUCTION

The x-ray pulsar 1E2259+586 is unique in that it is located at the geometric center of curvature of a large semi-circular shell of diffuse x-ray emission (Gregory and Fahlman 1980, Fahlman and Gregory 1981). The shell has a radio counterpart and has been identified as a supernova remnant (SNR), designated by us as G109.1-1.0. Its distance and age are estimated as 3.6 ± 0.4 kpc and ~10^4 yrs. The compact x-ray source appears to be connected to the diffuse shell by a long curving arc of x-ray emission. This, together with its central location is compelling evidence that 1E2259+586 is the stellar remnant of the supernova explosion responsible for G109.1-1.0.

The x-ray spectrum of the compact source is harder than that of the diffuse emission. Acceptable fits to the output of the IPC pulse height analyser are obtained for thermal spectra with kT>1.0 kev and low energy absorption equivalent to a line of sight column density of $N_H \simeq 0.8 \times 10^{22}$ cm^{-2}. This column density is consistent with our estimate of the distance to the diffuse remnant and leads to a luminosity in the IPC band (0.5-4.0 kev) of $L_x \simeq 2 \times 10^{35}$ ergs S^{-1}. This luminosity is consistent with the object being a neutron star accreting mass.

The pulse period derived from an IPC observation in June 1980 was found to be 3.4890 sec (Fahlman and Gregory 1981). Subsequent observations, discussed below, indicate that the pulse period should be

J. Danziger and P. Gorenstein (eds.), Supernova Remnants and their X-Ray Emission, 445–453.
© *1983 by the IAU.*

double this value. In any event the period is long enough to indicate
that the object is not a classical pulsar powered by the rotational
energy of the neutron star. It appears to be an accretion powered pulsar and is probably a member of a binary star system.

We have spent considerable effort in a search for an optical counterpart and had suggested an identification with a faint, B ≈ 22 mag.
star (Fahlman et al. 1982). On the basis of recent optical photometry,
carried out at KPNO in Sept. 1982 by J. Middleditch and G. Fahlman, it
appears that the counterpart is actually the even fainter star (B ≈ 23.5
mag.), located just 3 arc sec. SE of the earlier candidate (see Figure
1 of Fahlman et al. 1982). This work will be discussed in more detail
elsewhere. The point to note here is that the optical counterpart is
very faint and cannot be a massive star. Hence, the postulated binary
containing 1E2259+586 is very dissimilar to the well known binary SS 433.
The 1E2259+586 system apparently belongs to the group of low mass x-ray
binary systems (see van Paradijs, 1981 for a discussion of their properties) and is the only such object associated with a diffuse remnant.

The existence of a compact interacting binary system at the center
of a relatively young SNR is unexpected (van den Heuvel 1981 a,b;
Tutukov 1981) and, clearly, the properties of this system are of more
than usual interest because of this. Here we will present the main results from our analysis of the x-ray photon arrival times taken from
observations made at two epochs with the Einstein Observatory.

2. Results from the Photon Arrival Time Analysis

The data discussed here is taken from the Einstein IPC observations
described in Table 1. The photon arrival times were corrected for earth
and satellite motion using the ephemeris program installed at the Harvard-Smithsonian Center for Astrophysics (R.F. Harnden, private communication).

Table 1
IPC Observation Summary

DESIGNATION	I8102	I9984	I9985	I9986
DATE (m/d/y)	7/7/80	1/23/81	1/24/81	1/25/81
DURATION (sec)	7864	4588	3318	5243
SOURCE ON TIME (sec)	5205	3079	3282	3217
Mean Count Rate	0.90	0.79	0.81	0.83

2.1. The Pulsar Light Curve

The second epoch observations (I9984, -5, -6) were combined to
define the dominant pulse period via a fine grained folding analysis.
The period derived in this way was P = 3.48931 ± 0.000007 sec; the same
to within the error as the period derived from the first epoch data
(I8102), P = 3.48932 ± 0.000015 sec. Unexpectedly, a power spectrum

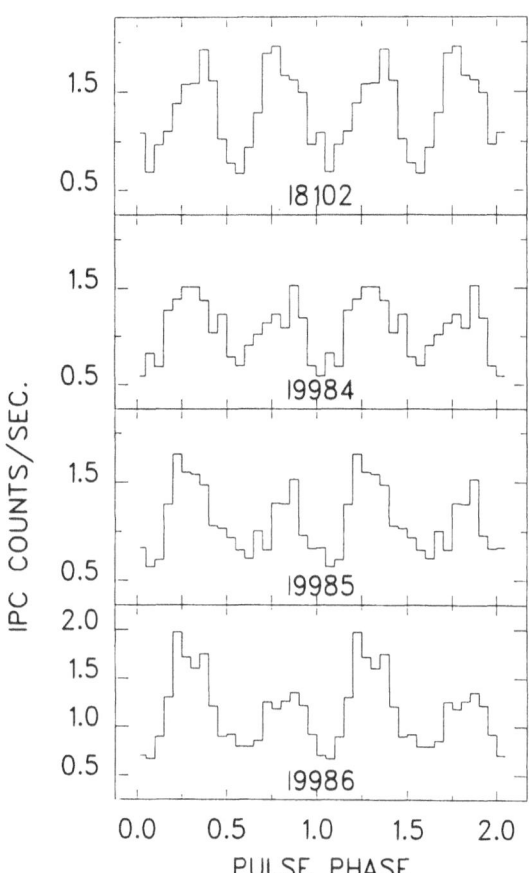

Figure 1: Pulsar light curves. Two cycles of the P = 6.978632 sec.
pulse are shown with 20 bins per cycle. Zero phase is set arbitrarily
but all four data sets have been aligned in phase. Note that the modulated signal is sitting on top of a constant flux of approximately 0.5
counts/sec. The data shown is that in a box with a length of 4 arc
min. per side centered on the pulsar. The background from the diffuse
source is expected to contribute no more than 0.05 counts/sec in this
field.

of the continuous data in I9985 showed a significant peak at the frequency corresponding to double the best folding period, i.e., at 143.3
mHz (the fundamental). The power in the first harmonic (286.6 mHz) was
almost an order of magnitude larger than that in the fundamental. A
power spectrum of a comparable continuous data segment in the I8102
data did not show significant power at 143.3 mHz. At this epoch then,
the two poles of the pulsar were statistically indistinguishable.

The light curves obtained by folding our data at a period of
6.978632 sec are shown in Figure 1. It is quite clear that a change in
the observed character of the pulses has occurred and that the true
pulsar period is 6.978632 sec.

Since pulsars in general are thought to be oblique rotators, the
most direct way of accounting for the difference in the light curve is
to postulate a change in the inclination angle of the rotation axis of
the neutron star relative to our line of sight. In other words, the
neutron star may be precessing.

The change in the pulsar light curve introduces some ambiguity in
phasing the two epochs together. If we adopt the zero phase time and
pulse period for the second epoch and count cycles back to the first
epoch we find a phase discrepancy which has two possible values depending on which half of the 6.978632 sec period is used to match phases.
The best fitting period for both epochs is either $P = 6.97863313$ or
$P = 6.97863173$ with an error of $\pm 0.14 \times 10^{-6}$ sec. Of course this period is strictly nominal because whole cycles lost between the two epochs would not be detectable. If we interpret the phase discrepancy as
a period change then the pulsar is speeding up with an average $\dot{P} \simeq -7 \times
10^{-14}$ or slowing down with $\dot{P} \simeq 2 \times 10^{14}$. Following Rappaport and Joss
(1977), we might have expected spin up to occur at a rate of $\dot{P} \sim -2 \times
10^{-12}$. The difference between the predicted and 'observed' $\dot{P}$ can be
eliminated if some 5 cycles were lost in the six month interval. If we
simply use the best period available at the two epochs, we can place an
upper limit to the period change of $|\dot{P}| \lesssim 2 \times 10^{-11}$. Evidently third
epoch x-ray data is needed to pin down any period change.

2.2. Orbital Motion

The three consecutive observations in January 1981 were used in a
modified pulse arrival time delay analysis to look for evidence of
orbital motion. The procedure used was as follows:
(i) assume a trial orbital period P_0,
(ii) bin the photon arrival times according to orbital phase,
(iii) fold in each orbital bin at the dominant pulse period (3.489316
sec),
(iv) cross-correlate these pulses with a suitable template and determine
the position of the maximum in the cross-correlation function,
(v) the variation on the position of the maximum of the cross-correlation function measures the time delay in the pulse arrival relative to
a fixed epoch; it is examined for evidence of Keplerian motion (essen-

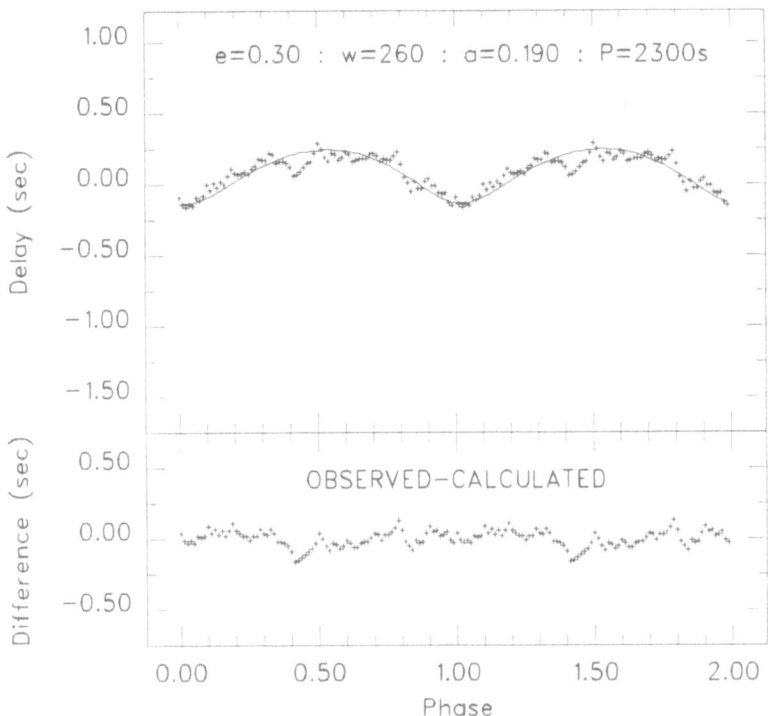

Figure 2: Orbital Motion. The upper panel shows the time delay points
determined by the method described in the text. The solid line is the
model with the parameters shown: e is the eccentricity, w is the argument of periastron, a is the projected semi-major axis of the pulsar
orbit around the center of mass of the system measured in light seconds.
Two complete cycles of the orbit are shown. The lower panel shows the
residuals from the model fit on the same scale as the upper panel. It
is apparent that the residuals are dominated by a distinct minima centered at phase ~0.4. This may well be an artifact of the method and
should not be considered significant at this time.

tially one looks for a smooth variation with orbital phase).

A systematic search for orbital periods between 1000 and 10,000 sec was made. The large gaps in the data prevented all independent periods in this range from being checked and generally complicated the interpretation of the results obtained from an automated search program. Moreover, to achieve adequate resolution, we settled on a minimum number of bins (orbital + pulse) of 80 which resulted in an effective (modulated) signal to noise ratio in each bin of between 3:1 and 5:1.

Within the range searched, the only reasonably coherent variation was found for orbital periods near 2300 sec. In Figure 2, we show a typical solution for the variation seen at $P_0 = 2300$ sec. Similar phase modulations are seen for the trial orbital periods within about 30 sec of this value but, in view of the gaps in the data the real uncertainty in the period is perhaps two or three times greater. To obtain the data points shown in Figure 2, overlapping orbital phase bins were used; there are actually only eight completely independent orbital phase points.

The Keplerian orbit fitted to the data points requires the specification of four parameters: ω, the argument of periastron; e the eccentricity; a_x sini (designated simply as a on the plot), the projected semi-major axis measured in light seconds and ϕ_0, a parameter to determine the epoch of periastron. Given trial values of e and ω, a_x sini and ϕ_0 were found by fitting a normalized model curve to the data; the best fit being determined by a least squares criterion. Provision was made for some manual adjustment of these latter two parameters because the residuals are dominated by systematic effects. The complexity of projected Keplerian motion allows virtually any reasonably smooth curve to be fitted with some combination of parameters. The particular fit shown is representative of the best solutions obtained. Similar results are obtained for a fairly wide range of parameters:

$$0.16 \lesssim a_x \text{ sini} \lesssim 0.21; \quad 200° \lesssim \omega \lesssim 280°; \quad 0.15 \lesssim e \lesssim 0.55 \qquad (1.)$$

In the case of the eccentricity, a wide range gives formally acceptable fits but values below about 0.2 generally show objectionable (even by the standards used here!) systematic departures from the data.

From a purely statistical point of view, none of the fits is particularly impressive - the rms residual after fitting a finite amplitude model are typically only a little more than a factor of two better than 'fitting' a zero amplitude model. The I8102 data covers slightly more than three of these orbits but a similar analysis of that data set did not reveal orbital phase modulation. However, for this data set the signal to noise in a given phase bin is so low (~3:1 on average) that any small amplitude phase modulation is expected to be hidden by the errors in determining the pulse phase.

It should be evident from the forgoing remarks that the results suggesting orbital motion must be regarded with caution. Nevertheless, the 2300 sec orbit is consistent with the 1 mHz downshift (at the 3.5 sec period) in the infrared observations noted in Fahlman et al. (1982) and discussed more thoroughly by Middleditch, Pennypacker and Burns (1982, preprint). A similar downshift was noted by Middleditch and Fahlman in their optical observations mentioned earlier. The available x-ray data has been all but exhausted and further progress in studying the binary nature of the source will undoubtably come by pursuing the optical and infrared observations already initiated.

3. Comments on the Nature and Origin of the Binary System

The spectroscopic mass function obtained from the orbital solutions discussed here is

$$m_2^3 \sin^3 i \,/\, (m_x + m_2)^2 = 0.008 \pm 0.0002 \; M_\odot \quad . \qquad (2.)$$

For a reasonable pulsar mass $m_x = 1.0-1.4 \; M_\odot$, the mass of the secondary is $m_2 \gtrsim 0.2 \; M_\odot$. If the companion is a normal main sequence star, its radius would exceed the characteristic radius of its Roche lobe by a factor of at least two. The relatively low x-ray luminosity is inconsistent with a high rate of mass transfer and the extreme faintness of the companion shows that an optically bright accretion disk is absent. Consequently, the companion is unlikely to be filling its Roche lobe (except perhaps near periastron) let alone exceeding it by a factor of two or more. We are forced to conclude that the companion is not a normal low mass main sequence star. The possibility that it is a degenerate star can be entertained. Here, the difficulty is just the opposite - a normal degenerate star with the minimum mass of 0.2 $M_\odot$ and hence the largest radius, would underfill its Roche lobe by a factor of about 5, even at periastron (for $e \simeq 0.3$). The system would be fully detached and presumably quiescent. It would be possible to induce some mass flow if the eccentricity is sufficiently high, $e \gtrsim 0.8$. Such high values are outside the range of the good fits to our data but cannot be totally ruled out since our analytic method would tend to underestimate the true eccentricity of such nearly rectilinear orbits. Further study on this point is needed. Continued observations of the infrared and optical pulses from the system will likely put better constraints on the secondary than the x-ray data.

The supernova configuration must have been even more massive and compact than the current system. Assuming an initially circular orbit and adopting a current eccentricity of 0.3, we can estimate the mass ejected by the explosion (Wheeler, Lecar and McKee, 1975). Ignoring the possible momentum transfer to the companion from incident ejecta (Fryxell and Arnett, 1981), we estimate an ejected mass of ~0.2-0.3 $M_\odot$ which implies that the original orbit was some 30% smaller than the observed now. Such a system would be much more compact than any of the known cataclysmic variables which might otherwise be considered as plausible precursors to 1E2259+586. In effect, this is also an argument against the companion being a non-degenerate dwarf.

An alternate possibility is that the precursor consisted of two degenerate dwarfs. Two spectacular examples of twin degenerate inter-acting binaries are HZ 29 (Patterson et al. 1979) and G61-29 (Nather, Robinson and Stover, 1981). Since both of these systems are thought to have extremely small mass ratios, they are probably not examples of the immediate percursor to 1E2259+586. Rather it is some variant of the parent to those systems which might be imagined as the precursor to 1E2259+586. An interesting discussion of the evolutionary paths leading to twin degenerate systems is given by Nather, Robinson and Stover (1981). One difficulty with this picture is that the mechanism which triggered the supernova is not clear. The obvious possibility is cata-strophic mass loss from the smaller to the larger star but, given a companion mass of ~0.2-0.3 $M_\odot$ and the assumption that it must fill its Roche lobe, this would imply such a small initial orbit that the cur-rent orbit would be unattainable through impulsive mass loss associated with the supernova explosion.

Until the nature of the secondary is known, i.e., whether it is degenerate or not, one has little basis for further speculation on the precursor system. If the secondary is degenerate then one might hope to observe the relativistic periastron advance of the orbit (Will, 1975):

$$\dot{\omega}_R = 6\pi \frac{(GM)}{C^2} \left[P_0 \, a(1 - e^2) \right]^{-1} \tag{3.}$$

For reasonable system parameters, $a_x \sin i = 0.175$ light-seconds and e = 0.3, we find $\dot{\omega}_R \simeq 100° \, (M/M_\odot)^{4/3} \, yr^{-1}$, where M is the total mass of the system. If the companion is not degenerate, then a large quadruple component to the periastron shift will be added to this value. Hence, an observation of the periastron shift, could provide an important con-straint on the secondary. Further optical/infrared studies could pro-vide this data.

ACKNOWLEDGEMENT

Our work is supported through grants from the Natural Sciences and Engineering Research Council of Canada. We are very grateful to the staff at the Einstein Observatory data center, in particular Rick Harnden, for their assistance.

REFERENCES

Fahlman, G.G. and Gregory, P.: 1981, Nature 293, 202.
Fahlman, G.G., Hickson, P., Richer, H.B. and Middleditch, J.: 1982, Astrophys. J. 261, L1.
Fryxell, B.A. and Arnett, W.D.: 1981, Astrophys. J. 243, 994.

Gregory, P.C. and Fahlman, G.G.: 1980, Nature 287, 805.
van den Heuvel, E.P.J.: 1981a, in "Fundamental Problems in the Theory
    of Stellar Evolution" (D. Sugimoto, D.Q. Lamb and D.N. Schramm, eds.),
    Reidel: Dordrecht, pp. 155-175.
van den Heuvel, E.P.J.: 1981b, Space Sci. Rev. 30, 623.
van den Paradijs, J.: 1981, Astron. Astrophys. 103, 140.
Patterson, J., Nather, R.E., Robinson, E.L. and Handler, F.: 1979,
    Astrophys. J. 232, 819.
Rappaport, S. and Joss, P.C.: 1977, Nature 266, 683.
Tutukov, A.V.: 1981, in "Fundamental Problems in the Theory of Stellar
    Evolution" (D. Sugimoto, D.Q. Lamb and D.N. Schramm, eds.), Reidel:
    Dordrecht, pp. 137-154.
Wheeler, J.C., Lecar, M., and McKee, C.F.: 1975, Astrophys. J. 200, 145.
Will, C.M.: 1975, Astrophys. J. 196, L3.

RADIO OBSERVATIONS OF THE SUPERNOVA REMNANT CTB109 (G109.2-1.0)

V.A. Hughes[1], R.H. Harten[2], C. Costain[3], L. Nelson[1] and M.R. Viner[1]
[1] Astronomy Group, Queen's University at Kingston, Ontario, Canada.
[2] Netherlands Foundation for Radio Astronomy, Dwingeloo, Netherlands.
[3] Dominion Radio Astrophysical Observatory, Penticton, B.C., Canada.

The supernova remnant G109.2-1.0 was discovered at λ49cm by Hughes, Harten and van den Bergh (1981) during a survey of part of the Galactic plane. The northern part of it had been detected previously as the non-thermal radio source CTB109 by Wilson and Bolton (1960), and by Raghava Roa et al (1965), but the extended low brightness of the source and its close proximity to the very strong source Cas A, from which it is separated by ~5', excluded it from any further detailed study. It was discovered independently at X-ray wavelengths by Gregory and Fahlman (1980). Recently, the original WSRT radio observations have been found to be in error as a result of applying the CLEAN procedure to an extended source, and since the object appears to contain an X-ray pulsar (Fahlman and Gregory, 1981), it was decided to carry out a more detailed and extensive mapping of the remnant using different antenna arrays and frequencies. This paper describes the results obtained at λ49cm and λ21cm using the Westerbork Synthesis Radio Telescope (WSRT), at λ21cm using the aperture synthesis array at the Dominion Radio Astrophysical Observatory (DRAO) and at λ4.6cm using the 46m telescope of the Algonquin Radio Observatory (ARO). Thus, data has been obtained from three completely independent telescopes, using completely independent data reduction systems. Of importance is the fact that not only have wavelengths been chosen such that the larger dimensions of the array give a reasonable angular resolution of <1', but also that the smallest spacing enables the larger angular dimensions of the remnant to be observed. This paper presents some of the results and a brief interpretation.

Figure 1 shows the revised map of the SNR, convolved to a beamwidth of 100", using the WSRT at λ49cm. The map shows a half-shell of emissions with a number of radio condensations. There is very clear indication of a "hole" in the emission, offset from the centre of the remnant. Also indicated are the positions of two compact sources, and the position of the X-ray pulsar (Fahlman and Gregory 1981). It is clear that the pulsar is not coincident with one of the compact sources, or with the peak of the nearby region of extended emission. The overall

455

J. Danziger and P. Gorenstein (eds.), Supernova Remnants and their X-Ray Emission, 455–458.

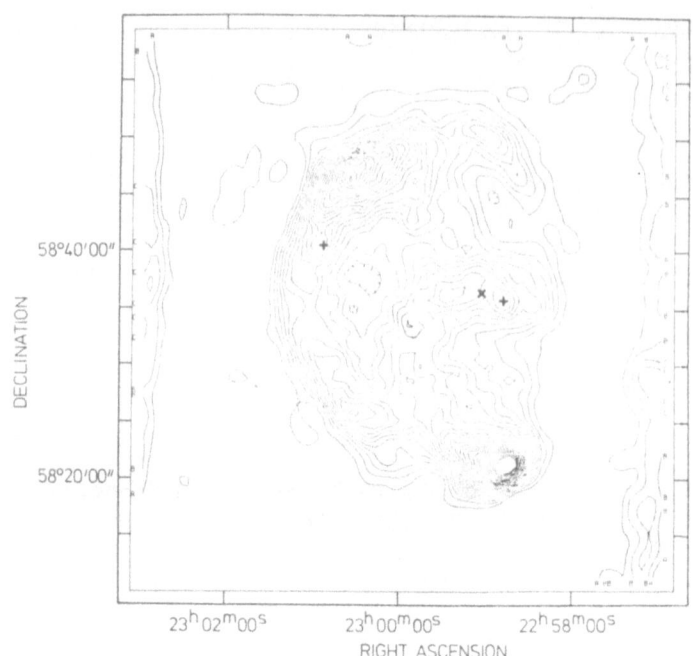

Figure 1. WSRT λ49cm map of CTB109, convolved to a resolution of 100".
Marked by a + are the positions of two compact sources; X marks the
position of the X-ray pulsar.

dimensions of the SNR are ~36′ and ~26′ in the N–S and E–W directions,
respectively.

        In an attempt to detect possible radio emission from the
pulsar, observations were made with the WSRT at λ21cm. In this case,
the minimum antenna spacing of 36m (171λ) did not receive the wide angle
components associated with the SNR, but enabled the detection of sources
as small as 20" with a flux density limit of 0.5 mJy. It was quite
clear that though there were about 13 compact sources observed,
corresponding to the number of extra-galactic sources expected, no radio
source at this level was detected within 2′ of the position of the
pulsar. Thus, we see no radio evidence for any obvious small diameter
remnant from the supernova explosion.

        In Figure 2 we show a comparison of the WSRT λ49cm map with
resolution in the E–W direction of 56" and the DRAO map at λ21cm with
resolution of 60". The minimum spacings are 36m (73λ) and 17.14m (82λ)
respectively. Contours are drawn in units of 1, 3, 5, 7....... in each
case, the unit values being 5 mJy for the WSRT map and 4 mJy for the
DRAO map. As can be seen, the remarkable similarity between the two
maps indicates no detectable variation in the spectral index of the
radiation across the map, and when account is taken of the difference in

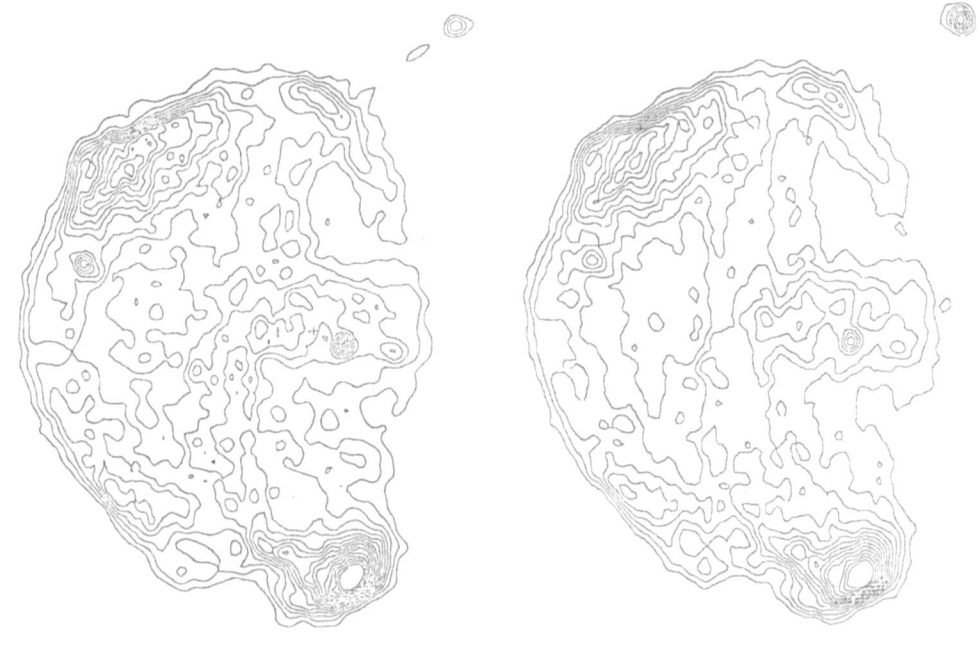

WSRT 49 CM   **DRAO 21 CM**

Figure 2. WSRT λ49cm map of CTB109, resolution of 56", and DRAO λ21cm
map, resolution 60". Contour units in each case are 1,3,5,7,····· The
unit for the WSRT map is 5 mJy and that for the DRAO map is 4 mJy.

angular resolution, leads to a uniform spectral index over the maps of
$\alpha = 0.45$ where $S \propto \nu^{-\alpha}$.   Such a value for α is typical for well
established SNR's, and shows no evidence for any component of radiation
that could be associated with the pulsar.

        Finally, the DRAO λ21cm map was convolved to the beamwidth
of 4.8′, equal to the resolution of the ARO map at λ4.6cm. A comparison
showed that at this resolution, again no change in spectral index was
apparent over the source. The total flux of 13 Jy at λ21cm and 6.6 Jy
at λ4.6cm gives a value of $\alpha = 0.45$ between these two wavelengths,
similar to the value obtained above between λ21cm and λ49cm.

        From the above data we have derived a number of properties
of the SNR. Our previous distance of 4.3 - 5.2 kpc, obtained using the
Σ-D relationship, has been revised to 5 - 6 kpc in the light of the new
flux densities. Since the Perseus arm bifurcates in this direction with
one part being at an estimated distance of 5 kpc, we would place the
remnant in or close to this outer arm. However, the uncertainties
involved in the use of the Σ-D method, as discussed at this Symposium,
do not allow a more accurate distance to be derived. Noting that the
optical filaments as seen by Hughes, Harten and van den Bergh (1981)
have not moved by more than 1" in the 27 years since the PSS plates were
taken, we obtain the maximum transverse speed of these filaments as

<900 km/sec, though it is not clear how this speed is related to the expansion speed of the SNR.

With regards to the band of X-ray emission in an approximately E-W direction across the centre of the radio SNR, it is clear that this does not have a counterpart in the radio emission and in fact passes through a radio "hole". In addition, there is no apparent change in the radio spectrum that we can associate with it. The angular extension of the apparent jet is ~16' or 23 pc, so that if it is the result of electrons being emitted from the pulsar at the speed of light, it would take them about 100 years to travel its length. This puts rather stringent limits on the mechanism for producing the jet; for instance, it is difficult to build models of synchrotron emission at X-ray wavelengths when the electron decay time is <100 years. More likely, the X-ray emission is due to bremsstrahlung from a hot plasma situated behind the expanding shock front. Models are at present being developed and will be published at a later date.

The Westerbork Synthesis Radio Telescope is operated by the Netherlands Foundation for Radio Astronomy (SRZM) with financial support of the Netherlands Organization for the Advancement of Pure Research (ZWO). The Dominion Radio Astrophysical Observatory and the Algonquin Radio Observatory are operated by the Herzberg Institute of Astrophysics of the National Research Council of Canada. Some of the work was supported by an Operating Grant from the Natural Sciences and Engineering Research Council of Canada.

REFERENCES

Fahlman, G.C. and Gregory, P.C. 1981, Nature, **293**, 202.
Gregory, P.C. and Fahlman, G.C. 1980, Nature, **287**, 805.
Hughes, V.A., Harten, R.H. and van den Bergh, S. 1981, Astrophys. J., **246**, L127.
Raghava Roa, R., Medd, W.J., Higgs, L.A. and Broten, M.W. 1965, Mon. Not. Roy. Astron. Soc., **129**, 159.
Wilson, R.W. and Bolton, J.G. 1960, Pub. Astron. Soc. Pac., **72**, 331.

DISCUSSION

BLAIR: Bob Kirshner and I obtained spectra of one of the outer optical filaments in this object. Because of the faintness of the filament, the quality of the spectrum is not high, but it does indicate a large amount of reddening and an otherwise normal SNR spectrum (i.e. strong [SII] and [NII] in comparison to Hα).

WEILER: I think that you have shown very strong evidence for removing CTB 109 as a possible member of the Class-C combination SNR's. I don't believe any doubt remains.

X-ray Studies of SS433

J.E. Grindlay, D. Band, F. Seward, and L. Stella
Smithsonian/Harvard Center for Astrophysics
and
M. Watson
University of Leicester

Results of Einstein Observations of SS433 are discussed which address both the nature of the diffuse X-ray lobes and the relationships between SS433 and W50 as well as the time variability and nature of the central X-ray source. The diffuse X-ray lobes extend out to the quasi-spherical shell seen in the radio maps of W50 and suggest that the X-ray lobes are powered by the interaction of shock-heating from the SS433 jets and the denser material in the W50 shell. The central X-ray source in SS433 is time variable but only on timescales $\gtrsim$ 500-1000 sec. Flares, in which the non-thermal spectrum hardens, are detected at two preferred phases in the 13.08 day binary orbit. Constraints on the central X-ray source size as well as a possible eclipse by the companion star suggest the compact object in SS433 may be an ~10 $M_\odot$ black hole.

1. Introduction

The peculiar object SS433 was observed in some 37 separate pointings with the Einstein X-ray Observatory (Giacconi et al. 1979) between March 1979 and October 1980. Observations were made with the two imaging instruments, the HRI and IPC, which showed the previously known X-ray source (Seward et al. 1976, Marshall et al. 1979) was surrounded by diffuse X-ray emission in two lobes (Seward et al. 1980, Watson et al. 1983). The MPC data from all the Einstein observations have been used to study the spectrum and variability of the central source (Grindlay et al. 1983). We shall describe both the diffuse and compact source results here and discuss a common model.

2. Diffuse X-ray Lobes

A ~1° x 2° field was mapped out around SS433 in three separate IPC exposures, each with ~15 ksec exposure time. The common X-ray image is shown in contour form in Figure 1. The X-ray lobes are striking direct confirmation of the existence of the relativistic jets in SS433. They define the same position angle (99°) on the sky as the jets and they have brightened "ends" at angular distances (from SS433) of 35-40 arcmin where they intersect the bright radio shell of W50 (Watson et al. 1983).

459

J. Danziger and P. Gorenstein (eds.), Supernova Remnants and their X-Ray Emission, 459–463.
© 1983 by the IAU.

The luminosity in each lobe is ~6 x 10^{34} erg sec^{-1} in the 0.5-4.5 keV band, and the spectra can be fit with thermal bremsstrahlung models with kT ≃ 1.5 keV. There is some evidence that the diffuse emission becomes harder towards the central source.

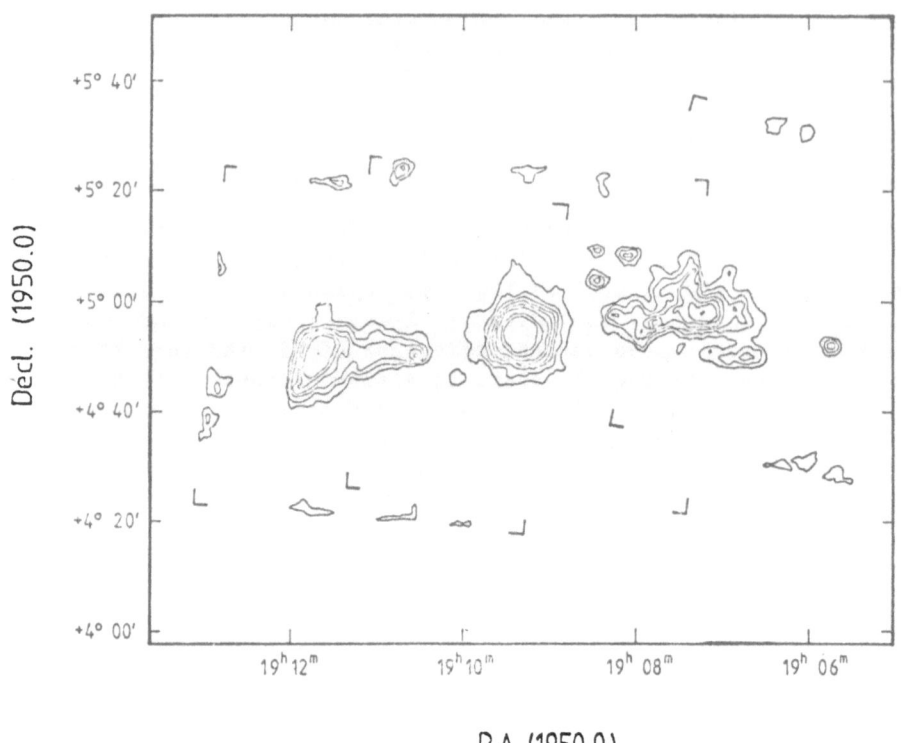

R A. (1950.0)

Figure 1 – Contour map of the (three) combined IPC images of SS433 in the ~0.7-4 keV band. The lowest contour is at ~7σ above background and corresponds to 2.3 x 10^{-3} IPC cts sec^{-1} arcmin^{-2}; subsequent contours are in steps of ~2.5σ (from Watson et al. 1983).

Perhaps the most striking feature of the lobes is their axial symmetry and center-filled morphology. The angular width subtended by the lobes is only ~15° (FWHM) although faint diffuse emission extends out from the lobe axis with full widths of ≳50°. The angular width of the lobe is significantly narrower in the hard X-ray band (1-2.5 keV) than in the soft band (0.5-1.0 keV) of the IPC. However, even in the soft band the FWHM is significantly narrower than the ±40° cone expected from the precession cone angle of the jets as inferred from the optical data and kinematic model (Margon 1982). If the X-ray lobes are due to the direct interaction of the beams with the material in W50 limb-brightened X-ray lobes would be expected. The lobe morphology may be used to distinguish between non-thermal and thermal models for the emission.

In a non-thermal model, the diffuse X-rays could be due to synchrotron radiation from electrons with energy ~6 x 10^4 GeV in a (equipartition) magnetic field of B ~4 μg. These electrons with lifetimes of ~10^4 year, could be produced by a first-order Fermi acceleration process in which ~40 collisions of an electron with material in the relativistic beams would boost the energies by the factor ~10^4 required from the initial ~5 GeV radio-emitting electrons. The center-filled morphology of the lobes would require the diffusion timescale (limited by the Alfven speed) for particles be less than the radiation lifetime. This is not possible for reasonable combinations of B and particle density, and so the precessing jets would have to fill the entire cone.

A thermal emission model is consistent with the lobe surface brightness and approximate spectrum profiles. The lobe spectra are fit by bremsstrahlung spectra with kT $\simeq$ 1-3 keV, increasing towards the cone axis and central object. The particle densities required range from ~0.2 cm^{-3} in the faint emission plateau at λ 45 arcmin from SS433 to ~0.7 cm^{-3} in emission peak at radius ~35 arcmin. The emission peak in the lobes is probably due to the density peak in the shell (at ~35 arcmin) of W50 as evident in both the radio (Geldzahler et al. 1980) and optical (Zealey et al. 1980) maps. The center-filled lobe morphology and possible temperature gradient suggests the actual heating mechanism of the gas is by shocks propagating perpendicular to the beams and intersecting along the cone axis. This requires heating timescales long enough (~10^4 yr) for shock propagation (at ~300 km sec^{-1}) to the cone axis.

The total thermal energy content of the lobe is 1.2 x 10^{51} ergs and the cooling time is ~3 x 10^8 years. This requires a rate of energy supply (from the beams) of $L_{beams} \simeq$ 4 x $10^{40}/(\varepsilon\ T_3)$ erg s^{-1}, where ε is the conversion efficiency from beam kinetic energy to thermal energy and T_3 is the timescale for energy supply in units of 10^3 years. Thus, for $T_3 \gtrsim$ 1 (corresponding to the minimum propagation times at 0.26 c of the material in the beams out to the lobes), then $L_b \lesssim 10^{41}$ erg s^{-1}.

3.   X-ray Emission from the Central Region of SS433

The central object in SS433 is, of course, especially interesting. Its X-ray emission is apparently non-thermal and characterized by flares with ~1 day timescales (Seaquist et al. 1982, Grindlay et al. 1983). In the latter study, we have shown that since there is no evidence for variability on timescales $\lesssim$500-1000 sec, a plausible lower limit on the source size is ~$10^{12}-10^{13}$ cm. The central source appears to have a relatively hard power law spectrum which may be eclipsed briefly once per 13.08 day orbit. The possible eclipse is evident at phase 0.3 in the X-ray light curve plotted in Figure 2. The spectral index, as measured in both the MPC and SSS instruments, steepens significantly in the dip, and there is a marginal increase in the low energy absorption. The X-ray luminosity plotted in Figure 2 also shows two phases (~0.1 and ~0.6) of the 13.08 day period where flares occur. These are characterized by pronounced hardening of the power law spectrum, with even positive (energy) spectral indices observed in some flares.

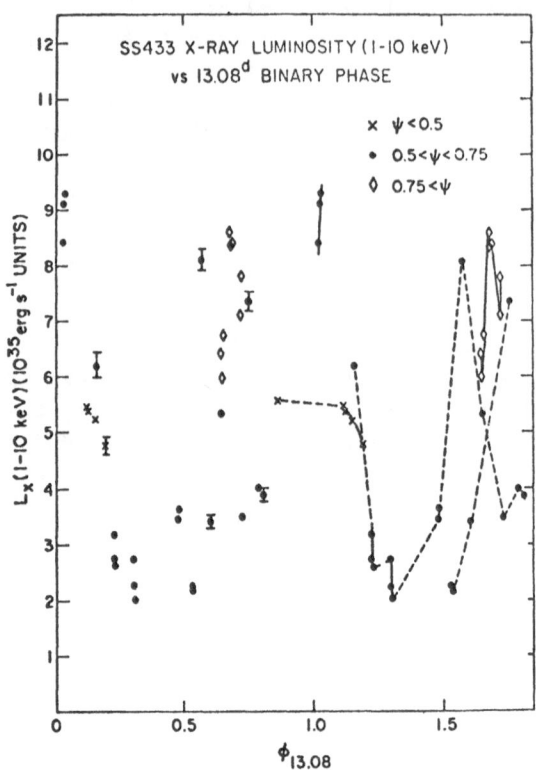

Figure 2 – Apparent modulation of the ~1-10 keV X-ray luminosity of SS433 at the 13.08 day binary period (plotted twice, for clarity). Three different intervals of 164 day phase, Ψ, are indicated. Solid lines connect adjacent observations (within 0.75 day) and dashed lines connect points observed in the same sequence by Einstein. A possible partial eclipse is evident at phase ~0.3, while flares are evident primarily at phases ~0.1 and ~0.6 (from Grindlay et al. 1983).

The flare modulation is approximately synchronous with both the velocity maxima/minima in the "stationary" HeII λ4686 line emission (Crampton and Hutchings 1981) as well as the photometric (continuum) variations in a similar range of 164 day phase, $0.5 < Ψ < 0.6$ (Kemp et al. 1980). The X-ray and optical maxima at quadrature phases suggest the accretion disk is comparable in optical brightness to the companion star and that the flares are triggered by enhanced accretion into the disk plane. If the companion star also has inclined spin, as in the slaved disk models of Katz et al. 1982 which give rise to the nodding motions of the disk, then the Roche lobe radius of the companion star will undergo two maxima/minima per orbit provided the orbit is nearly circular (Avni and Schiller 1982). This could provide the required modulation of the accretion rate. Alternatively, if internal damping has aligned the companion star (Papaloizou and Pringle 1982), the precessing disk may still undergo enhanced accretion twice per orbit when its mid-plane crosses the orbital plane. In either case the phase of the X-ray maxima is expected to advance with 164 day phase if the disk precesses with the beams.

The possible X-ray eclipse and yet finite X-ray source size can constrain the source mass. To (partially) eclipse the X-ray source (with radius $\sim 10^{12}-10^{13}$ cm), the stellar companion must have a comparable radius. A $\sim 20-30$ $M_\odot$ B giant (with radius $\sim 2 \times 10^{12}$cm) would be consistent as has been suggested (Zealey et al. 1980) on the basis of the inferred mass loss in the system. The mass function of SS433 derived by Crampton and Hutchings (1981) would then require the compact object to be ~ 10 $M_\odot$ and thus presumably a black hole.

4. Conclusions

X-ray studies of SS433 constrain both the nature of the jets (with age $\gtrsim 10^3-10^4$ years and total kinetic energy $\lesssim 10^{41}$ erg s^{-2}) and ISM into which they expand (which appears to be consistent with a SNR). They constrain also the nature of the central binary system, which appears to contain a B giant, a massive (~ 10 $M_\odot$) compact object (black hole) and tilted accretion disk in which variable accretion occurs. Particle acceleration (to $\gamma \gtrsim 10^2-10^4$) must occur in the disk or disk corona to explain the central non-thermal X-ray source which is eclipsed. The acceleration region may be in the inner ends of the beams, with blobs of relativistic plasma then entrained in the jets and moving out to give the synchrotron radio emission (and an additional inverse Compton X-ray component) at distances of $\sim 10^{14}-10^{15}$ cm as envisioned in the model of Seaquist et al. (1982).

References

Avni, Y. and Schiller, N. 1982, Ap.J., 257, 703.
Crampton, D. and Hutchings, J. 1981, Ap.J., 251, 604.
Giacconi, R., et al. 1979, Ap.J., 230, 540.
Grindlay, J., Band, D., Seward, F., Leahy, D., Weisskopf, M., and Marshall, F. 1983, Ap.J., submitted.
Margon, B. 1982, Science, 215, 247.
Papaloizou, J. and Pringle, J.E. 1982, MNRAS, 200, 49.
Seaquist, E., Gilmore, W., Johnston, K., and Grindlay, J. 1982, Ap.J., 260, 220.
Seward, F., Page, C., Turner, M., and Pounds, K. 1976, MNRAS, 175, 39.
Seward, F., Grindlay, J., Seaquist, E., and Gilmore, W. 1980, Nature, 287, 806.
Watson, M., Willingale, R., Grindlay, J., and Seward, F. 1983, Ap.J., submitted.
Zealey, W., Dopita, M., and Martin, D.F. 1980, MNRAS, 192, 731.

STRUCTURE OF EXTENDED RADIO EMISSION IN W50 AND ITS RELATION TO SS 433

T. Velusamy and A. Pramesh Rao
Radio Astronomy Centre(TIFR), Ootacamund, India

The association of SS 433 with SNR W50 suggests several
interesting scenerios for the complex emission features
observed in W50. Since W50 lies close to galactic plane
(b = -2°.5), it is possible some of these features are
just chance superpositions. Several of these are seen in
association with diffuse X-rays (Seward et al. 1982) and
optical filaments (Zeally et al. 1980). At radio wavelengths W50 has been well studied, at 11 cm (Velusamy and
Kundu, 1974; Geldzahler et al. 1980) and at 18 cm (Downes
et al. 1981; 1982). In this paper we present new observations at 92 cm (327 MHz).

The observations at 327 MHz were made during June-
July 1982, using the partially completed Ooty Synthesis
Radio Telescope (OSRT). We have mapped a 3°x2° region
containing SNR W50 and SS 433, using 3 antennas of OSRT,
with maximum spacing of 1.2km. The primary beam of the
largest antenna (ORT) is 170'x8' ; while that of others is
360'x360'. With the usual observing procedure we can map
a field of view of only 170'x8'. However, we have overcome this limitation by using a technique of rapid switching in declination in order to derive the complex visibility corresponding to a much larger field in the North-
South direction (120' in the case of W50). Maps are
produced by standard techniques of Fourier inversion and
cleaning. The shortest spacing available was 100 wavelengths. We have added the visibility close to u=0, using
the total power scans across W50 made with the ORT.

In Fig.1 is shown the 327 MHz map of W50. The restoring beam used for this map was 7'.5x7'.5 which is the same
as for the 1.7 GHz map of Downes et al. (1982). The overall structure of W50 is similar at these wavelengths. At
327 MHz a weak radio shell is seen at a radius of 20'-30'
from SS 433. It should be noted that the broader components

J. Danziger and P. Gorenstein (eds.), Supernova Remnants and their X-Ray Emission, 465–467.

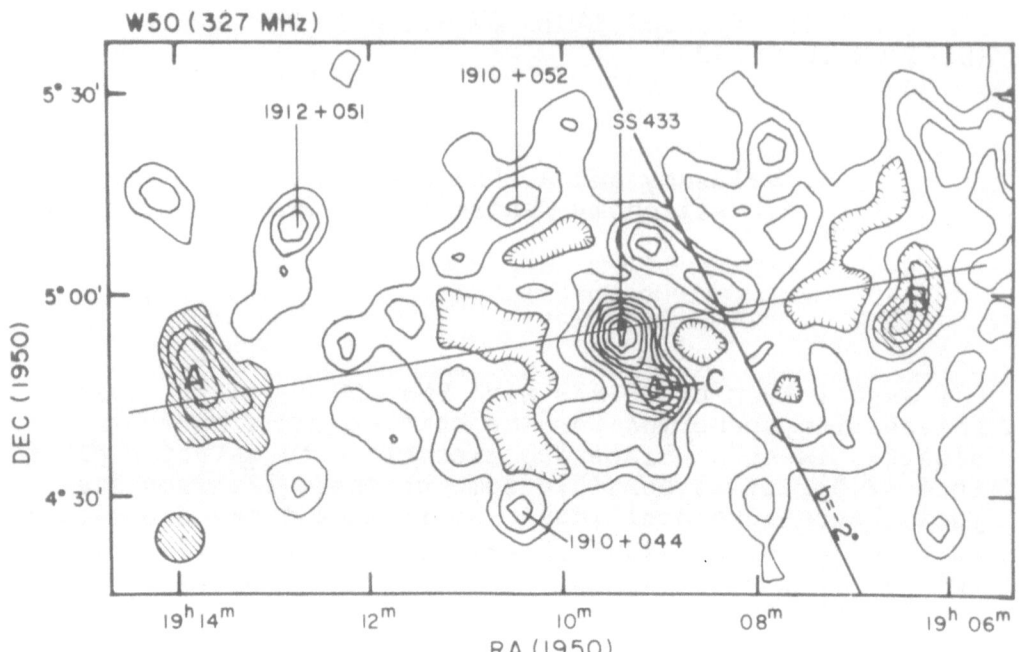

Fig.1. Map of W50 at 327 MHz. Restoring beam is
       7ʹ.5x7ʹ.5. Contour interval is 0.15 Jy/beam.
       The axis of the jet is also marked.

of this shell are not seen prominently in this map, as they
suffer from the effects of missing spacings from u ∼ 10 to
100 wavelengths. In the west, the map is confused by the
steep gradients in the galactic emission close to the plane.
A comparison of the 92cm map with the centimeter maps shows
that at meterwavelength the emission is less pronounced over
the parts of the shell where linear polarization is minimum
(e.g. along the northern shell) indicating considerable
mixture of thermal and nonthermal emission over W50 complex.

     The hatched areas marked 'A' and 'B' in Fig.1, lie
along the jet axis at PA 100⁰ (Hjellming and Johnston,
1981), and are separated by 2⁰. Their apparent resemblance
to 'Double lobes' in extragalactic sources is remarkable.
It is conceivable that these lobes are indeed parts of the
shell which have acquired excess expansion velocities due
to the dynamical effects of the beams. In X-ray also a
double lobe structure is seen (Seward et al.1980). However
the X-ray lobes are separated only by 70'. The sharp outer
boundaries of the X-ray emission suggest that the beams do
not propagate beyond about 40' from SS 433. This implies
that the radio lobes are no longer being accelerated by the
beams.

Another interesting feature near the center is the jet-like extension (C in Fig.1) towards the southwest direction. This feature is also clearly seen in the 11 and 18 cm maps (Downes et al. 1982). In our highest resolution (3'x6') map also, it appears to be connected to SS 433. It seems to have a relatively steeper spectrum than SS 433. It is unlikely that this extension is the result of the relativistic beams from SS 433, since existence of the feature 'A' and 'B' imply the beams have been well collimated and stable. It could have resulted from transient activity in SS 433. Clearly, observations with still higher resolution will be needed to confirm its association with SS 433.

We thank the staff of the Radio Astronomy Centre but for whom these observations would not have been possible.

REFERENCES

Downes,A.J.B., Pauls,T., Salter,C.J., 1981, Astron. Astrophys., 97, 296.

Downes,A.J.B., Pauls,T., Salter,C.J., 1982, Astron. Astrophys. (In press).

Geldzahler,A.J.B., Pauls,T., Salter,C.J.,1980,Astron. Astrophys., 84, 237.

Hjellming,R.M., Johnston,K.J., 1981, Astrophys.J.(Lett.), 246, L141.

Seward,F., Grindlay,J., Seaquist,E., Gilmore,W., 1980, Nature, 287, 806.

Velusamy,T., Kundu,M.R., 1974, Astron. Astrophys., 32, 375.

Zealey,W.J., Dopita,M.A., Malin,D.F., 1980, Mon. Not. R. astr. Soc., 192, 731.

RADIO INTERFEROMETER OBSERVATIONS OF COMPACT SOURCES IN SUPERNOVA REMNANTS

R.E. Spencer & R. Haley
University of Manchester,
Nuffield Radio Astronomy Laboratories, Jodrell Bank
Macclesfield, Cheshire, England

The remains of the progenitor stars of supernovae are likely to have compact radio structure and steep radio spectra, as in the cases of pulsars and SS433. An instrument which is sensitive to compact structure at low frequencies is therefore suitable for a search for new objects. The radio-linked interferometer comprising the MK IA 76m telescope and the Defford 25m telescope situated 127 km to the south has a resolution of $\sim$1 arcsec at 408 MHz and is ideal for such a task.

A compact radio source within the field of view of $\sim$1 deg^2 of the instrument gives rise to an output fringe at a delay and frequency which depends on the source position. On-line data processing produces plots of fringe amplitude versus delay and fringe frequency from which the position of sources in the sky can be measured (Lyne et al. 1982). The positional accuracy is limited by delay resolution and ionospheric variations to $\sim$1 arc min. Sources emitting $\geq$100 mJy in components $\leq$1 arc sec in angular size can be readily detected and the instrument is insensitive to extended, low brightness emission such as that from supernova remnants themselves.

Areas of sky in and around 27 remnants selected from the Milne (1979) catalogue were searched in May 1980. The larger remnants were covered by several overlapping search areas. In total 27 sources were detected with an average surface density of 0.86 $\pm$ 0.18 deg^{-2} for those stronger than 100 mJy. A proportion of the detected objects are likely to be background extragalactic sources, and so 35 areas at high ($>$10^0) galactic latitude were selected at random and searched for compact sources in the same way. The surface density of these presumably extra-galactic sources was 0.69 $\pm$ 0.16 deg^{-2} and so the slight excess of sources found in or near supernova remnants is not significant. However there may be one or two sources which are indeed galactic and associated with remnants. A list of sources which are interesting either because of unusual spectra (from previously published data) or by virtue of their position in the remnant, is given in the table.

J. Danziger and P. Gorenstein (eds.), Supernova Remnants and their X-Ray Emission, 469–470.
© *1983 by the IAU.*

Name		Position (1950)	
		R.A.	Dec
G69.0+2.7	CTB80	19 50 22	32 42
G74.9+1.2	CTB87	20 13 32	37 01
G93.6-0.2	CTB104A	21 30 56	50 50
		21 29 29	49 58
G94.0+1.0		21 24 56	51 47
G119.5+1.0	CTA1	00 14 34	72 53
		00 16 54	73 11
G160.5+2.8	HB9	05 02 14	46 50
		05 02 50	46 53
		05 03 50	46 49
		05 03 39	46 40
G166.0+2.5		05 12 19	41 48
G166.0+4.3		05 25 48	43 15
		05 27 01	43 04
		05 22 31	42 34

REFERENCES

Lyne, A.G., Anderson, B. & Salter, M.J., 1982. Mon.Not.R.astr.Soc., in the press.
Milne, D.K., 1979. Aust.J.Phys., 32, 83.

X-RAYS FROM RADIO PULSARS: THE PORTABLE SUPERNOVA REMNANTS

David J. Helfand
Columbia Astrophysics Laboratory, Columbia University

ABSTRACT

Neutron stars are the longest-lived remnants of supernova explosions. As a reservoir of thermal energy remaining from the explosion and generated by frictional coupling between core and crust, as a storehouse of magnetic and rotational kinetic energy which allows the star to act as a high energy particle accelerator, and as the source of a deep gravitational potential which can generate heat from infalling matter, neutron stars remain capable of producing high energy radiation for a Hubble time. We review here the results of an extensive survey of supernova remnants and radio pulsars with the imaging instruments on board the Einstein Observatory and discuss the implications of these results for pulsar physics and for the origin and evolution of galactic neutron stars.

I. INTRODUCTION

The concept of a neutron star was first discussed over fifty years ago. Confirmation that such stars exist is dated from the announcement in 1968 of the discovery of the first four pulsars by radio astronomers working at Cambridge University (Hewish et al. 1968). Over the past 15 years, the number of known radio pulsars has increased nearly a hundredfold, and a wealth of data on pulse intensities, polarization, spectra, and repetition rates has been accumulated. These data have been used to model the pulsar magnetosphere, probe the stellar interior, and study the origin and evolution of the galactic population of neutron stars. Yet the radio luminosities of most pulsars represent $\lesssim 10^{-5}$ of the stars' energy loss rate as measured from their increasing periods. The remaining rotational kinetic energy is converted to low frequency dipole radiation at the pulsar's spin frequency, higher frequency electromagnetic waves, and high energy particles in unknown proportions. We discuss herein a new diagnostic for the properties of isolated neutron stars: X-ray emission from radio pulsars.

X-ray astronomers had in fact been observing neutron stars as X-ray binaries for several years when the discovery of radio pulsars was announced, although it was not until the launch of the first X-ray satellite, Uhuru, that the description of these systems as a

471

J. Danziger and P. Gorenstein (eds.), Supernova Remnants and their X-Ray Emission, 471–486.

neutron star accreting matter from a normal companion became accepted. The first radio pulsar to be detected as an X-ray source was the youngest such object known - the pulsar at the center of the Crab Nebula (Fritz et al. 1969). Its X-rays are pulsed at the stellar rotation frequency and the pulse morphology is similar to that in the radio, optical, and gamma-ray spectral regimes; the X-ray to radio luminosity ratio $L_x/L_R \sim 10^6$. The only other radio pulsar detected as an X-ray source in the ensuing decade was the second-youngest pulsar known, PSR 0833-45 in the Vela supernova remnant (Harnden and Gorenstein 1973; Moore et al. 1974). Recent imaging observations have shown that the emission consists of an unpulsed X-ray point source centered on the pulsar surrounded by a ~2' diffuse X-ray nebula: $L_x/L_R \sim 3 \times 10^4$ (Harnden, this volume). Prior to 1980, the Crab and Vela were the only two radio pulsars detected outside the radio band, and they remain the only objects detected at optical and gamma-ray wavelengths.

The launch of the Einstein Observatory provided X-ray astronomers with a thousandfold increase in sensitivity to point sources emitting in the 0.1 to 4.0 keV band. With this new capability in mind, Helfand, Chanan, and Novick (1980) reviewed the prospects for detecting X-ray emission from isolated neutron stars and initiated a program of Einstein observations designed to further our understanding of radio pulsar interiors, magnetospheres, and evolution. In addition, a number of observers were surveying the X-ray emission from SNR within which young neutron stars might be found, and theorists, working on such problems as neutron star cooling, crust-core coupling, and magnetospheric emission mechanisms, were reassessing the levels of X-ray emission expected. The following is a preliminary review of the result of all the activity during the past three years.

2.   X-RAYS FROM YOUNG PULSARS

2.1   Initial Cooling - Known Radio Pulsars

At the time of its creation in the supernova event, a neutron star is likely to have a temperature $T > 10^{10}$ K. It subsequently cools via both neutrino emission from the interior and photon emission from the stellar surface. The cooling rate is sensitive to a number of factors including the central density and composition (e.g., quarks, pions, or neutrons), interior superfluidity, magnetic field strength, and crustal properties. Thus, the temperature history of a young neutron star is a sensitive probe of stellar structure and, throughout the interval 10^{-3} yr $< t < 10^4$ yr, the temperature must be measured in the X-ray band.

Among the first theoretical work on neutron star cooling was that of Bahcall and Wolf (1965) who calculated expected surface temperatures for the purpose of assessing the likelihood that the newly discovered, luminous celestial X-ray sources were in fact hot neutron stars. They concluded that temperatures would remain in the range of 10^7-10^8 K for too short a time to explain the observed population of sources. The discovery of radio pulsars stimulated considerable further work in the following decade (see Tsuruta 1979 for a comprehensive review). These results may be summarized as a prediction that, barring exotic interior compositions, neutron star surface temperatures should remain above a few million degrees for a few thousand years after birth. The only observatinal datum was an upper limit to the nonpulsed radiation from the Crab pulsar obtained during lunar occultation (Wolff et al. 1975; Toor and Seward 1977); it led to an implied temperature of $T \lesssim 3 \times 10^6$ K.

Stimulated by the prospect of considerable observational progress offered by the imaging X-ray instruments on Einstein, (Giacconi et al. 1979) several groups have readdressed the problems of neutron star cooling over the past three years (Glen and Sutherland 1980; Richardson 1980; van Riper and Lamb 1981; van Riper, this volume; Nomoto and Tsuruta 1981 and this volume). These recent calculations have included full general relativistic thermodynamics, the effect of a finite thermal conduction timescale, a more realistic treatment of the structure of the thin envelope across which the temperature gradient is steepest, new equations of state, improved treatment of superfluid and magnetic field effects and the recalculation of accelerated cooling from quarks and pions. While the predictions disagree in some of their details, the general result has been lower expected surface temperatures. A star with standard composition is expected to have $T \lesssim 2.5 \times 10^6$ by $t \sim 100$ yr and to have cooled to $T < 1.5 \times 10^6$ at $t \sim 10^4$ yr. This represents a decrease in the expected X-ray luminosity for a young star of a factor $\gtrsim 50$ and it should be emphasized that enough remaining theoretical uncertainties have been identified that a further readjustment by a similar factor is not excluded.

The chief value of the Einstein observations with regard to the neutron star cooling question has been to stimulate the aforementioned theoretical work; no definitive examples of a cooling young neutron stars have been identified, although a few noteworthy observations have been achieved. Four known neutron stars have been observed to emit X-rays, although the contribution to this emission of surface thermal radiation remains unclear. From a survey of over 70 SNR in the Galaxy, a single candidate hot neutron star has been observed and a number of significant upper limits have been set. We review these results below.

For the case of the Crab pulsar, high resolution imager (HRI - 4" spatial resolution, $\Delta E = 0.1-3.5$ keV) data show that the pulsar fades to ~1% of its peak value at phase ~0.85 from the main pulse (Harnden, this volume), a level similar to that seen in the nonthermal optical emission profile (Smith 1981). Interpreting this as an upper limit to emission for a 15km blackbody leads to a value of $T < 2.5 \times 10^6$ K, consistent with current theory. It should be noted, however, that a hot young neutron star is most unlikely to be radiating as a uniform temperature blackbody. The optical properties of the surface as well as the strong magnetic field, are likely to cause significant departure from a Planck function spectral form (Brinkman 1980); for realistic parameters, these departures should be a factor ~2 (Cheng and Helfand 1983). In addition, the magnetic field introduces an anisotropic thermal conductivity in the surface layers of the star leading to a surface temperature distribution significantly hotter at the magnetic poles than at the magnetic equator (Greenstein and Hartke 1982). This will introduce a modulated component to even a purely thermal X-ray flux, further complicating the interpretation of the data. The most conservative expression of the Crab results, then, is that they are not inconsistent with current predictions from cooling theory.

As noted in the introduction, a point source of X-rays has now been seen to be coincident with the Vela pulsar. Interpreting the total point source flux as surface blackbody emission yields a temperature of ~9 × 10^5 K for this ~10^4 year old star, again consistent with current theoretical estimates. However, this emission is not modulated at the pulsar rotation period to less than the 1% level. For the reason cited above, this is difficult to understand if the emission arises from the stellar surface. Using the modulated flux to establish a temperature upper limit leads to a value of $T \lesssim 3 \times 10^5$ K ($L_x \lesssim 10^{31}$ ergs s^{-1}), substantially below that expected unless rapid early cooling from a pion condensate core is important. A critical observation for future X-ray missions will be the determination of the spectrum of the Vela point source emission over as wide an energy range as possible to help separate nonthermal and thermal components and establish the actual temperature of the pulsar surface.

The only other known radio pulsar associated with a SNR is the source first discovered as a 150 msec X-ray pulsator in MSH 15-52 (Harnden and Seward 1982). The age of this object is somewhat uncertain, since the SNR age (~10^4 yr) and pulsar spin down time (~1500 yr) are in substantial disagreement. Nonetheless, the object is clearly young and provides another important test of neutron star cooling theory. The X-ray luminosity in the Einstein band is $L_x \sim 5 \times 10^{34}$ ergs s^{-1} and the signal is ~100% modulated with a single pulse of FWHM ~25% each period. The single radio pulse is considerably narrower, with a FWHM ~9% (Manchester et al. 1982); the relative phase of the pulses has not yet been determined. This situation is in marked contrast to the Crab pulsar for which the radio and X-ray pulse profiles are quite similar. Interpreting the total flux from this source as surface thermal emission yields $T \sim 2.5 \times 10^6$ K. Greenstein and Hartke (1982) have attempted to model the X-ray pulse profile with an inhomogeneous surface temperature distribution. They require a magnetic polar cap temperature of $T \sim 9 \times 10^6$ K and a (perhaps implausibly) high pole-to-equator temperature ratio to fit the data. If this interpretation is correct, the pulse width and modulation fix the angle between the magnetic and rotation axes at ~40°. The high temperature required, however, may imply a source of polar cap heating (§2.1) or a nonthermal emission component (§1.3) and the implications of this object for neutron star cooling theory remain unclear. Again, spectral and polarimetric data, attainable with an AXAF-scale facility, are crucial for further progress.

2.2 Initial Cooling – A SNR Search

The galactic supernova rate and pulsar birthrate are now in approximate agreement (Lyne, Manchester, and Taylor, this volume), suggesting that a significant fraction of supernovae leave behind a neutron star. Most, if not all, of these objects must be radio pulsars, although perhaps only 1 in 5 will be detectable at radio frequencies owing to the narrowly beamed emission. One anticipated result of the Einstein survey of SNR was that a number of such hot young neutron stars would be discovered. This has not turned out to be the case.

A total of 72 galactic SNR were mapped with the imaging instruments on Einstein; 33 of these were within 5 kpc and any central neutron star with $T \gtrsim 1.5 \times 10^6$ K would have been easily seen (NB – A high neutron star velocity will not remove it from the remnant's center for ~10^4 yr since the average remnant expansion velocity over this time (~500 km s^{-1}) is considerably greater than the average measured pulsar transverse velocity of ~150 km s^{-1}). A total of twelve X-ray point sources associated with these remnants were detected (see Helfand and Becker 1982 for a complete review of these data). Briefly, four are in Crab-like remnants (the Crab, Vela, 3C58, and CTB80), two are X-ray binaries (SS433 and G109.1-1.0), one is MSH15-52, discussed above, and three are seen in large angular diameter remnants and are likely to be chance superpositions of foreground or background objects (W28, PKS1209-56, and G127.1+0.5). This leaves two sources, one of which is very distant and about which little is known (Kriss, private communication), and the other of which is in RCW 103 (Tuohy and Garmire 1980). This latter source is located precisely at the center of a 1-3000 yr old remnant and has no optical counterpart brighter than m_v ~ 22, implying $L_{opt} < 10^{-2.5}$ that of the Crab pulsar optical luminosity (Tuohy et al. 1982). The limit on radio pulses is similarly ~10^{-2} that of the Crab. The source is too weak for an effective search for X-ray pulsations. A blackbody interpretation leads to a surface temperature estimate of ~2×10^6 K, consistent with standard cooling theory.

Apart from the two binary sources, RCW103 contains the only X-ray point source seen in any young (<10^4 yr) shell-type SNR. This lack of detections may be the most significant datum to emerge from the search for cooling neutron stars. Of the seven historical remnants (Cas A, Kepler, Tycho, 3C58, the Crab, SN1006, and SN185-

RCW86), only the two Crab-like objects contain point sources. For
SN1006, the limit on thermal emission is particularly severe;
$T < 5 \times 10^5$ K for a neutron star anywhere within the inner 20' of
the remnant (Winkler, private communication). For the other objects,
the limits range from $1-2 \times 10^6$ K. We are forced, then, to one of
three conclusions: 1) current cooling calculations are in error for
standard neutron star models; 2) the neutron stars in these remnants
contain exotic species which lead to faster early cooling; or 3) that
a substantial fraction of supernova do not leave neutron stars. For
the putative Type I remnants of Kepler, Tycho, and SN1006, the latter
conclusion is in agreement with the majority view amongst modellers
of SN explosions (see Wheeler 1982 for a review). For Cas A and
others of the dozens of remnants surveyed, however, conclusion 3
again raises the question of whether or not there are enough classical SN in the Galaxy to produce the observed number of radio pulsars. Further evidence on this point is presented in §4. A decision
on whether conclusion 1 or 2 are important will have to await further
observations of the known young neutron stars discussed above.

2.3 Magnetospheric Emission

An X-ray image of the Crab Nebula is dominated by the bright
point source centered on the pulsar. These X-rays are ~100%
modulated at the stellar rotation frequency and the pulse morphology
is generally similar in form and phase to the emission at radio wavelengths. Important differences exist, however, between the radio and
higher energy pulses, suggesting a different emission mechanism is
responsible. There is a break in the spectrum between the radio and
the optical/X-ray pulses and the brightness temperature of the latter
is low enough that incoherent processes suffice to explain the observed intensity. In addition, unlike the radio pulses, the higher
frequency radiation is not time variable and there is no observed
counterpart to the "giant pulses" outside the radio band.

Shklovskii (1970) was the first to suggest that incoherent synchrotron radiation far from the stellar surface was responsible for
the optical (and higher energy) pulsed radiation. Adopting this
idea, Pacini (1970) pointed out from fairly general considerations
that a pulsar's optical luminosity should scale as
$L \quad B_0^4 \; P^{-10} \quad P^{-8} \; \dot{P}^2$, where B_0 is the star's surface magnetic field
strength derived assuming that rotational braking results exclusively
from magnetic dipole radiation and a purely dipolar field geometry
applies. The discovery of 25th magnitude optical pulses from the
Vela pulsar (Wallace et al. 1977) supported this conclusion. The
ratio of the Crab (P $= 0.033$s, $\dot{P} = 4.21 \times 10^{-13}$ ss^{-1}) pulsed X-ray
luminosity to the point source component of the Vela (P $= 0.089$ s,
$\dot{P} = 1.25 \times 10^{-13}$ ss^{-1}) X-ray emission also approximately agrees with
this scaling law; the problem, of course, is that the Vela X-rays are
not pulsed. For the other young pulsar in MSH 15-52 (P $= 0.15$ s,
$\dot{P} = 1.5 \times 10^{-12}$ ss^{-1}), the X-ray emission does not scale to the
Crab X-ray luminosity following this relation, exceeding the nominal
ratio by a factor of $\sim 10^3$, although the fraction of the rotational

kinetic energy loss that emerges as pulsed X-rays is approximately the same for the two sources ($L_x/\dot{E} \sim 10^{-4}$). Unlike the Crab, however, the X-ray pulses from MSH15-52 have a shape that differs substantially from the radio pulse morphology and, as we have suggested above, may contain a significant thermal component.

The Crab pulsar, then, remains the only clear example of non-thermal magnetospheric emission in the X-ray band. Other candidates from the list of SNR/point source associations cited above include the centrally located objects in the Crab-like remnants 3C58 and CTB80, and RCW 103. All are too weak to perform a meaningful pulsation analysis; Tuohy et al. (1982) have argued against this interpretation for RCW 103 on the basis of an L_x/L_{opt} ratio which differs from the Crab by a factor of $>10^2$. It is interesting to note that in the other historical SNR, the limits on pulsar emission expressed as a fraction of the Crab luminosity are now more stringent in the X-ray band than at radio frequencies, reaching $<10^{-5} L_{Crab}$ for SN1006. Limits for the remaining sample of older remnants are typically $\sim 10^{-3} L_{Crab}$. However, the much steeper period dependence of the high energy emission suggests that this is not a very serious constraint on the population of young pulsars in SNR.

3.   X-RAYS FROM OLD PULSARS

After 10^4-10^5 yr a pular is expected to have cooled to less than a few hundred thousand degrees and to have spun down to the point where high energy magnetosphere emission is insignificant. Several other processes, however, still operate to keep the star warm enough to emit X-rays. A measurement of, or limit on, the level of this emission can thus be an important constraint on models for pulsar magnetospheres and neutron star interiors.

3.1 An X-Ray Survey of Radio Pulsars

To exploit this diagnostic, we conducted a survey of nearly two dozen radio pulsars using the imaging instruments on Einstein. The objects were chosen to be nearby ($\lesssim 1$ kpc) and/or to have properties expected to maximize a particular X-ray heating effect (e.g., low period, high magnetic field, etc.). The radio position errors were all <30" and typically $\lesssim 1$". All sources were first observed with the imaging proportional counter (IPC) for exposures of $\sim$2,000 to $\sim$20,000 s, implying flux thresholds in the 0.10-3.5 keV band of $\sim 10^{-13}$ ergs cm^{-2} s^{-1}. The log N-log S relation for IPC sources predicts $\sim$2 sources per square degree above this level. Taking a generous 90" error circle radius for the X-ray positions, we have thus searched a total of $\lesssim 0.05$ square degrees in the survey, implying that $\lesssim 1$ serendipitous X-ray source should have appeared coincident with a pulsar position by chance. A total of eight sources were detected. The one with the largest position error was subsequently identified as an AGN (Margon, Chanan, and Downes 1981) (the one statistically expected spurious source); we are confident that the remaining seven sources represent the detection of X-ray emission from radio pulsars. In addition to the three objects cited above (Crab, Vela, and MSH15-52), they include PSRs 0355+54, 0950+08, 1055-52, 1642-03, and 1929+10; their IPC counting rates are, respectively, ~0.01, 0.008, 0.10, 0.01, and 0.006 ct s^{-1}. The five older objects have spin down ages ranging from ~5 × 10^5 to ~10^7 yr; all have periods less than 0.4 s, more than a factor of 2 below the median for all pulsars.

Three of the objects were subsequently observed with the HRI to obtain higher spatial resolution and better positional accuracy. For PSR 1929+10, a source with a counting rate of 0.0007±0.0002 ct s^{-1}, centered on the radio position (±2"), was detected. The resulting IPC/HRI ratio of 8±3 was consistent with the moderate-energy IPC pulse height spectrum. We conclude that this pulsar is a point source of X-ray emission and discuss the implications below. In the case of PSR 1642-03, no source was detected in either of two long HRI exposures totally >30,000 s, one of which was taken within hours of an IPC pointing that showed a 5 σ source coincident with the radio position and unchanged in intensity from an observation 18 months earlier. The implied IPC/HRI ratio for a point source, >25, is inconsistent with the soft IPC pulse height spectrum in which $\gtrsim$ 75% of the counts fall below 1.5 keV. In this case we conclude that the source is extended on a scale of ~90", implying a surface brightness too low for detection with the HRI but consistent with a point source for the IPC. The HRI pointing for PSR1055-52 also showed evidence of extent on a scale of ~10" (Cheng and Helfand 1982). Finally, the IPC source associated with PSR 0355+54 consisted of an elongated region of emission 5' long with the pulsar at one end.

A search for pulsations was carried out for three of the sources. Barycentric corrected photon arrival times were folded at the radio pulsar period; in each case, errors in the measured values of P and $\dot{P}$ led to a phase drift of <10^{-2} P over the length of the observation when extrapolated to the epoch of the X-ray data. The limits on any modulation are <30% and <3% for PSRs 1642-03 and 1055-52, respectively (the two extended sources), and <25% for the point-like source PSR 1929+10.

In addition to the detections, we obtained stringent upper limits on ~20 other pulsars with the IPC. In particular, observations of 10 sources lying within 500 pc led to X-ray luminosity limits of between 0.5 and 5 × 10^{30} ergs s^{-1}, implying blackbody temperature limits of between ~3 and 5 × 10^5 K. Below, we outline briefly the implications of these point source limits and the detections of PSR 1929+10 and Vela for various pulsar heating models.

3.2 Heating Mechanisms

The radio pulse emission mechanism and the structure of the

pulsar magnetosphere are problems which have proven remarkably resistant to solution over the past 15 years. A currently popular class of models predicts significant heating of the magnetic polar cap by electrons or positons flowing back toward the star after having been accelerated in the pair production discharge of the outer magnetosphere (Cheng and Ruderman 1980; Arons 1981 [hereafter CR and AS]). The predicted X-ray luminosities from the hot polar caps ($\sim 10^{10}$ cm^2) range from 10^{26}-10^{31} ergs s^{-1} depending on such parameters as the pulsar's period, radius, and surface magnetic field strength. This emission should be modulated at the pulsar rotation period, although the depth of the modulation depends on the thermal conductivity of the stellar surface and the relative orientation of the rotation and magnetic axes (Greenstein and Hartke 1982). For PSR 1929+10, the models predict $L_x \sim 3 \times 10^{28}$ (CR) and $L_x \sim 8 \times 10^{28}$ (AS) ergs s^{-1}. Adopting a distance of 65 pc, the X-ray data imply a luminosity of $L_x \sim 6 \times 10^{28}$ ergs s^{-1}; the limit on the modulated flux yields $L_x \lesssim 2 \times 10^{28}$ ergs s^{-1}. These values are not inconsistent with the predictions (which have come down by 2-3 orders of magnitude over the past few years), but imply that little or no pair production can occur in the outer electrostatic gap of the AS picture. For Vela, the predictions are $L_x \sim 2 \times 10^{31}$ (CR) and 2×10^{30} (AS) ergs s^{-1}. The observed luminosity from the point source is $\sim 3 \times 10^{32}$ ergs s^{-1} although the modulated component is $\lesssim 1\%$ of this value. The RS model requires synchrotron cooling of the backflowing electrons to meet this latter limit. Perhaps the most difficult problem these models face for Vela, however, is the fact that the modulated X-ray flux is $< 10^{-4}$ of the gamma-ray flux which is thought to arise from one sign of the cascading pairs as the other flows back to heat the stellar surface. A solution may be found either through alternative magnetospheric models or from a detailed analysis of the heat transport in the stellar crust which might spread the energy of the cap's bombardment more evenly over the star, reducing the X-ray modulation.

The magnetic field of a neutron star is tied to the solid crust, and the electromagnetic braking torque acts directly on this outer component to slow the rotation. The core is largely superfluid neutrons, and the details of the boundary between these two components has been a matter of considerable interest. The strength of the coupling will determine the degree of differential rotation expected and, in turn, the amount of frictional heating which will occur at the crust/core interface. One critical question is whether or not the superfluid vortices of the core pin to crustal nuclei. A calculation of heating under the assumption of no pinning has been carried out by Harding et al. (1978). The predicted temperatures are primarily a function of stellar mass owing to the changing ratio of crust and core mass fractions; less massive stars have much thicker crusts implying lower heating rates from the more rapidly spinning core. The X-ray flux from PSR 1929+10, if interpreted as emission from a blackbody of radius 15 km, implies a temperature of $\sim 2 \times 10^5$ K (the spectrum is inconsistent with such a low temperature, however). This limit on the thermal luminosity of this source as well as the undetected objects, would imply masses of <0.5 $M_\odot$ for these pulsars. This is to be contrasted with the observed neutron star masses in a number of X-ray and radio binary pulsars which cluster around 1.4 $M_\odot$. A more likely conclusion is that some vortex pinning does occur. Recently, Alpar et al. (1983) have investigated the question in detail and have calculated heating rates that depend on P, $\dot{P}$ and the properties of the boundary layer. The only two pulsars in our survey that should be detectable under these assumptions of fairly strong pinning are PSR 0950+08 and PSR 1929+10, both of which were seen; these were also the only two sources for which no evidence of spatial extent are adduced. The X-ray luminosity of these sources constrains the details of the pinning model, offering an important new diagnostic for the interior dynamics of neutron stars (Helfand 1983).

Another potential source of heat in all neutron stars is the release of energy accompanying the sudden speed up in rotation rate observed as a "glitch." Vela is by far the most active source in this regard with a total of six events ($\Delta P/P \sim 10^{-6} - 10^{-7}$) in the last 12 years. The Crab has undergone two or three much smaller spinups ($\Delta P/P \sim 10^{-9}$) and Vela-sized glitches have been observed once each in three older sources (see Alpar and Ho 1983 for a recent summary). In addition, nearly all pulsars exhibit some degree of "timing noise" indicating small irregular changes in rotation rate (e.g., Helfand et al. 1980). The amount of energy released in these events depends critically on the mechanism responsible. An early model for the Crab (and Vela) glitches postulated sudden cracks in the neutron star crust (or core) with an accompanying release of gravitational and strain energy amounting to $10^{40} - 10^{45}$ ergs per event (Ruderman 1969; Baym and Pines 1971; Pines et al. 1972). More recently, a scenario involving the sudden unpinning of the superfluid vortex lines has been advanced (Anderson and Itoh 1975; Alpar et al. 1981); such an event would release considerably less energy than a starquake (only $\sim 10^{39}$ ergs for a Vela sized glitch).

To the extent that energy released in a glitch is thermalized and eventually radiated away in photons from the stellar surface, the temperature limits imposed by these X-ray data constrain glitch models. For example, in PSR 1929+10, the observed luminosity of $\sim 6 \times 10^{28}$ ergs s^{-1} implies an energy input from glitches of <2×10^{36} ergs yr^{-1}. If each event deposits a total energy of $\gtrsim 10^{39}$ ergs, the events must be separated by $\gtrsim 500$ yrs. This is approximately the observed rate for old pulsars (Alpar and Ho 1983). For Vela, the X-ray data allow a choice between the two models. The total point source X-ray luminosity is $\sim 3 \times 10^{32}$ ergs s$^{-1} \sim 10^{40}$ ergs yr^{-1}. There has been one glitch every two or three years since the pulsar's discovery and thus, to balance the radiated energy there must be <3×10^{40} ergs per event released as thermal energy in the star. This is only 10^{-4} of the total energy associated with a corequake but is consistent with the unpin-

ning picture. Coupled with the evidence in favor of a pinned superfluid component derived from the crust-core coupling arguments
presented above, these data offer strong support to the emerging
picture of the superfluid behavior in pulsar interiors and the interaction of this component with the remainder of the star.

4.   X-RAY SYNCHROTRON NEBULAE

Left to consider is the detection of extended X-ray emission
from radio pulsars. The Crab Nebula is, of course, the archetype of
an extended region of synchrotron radiation driven by particles accelerated by a central pulsar. Before the launch of Einstein, it was
the only such object known to emit X-rays, although several other
radio supernova remnants had been identified as analogous objects
from their centrally peaked brightness distributions, flat radio
spectra, and high linear polarization. Several of these Crab-like
remnants have now been detected as X-ray sources (Wilson 1980; Becker
et al. 1982) and in two cases, a central point source (presumably the
driving pulsar) is seen. In addition, we have detected such nebulae
around six radio pulsars including the Crab, Vela, MSH15-52, PSRs
1055-52, 1642-03, and 0355+54. They range in age from ~10^3 to
~10^6 years, and in luminosity from 2×10^{37} ergs s^{-1} to
6×10^{31} ergs s^{-1}. Seward (the volume) has discussed the younger
three sources and a detailed argument concerning the case for synchroton nebula around PSR1055-52 has been given by Cheng and Helfand
(1983).

A general scenario for the origin and evolution of these nebula
can be found in Helfand (1983). Briefly, we adopt the pair production discharge magnetospheric models cited above and use the observed
intensity and spectrum of the pulsed gamma-rays from the Vela pulsar
to define the parameters of the particles injected into the surrounding magnetic field. The nebular field strength is constrained by two
considerations: 1) the maximum potential drop (and, thus, the
highest particle energy) which can occur in the magnetosphere coupled
with limits on diffuse gamma emission from the vicinity, and 2) by
the reflection of the turnover of the particle spectrum implied by
the lack of pulsed X-ray emission in the turn over in the diffuse
radiation spectrum between the X-ray and optical regimes (i.e., no
optical nebula is seen). The size of the nebula is derived from the
requirement of pressure balance between the expanding bubble of
particles moving outward at the Alfven speed and the thermal pressure
of the surrounding medium (Blanford et al. 1973). The observed
properties of the Vela nebula are well matched by this picture;
extrapolating to lower fields and particle luminosities can yield
substantial agreement with the sizes and luminosities of the older
pulsar nebulae as well.

The detection of X-ray synchrotron nebulae around a number of
radio pulsars provides important new information on magnetospheric
particle acceleration models, gamma-ray emission mechanisms, evolu-

tion of the particle flux, and the contribution of pulsars to the electron cosmic ray and diffuse gamma-ray backgrounds. Most importantly, however, the detection of every object above the value of BP^{-2} at which the outer gap particle acceleration mechanism is supposed to cease (and the nondetection of every source below this value) strongly suggests that X-ray synchrotron nebulae are a necessary consequence of an active young pulsar. This conclusion holds important implications for the pulsar/SNR association problem. These nebulae are almost certainly isotropic and, as such, should be visible within any SNR which contains an active pulsar. The nebula surrounding PSR 1055-52 (a $\sim 5 \times 10^5$ yr old source) is sufficiently luminous to have been detected in any of the many remnants surveyed out to 4 kpc, and a Vela-type nebula ($t \sim 10^4$ yr) could have been seen to twice that distance. The only new synchrotron nebula found, however, was the one surrounding the newly identified pulsar in MSH15-52. We are thus forced to conclude either 1) that the majority of SN which produce remnants do not create neutrons stars, or 2) that the pulsar phenomenon does not turn on in at least some sources for $>10^4$ yr (i.e., after the SNR is dissipated or the pulsar has moved away). If we adopt 1) we must search for new sites of neutron star formation; if 2) is correct, it could tell us much about the pulsar emission process (Radhakrishnan and Srinivasan 1980) and/or the generation of neutron star magnetic fields (Blanford et al. 1982).

5.   SUMMARY

We have presented a review of the X-ray data currently available on the longest-lived remnant of a supernova explosion - the neutron star - and have summarized the implications these data hold for problems in pulsar physics and neutron star evolution. There is still no definitive detection of thermal emission from the surface of a young neutron star, although the few candidate examples and stringent upper limits now available have led to significant advances in cooling theories. The Crab pulsar remains the only clear example of nonthermal magnetospheric X-ray emission and the bewildering array of behavior seen in other young pulsars and Crab-like remnants suggests further work on the high energy pulse emission process is needed. Several older radio pulsars have been detected for the first time outside the radio band; the pulsar survey observations offer important new constraints on models of polar cap heating, crust-core coupling and glitches. The unexpected discovery of synchrotron nebulae around a number of objects offers new information on the evolution of particle acceleration in the magnetosphere and holds important implications for the origin of galactic neutron stars.

The author acknowledges the support of the National Aeronautics and Space Administration under Contract NAS 8-30753. This is Columbia Astrophysics Laboratory Contribution No. 235.

6. REFERENCES

Alpar, A., Anderson, P. W., Pines, D., and Shaham, J. 1981,
     Astrophys. J. (Letters), 249, p. L29.
Alpar, A., Anderson, P. W., Pines, D., and Shaham, J. 1983,
     Astrophys. J., submitted.
Alpar, A., and Ho., C. 1983, Mon. Not. R. Astron. Soc., in press.
Anderson, P. W., and Itoh, N. 1975, Nature, 256, p. 25.
Arons, J. 1981, Astrophys. J., 248, p. 1099.
Bahcall, J.N., and Wolf, R.A. 1965, Astrophys. J., 142, p. 1254.
Baym, G., and Pines, D. 1971, Ann. Phys., 66, 816.
Becker, R.H., Helfand, D.J., and Szymkowiak, A.E. 1982, Astrophys.
     J., 255, p. 557.
Blandford, R.D., Applegate, J.H., and Hernquist, L. 1982, preprint.
Blandford, R.D., Ostriker, J.P., Pacini, F., and Rees, M.J. 1973
     Astron. Astrophys., 23, p. 145.
Brinkman, 1980, Astron. Astrophys., 82, p. 352.
Chanan, G.A., Margon, B., and Downes, R.A. 1981, Astrophys. J.
     (Letters), 243, p. L5.
Cheng, A.F., and Helfand, D.J. 1983, Astrophys. J., in press.
Cheng, A.F., and Ruderman, M.A. 1981, Astrophys. J., 235, p. 576.
Fritz, G., Henry, R.C., Meekins, J.F., Chubb, T.A., and Friedman, H.
     1969, Science, 164, p. 709.
Giacconi, R. et al. 1979, Astrophys. J., 230, p. 540.
Glen, G., and Sutherland, P. 1988, Astrophys. J., 239, p. 671.
Greenstein, G., and Hartke, G.J. 1983, Astrophys. J., submitted.
Harding, D., Geyer, R.A., and Greenstein, G. 1978, Astrophys. J.,
     222, p. 991.
Harnden, F.R., and Gorenstein, P. 1973, Nature, 241, p. 107.
Harnden, F.R., and Seward, F.D. 1982, Astrophys. J. (Letters), 256,
     p. L45.
Helfand, D.J. 1983, in preparation.
Helfand, D.J., and Becker, R.H. 1982, Nature, submitted.
Helfand, D.J., Chanan, G. A., and Novick, R. 1980, Nature,
     283, p. 337.
Hewish, A., Bell, S.J., Pilkington, J.D.M., Scott, P. F., and
     Collins, R.A. 1968, Nature, 217, p. 709.
Manchester, R.N., Tuohy, I.R., and D.'Amico, N. 1982, Astrophys. J.
     (Letters), 262, p. L31.
Moore, W.E., Agrawal, P. C., and Garmire, G. 1974, Astrophys. J.
     (Letters), 189, p. L117.
Nomoto, K., and Tsuruta, S. 1981, Astrophys. J. (Letters), 250,
     p. L19.
Pacini, F. 1971, Astrophys. J. (Letters), 163, p. L17.
Pines, D., Shaham, J., and Ruderman, M.A. 1972, Nature, Phys. Sci.
     237, p. 83.
Radhakrishnan, V., and Srinivasan, G. 1980, J. Astron. Astrophys.
     1, p. 25.
Richardson, M.B. 1980, Ph.D. dissertation, SUNY, Rochester.
Ruderman, M.A. 1969, Nature, 223, p. 447.
Shklowskii, I.S. 1970, Astrophys. J. (Letters), 159, p. L77.

Smith, F.G. 1981, Pulsars, ed. W. Sieber and R. Wielebinski, IAU
    symposium 95, (Reidel) p. 221.
Toor, A., and Seward, F.D. 1977, Astrophys. J., 216, p. 560.
Tsuruta, S. 1979, Physics Reports, 56, p. 237.
Tuohy, I.R., and Garmire, G. 1980, Astrophys. J. (Letters), 239,
    p. L107.
Tuohy, I.R., Garmire, G.P., Manchester, R.N., and Dopita, M.A.
    1982, preprint.
van Riper, K.A., and Lamb, D.Q. 1981, Astrophys. J., 244, p. L13.
Wallace, P. T. et al. 1977, Nature, 266, p. 692.
Wheeler, J.C. 1982, in Supernovae: A Survey of Current Research,
    ed. M.J. Reese and R.J. Stoneham (Dordrecht: Reidel), p. 167.
Wilson, A. S. 1980, Astrophys. J. (Letters), 241, p. 19.
Wolff, R.S., Kestenbaum, H. L., Ku, W. H.-M., and Novick, R. 1975,
    Astrophys. J. (Letters), 202, p. L77.

DISCUSSION

WINKLER: How bright would you expect the synchrotron nebula around the Vela pulsar to be optically? Shouldn't it be detectable?

HELFAND: Extrapolating from 0.2 keV to 2 eV using a Crab-like spectrum (similar to the observed X-ray power law spectral index) yields an integrated magnitude of $m_v \sim 17$ or $m_v \sim 27$/sq arcsecond. However, a turnover in the injected electron spectrum may well leave it substantially below this value. We have proposed for time at Cerro Tololo to observe several of these X-ray nebulae for optical counterparts.

BENFORD: Your analysis is thoroughly married to the Cheng-Ruderman and/or Arons-Scharlemann models. These have had to recant and adjust their pulsed X-ray luminosity by 1,000 to escape the constraints of observation. Also, it seems outer gaps might have trouble giving sharp pulses in X-ray and gamma-ray. In view of this, shouldn't you consider other ways to connect with the nebular parameters - say a larger B?

HELFAND: Yes, but I think the essential point is that particules accelerated by the pulsar are producing an X-ray synchrotron nebula. I have presented one possible more detailed scenario, but I am sure there are others. The one point in favor of gap models (or at least consistent with them) is that pulsars with values of $B_{12}P^{-2} \lesssim 10$ do not produce such nebulae whereas all those with $B_{12}P^{-2} > 10$ do, as the theory would predict.

SALVATI: a) In the Vela X-ray Nebula, the emitting particles must be reaccelerated. Then one should expect no relation between the nebular spectrum and the gamma-pulsar spectrum.
b)   As for PSR 0355, you postulate very high energy particles, capable of emitting in the gamma-ray range before leaving the pulsar neighborhood. Has the gamma-ray emission been observed?

HELFAND: a) I am not sure there should be no relation, but I agree I have oversimplified things considerably.
b)   My scenario does indeed require that all of these pulsars with nebulae should be gamma-ray sources and none but the Crab and Vela have been detected. The predicted fluxes are a factor of ~10-100 below the COS-B upper limits for PSR 1055-52, PSR 0355-154 and PSR 1642-03 but should be detectable with GRO.

MANCHESTER: Some models suggest that the gamma-ray pulses from the Vela pulsar are generated by the inverse Compton radiation. If this is true, what would be the implications for your model?

HELFAND: I have been chastened before by one of my colleagues to refer to my description of these nebulae as a "scenario" rather than a model and have taken the suggestion to heart. As Dr. Salvati has

already mentioned, the requirement for particle reacceleration in the
Vela Nebula (but not, I might note, for the other sources), may well
destroy any relationship between the pulsed gamma-ray and X-ray spectra; an alternative to curvature radiation as the production mechanism for the gamma-ray pulses would do the same.

ON THE ASSOCIATION BETWEEN PULSARS AND SUPERNOVA REMNANTS

V. Radhakrishnan and G. Srinivasan
Raman Research Institute
Bangalore-560080, India

The true nature of the association between pulsars and supernova remnants has remained an intriguing and poorly understood problem even after all these years of research on them. We attempt in this review to marshal all the evidence one has on this question, and to see what conclusions we can draw.

The idea that there should be an association at all owes its origin to Baade and Zwicky (1934) who advanced the view that supernovae represent the transitions from ordinary stars into neutron stars - a new and revolutionary concept at that time, just two years after the discovery of the neutron. Following their suggestion, all theories of supernovae for many years afterwards involved a neutron star as an essential member of the cast, and one capable in principle of releasing up to 10^{53} ergs of gravitational binding energy at the time of its formation. But the details of how a part of this energy could be coupled to the infalling envelope of the star to arrest its collapse, and to accelerate it outwards to velocities of thousands of km s^{-1}, have remained a major problem. It is only now that plausible scenarios are being advanced in which the neutron star collapses to greater than nuclear densities and rebounds transmitting energy to the envelope (see Brown et al. 1981 and references therein).

After pulsars were discovered, and when it was suspected that they had to be spinning neutron stars, it was predicted (Woltjer 1968) that they should be found in SNRs, and in particular that there should be one in the Crab Nebula. It should be remembered, however, that in the case of the Crab Nebula, the prediction was motivated at least as much by the need to have a central engine to explain its continuing activity (Pacini 1968), as to find a natural birth place for neutron stars. The discovery of a pulsar in the Vela supernova remnant within months (Large et al. 1968) and one in the Crab soon after (Staelin and Reifenstein 1968), seemed at the time to have answered several questions all at once.

J. Danziger and P. Gorenstein (eds.), Supernova Remnants and their X-Ray Emission, 487–494.
© 1983 by the IAU.

The extraordinary thing is that the total number of firm associations with supernova remnants remained at these two for over a decade or
so although the total number of pulsars and supernova remnants mounted
up with time. The fact that no further associations were found in spite
of repeated searches was attributed to selection effects like distance,
dispersion, interstellar scattering, multi-path propagation, small
beaming angle, etc., all of which when put together, "satisfactorily"
explained why no more pulsars were seen in SNR. (see Manchester and
Taylor 1977). This can also be seen by studying the distribution of
observed pulsars and supernova remnants with distance from the Sun. All
of the selection effects can be thought of as combining to cut down the
distance within which pulsars will be picked up in searches over areas
of the sky which may include supernova remnants. Assuming that all SNRs
have a pulsar in them, it can be shown that the number of these pulsars
expected to be found is of the order of one, quite consistent with the
associations we have had.

But apart from the absence of more associations, there are important differences even between the two known cases, the significance of
which needs to be understood. We mentioned earlier that in most theories
of supernovae, it was the binding energy of the neutron star that was
somehow responsible for the motion of the ejecta which subsequently
formed the remnant. In the case of the Crab, the optical filaments
appear to be the only ejecta ($\sim 1\ M_\odot$) whose present velocities - with
some little acceleration since 1054 AD - were mainly acquired at the
time of the explosion. These filaments seem to lie on the periphery of
a bright and centrally concentrated nebula now believed by all to have
been created by the pulsar over a period since its formation. There is
no canonical shell around this object although such shells are commonly
found in both younger and older remnants, most of which have a hollow
interior as seen in the radio.

If it is really part of the binding energy of the neutron star
which powers the supernova remnant, then there are clearly two completely different ways in which this happens. If energy is imparted to the
envelope at the time of the explosion, then it presumably results in a
shell remnant whose characteristics will depend on the mass and initial
velocity, and whose development will be governed by the density of the
medium into which it is expanding. If the energy stored in the rotation
of the neutron star is released subsequently through its functioning as
a pulsar, then a filled remnant similar to the Crab nebula will be
formed, whose characteristics will depend upon the initial spin rate and
field of the pulsar as shown by Pacini and Salvati (1973), and whose
evolution must also be affected by the interstellar medium.

That the Crab Nebula, although special, is not unique, has been
argued by Weiler (and others) in a series of papers over many years.
There are a handful of other such objects in the galaxy, with similar
characteristics, all presumably powered by a fast spinning pulsar
inside (see e.g., Wilson and Weiler 1976; Weiler and Wilson 1977;

Weiler and Shaver 1978; Weiler and Panagia 1978; Weiler 1978; Caswell 1979; Weiler and Panagia 1980; and Wilson 1980).

When the Vela pulsar was discovered by Large et al. (1968) it was immediately associated with the large shell remnant whose age as deduced by the standard method was in reasonable agreement with the characteristic age of the pulsar ($P/2\dot{P} \sim 12,000$ years). The rough agreement in the rotation measures obtained for both pulsar and SNR was taken to add further support to the association although the dispersion measure of this pulsar when coupled with the average electron density (0.03 cm^{-3}) in the interstellar medium would have led to a distance (2.3 kpc) four to five times that estimated independently for the SNR. This discrepancy is now understood as due to the presence of a large HII region in the line of sight.

The major difference from the case of the Crab is the form of the associated SNR which clearly calls for an explanation. The position of the pulsar, unlike that in the Crab, is substantially displaced from the apparent centre of the large shell remnant and is in fact located in a centrally concentrated nebulosity at the edge of the shell. It has been pointed out by Weiler and Panagia (1980) that the spectral index, general form and polarization characteristics of this nebulosity were reminiscent of the Crab nebula, and very different from that of the shell. It should also be noted that if the pulsar had been born close to the centre of the present shell, it would require an unlikely space velocity in excess of 1000 km s^{-1} to reach its present position, or the shell would have had to expand in a very non-uniform way. That the nebulosity at the position of this pulsar is not just a piece of the shell against which the pulsar is superposed on the sky, is confirmed by the non-thermal X-radiation from it, which resembles that of the Crab nebula. Thus the association of the pulsar with the radio and X-ray source surrounding it seems beyond doubt, while the connection with the shell remnant is less well established. (Weiler and Panagia 1980; Radhakrishnan and Srinivasan 1980).

If we assume that the Vela *shell remnant* is also definitely associated with the pulsar, this will imply that both features of this remnant can be produced by the same neutron star, with the energy to one being delivered at birth, and to the other over a characteristic time determined by the initial period and field of the pulsar. Observational evidence in support of such duality has been (until recently) very meagre however. Only one other shell remnant (G326.3-1.8) shows a radio concentration within the shell, and the half-a-dozen or so remnants like 3C58 display the fried-egg morphology of the Crab, and a similar lack of any shell around them. This led to the hypothesis advanced by Radhakrishnan and Srinivasan (1980) that the two types of remnants were perhaps exclusive. Following Ostriker and Gunn (1971), we proposed that the energy for the expansion of shell-type remnants came from the rotational energy of those neutron stars which (for some reason) did not function as pulsars at birth; if they did, they would produce pulsar - nebulae (plerions) like the Crab.

Our contention was that the neutron stars that should be there in the young classical shell remnants like Cas A, Kepler and Tycho, did not function as pulsars at all. The absence of emission from a central nebula in either radio or X-ray supported our hypothesis, which however left several awkward questions unanswered. One of them was why very little ejected mass is observed associated with bright plerions like the Crab and 3C58. This is not easily explained, unless strong pulsar activity at birth is itself associated with a very low mass for the envelope. We shall return to this possibility at the end of this review.

While the absence of X-ray nebulae within most shells supported the picture mentioned above, the absence of point X-ray sources in most SNRs was a source of embarrassment. If neutron stars were left behind by every supernova explosion as we had assumed, some fraction of them should have been detectable by their thermal emission in X-rays. The fact that very few such compact X-ray sources were found, either meant that there were no neutron stars in most remnants, or that neutron stars cooled much faster than generally believed as Radhakrishnan and Srinivasan (1980) had suggested. (See discussion by Helfand in this volume).

The former point of view was taken by supporters of theories of supernova explosions which did not leave behind any neutron star. It was claimed that given the right conditions, some stars could suffer thermo-nuclear detonation in their degenerate carbon cores and disintegrate completely leaving no collapsed remnant. It was suggested that these were type I supernovae associated with stars in the mass range 1.4-6 $M_\odot$; stars of greater mass caused type II supernovae which left behind neutron stars which could function as pulsars. (See Tinsley 1977 and references therein). Lack of the expected amount of iron peak elements in the spectra of type I supernovae, was one of the observational facts that did not support this point of view. Another was the high mass ($\sim$ 15 $M_\odot$) estimated for the shell of Cassiopeia (Fabian et al. 1980) which apparently has no pulsar in it, and the low mass ($\sim$ 1 $M_\odot$) found around the Crab pulsar.

A different point of view from which this problem can be looked at, is in regard to the birthrates of supernovae, supernova remnants and pulsars. Starting with SN, the best estimate from extragalactic observations is one in 20 years (Tammann 1977) with an uncertainty of a factor of two. A similar estimate has also been made by Clark and Stephenson (1977) who studied historical records and arrived at a not too different figure of 1 in 30 years or less. One can also derive a birthrate for SNRs, which should be the same, unless a fraction of SN did not leave remnants. This exercise has been done in various ways, leading to estimates of one in 50 $\pm$ 25 y (Ilovaisky and Lequeux (1972), $\sim$80 y (Caswell and Lerche 1979), and $\sim$ 150 y (Clark and Caswell 1976). A recent attempt by Srinivasan and Dwarakanath (1982a) gives 1 in 25 y, and a similar number has been derived by Higdon and Lingenfelter (1980). Both papers have argued that the ages of shell SNRs have been much overestimated previously. (Even more recently, Mills - this volume - has derived a birthrate of 1 in 30 years). It is seen

that within the errors, all these recent estimates of the birthrate of
SNRs agree with the occurrence rate of SN, and for our purposes we shall
assume a birthrate of 1 in 25-30 years for both.

The birthrate of pulsars is a more touchy number, involving as it
does, numerous selection effects both known and unknown. Among the
well-known estimates in the literature are those of Taylor and Manchester
(1977), Phinney and Blandford (1981) and Lyne (1981) all of the order of
1 in 10 years, or more frequently. However, a more recent estimate
obtained by studying the current in the P-$\dot{P}$ diagram and allowing for
luminosity selection effects gives 1 in 21^{+6}_{-5} y (Vivekanand and Narayan
1981), where the errors are 95% confidence limits. This last estimate
of the pulsar birthrate is significantly lower than previous ones and
is, within the errors, compatible with the SNR birthrate adopted above.

All of the above estimates of the pulsar birthrate assume a beaming
factor of 5, and this has long been a weak point frequently pounced upon
by theorists. The observational astronomers, working with longitude
scans, and resigned to the impossibility of obtaining a latitude scan
through pulsar beams, have become accustomed to thinking of them as
circular. Very recently, however, the equivalent of a latitudinal scan
has been achieved by a statistical analysis (Narayan and Vivekanand
1982) of the polarization angle sweep obtained from new high quality
observations (Backer and Rankin 1980). The surprising result of this
analysis is that pulsar beams are found to be elongated in latitude by a
factor of $3^{+0.5}_{-0.4}$ over the longitudinal width, but that the elongation
appears to be a function of the pulsar period. Because of this
dependence on period, the effective value of the average beaming factor
is somewhat uncertain; but whatever the magnitude of the change from
the previously assumed value, it can only decrease the birthrate.
Incorporating a new determination of the distribution of dispersing
electrons (Vivekanand and Narayan 1982), the birthrate estimate is
expected to be even further reduced.

While all these developments reduce the urgency of manufacturing
neutron stars in less spectacular events than supernovae, it would still
appear that practically every SNR has a neutron star in it. In any case,
even if only some SNRs have neutron stars inside, why do we not see
evidence of their activity as radio and X-ray nebulae around them? As
pointed out by Radhakrishnan and Srinivasan (1980) this is independent
of beaming angle and viewing geometry. We are thus forced to conclude
that the majority of neutron stars that may exist in shell remnants are
not functioning like the Crab and Vela pulsars. This could be due to an
absence of particle and field production as we conjectured, or perhaps
because of a greatly reduced output due to slower initial rotation of
the pulsar. The possibility that shell-type remnants leave behind a
slow pulsar has also been suggested by Weiler (1978) in a discussion
which however involved the different types of supernovae.

The most interesting new pulsar-supernova association is the
recently discovered PSR 1509+58 in the shell remnant MSH 15-52 (Seward

and Harnden 1982). These authors discovered both X-ray pulses from this object and also an X-ray nebula around it. Radio measurements (McCulloch et al., 1982; Manchester et al. 1982) have confirmed that this is indeed a regular pulsar, and not an accreting X-ray system, thus making it the third firm PSR-SNR association. From the strength of the X-ray nebula around the pulsar it has recently been concluded that this pulsar must have had an "intermediate" value of initial period of $\gtrsim 70$ ms (Srinivasan et al. 1982). Otherwise, given the exceptionally high magnetic field of this pulsar (~ 4 B_{Crab}), a short initial period would have resulted in an optical and X-ray nebulosity even more spectacular than the Crab nebula.

That objects like the Crab nebula are rare, with a birthrate of ~ 1 in 350 years, can be shown simply by considering the age, luminosity and expected evolution of the nebula, together with the number of such objects seen in the galaxy (Srinivasan and Dwarakanath 1982b). But because it represented 50% of the known associations for over a decade, the temptation to use it as a proto-type to understand the properties of pulsars in supernova remnants has been irresistible. Much of the difficulty we have had in understanding these objects can, we believe, be attributed to this reason.

Completely independent evidence, not connected with SNRs in any way, that the initial periods of the majority of pulsars may be much longer than generally believed was provided by the same analysis of the current in the P-$\dot{P}$ diagram referred to earlier (Vivekanand and Narayan 1981). This analysis suggested that the majority of pulsars were either born with, or turned on at, a period of around 0.5 seconds. This estimate of the "injection" into the P-$\dot{P}$ diagram has been refined somewhat taking into account new selection effects (Vivekanand et al. 1982) and the latest figure is $\gtrsim 50\%$. This result is certainly in accord with the absence of central concentrations in shell-type remnants, even though there may be neutron stars within. But it cannot be used to choose unambiguously between a long period for the neutron stars at birth, or an interval of quiescence after birth until they have slowed down through dipole radiation.

We return finally to a question concerning the Crab which was touched upon earlier but not elaborated. It was mentioned that one of the extraordinary features of this nebula is that there is no shell around the object. The possibility that a greater amount of mass (than seen in the filaments) was ejected, but is invisible because of expansion in a very low density medium, has been suggested by several authors (Chevalier 1977; Murdin and Clark 1981). The latter authors have found a weak optical halo surrounding the nebula, and have argued that this may originate in such a shell. However, Wilson and Weiler (private communication, Weiler) find no radio emission to support this interpretation. Pending X-ray observations of the optical halo found by Murdin and Clark (1981), it appears to us that the evidence for any substantial amount of ejected matter beyond the present boundary is weak.

If the mass ejected were in fact low, it is very tempting to think that it might be connected with the powerful activity of the pulsar. We have already discussed the need for a short spin period at birth to produce the particles and field that would form a bright nebula. If a high spin-rate at birth was, say, a consequence of a low mass for the envelope to which the core would be magnetically connected during collapse, we might have an explanation.

To speculate further, it should be noted that if such a mechanism could be shown to work, the envelope would extract a fraction of the rotational energy of the neutron star, which would otherwise reside in it for later use; also, the more massive the envelope, the greater the fraction. It is not clear whether a low-mass envelope accelerated in this fashion would fragment, as the mass in the Crab nebula seems to have done; but the kinetic energy can certainly be accounted for without requiring any 'bounce' at all. As an illustration, if angular momentum conservation had given the Crab pulsar an initial period of 10 ms, then slowing it down to 16 ms provides the $\sim 10^{50}$ ergs of kinetic energy now found in the ejecta.

To summarize, all the facts put together seem to point to the following conclusions:

1. In spite of the errors in all such estimates, it seems reasonable to believe that the birthrate of pulsars is not higher than the birthrate of SNRs.

2. If all SN leave behind neutron stars, there is no need to look for alternative ways of producing them.

3. If there are neutron stars inside shell remnants, the majority were either born spinning slowly, or are not functioning as pulsars even if spinning fast.

4. The Crab phenomenon is relatively rare and not a prototype of either young pulsars or young supernova remnants.

REFERENCES

Baade, W., and Zwicky, F. 1934, Proc. natn. Acad. Sci. U.S.A., 20, 254.
Backer, D.C., Rankin, J.M. 1980, Astrophys. J., 42, 143.
Brown, G.E., Bethe, H.A., Baym, G. 1981, preprint (NORDITA-81/17).
Caswell, J.L. 1979, Mon. Not. R. astr. Soc., 187, 431.
Caswell, J.L., Lerche, I. 1979, Mon. Not. R. astr. Soc., 187, 201.
Chevalier, R. 1977, Supernovae, ed. D. Schramm, Reidel, Dordrecht, p.53.
Clark, D.H., Caswell, J.L. 1976, Mon. Not. R. astr. Soc., 174, 267.
Clark, D.H., Stephenson, F.R. 1977, Mon. Not. R. astr. Soc., 179, 87p.
Fabian, A.C., Willingale, R., Pye, J.P., Murray, S.S., Fabbiano, G.
    1980, Mon. Not. R. astr. Soc., 193, 175.
Higdon, J.C., Lingenfelter, R.E. 1980, Astrophys. J., 239, 867.

Ilovaisky, S.A., Lequeux, J. 1972, Astron. Astrophys., 20, 347.
Large, M.I., Vaughan, A.H., Mills, B.Y. 1968, Nature, 220, 340.
Lyne, A.G. 1981, Pulsars, IAU Symposium No. 95, Eds. W. Sieber and
    W.R. Wielebinski, Reidel, Dordrecht, p. 423.
Manchester, R.N., Taylor, J. 1977, Pulsars (Freeman).
Manchester, R.N., Tuohy, I.R., D'Amico, N. 1982, RPP No. 2639.
McCulloch, P.M., Hamilton, P.A., Ables, J.G. 1982, IAU Circular No.3704,
    June 15.
Murdin, P., Clark, D.H. 1981, Nature, 294, 543.
Narayan, R., Vivekanand, M. 1982, submitted to Astron. Astrophys..
Ostriker, J.P., Gunn, J.E. 1971, Astrophys. J. 164, L95.
Pacini, F. 1968, Nature, 219, 145.
Pacini, F., Salvati, M. 1973, Astrophys. J. 186, 249.
Phinney, E.S., Blandford, R.D. 1981, Mon. Not. R. astr. Soc., 194, 137.
Radhakrishnan, V. and Srinivasan, G. 1980, J. Astrophys. Astr. 1, 25.
Seward, F.D., Harnden, Jr., F.R. 1982, Astrophys. J., 256, L45.
Srinivasan, G., Dwarakanath, K.S. 1982a, to appear in J. Astrophys. Astr.
Srinivasan, G., Dwarakanath, K.S. 1982b, to be published.
Srinivasan, G., Dwarakanath, K.S., Radhakrishnan, V., 1982, Curr. Sci.
    51, 596.
Staelin, D.H., Reifenstein, III, E.C. 1968, IAU Circular 2110.
Tammann, G.A. 1977, Ann. N.Y. Acad. Sci., 302, 61.
Taylor, J.H., Manchester, R.N. 1977, Astrophys. J., 215, 885.
Tinsley, B.M., 1977, Supernovae, ed. D. Schramm, Reidel, Dordrecht,
    p.117.
Vivekanand, M., Narayan, R. 1981, J. Astrophys. Astr., 2, 315.
Vivekanand, M., Narayan, R. 1982, submitted to J. Astrophys. Astr.
Vivekanand, M., Narayan, R., Radhakrishnan, V. 1982, to appear in
    J. Astrophys. Astr.
Weiler, K.W. 1978, Mem. Soc. Ital. Astron. 49, 545.
Weiler, K.W., Panagia, N. 1978, Astron. Astrophys., 70, 419.
Weiler, K.W., Panagia, N. 1980, Astron. Astrophys. 90, 269.
Weiler, K.W., Shaver, P.A. 1978, Astron. Astrophys. 70, 389.
Weiler, K.W., Wilson, A.S. 1977, Supernovae, ed. D.N. Schramm,
    Reidel, Dordrecht, p.67.
Wilson, A.S., 1980, Astrophys. J., 241, L19.
Wilson, A.S., Weiler, K.W., 1976, Astron. Astrophys. 49, 357.
Woltjer, L. 1968, Astrophys. J., 152, L179.

A SEARCH FOR PULSARS ASSOCIATED WITH SUPERNOVA REMNANTS IN THE GALAXY AND THE MAGELLANIC CLOUDS

R.N. Manchester
Division of Radiophysics, CSIRO, Sydney, Australia
I.R. Tuohy
Mount Stromlo and Siding Spring Observatories,
Research School of Physical Sciences, Australian National
University, Canberra
N. D'Amico
Istituto di Fisica dell'Università, Palermo, Sicily

It is widely accepted and almost certainly true that both pulsars and supernova remnants (SNRs) are products of the collapse of a star at the end of its evolution. Given this, it is a considerable puzzle why, of the more than 120 known SNRs in the Galaxy, only two have unambiguously associated pulsars. Beaming of the pulsar emission probably accounts for the absence of detectable pulsars in up to 80% of the SNRs; however, this still leaves 20-30 SNRs in which one should be able to detect a pulsar. Vivekanand and Narayan (1981) show that there is a deficit of pulsars with periods $\lesssim 0.5$ s and suggest that a majority of pulsars do not become active for a time $\sim 10^4$ years after their birth. This would account for the lack of pulsar-SNR associations. It is however possible that the observed lack of short-period pulsars is simply due to observational selection. In the past, most pulsar searches have been made at relatively low radio frequencies, typically close to 400 MHz. At these frequencies SNRs are bright and the effects of interstellar scattering are significant, especially for distant, short-period pulsars. Further, most of these searches have used a relatively long sampling interval, typically about 20 ms, which further reduces the sensitivity for short-period pulsars.

We have undertaken a survey specifically designed to avoid these selection effects. The Parkes 64-m telescope was used in two sessions, November 1981 and March 1982, with a dual-channel cooled FET receiver operating at a frequency of 1.4 GHz. Each channel was divided into four adjacent 5 MHz bands and the detected output of each band sampled at 2-ms intervals. In off-line analysis the data were de-dispersed and searched for periodicities. For typical observation times of 20 min, the limiting sensitivity was ~ 1 mJy for pulsars with periods in the range 20 ms to 4 s and dispersion measures $\lesssim 1800$ cm^{-3} pc. A total of 55 galactic SNRs, 30 Magellanic Cloud SNRs, three COS-B error circles (Swanenburg et al., 1981) and about 20 miscellaneous objects were searched.

J. Danziger and P. Gorenstein (eds.), Supernova Remnants and their X-Ray Emission, 495–497.
© *1983 by the IAU.*

R. N. MANCHESTER ET AL.

TABLE 1 - Parameters of pulsars detected at 1.4 GHz..

PSR	R.A. (1950)	Dec. (1950)	P (s)	DM (cm^{-3} pc)	Association
Definite					
1338-62	$13^h 38^m 30^s \pm 20^s$	$-62°08' \pm 2'$	0.1932024	885±45	G308.7+0.0
1509-58	$15^h 09^m 59^s.5^*$	$-58°56'57''^*$	0.15021718	235±25	G320.4-1.4 X-ray pulsar
1758-23	$17^h 58^m 15^s \pm 30^s$	$-23°05' \pm 3'$	0.4157644	1050±100	W28, CG006-00
Probable					
1758-24	$17^h 58^m 29^s \pm 45^s$	$-24°50' \pm 10'$	0.1248315	260±60	G5.3-1.0
1802-23	$18^h 02^m 45^s \pm 45^s$	$-23°22' \pm 10'$	0.1125343	400±100	CG006-00

*X-ray position (Seward and Harnden, 1982).

The results of the search are summarized in Table 1. Three pulsars were detected and confirmed; for two other objects, periodic signals were detected but not confirmed. All five detections are associated with galactic objects. The most interesting detection is that of PSR 1509-58, which is also observed as an X-ray pulsar (Seward and Harnden, 1982). This pulsar, the radio detection of which has been previously reported by the present authors (Manchester et al., 1982), is close to the centre of the SNR G320.4-1.4 (MSH 15-52); it is most probably associated with this SNR and hence joins the Crab and Vela examples as a third relatively unambiguous pulsar-SNR association.

The two other confirmed pulsars are interesting in that their dispersion measures are very high - in the case of 1758-23, more than twice the previously highest value (Manchester and Taylor, 1981). PSR 1338-62 is close to but outside the radio contours of the SNR G308.7+0.0 (Caswell et al., 1981) and is probably not associated with this SNR. Similarly PSR 1758-23 is outside the radio contours of W28 (Shaver and Goss, 1970) but within the error circle for the COS-B γ-ray source CG006-00 and is probably not associated with either object. The two unconfirmed detections are both short-period pulsars, and it is important to confirm their reality and (if confirmed) to improve their positions in order to establish any association. A further result from this search is an upper limit on the pulsed flux from the central object in RCW 103 (Tuohy et al., 1982), which strengthens the case for interpreting the X-ray emission from this object as thermal emission from the surface of a neutron star.

REFERENCES

Caswell, J.L., Milne, D.K., and Wellington, K.J.: 1981, *Mon. Not. R. Astron. Soc.* 195, p. 89.

Manchester, R.N., and Taylor, J.H.: 1981, *Astron. J.* 86, p. 1953.

Manchester, R.N., Tuohy, I.R., and D'Amico, N.: 1982, *Astrophys. J. (Lett.)* 262 (in press).

Seward, F.D., and Harnden, F.R.: 1982, *Astrophys. J. (Lett.)* 256, p. L45.

Shaver, P.A., and Goss, W.M.: 1970, *Aust. J. Phys. Suppl.* No. 14, p. 77.

Swanenberg, B.N. et al.: 1981, *Astrophys. J. (Lett.)* 243, p. L69.

Tuohy, I.R., Garmire, G.P., Manchester, R.N., and Dopita, M.A.: 1982, *Astrophys. J.* (submitted).

Vivekanand, M., and Narayan, R.: 1981, *J. Astrophys. Astron.* 2, p. 315.

STRONG WAVES FROM PULSARS AND MORPHOLOGIES IN SNRS

Gregory Benford
University of California, Irvine

Attilio Ferrari and Silvano Massaglia
Istituto di Fisica Generale dell'Universita', Torino

1.   INTRODUCTION

Canonical models for pulsars predict the emission of low–frequency waves of large amplitudes, produced by the rotation of a neutron star possessing a strong surface magnetic field. Pacini (1968) proposed this as the basic drain which yields to the pulsar slowing–down rate. The main relevance of the large amplitude wave (LAW) is the energetic link it provides between the pulsar and the surrounding medium. This role has been differently emphasized (Rees and Gunn, 1974; Ferrari, 1974), referring to absorption effects by relativistic particle acceleration and thermal heating, either close to the pulsar magnetosphere or in the nebula. It has been analyzed in the special case of the Crab Nebula, where observations are especially rich (Rees, 1971). As the Crab Nebula displays a cavity around the pulsar of dimension $\sim 10^{17}$ cm, the function of the wave in sweeping dense gas away from the circumpulsar region is widely accepted. Absorption probably occurs at the inner edges of the nebula; i.e., where the wave pressure and the nebular pressure come into balance. Ferrari (1974) interpreted the wisps of the Crab Nebula as the region where plasma absorption occurs, damping the large amplitude wave and driving "parametric" plasma turbulence, thus trasferring energy to optical radiation powering the nebula. The mechanism has been extended to interpret the specific features of the "wisps" emission (Benford et al.,1978). Possibly the wave fills the nebula completely, permeating the space outside filaments with electromagnetic energy, continuously accelerating electrons for the extended radio and optical emission (Rees, 1971).

The details of the interaction of a low–frequency wave with magnetospheric and circumstellar plasmas has been discussed by Dobrowolny and Ferrari (1976) and by Benford et al. (1978). Benford et al. (1982) propose a general scenario for the interaction between the strong electromagnetic wave and the surrounding medium. The aim is to derive which morphologies should be expected for a range of the physical parameters corresponding to different stages of remnant's evolution or location in the Galaxy. In particular they attempt to define which observations would allow indirect detection of a spinning neutron star in a supernova remnant, where the geometry is not favourable to direct observation of pulses. Following this analysis, it turns out that two scenarios can be envisioned, depending on the density of the circumpulsar region: i) low–density magnetosphere case (LDM), which leads to morphologies typical of Crab–like supernova remnants, and ii) high density magnetosphere case (HDM), which yields to objects like

J. Danziger and P. Gorenstein (eds.), Supernova Remnants and their X-Ray Emission, 499–501.

SS433 and compact, high–brightness sources. We summarize here the main results of this model in connection with possible observational checks.

2.   CRAB–LIKE SNRS

Two classes of SNRs are usually recognized: shell–type SNRs (possibly caused by Type I SNs) and Crab–like SNRs or "plerions" (possibly due to Type II events) (Weiler and Panagia, 1978, 1981). Plerions may be supported by the activity of young, fast pulsars, and should be short lived ($\sim 2 \times 10^4$ yr) from heavy braking. In most observed plerions the central object has not been detected as a pulsar. Beaming away from us may account for this; instead, energy loss occurs mostly through large amplitude wave mecha–nisms within a parsec of the object. If so, the only method of detection may be through non–thermal X–ray emission or coherent (but unpulsed) radio radiation.

To illustrate the range of parameters needed, Benford et al. (1982) consider a sample of plerions detected in the radio with luminosity L_r. The total power emitted by a spinning magnetic dipole is

$$L_d \approx 5 \times 10^{31} \, a_6^6 \, B_{12}^2 \, P^{-4} \text{ erg/sec} \tag{1}$$

By equating the observed L_r to L_d one obtains a crude upper limit on the period, P. A minimum distance at which parametric instabilities can occur in a circumstellar plasma is then found. This is roughly the size of a non–thermal X–ray feature arising from synchro–Compton processes as the wave decays. The expected angular sizes fall within the angular resolution of forthcoming X–ray satellites. To estimate X–ray luminosities we scale from L_r to L_x using the same ratio as the Crab in 2–6 keV range. Benford et al.(1978) found the synchro–Compton radiation from the Crab in this range to be given by $W_x \cong 10^{37} \, n_x/n_w$ where n_x/n_w is the ratio of electrons at $\gamma \approx 10^8$ to the bulk density carried by the wave, n_w. Setting $W_x \cong L_x$, $n_x/n_w \sim 10^{-4} \div 10^{-3}$. Parametric decay of the wave produces particle bunching, leading to coherent, antenna–like emission (Leboeuf et al., 1982). Thus some coherence may occur far from the central object, though the plasma frequencies in this region may be so low that the emission is at frequencies below the radio. It seems likely that coherence is best observed close to the central object.

As an interesting case, we finally recall that Seward and Harnden (1982) found pulsed X–ray emission coming from a point source MSH 15–52. The observed period is 0.150 sec. The source appears to be near a small region of non–thermal diffuse X–ray emission, probably powered by the pulsar itself, as it is typical of Crab–like SNRs. The diffuse X ray luminosity is $L_x \simeq 2 \times 10^{35}$ erg/sec. If we suppose this is a pulsar, rather than a binary neutron star source, we can apply the method above to get an upper limit on the period in terms of the X–ray luminosity. This yields $P \cong 0.126 \, a_6^{3/2} \, B_{12}^{1/2}$ sec, quite close to the observed X–ray period.

3.   HIGH–BRIGHTNESS TEMPERATURE REMNANTS

If the plasma beyond the light cylinder is too dense ($\gtrsim 10^{-8}/P^2$ cm^{-3}), the wave will be absorbed in a few light cylinder radii, as the wave pressure does not exceed the plasma pressure (Dobrowolny et al., 1976). Computer simulations of Lebouef et al. (1982) show that, regardless of waveform, the large amplitude wave rapidly decays into a mixture of ultrarelativistic particles and waves. What is more, localized density spikes appear at the extrema of the wave magnetic field. Bunched electrons can radiate coherently.

We may expect that many spinning neutron stars will be surrounded by dense plasmas. This will be the case particularly if the pulsar has high linear speed and thus confronts fresh interstellar plasma constantly, so that it cannot evacuate a cavity. Such smothered magnetospheres should display the effects of wave decay beyond their light cylinder radii, and if the parametric decay involves particle bunching, a steady coherent emission should appear. This radiation will generally not be pulsed, since the bunches are not being swept around by a rigid magnetosphere. We can crudely estimate the brightness temperature of a light electron wind oscillating in the wave troughs. For the typical pulsar parameters, even if the wave decays parametrically in a distance L, it will radiate in radio wavelengths with brightness temperature

$$T_b \approx 10^{19} \; L_{10} \; \nu_9^{-3} \; \gamma_4^2 \; n_6 \; B_{12}^2 P_{-1}^{-6} \; °K \tag{2}$$

where γ is the wind Lorentz factor, n its plasma density, and B, a, P are, respectively, the surface magnetic field, the radius and the period of the spinning object. Such bright compact, non–pulsing objects should be sought by scintillation techniques (see Cordes and Dickey 1979, for a possible discovery of one of this kind of objects). If a dispersive smearing of a pulsed component can be eliminated, such a source will provide a signature of the large amplitude wave.

REFERENCES

Benford,G., Bodo,G. and Ferrari,A., (1978) Astron. Astrophys. 70, 213.

Benford,G., Ferrari,A. and Massaglia,S., (1982) Ap. J., submitted.

Cordes,J.M. and Dickey,J.M., (1979) Nature 281,24.

Dobrowolny,M. and Ferrari,A., (1976) Astron. Astrophys. 47, 97.

Ferrari,A., (1974) "Supernovae and Supernova Remnants", D.Reidel, 375.

Leboeuf,J.N., Ashorn–Abdalla,M., Tajima,T., Dawson,J.M., Coroniti,F.V. and
        Kennell,C.F., (1982), to be published.

Pacini,F., (1968) Nature 219, 145.

Rees,M.J., (1971) Nature 230, 55.

Rees,M.J. and Gunn,J.E., (1974) Mon. Not. R. astr. Soc. 167, 1.

Seward,F.D. and Harnden,F.R., (1982) Ap. J., in press.

Weiler,K.W. and Panagia,N., (1978) Astron. Astrophys. 70, 419.

Weiler,K.W. and Panagia,N., (1980) Astron. Astrophys. 90, 269.

THREE NEW PULSARS NEAR SUPERNOVA REMNANTS

D.K. Mohanty
Radio Astronomy Centre(TIFR),Ootacamund,India

A search for pulsars in the directions of twenty Galactic supernova remnants (SNRs) was carried out at 327 MHz with the Ooty Radio Telescope (ORT). We have measured the P for one of the pulsars which is just outside the shell of SNR W44 and discuss here the possibility of its association with the HI shell around W44. The parameters of the remaining two pulsars have not yet been determined with sufficient accuracy to suggest association with the respective SNRs.

An area of $2^o \times 2^o$ in the direction of each of the SNR was observed for 50 minutes with the ORT. A multi-channel receiver system (channel width 300 kHz) with an overall bandwidth of 4 MHz was used with each of the 12 beams of the ORT (covering 36' in declination) and the receiver outputs were sampled at 16 msec. Data were de-dispersed and folded using a Fast Folding Algorithm for periods between 0.056 s and 1.58 s. Sensitivities in the range of 4-15 mJy was obtained in various directions. Preliminary results are summarized in Table 1.

The parameters of the pulsars in W28 and G7.7-3.7 have large errors.Since we are still in the process of improving

TABLE 1

NAME	RA(1950) h m	DEC(1950) o '	PERIOD sec	DM cm^{-3}pc	FLUX mJy	FIELD
1853+00	18 53.5±.3	+00 55±0.5	0.3569290	90±10	8	W44
1757-23	17 57±4	-23 43±1.5	1.03082±2	280±40	6	W28
1814-23	18 14±4	-23 13±1.5	0.62547±2	240±40	6	G7.7-3.7

J. Danziger and P. Gorenstein (eds.), Supernova Remnants and their X-Ray Emission, 503–504.

these parameters, in the subsequent discussions we will confine ourselves to the pulsar near W44.

Barycentric arrival time of the pulses from W44 was computed using ephemeris supplied by L.I.Shapiro of MIT. The $\dot{P}$ was estimated to be $0.3\pm0.15\times10^{-15}ss^{-1}$. The pulsar is located about 5' south of the shell of W44. This pulsar is also inside an expanding HI shell detected by Knapp and Kerr (1974). The parameters of the pulsar, SNR and HI shell are summarized in Table 2.

TABLE 2

	PULSAR	SNR	HI SHELL
Distance (kpc)	3	3.1	3.1
Diameter (pc)	-	26	80
Distance from pulsar(pc)	-	17	20
Age (yrs)	2×10^7	4000	7×10^6

Considering the ages of the three objects listed above, the pulsar is more likely to be associated with the HI shell rather than the SNR W44. The true age of the pulsar, which is about 1/4th of its characteristic age (Manchester and Taylor 1977) agrees with that of the HI shell. The estimated energy in the shell indicate that the shell was perhaps formed by a SN II event which is supposed to yield neutron star. The evidence suggests a scenario where, the supernova that yielded the pulsar was also responsible for the expanding HI shell. It is possible that the SNR W44 is the remnant of a second generation star. Stacy and Jackson (1982) have also proposed the association of an expanding HI shell with PSR 0740-28. It is of interest to measure the proper motion of these pulsars.

I thank Dr. T. Velusamy for valuable discussions and for assistance with observations. I also thank the observing staff of the ORT for their help with data reduction.

REFERENCES

Manchester,R.N. and Taylor,J.N., 1977, Pulsars,Freeman Co.

Knapp,G.R. and Kerr,F.J.,1974, Astron. Astrophys.,33, 463.

Stacy,J.G. and Jackson,P.D., 1982, Nature, 296, 42.

X-RAY AND GAMMA-RAY RADIATION FROM THE SWITCHED-OFF RADIOPULSARS

A.I. Tsygan
A.F. Ioffe Institute of Physics and Technology,
Leningrad, USSR

It is shown that pulsars that have ceased to generate electron-positron pairs (switched-off radiopulsars) may be the sources of X-ray and γ-ray radiation. The magnetic dipole radiation from these rotating neutron stars is transformed near the "light radius" into hard radiation by the plasma that is created due to ionization of interstellar neutral hydrogen.

A radiopulsar with a period P that obeys

$$P = P_o \simeq \beta \left(\frac{B_o}{10^{12}}\right)^{8/13} (\cos\alpha)^{6/13} \qquad (1)$$

stops producing an avalanche of electron-positron pairs. Here α is the angle between the angular velocity Ω and magnetic field B_o in the avalanche region, and β is a numerical factor ~ 1. At $\cos\alpha \sim 1$ the period $P_o \sim \beta(B_o/10^{12})^{8/13}$ (Sturrock, 1971; Ruderman and Sutherland, 1975). For an orthogonal rotator ($\Omega \perp \mu, \mu$ is the magnetic moment of the star) one should put $\alpha \sim (\pi/2 - \sqrt{\Omega a/c})$ in equation (1), i.e., $P_o \sim 0.2\beta^{13/16}$ $(B_o/10^{12})^{1/2}$, where a is the neutron star radius. Let us consider the pulsars with high radio emission factors $\eta = L_r/\dot{E} > 10^{-4}$, L_r is the radioluminosity and $\dot{E}$ the total energy loss rate. A separation of pulsars into two groups, with $\eta > 5 \times 10^{-6}$ and $\eta < 5 \times 10^{-6}$, respectively, was proposed by Vladimirsky (1981). Figure 1 shows the pulsars with $\eta > 10^{-4}$ (dots) and with $\eta < 3 \times 10^{-7}$ (crosses), according to Manchester and Taylor (1977), Smith (1977). We assume that all the pulsars displayed between the lines I and II are passing the switch-off stage and differ only by the angle between Ω and μ. The line I is quite well fitted by equation (1) with $\beta = 2$ at $\cos\alpha \sim 1$ (there should be two "switch-off lines", for $\alpha \to 0$ and $\alpha \to \pi$, which are likely to appear in figure 1). Line II corresponds to switching off pulsars with $\Omega \perp \mu$. It is probable that the avalanche generation has an oscillatory character, i.e., the plasma is being ejected with the bunches. The latter yields the high values of the parameter η.

J. Danziger and P. Gorenstein (eds.), Supernova Remnants and their X-Ray Emission, 505–507.
© 1983 by the IAU.

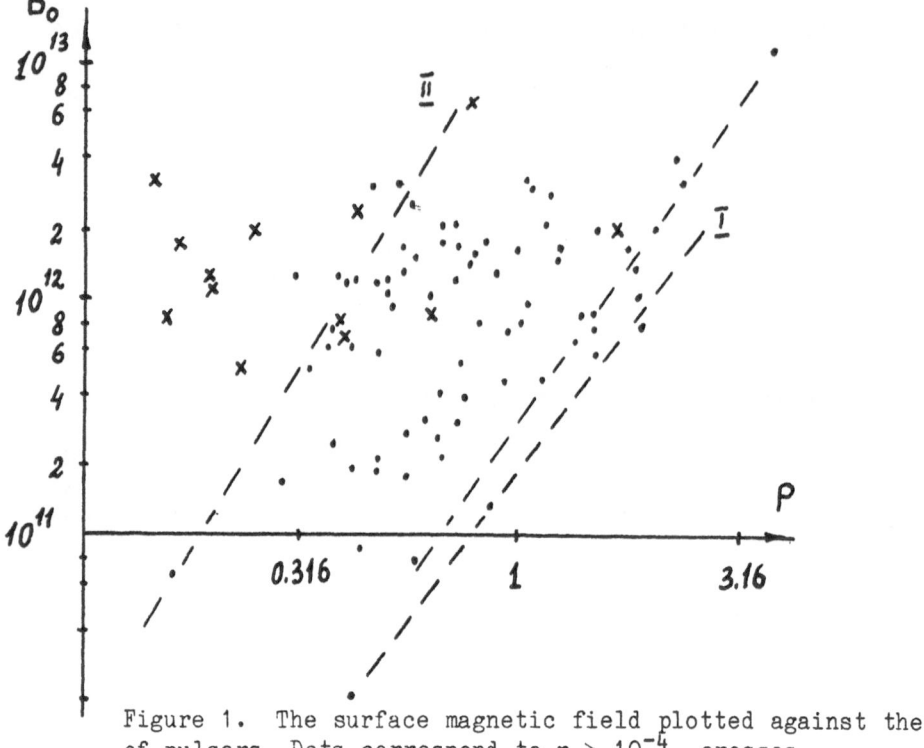

Figure 1.   The surface magnetic field plotted against the period
of pulsars. Dots correspond to $\eta > 10^{-4}$, crosses
to $\eta < 3 \times 10^{-7}$.

        Let us consider a radiopulsar in the switch-off stage $(P > P_0)$ which
ejects no electron-positron plasma. Magnetic dipole radiation from the
rotating neutron star pushes away the interstellar plasma at a distance
$\sim 10^{15}-10^{16}$cm. Interstellar neutral hydrogen is, however, captured by
the neutron star gravitational field, and is then ionized by the star's
thermal radiation. For a surface temperature of the star $T_s = (2-3)$
$\times 10^4$K (the cooling time $\sim 10^6$ years (Glen and Sutherland, 1980)) the
radius of photo-ionization of neutral hydrogen is given by $R = 10^8-10^{10}$
cm. The cross section of neutral hydrogen capture by the star is
$\sigma = \pi R^2(1+2GM/RV_\infty^2)$, $V_\infty \sim 10^7$cm s^{-1} being the star's velocity relative
to the neutral hydrogen. The hydrogen number density in the ionized
region equals $n(R) = n_\infty V_{ff}(R)/4V_\infty$, where $V_{ff}(R) = \sqrt{2GM/R}$ and n_∞ is the
number density of neutral hydrogen. In the quasistatic zone, within the
"light radius" sphere $(\Gamma_L = c/\Omega)$, the electromagnetic field accelerates
some fraction of the charged particles towards the neutron star (Tsygan,
1980). They produce two "hot spots" with radii $R_0 = a\sqrt{\Omega a/c}$ at the bottom
of the "open" magnetic field lines on the surface of the neutron star.
While cooling down to a temperature $T_s < 2 \times 10^4$K, the neutral hydrogen
will be ionized in the vicinity of the "light radius" by X-ray radiation
from the "hot spots". At $P = 0.5-1.5$s and $n_\infty = 0.1$cm^{-3} the temperature of
the "spots" is $T \sim 10^6$K and their luminosity is $2 \times 10^{27}-10^{29}$ erg s^{-1}.

At the "light radius" the electric field component along the magnetic field equals $E_{\parallel} \sim 0.5B_0 (\Omega a/c)^3$, and accelerates electrons along the magnetic field up to a γ-factor given by $\gamma = eE_{\parallel}\Gamma_L/mc^2$ or $\gamma = (3E_{\parallel}\Gamma_L^2/2e)^{1/4}$ if the radiative losses are significant (Ferrari and Trussoni, 1974). For a switched-off pulsar with $\Omega\parallel\mu$, $B_0 = 10^{12}$ G and P = 0.5s (P $\lesssim$ P_0 according to equation (1)) electrons are accelerated up to $\gamma = 2.8 \times 10^7$ and emit quanta (curvature radiation) of energy $\hbar\omega \sim \hbar c\gamma^3/2\Gamma_L = \hbar\Omega\gamma^3/2 = 1.4 \times 10^{-4}$ erg (70 Mev). In this case the γ and X-ray ($\hbar\omega \sim 100$ Kev) luminosities are about $L_\gamma = 2.5 \times 10^{31}$ erg s^{-1} and $L_X = 4 \times 10^{27}$ erg s^{-1}, respectively.

REFERENCES

Ferrari A., Trussoni E., 1974. Astronomy and Astrophysics, 36, 267.
Glen G., Sutherland P., 1980. Astrophys. J., 239, 671.
Manchester R.N., Taylor J.H., 1977. Pulsars. (San Francisco: Freeman).
Ruderman M.A., Sutherland P.G., 1975. Astrophys. J., 196, 51.
Smith F.G., 1977. Pulsars. (Cambridge: Cambridge University Press).
Sturrock P.A., 1971. Astrophys. J., 164, 529.
Tsygan A.I., 1980. Sov. Astrophys. J., 57, 73.
Vladimirsky B.M., 1981. Conf. Papers of 17th Internat. Cosmic Ray
    Conference, Paris, Vol. 1, 38.

EVOLUTION OF NEUTRON STARS IN YOUNG SUPERNOVA REMNANTS

K. Nomoto[+] and S. Tsuruta[++]
Max-Planck-Institut für Physik und Astrophysik,
Garching bei München, West Germany

The exciting observational developments in recent years (see Seward, Helfand, Harnden, Becker, etc., in this volume) have made it worthwhile to reexamine neutron star cooling theories. Here we shall give an intermediate report on our work.

PHYSICAL INPUT

The basic stellar structure equations we used are the fully general relativistic version by Thorne (1977). These equations have been solved in the manner described by Nomoto and Tsuruta (1981, 1982), using the same physical input. In order to see the effect of nuclear inter-actions in the central core, we chose (i) a stiff equation of state by Pandharipande and Smith (1975) called PS model and (ii) a soft equation of state of Reid type called BPS model (Baym, Pethick and Sutherland 1971), because they give boundaries to relatively "realistic" equations of state. The neutrino emissivities included are: modified Urca, neutron-neutron and neutron-proton bremsstrahlung, electron-ion bremsstrahlung, and plasmon, pair and photo-neutrino emissions. The table by Huebner et. al. (1977) was used for radiative opacity and the work of Flowers and Itoh (1976) for conductive opacity.

RESULTS

Typical cooling curves for a stiff model are shown in Figure 1. The gravitational mass $M_G = 1.3 M_\odot$, central density $\rho^c = 4.0 \times 10^{14}$ gm cm^{-3}, and radius R = 16.0 km.

Recently the work of Flowers and Itoh was criticised by Yakovlev and Urpin (1980). We carried out envelope calculations by using both the conductive opacity of Yakovlev-Urpin (κ_c(YU)) and that of Flowers-Itoh (κ_c(FI)). Our results are shown in Figure 2. T_b is the temperature at the boundary of the isothermal core and T_e is the surface temperature. (Here we assumed that the isothermal state has been reached already.) We set the surface gravity $g_s = 10^{14}$ cm sec^{-2}, which corresponds roughly to PS model. The solid curve refers to κ_c(FI) while the dashed curve is for κ_c(YU). We note that the difference is negligible. It is somewhat larger as g_s increases, but it is less than a factor of ~ 1.2 in cases of our interest ($\sim 100 - 10^4$ years).

(+) On leave from Department of Earth Science and Astronomy,
    University of Tokyo
(++) On leave from Department of Physics, Montana State University

J. Danziger and P. Gorenstein (eds.), Supernova Remnants and their X-Ray Emission, 509–512.
© 1983 by the IAU.

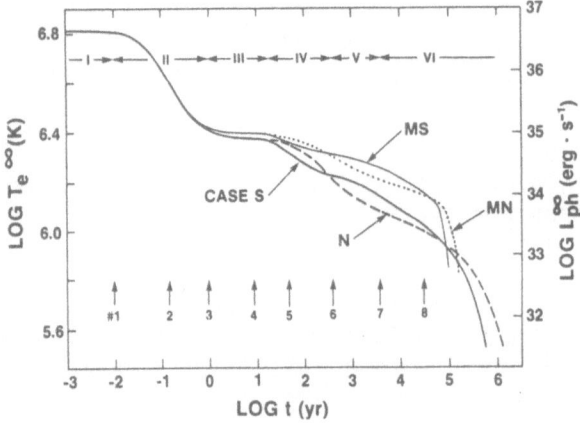

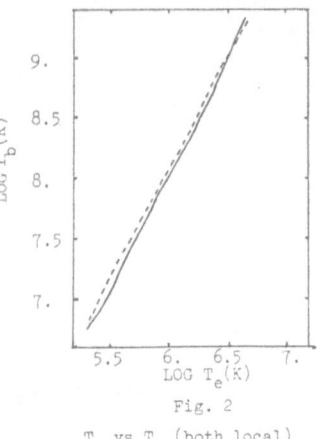

Fig. 2

T_b vs T_e (both local)

Fig. 1

The surface temperature T_e^∞ (left) and photon luminosity L_{ph}^∞ (right) as a function of t (all measured at infinity), for PS model. The surface magnetic field H = 0 for Case N (no superfluid) and Case S (with superfluid); and H = 5 x 10^{12} gauss for Case MN (no superfluid) and Case MS (with superfluid).

Some characteristics of compact sources in young supernova remnants (SNR's) are shown in Table 1 (see reports by Seward, Helfand, Harnden, Becker, etc., in this volume). The names of SNR's are shown in the first column. The next three columns give the estimated distances, diameters and ages whenever known. Point sources are detected in six of these SNR's, which are marked as (*). For these sources, the last column gives the surface temperature (observed at infinity) assuming that all of the non-pulsed portion of the point source radiation comes from the neutron star surface. For the other sources, the last column gives the 3σ level upper limits to the surface temperature. These values are obtained by assuming blackbody radiation and for R = 7 - 16km, which corresponds to a variety of "realistic" equations of state. Measured values of interstellar

SNR	Distance (Kpc)	Diameter (pc)	Age (years)	Temperature (x 10^6 K)
Cas A	2.8	2	300	1.5 - 1.7
Kepler	8.0	–	375	1.8 - 2.2
Tycho	3.0	6	407	1.2 - 1.5
Crab(*)	1.7- 2	1	925	2.0 - 2.4
SN 1006	1.2	4.4	973	0.65 - 0.8
RCW 86	2.5	–	1794	1.4 - 1.8
RCW 103(*)	∿ 2	–	{∿1000 2000	1.7 - 2.2
MSH15-52(*)	4.2	10	{∿1600 2100	?
W 28	2.3	10	3400	1.6 - 2.0
G350.0-18	4.0	35	∿8000	1.8 - 2.2
G32.7-0.2	4.8	35	∿10000	2.0 - 2.4
Vela X(*)	0.4	20-40	∿12000	1.2 - 1.5
3C 58(*)	2.6	10-15	?	2.0 - 2.4
CTB 80(*)	3.0	–	?	1.2 - 1.6

Table 1

Characteristics of compact sources in young SNR's

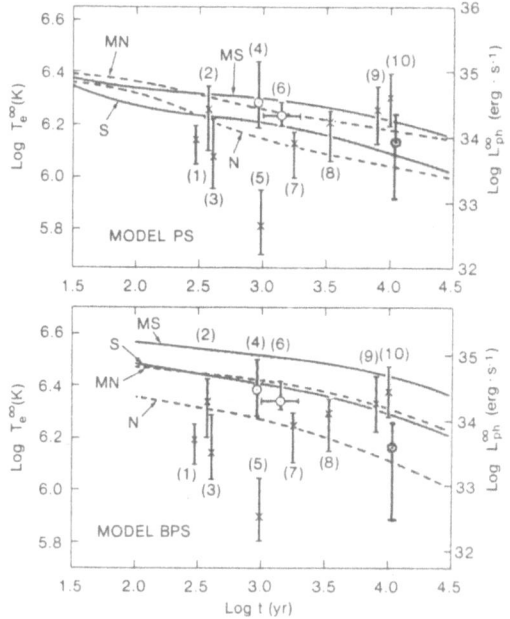

Fig. 3

Comparison between the observational data
and the "standard" cooling curves of PS model
(top) and BPS model (bottom). T_e^∞ and L_{ph}^∞
are shown as a function of t. The numbers
are: (1) Cas A, (2) Kepler, (3) Tycho,
(4) Crab, (5) SN1006, (6) RCW103, (7) RCW86,
(8) W28, (9) G350.0-18, (10) G22.7-02, and
(11) Vela. For BPS model, M = 1.3M$_\odot$,
ρ^c = 3.2 x 10^{15}gm cm^{-3}, and R = 8.0 km.

density are used whenever known.
Otherwise it is assumed to be 0.3
- 1 cm^{-3}. The origin of the point
source emission remains unclear.
However, it is already interesting
in the sense that they strongly
suggest the existence of an active
neutron star in these remnants.

Our results are compared
with the data from the Einstein
Observatory in Figure 3, for typical
stars of M$_G$ = 1.3M$_\odot$. The crosses
indicate the observed upper limits
to the surface temperature of various SNR's, assuming the presence of
neutron stars in these remnants.
The circles refer to possible
"detections" of point sources.
These are accompanied by the estimated error bars which are mostly
due to uncertainties in interstellar
absorption.

CONCLUSION

Major points of interest
are summarised below:
(a) - Some of the observed upper
limits (Cas A, SN 1006, Tycho) are
below the "standard" theoretical
curves (with no drastically fast
cooling agents), while they are above the "non-standard" cooling curves
with charged pion condensates and quarks. A natural interpretation is
that there are no neutron stars in these SNR's, or that these stars contain a substantial amount of such exotic particles as charged pion condensates or quarks.
(b) - If some of the "detections" (e.g. RCW 103, Vela) indeed refer to the
surface emission, these "detection" points are consistent with "standard"
cooling, while they are inconsistent with charged pion and quark cooling.
(c) - The effect of general relativity on thermal properties such as energy
transport and energy balance is negligible.
(d) - The effect of equations of state on observable temperatures is relatively small, and the use of a better equation of state should not change
our conclusion (a) and (b).
(e) - The above points confirm our major conclusion reached earlier
(Tsuruta 1980, 1981).
(f) - The effect of the finite time scale of thermal conduction is evident
for PS model, for a few to several thousand years (see Fig. 1). Many young
SNR's are as old as or younger than these ages. That means the isothermal
approach is not valid for a stiff model such as PS model. On the other
hand, it is adequate for a soft model such as BPS model.

(g) - The use of $\kappa_c(YU)$ instead of $\kappa_c(FI)$ should not change our conclusion, because the difference is negligibly small (Fig. 2). Ichimaru et. al. (1982) recently obtained conductive opacity by treating the problem more accurately than Yakovlev-Urpin (YU) and Flowers and Itoh (FI) did. The use of their better opacity, however, should not change our conclusion, because their values generally lie between those of YU and FI.
(h) - If the surface is composed of elements lighter than iron (possible for very young stars), the surface temperature will be higher (at a given age). This is because the conductive opacity will then be decreased (e.g. by a factor of ∿20 for Helium in the critical liquid region). This will further enhance the trend noted in (a). This may be the case e.g. in Cas A.
(i) - The state of the surface matter (gas, solid,etc.) may turn out to be important, and we are currently investigating its effect on cooling.
(j) - If neutral pions are present (a realistic possibility), the surface temperature will be higher again (Tamagaki 1981, Takatsuka 1981), and the trend noted in (a) will be enhanced further.
(k) - The use of better theory of magnetic properties will not change our conclusion (a), because the lower boundary to theoretical temperatures will be determined by a zero-field curve. However, the upper boundary will be determined by a strong field curve. Therefore, the effect of a more "realistic" treatment of magnetic properties should be important if "detections" of the surface temperatures be confirmed.

A major part of the work reported here has been carried out in Max-Planck Institute in München, Institute of Space and Astronautical Science in Tokyo, and Institute of Astronomy in Cambridge. We are grateful to Drs. R. Kippenhahn, G. Börner, W. Hillebrandt, H.-C. Thomas, M. Oda, Y. Tanaka, M.J. Rees, and the others in these institutions, for their hospitality and help.

REFERENCES

Baym, G., Pethick, C.J., and Sutherland, P.G.: 1971. Ap. J. 170, p. 299.
Flowers, E.G., and Itoh, N.: 1976. Ap. J. 206, p. 218.
Huebner, W.F., et. al.: 1977. Astrophysical Opacity Library (Rep. No. LA 6760M, Los Alamos Scientific Laboratory).
Ichimaru, S., Itoh, N., Iyetomi, H., and Mitake, S.: 1982. in preparation.
Nomoto, K., and Tsuruta, S.: 1981. Ap. J. Letters 250, p. L19.
Nomoto, K., and Tsuruta, S.: 1982. in preparation.
Pandharipande, V.R., and Smith, R.A.: 1975. Nucl. Phys. A237, p. 507.
Takatsuka, T.: 1981. Proceedings of the 2nd International Conference on "Recent Progress in Many Body Theories", Oaxtepec, Morelos, Mexico.
Tamagaki, R.: 1981. Proceedings of the 2nd International Conference on "Recent Progress in Many Body Theories", Oaxtepec, Morelos, Mexico.
Thorne, K.S.: 1977. Ap. J. 212, p. 825.
Tsuruta, S.: 1980. X-Ray Astronomy (ed. R. Giacconi and G. Setti, Dordrecht: Reidel), p. 73.
Tsuruta, S.: 1981. Pulsars (ed. W. Sieber and R. Wielebinski, Dordrecht: Reidel), p. 331.
Yakovlev, D.G., and Urpin, V.A.: 1980. Sov. Astron. 24, p. 303.

NEUTRON STAR COOLING: EFFECTS OF ENVELOPE PHYSICS

Kenneth A. Van Riper
Computational Physics Group
Applied Theoretical Physics Division
Los Alamos National Laboratory
Los Alamos, New Mexico 87545, USA

ABSTRACT

Neutron star cooling calculations are reported which employ improved physics in the calculation of the temperature drop through the atmosphere. The atmosphere microphysics is discussed briefly. The predicted neutron star surface temperatures, in the interesting interval $300 \leq t \ (yr) \leq 10^5$, do not differ appreciably from the earlier results of Van Riper and Lamb (1981) for a non-magnetic star; for a magnetic star, the surface temperature is lower than in the previous work. Comparison with observational limits show that an exotic cooling mechanism such as neutrino emission from a pion-condensate or in the presence of percolating quarks, is not required, unless the existence of a neutron star in the Tycho or SN1006 supernova remnants is established.

A neutron star cooling model calculates the evolution of the star's temperature by balancing the energy lost, through volume neutrino emission and surface photon radiation, to the change in the thermal energy of the star. The model assumes the interior ($\rho > 10^{10}$ g cm^{-3}) is isothermal. (See Nomoto and Tsuruta, this volume, for an evolutionary model where the isothermality assumption is relaxed; this model does approach isothermality after several hundred years. The differences between their soft cooling model and our soft cooling model are much greater than the differences between their cooling model and their evolutionary model.) The thermal content of the star depends on the interior temperature T_m. The observable surface temperature T_s is related to T_m by an atmosphere (or envelope) calculation, which solves the coupled equations of atmosphere structure and heat transport by electron conduction and radiation diffusion. The atmosphere integration requires an equation of state, a radiative opacity, and a thermal conductivity. (Each of these microphysics relations is a theoretical construct.)

Previous work (Tsuruta 1979; Glenn and Sutherland 1980; Van Riper and Lamb 1981) has relied on the thermal conductivities

513

J. Danziger and P. Gorenstein (eds.), Supernova Remnants and their X-Ray Emission, 513–516.
© *1983 by the IAU.*

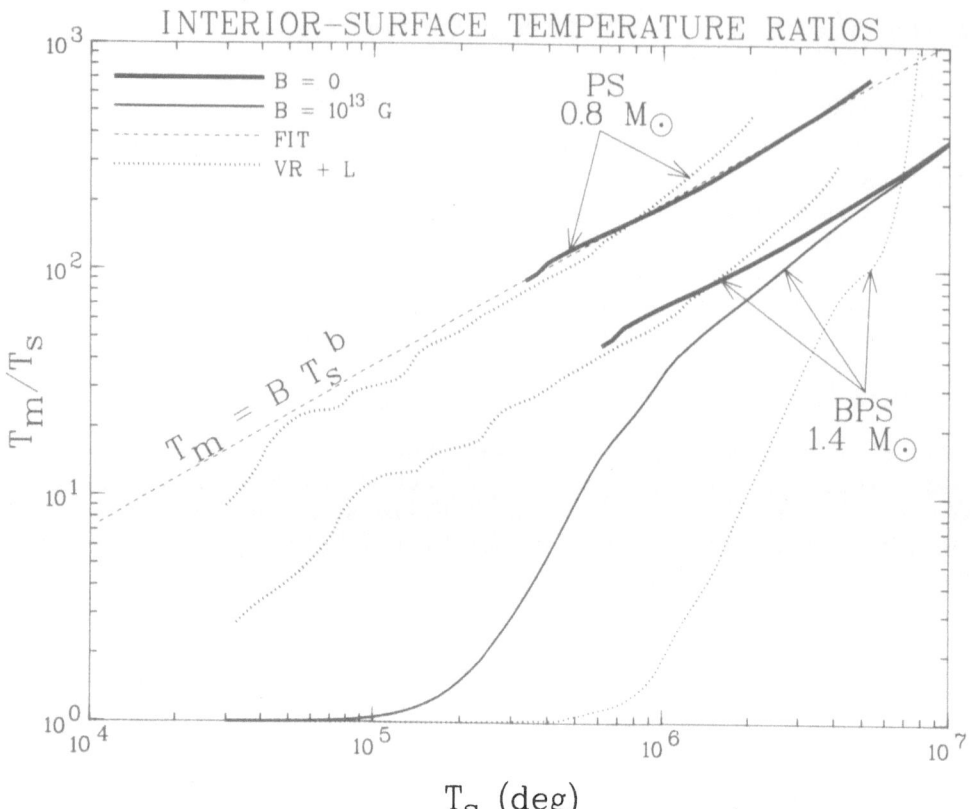

$$T_S \ (deg)$$

Figure 1. Ratio of the temperature at the edge of the isothermal core T_m to the surface temperature T_S, as a function of the surface temperature. The central temperature T_c is higher than T_m by the ratio of redshifts $e^{\phi}m/e^{\phi}c$. T_S is the local surface temperature, not the apparent temperature $T_\infty = e^{\phi}sT_S$ (T_∞ is the temperature an observer would infer from a thermal spectrum). The solid lines are from the current calculations. The current zero field calculations do not extend below $T_S \approx 2 \times 10^5$ deg because integrations at the lower tempertatures encounter a regime where the (negative) Coulomb correction dominates the total pressure. In the cooling calculations, these atmosphere relations were extrapolated with the slope of the straight dashed line (the extrapolated regions are shown by dashed lines in Fig. 2). The dotted lines are from Van Riper and Lamb (1981). The two curves that reach $T_m/T_S = 1$ are for an atmosphere with a magnetic field $B \approx 10^{13}$ G; the other cases are for $B = 0$. The dashed line is a first order fit, of the form shown, to the zero field case. A fourth order fit to the (solid) magnetic case has been plotted. The temperature ratios depend only on the neutron star surface gravity $g = GM/[R^2(1-2GM/c^2R)^{\frac{1}{2}}]$. Two extreme cases are shown: A low mass star with a stiff interior equation of state [Pandharipande and Smith (1975)], for which log g = 13.614 (cm s^{-2}), and a soft [Bayn, Pethick, and Sutherland (1971) EOS] star with cannonical mass, for which log g = 14.712.

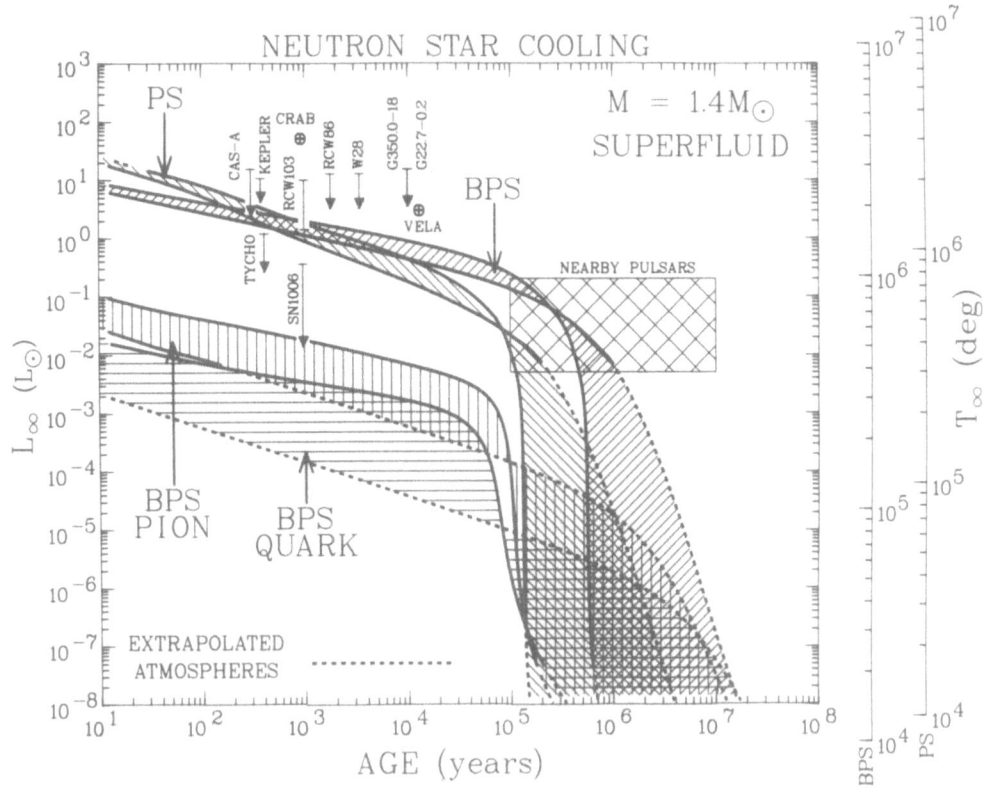

Figure 2. Apparent luminosity L_∞ of cooling neutron stars as a function of the stars age. All cases shown are for a 1.4 $M_\odot$ neutron star with neutron and proton superluidity taken into account. The label "PS" stands for the stiff equation of state of Pandharipande and Smith (1975), while "BPS" stands for the soft EOS of Baym, Pethick, Sutherland (1971). The presence of a pion-condensate (Maxwell, et al 1977) or of percolating quarks (Kiguchi and Sato 1981) results in greatly enhanced neutrino emission and accelerated cooling, as shown. For each case, a shaded region is delimited by two cooling curves, corresponding to a surface magnetic field B ≈ 10^{13} G and to no surface field. In the early, neutrino dominated cooling era (t < 10_5 yr, shallow cooling curves) the magnetic curve lies above the B = 0 curve. In this era, the evolution of the interior temperature, which is determined by volume neutrino emission, is independent of the surface field; the displacement of the curves reflects the different T_m/T_s ratio. At later times, surface photon emission is the dominant cooling mechanism. The magnetic curve then lies below the B = 0 curve, reflecting the lower opacities in the former. The experimental data is discussed in the text, and the atmosphere extrapolations are explained in Fig. 1 and its caption.

calculated by Flowers and Itoh (1976). Subsequent calculations of the conductivity by Yakovlev and Urpin (1980), who use a better plasma structure factor, find that the conductivity may differ from the Flowers and Itoh result by as much as a factor of three. The importance of this was first pointed out by Gudmundsson, Pethick, and Epstein (1982). The present calculations use the Yakovlev and Urpin conductivities in (electron) degenerate matter as long as the density $\rho \geq 10^4$ g cm^{-3}. In non-degenerate matter, conductivities calculated from the Hubbard-Lampe (1969) model are used. This leaves, unfortunately, a large region in which the conductivity must be interpolated. The radiative opacity is, as in all recent work, taken from a table supplied by Los Alamos (Huebner, et. al., 1977). This table also contains the Hubbard-Lampe conductivities, an equation of state, and the ionization level of the iron atoms (the atmosphere is assumed to consist solely of this species). Rather than use this equation of state directly, as did Van Riper and Lamb, an equation of state is computed from the number of free eletrons; a coulomb correction to the pressure is included. For stars with a surface magnetic field, the radiative and conductive opacities are multiplied by a correction < 1 (see Tsuruta 1979). A mistake in Van Riper and Lamb's correction has been fixed.

The temperture changes through some represenative atmospheres are shown in Fig. 1. The zero field cases are fit well by $T_m = (T_s/T_b)^b$, where the fitting parameters depend only on the surface gravity g: log $T_b = -2.3$ log g + 0.2562, b = 0.0022 log g + 1.694. The magnetic cases require a fourth order [log $T_s = \sum_i \log(T_s/s_i)^i$, i=0, 4] fit. Again, the parameters s_i depend only on g. Neutron star cooling curves are shown in fig. 2. The observational points are the same as those shown in Van Riper and Lamb (1981). The Vela point has been displaced slightly to the lower right. The Crab, Vela, and RCW 103 points should be regarded as upper limits to any thermal emission from the neutron star surface. The rectangle characterizes the upper limits that have been obtained for seven nearby pulsars.

Baym, G., Pehtick, C.J., and Sutherland, P.G. 1971, Ap. J., 170, pp.299.

Flowers, E.G. and Itoh, N. 1976, Ap. J., 206, pp.218.

Glenn, G. and Sutherland, P.G. 1980, Ap. J., 239, pp.671.

Gudmundsson, E.H., Pethick, C.J., and Epstein, R.I. 1982, Ap. J. (Lett.), 259, pp.L19.

Hubbard. W.B. and Lampe, M. 1969, Ap. J. Suppl., 18, pp.297.

Huebner, W.F., Merts, A.L. Magee, N.H. and Argo, M.F. 1977, Report LA-6760-M, Los Alamos National Laboratory.

Kiguchi, M. and Sato, K. 1981, Prog. Theor. Phys., 66, pp.725.

Maxwell, O.V., Brown, G.E., Campbell, D.K., Dashen, R.F., and Manassah, J.T. 1977, Ap. J., 216, pp.77.

Pandharipande, V.R. and Smith, R.A. 1975, Nucl. Phys. A., 208, pp.550.

Tsuruta, S. 1979, Phys. Rept., 56, pp.237.

Van Riper, K.A. and Lamb, D.Q. 1981, Ap. J. (Lett.), 244, pp.L13.

Yakovlev, D.G. and Urpin, V.A. 1980, Sov. Astr., 24, pp.303.

CHEMICAL ABUNDANCES IN THE INTERSTELLAR MEDIUM OF GALAXIES FROM
SPECTROPHOTOMETRY OF SUPERNOVA REMNANTS

S. D'Odorico[1] and M. Dopita[2]

[1] European Southern Observatory,
    8046 Garching bei München, West Germany

[2] Mount Stromlo and Siding Spring Observatory,
    Australian National University, Camberra, Australia

1. INTRODUCTION

In the recent past supernova remnants (SNR) have been identified in
external galaxies as far as 5 Mpc. These galaxies are: M 33(12), M 31(12),
NGC 6822(1), IC 1613(1), NGC 2403(2), NGC 300 (2), NGC 4449(1), where the
numbers of SNR confirmed by optical spectroscopy are given in parenthesis.
Data for objects in the Magellanic Clouds are discussed extensively
elsewhere in this volume.

With the exception of the SNR in NGC 4449, characterised by
broad [OIII]emission lines associated with a non-thermal radio source, all
the other SNR have been identified by their [SII]/Hα line intensity ratio.
An updated description of this technique and of its limitations is given by
Blair et al., (1981) and Dennefeld and Kunth (1981). By now it is also
clear that the SNR with a high[SII]/Hα ratio do not represent the global
SNR population in a galaxy. This can be verified from the data on the
Magellanic Clouds, where by combining the optical, radio and X-ray
observations a rather complete sample of SNR with different characteristics
has been obtained. The sulfur strong SNR are however the optically
brightest and serve well as probes of the chemical composition of the
interstellar medium of the galaxies they belong to.

In the next paragraphs we present the first spectrophotometric data for
four of the remnants mentioned above. We also show the emission line
intensity ratio trends for SNR in different galaxies and, by comparison
with theoretical models, interpret them in terms of variations in the
chemical composition.

J. Danziger and P. Gorenstein (eds.), Supernova Remnants and their X-Ray Emission, 517–524.
© 1983 by the IAU.

2. SPECTROPHOTOMETRY OF THE SNR IN IC 1613, NGC 6822 AND NGC 300

The candidates were taken from the list of D'Odorico et al. (1980) and include beside the single SNR identified in IC 1613 and NGC 6822, #2 and #5 in NGC 300. The four objects were observed both with the AAT 3.9m telescope in October 1979 and with the ESO 3.6m telescope in September 1980.

At the ESO telescope, the spectra were recorded with the IDS (Cullum 1979) attached to the Boller and Chivens spectrograph. Two separate spectral regions, λλ5400-7600Å and λλ 3600-5800Å, were observed at a resolution of 7Å. Figure 1 shows the ESO spectrum of IC 1613.

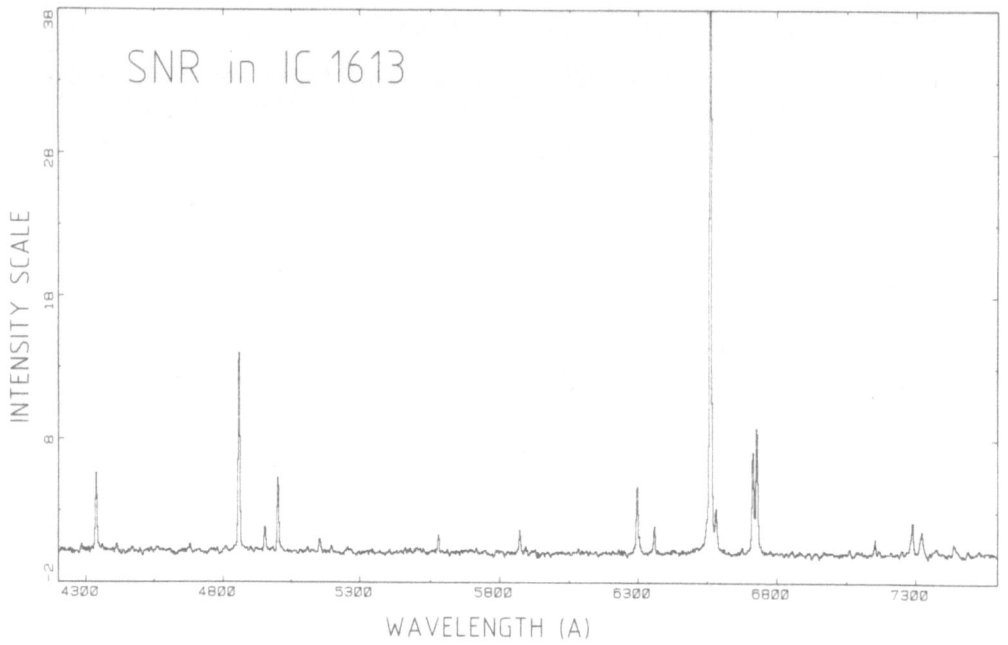

Figure 1 The combined blue and red ESO spectra of the SNR in IC 1613 are shown. The line identifications are given in Table 1.

The AAT spectra were recorded with the IPCS and cover the spectral region λλ3600-7000Å at a resolution of about 10Å.

Both sets of data have been corrected for the spectral sensitivity of the instruments by observations of standard stars. The ESO data have been combined with the AAT by using Hβ as a reference point for the blue spectra and Hα for the red ones. The agreement between the two sets of data was always within 20%, with the exception of NGC 300/#2 where however a contiguous HII region may have contaminated the spectra. The averaged

Table 1 – REDDENING CORRECTED AND MODEL PREDICTED LINE INTENSITIES IN SNR*

Object	$C_{H\beta}$	3727 [OII]	4100 Hδ	4340 Hγ	4959 [OIII]	5007 [OIII]	5876 HeI	6300 [OI]	6563 Hα	6583 [NII]	6717 [SII]	6731 [SII]
IC 1613/SNR^ø	0.6	174	24	43	12	34	8	30	266	20	36	45
		184	25	46	10	30	6	30	299	19	42	47
NGC 6822/SNR	0.6	377	–	41	24	73	–	74:	253	21	65	63
	–	406	25	46	23	66	7	46	300	17	49	54
NGC 300/#2	0.2	470	25	38	45	133	–	35	260	81	103	84
	–	572	25	46	34	100	7	55	300	99	87	95
NGC 300/#5	0.2	753	28	44	54	128	–	80	290	105	131	134
	–	723	25	46	46	136	7	44	303	102	120	125
$f(\lambda)^{+}$	–	0.298	0.195	0.132	-0.021	-0.032	-0.216	-0.281	-0.332	-0.335	-0.355	-0.357

* With respect to Hβ = 100. ø Other line intensities: 4069-76 [SII] = 20; 4363 [OIII] = 2.6;
4415 [FeII] = 3.4; 4686 HeII = 3.6; 5158 [FeII] = 5.5; 5198 [NI] = 2; 7155 [FeII] = 3;
7291 [CaII] = 12; 7325 [OII] = 9.
+ Adopted reddening function.

spectra were further corrected for reddening. The reddening constants $C_{H\beta}$ and the assumed reddening law are given in table 1. In the IC 1613 SNR the reddening of $C_{H\beta} = 0.6$ which fits well the Balmer decrement is much higher than the value of 0.13 derived by Humphreys (1980) from the blue stars in the galaxy, suggesting that dust must have been associated with the pre-supernova star.

The reddening corrected values are given in Table 1. From the scatter of values from spectra obtained on the two different instruments, we estimate the line intensities to be accurate to better than 30%.

3. SNR AS INDICATORS OF THE CHEMICAL COMPOSITION OF THE INTERSTELLAR GAS IN GALAXIES

It was evident from the beginning of the spectroscopic survey of SNR in external galaxies that chemical abundance differences were affecting in a systematic way the line intensity ratios in SNR. This stimulated the development of models for the emission of collisionally excited gas to interpret the spectra and determine the abundances. Results in this field have been obtained by Dopita et al. (1980), Blair et al. (1982), Dennefeld and Kunth (1981), Binette et al. (1982) where references to earlier works can also be found.

The new observations of the SNR in the metal poor galaxies IC 1613 and NGC 6822 reported here have been combined with the previous results in Fig. 2 and 3, which show the behaviour of the main line intensity ratios. In both figures two clear correlations emerge, which can be attributed to changes in the chemical abundances.The trend from high to low metal abundances goes from the SNR .in the most massive galaxies in the sample (M31, the Galaxy) through intermediate systems like M33, NGC 300 and the LMC to low mass galaxies like IC 1613, NGC 6822 and the SMC. This is in agreement with what is known on the chemical abundances of these galaxies from the HII regions and the stellar population. To put these results in a quantitative form, we can use a new self consistent model for the emission of shock ionized gas (Dopita in this volume and Binette, Dopita, Tuohy, 1982). The computations show that in the regime where hydrogen is fully preionised up to shock velocity larger than 300 Km s^{-1}, the intensity of the prominent optical forbidden lines with respect to Hβ vary little with velocity and the details of the preionisation. It is therefore possible to produce a set of models with constant shock parameters but variable abundances, to explore the effect of chemical composition on the line ratios. The results of these calculations are shown in detail elsewhere (Dopita, Binette, D'Odorico 1982). In Figure 2 and 3 we have superimposed on the observed points the theoretical curves for these line ratios. The [OIII] versus [OII] intensity diagram shows that the oxygen abundance for the SNR lie between 0.5×10^{-4} and 1.2×10^{-3}. It illustrates also the saturation effect which affects the [OII] lines at high oxygen abundances.

In Fig 3. we see that the 6548,6584/Hα ratio determines well the N/O abundance ratio once the oxygen abundance is known. The SNR in the galaxies

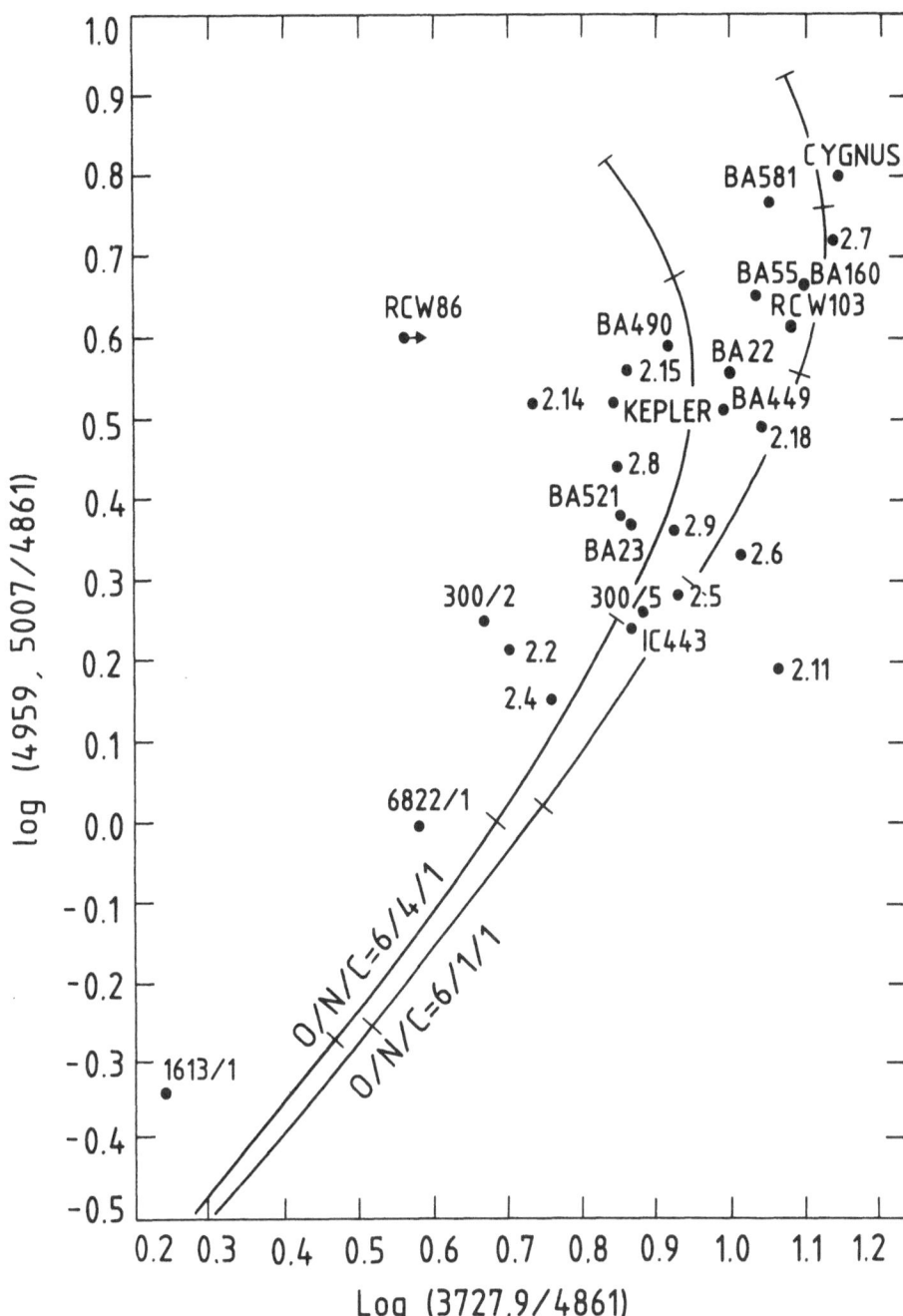

Figure 2. Data for the SNR in this figure are from this paper and from Dopita et al. (1980) for M33; Blair et al. (1982) for M31; Leibowitz and Danziger, (1982) for galactic objects. The ticks in the theoretical curves mark variations by a factor of two in the abundance of elements heavier than helium, the highest points corresponding to an oxygen abundance of 2.4 x 10^{-3}. All of the models used to build up the grid assume a preshock density of 10 cm^{-3} and a shock velocity of 106 km/sec.

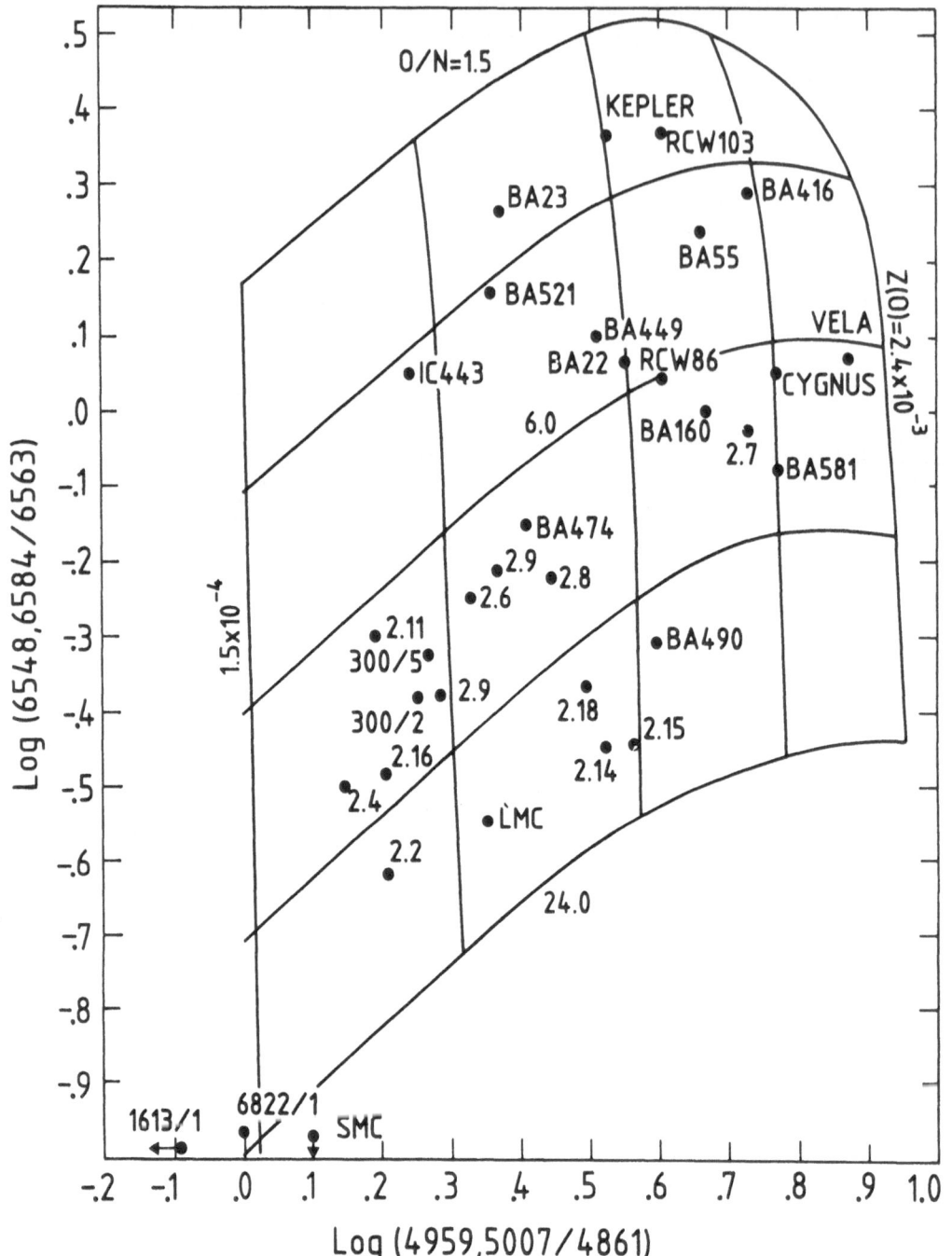

Figure 3. Data for the SNR in this figure are from the sources quoted in Fig. 2 and from Danziger, in this volume, for the LMC and SMC. The theoretical grid shows the effect of a change in the abundance of oxygen (vertical lines) and of the O/N ratio (diagonal lines). Both sets of lines are plotted at intervals corresponding to a factor of two change. Parameters of the shock as specified in Fig. 2.

IC 1613, NGC 6822 and the SMC have oxygen abundance lower than $2 \times 10^-$ and O/N larger than 24.

The theoretical grids give the possibility of understanding the gross variations of the line intensity ratios in terms of abundance, but more accurate, ad hoc models for a particular remnant can be found by an iterative process. Table 2 lists the shock parameters and the chemical abundances in number of atoms relative to H which give the best fit for the four remnants discussed in this paper. The predicted line intensities are given in table 1 and comparison with the unreddened values shows that the agreement is good.

Table 2 - Parameters of the Best Fit Models

Object	n_0 (cm^-)	Shock velocity	He	C	N	O	Mg	Si	S
IC 1613	20	131	8E-2	1.1E-5	5E-6	4E-5	2.3E-6	2.3E-6	1.1E-6
NGC 6822	10	131	8E-2	2E-5	4.5E-6	1E-4	2E-5	4E-6	1.6E-6
NGC 300/2	10	131	8E-2	3E-5	2.5E-5	1.4E-4	5E-6	5E-6	3E-6
NGC 300/5	10	131	8E-2	5E-5	3E-5	2.2E-4	1E-5	1E-5	6E-6

4. CONCLUSIONS

Six years ago, no supernova remnants were firmly identified beyond the Magellanic Clouds. Since then it has been possible to extend the iden- tifications to galaxies as far as 5Mpc, and to use the spectrophotometric data on SNR in combination with theoretical modelling of emission of the shock ionised gas to determine chemical abundances of the interstellar matter. This method can be applied to high metallicity systems without the shortcomings met in the study of HII region observations, (e.g. poor knowledge of the temperature). A new, powerful tool to investigate the chemical composition and evolution of galaxies has become available.

ACKNOWLEDGEMENTS

This work has been supported by the Italian National Research Council (C.N.R.).

REFERENCES

Binette L., Dopita M.A., D'Odorico S. and Benvenuti P. 1982, A.A. in press
Binette L., Dopita M.A. and Tuohy 1982, Ap. J. submitted
Blair W.P., Kirshner R.P. and Chevalier R.A. 1981, Ap.J. 247, 879
Blair W.P., Kirshner R.P. and Chevalier R.A. 1982, Ap.J. 254, 50
Cullum M. 1979, ESO Tech. Rept 11
Dennefeld M., and Kunth D., 1981, A.J. 86, 989
D'Odorico S., Dopita M.A. and Benvenuti P. 1980, Astr. Ap. Suppl.
Dopita M.A., Binette L. and D'Odorico S. 1982, Ap.J. submitted
Dopita M.A., D'Odorico S. and Benvenuti, P. 1980, Ap.J. 236, 628
Humphreys R.M., 1980, Ap.J. 238, 65
Leibowitz E.M. and Danziger I.J. 1982, MNRAS, in press

DISCUSSION

McKEE: I have two questions concerning the effects of ejecta on your
observations. First, the supernovae in irregular galaxies should eject as
much metals with the ambient gas as do SN in our galaxy. Since the ambient
metal abundance is much lower, however, the ejecta should be relatively
much more important. Do you see evidence for this? Second, Mike Dopita
indicated earlier in the conference that there is evidence for a large
oxygen abundance even in remnants as old as the Cygnus Loop. Do you see
such effects in your survey?

D'ODORICO: Our observations are representative of the emission from the
entire remnants in a stage where most of the emission comes from the swept
up material . We clearly reveal the differences in chemical abundances of
the IM in the different galaxies, but are not yet able to detect "second
order" effects. To answer the second point, but also in relation to the
first remark, it is only in galactic remnants (and possibly in the
Magellanic Clouds) that we can still reveal overabundances related to the
stellar ejecta. At which epoch and how well in a SNR the enriched material
is mixed with the IM is still an open question.

SHULL: Shocks of velocity 100km/s can destroy 50-80% of silicate grains
(see contribution by myself, Seab and McKee). Thus, part of the oxygen
abundance variations you report could be due to shock processing of
interstellar grains. Since N and S are probably undepleted, their abundance
variations are unaffected.

THE X-RAY PROPERTIES OF SUPERNOVA REMNANTS IN THE LARGE MAGELLANIC CLOUD

Knox S. Long
Physics Department, Johns Hopkins University

There are at least 25 supernova remnants (SNR) in the Large Magellanic Cloud (LMC) with X-ray luminosities exceeding 2 x 10^{35} erg s^{-1}. As many as 25 other SNR may be contained in the X-ray survey conducted with the Einstein Observatory of the LMC. The X-ray spectra of the 6 SNR observed with the Solid State Spectrometer (SSS) resemble their galactic counterparts; two SNR, N157B and 0540-69.3, may emit X-rays primarily by synchrotron radiation. The density of the medium in which SNR are expanding inferred from the X-ray data appears to decrease with SNR diameter; the density of the ISM inferred from the Balmer lines of 4 new SNR in the LMC is much lower than that inferred from X-ray observations. The apparent thermal energy content of LMC SNR evolves with diameter, peaking at $\sim$5 x 10^{50} ergs. The ratio of the densities of the X-ray and [SII] emitting plasmas is consistent with their being in pressure equilibrium. The SN rate in the LMC is $\sim$1 per 100-200 years. This is the number of SN expected from other considerations. The number diameter relation of LMC SNR is consistent with free expansion. The X-ray data are difficult to understand in terms of traditional Sedov models on SNR evolution; probably ejecta and multiphase ISM are required to explain the X-ray properties of LMC SNR.

The LMC is morphologically quite different from the Galaxy. An irregular barred galaxy, the LMC is roughly 1/10 as massive as the Galaxy. Nevertheless, because absorption along the line of sight to the LMC is small ($E_{B-V}\sim0.1$) and because all LMC objects are at essentially the same distance ($\sim$55 kpc), it may be better suited to studies of the generic properties of supernova remnants than the Galaxy. Recently, Long, Helfand and Grabelsky (1981) completed a study of the LMC with the imaging instruments on the Einstein Observatory. This survey consisted of more than 100 individual Imaging Proportional Counter (IPC) observations with times which averaged 2000 seconds. Most of the optically prominent portions of the LMC were covered, including the diffuse optical bar, the 30 Doradus nebula and surrounding HI cloud and much of the spiral arms. Sources were detected in all portions of the LMC; $\sim\frac{1}{2}$ of the sources lie superposed on the HI cloud which envelopes 30 Doradus, implying Pop I progenitors for the majority of the LMC

J. Danziger and P. Gorenstein (eds.), Supernova Remnants and their X-Ray Emission, 525–533.

sources. Approximately $\frac{1}{4}$ of the 97 sources in the survey are expected
to be foreground stars or active galaxies; active ground-based observational programs are underway to characterize the objects which were
detected.

A substantial number of sources in the survey were immediately
identifiable as SNR. Twelve of the 13 SNR which had been recognized
by Mathewson and Clarke (1973) on the basis of a non-thermal radio
spectrum and a large [SII]/H$\propto$ flux ratio were detected as X-ray sources;
the radio sources N157B and 0540-69.3 were also detected, apparently
confirming their identification as SNR despite the lack (at that time)
of optical counterparts. Five other objects were coincident with
entries in a list of SNR candidates which had been compiled by Davies,
Eliot and Meaburn (1976) solely on the basis of filimentary H$\propto$ emission.
Thus 17 objects previously suspected or identified as SNR were detected
as X-ray sources.

To further characterize the sources detected with the IPC, Long
Helfand and Grabelsky carried out High Resolution Imager (HRI) observations of many of the brighter sources. Of the twenty sources which had
no firm identification which were detected with the HRI, eight were
extended and thus determined to be SNR. In all 25 SNR have been detected
as X-ray sources in the LMC; about 15 other sources are known to be point
sources including LMC X-1, 2, 3, and 4. If a similar percentage of the
remaining sources are SNR, then the total number of SNR in the LMC
approaches 50.

A search to identify new SNR in the LMC based on the X-ray observations is being carried out by Mathewson, et. al. (1983); no new SNR
have been identified as yet, although optical emission associated with
all of the X-ray discovered SNR has been found. Four of these remnants
have spectra which are dominated by Balmer line emission similar to
Tycho's SNR and SN1006 (Tuohy et al 1982) which explains why they were
not recognized as SNR previously; the remainder appear to have typical
SNR emission characteristics.

SSS spectra of 6 SNR in the LMC were obtained by Clark et al. (1982).
Four of the remnants, N132D, N49, N49B and N63A, have emission-line
dominated spectra similar to most galactic remnants; two, N157B and
0540-69.3, have spectra which can be fit to a power law. These two
remnants, in addition to N103B, have centrally peaked X-ray morphologies
and may be examples of Crab-like SNR in the LMC.

The relationship between X-ray luminosity and SNR diameter for the
LMC remnants is shown plotted in figure 1. The diameters used here and
elsewhere are derived from a consideration of the X-ray and optical
morphologies documented by Mathewson, et al (1983). The SNR with optical
spectra dominated by the Balmer lines of hydrogen (▲) and the SNR with
centrally-condensed X-ray morphologies (■) are plotted separately.
Numerical models of SNR (such as those discussed by White and Long, 1983)
evolve from low to high luminosity as more IS material is shock heated,

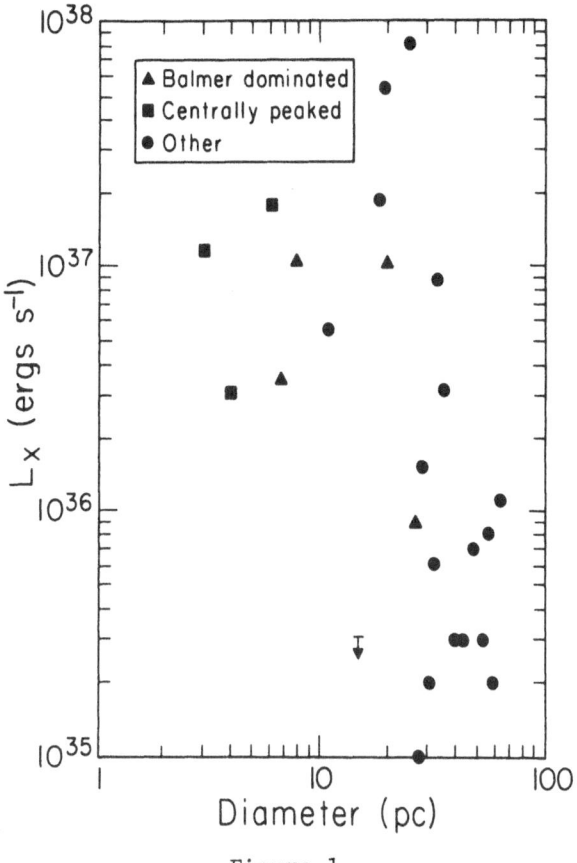

Figure 1.

and subsequently fade as the temperature of the material falls below that required to generate X-rays. The LMC SNR appear with considerable scatter to follow such an evolutionary track.

In principle, the luminosity L_x and diameter D of a SNR permit a determination of the density of the ISM into which a SNR is expanding if its X-ray emission is dominated by shocked gas from the ISM. Specifically, the ISM density

$$n = \left(\frac{6}{\pi}\right)^{\frac{1}{2}} \epsilon^{-\frac{1}{2}} f^{\frac{1}{2}} L_x^{\frac{1}{2}} D^{-3/2}$$

where ϵ is the specific emissivity and $f(=\langle n_e^2 \rangle / \langle n_e \rangle^2)$ is the fraction of the SNR volume filled with X-ray emitting material. If f and ϵ do not change greatly with diameter, the inferred ISM densities evolve strongly with diameter as shown in figure 2. Here f was taken to be $\frac{1}{4}$ and ϵ to be 3×10^{-23} erg cm^{-3} s^{-1}, appropriate for a strong shock and a collisionally-equilibrated plasma having cosmic abundance ratios and a temperature of 5×10^6 K (Raymond and Smith 1977). Possible explanations for this trend include the following: (1) Cloudlet evaporation.

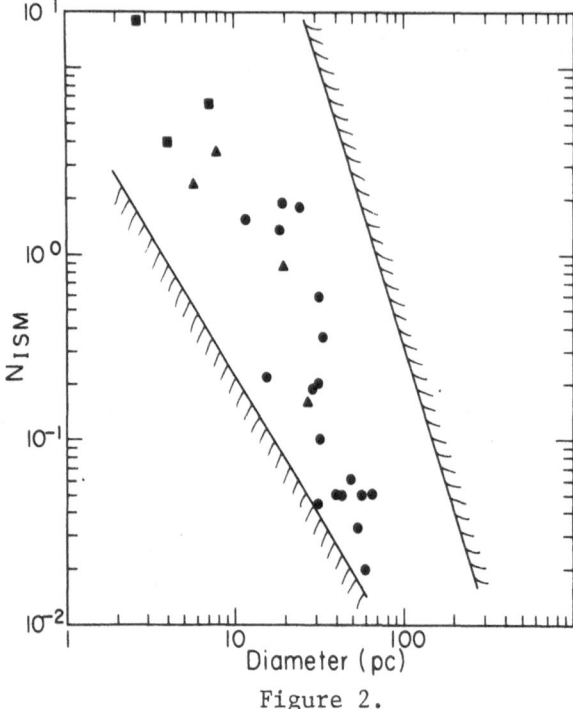

Figure 2.

The correlation of apparent density with SNR diameter was first noticed
in a small sample of galactic SNR by McKee and Ostriker (1977). They
used this correlation to support a multiphase model of the ISM in which
a pervasive tenuous gas is studded with dense cloudlets. After the
passage of a SNR shock which is carried by the tenuous medium, the
cloudlets are heated by either conduction or more slowly moving secondary shocks to X-ray temperatures. Because the disruption of cloudlets
is more rapid while the shock velocity is high, the apparent density
inferred using a single phase model of the ISM evolves with SNR diameter.
(2) Emission from SN ejecta. Substantial amounts of material are
ejected from the star in a SN explosion; if shock heated, either by
interaction with circumstellar material or the surrounding ISM, the
ejecta should emit at X-ray temperatures. Because this debris represents
a smaller and smaller portion of the total shocked gas as the SNR grows,
the apparent density, when ejecta are neglected, would appear to evolve.
Long, Dopita and Tuohy (1982) argue that if the ejecta is metal rich,
the density diameter relation can be understood. Analyses of SSS spectra
of young SNRs show deviations from cosmic abundance ratios of heavy
metals even when non-equilibrium effects are taken into account (Becker,
et al 1980 a,b). However, the SSS is insensitive to the ratio of metals
to H and He. SN ejecta are known to be important in some remnants.
Reverse shocks are observed in Cas A (Fabian, et al 1980), Tycho's SNR
(Seward, Gorenstein and Tucker 1982) and possibly N132D. In Balmer-
dominated SNR the density of the neutral ISM can be inferred from the
optical line intensities; for the LMC remnants, the neutral ISM density

is much lower than total density obtained from the X-ray data (Tuohy, et al 1982). (3) <u>Selection effects</u>. Finally, the X-ray survey of the LMC is luminosity limited. Until the post shock temperature falls below 10^6K, the X-ray luminosity scales as $n^2 D^3$. For a SNR to have been detected, the ISM density must exceed $0.2(D/10pc)^{-3/2}$. Furthermore, a SNR in a dense medium cools below 10^6K at a smaller diameter than does a SNR encountering more tenuous material. If the explosion energies of LMC SNR do not exceed 10^{51} ergs, observable SNR must be located in regions in which $n \lesssim 330 (D/10pc)^{-3}$. The excluded regions are shown shaded in Figure 2. Possibly all these effects contribute the apparent diameter density relation of the LMC SNR.

The thermal energy content of the LMC SNR can be inferred from the size, density and temperature of the X-ray emitting material. If we assume (in contrast to Gronenschild and Mewe 1982) that a line temperature of 5×10^6K derived from the SSS spectra under equilibrium assumptions is appropriate, then the thermal energy content evolves from 8×10^{48} ergs to 5×10^{50} ergs, as illustrated in figure 3. Blair, Kirshner and Chevalier have attempted to derive the thermal energy content of galactic and extragalactic SNR using the pressure inferred

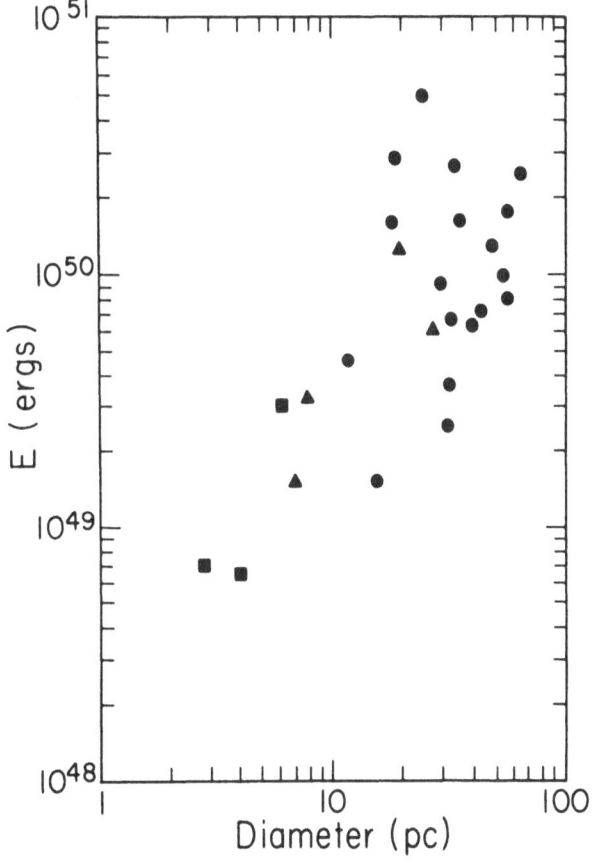

Figure 3

from [SII] line ratios, assuming that pressure in the clouds where SII
is formed is identical to that outside the clouds; Blair, Kirshner and
Chevalier (1981) also find that the thermal energy content in SNR increases
with radius. (See figure 5 of their paper.) The simplest interpretation
of these results is that energy of the initial explosion is evolving into
thermal energy and that this evolution is not complete until SNR are
quite large, at least 30pc in diameter. Blair, Kirshner and Chevalier
(1981) suggest this energy is stored in magnetic fields, but alternatively it may be stored in the kinetic energy of the ejecta. Such an
interpretation is probably inconsistent with single phase models of the
ISM where the evolution in expected to be much more rapid. In order for
the kinetic energy of the ejecta to remain high, it is necessary that
the medium which is slowing the ejecta be quite tenuous.

The densities derived from the X-ray observations and the densities
of the [SII] emitting regions as measured by Dopita[†] (1979) are correlated, as shown in figure 4. The ratio between densities of the shocked
cloudlets emitting in [SII] to the X-ray plasma is ~ 500 if the X-ray
filling factor is ¼. With this filling factor and a temperature of
~5 x 10^6K, there is approximate pressure equilibrium between the shocked
[SII] cloudlets which have a temperature of 10^4K and the post shock
X-ray plasma.

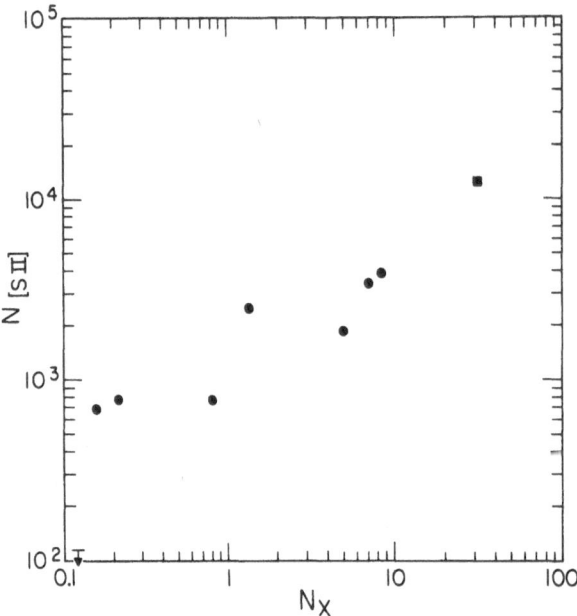

Figure 4.

[†]Although the post-shock densities are not given explicitly by Dopita,
they can be obtained using equation 8 of his paper.

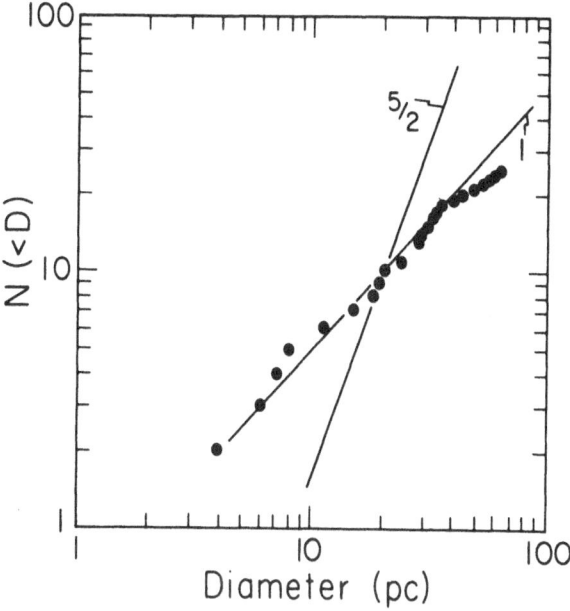

Figure 5.

If most of the LMC are expanding into a uniform density ISM and have swept up enough IS material so that their time evolution resembles that of a Sedov explosion, the number of SNR with diameters less than D should increase as $D^{5/2}$. This is not the case, as is evident from figure 5. Instead, the number-diameter relation is consistent with a power law index of 1. It is as if the SNR which we observe have not been decelerated at all by the ISM. This result is not new; Mathewson and Clarke in their original discussion noted that the number-diameter relation was much flatter than that expected from the Sedov relations. The result was discounted, however, because galactic SNR seem to obey a $N^{5/2}$ law, a result Mills (1983) has now shown to be due to the Σ -D relation assumed for galactic remnants. If SNR are still freely expanding at diameters of 30pc, high velocity material should exist within them; N132D is the only large diameter remnant in which material moving at velocities (> 1000 km s^{-1}) has been discovered (Lasker 1977). If SNR are freely expanding with a velocity which averages 5000 km s^{-1}, then the SN rate in the LMC lies between 1 per 200 (100) years assuming 25(50) SNR with diameters less than 50pc. The SN rate is not very dependent on the assumption of free expansion; Long, Helfand, and Grabelsky derived rates of 1 per 110 to 340 years from a Sedov analysis of SNR expansion. The SN rate in the LMC, for almost any reasonable set of assumptions, is consistent with the SN rate per unit mass in the Galaxy.

Although the importance of selection effects should not be under-estimated, the data on SNR in the LMC seem to present fundamental prob-lems for the traditional view that most SNR which we observe can be understood in terms of the Sedov solutions to a point explosion in a

uniform density medium. The observations do not currently resolve
whether most of the emission arises from SN ejecta or from cloudlets in
a multiphase ISM. It appears likely that the apparent density evolution,
the Balmer-line inferred densities, free expansion, multiple-temperature
SSS spectra, and thermal energy evolution can be accomodated within
either context. Probably both are partially correct.

        Acknowledgements: Much of this work was carried out in the context
of a collaboration on the LMC with DJ Helfand, IR Tuohy, MA Dopita and
DS Mathewson. Financial support was provided at Columbia and Johns
Hopkins by NASA contracts NAS 8-30753 and NAS 5-27000. Travel support
by the IAU is gratefully acknowledged.

REFERENCES

Becker, RH, Holt, SS, Smith, BW, White, NE, Boldt, EA, Mushotsky, RF
    and Serlemitsos, PF 1980A, ApJ (Letters) 235, L5.
Becker, RH, Szymkowiak, AE, Boldt, EA, Holt, SS and Serlemitsos, PJ
    1980B, ApJ (Letters) 240, L33.
Blair, WP, Kirshner, RP, and Chevalier, RA 1981, ApJ 247, 879.
Clark, DH, Tuohy, IR, Long, KS, Syzmkowiak, AE, Dopita, MA, Mathewson,
    DS, Culhane, JL 1982, ApJ 255, 440.
Davies, RD, Eliot, KH and Meaburn, J 1976, MNRAS 81, 89.
Dopita, MA 1979, ApJ Suppl 40, 456.
Fabian, AC, Willingale, R, Pye, JP, Murray, SS and Fabbiano, G 1980,
    MNRAS 193, 175.
Gronenschild, EHBM and Mewe, R 1982, Astr Ap Suppl 48, 305.
Lasker, BM 1977, PASP 89, 474.
Long, KS, Dopita, MA and Tuohy, IR 1982, ApJ 260, 202.
Long, KS and Helfand, DJ 1979, ApJ (Letters) 234, L77.
Long, KS, Helfand, DJ and Grabelsky, DA 1981, ApJ 248, 925.
Mathewson, DS and Clarke, JN 1973, ApJ 180, 725.
Mathewson, DS, Ford, VL, Dopita, MA, Tuohy, IR, Long, KS, and Helfand,
    DJ 1983, ApJ Suppl, in press.
McKee, CF and Ostriker, JP 1977, ApJ 218, 148.
Mills, BY 1983, this volume, p. 563.
Raymond, JC and Smith, BW 1977, ApJ Suppl 35, 419.
Seward, F, Gorenstein, P, and Tucker, W 1982, ApJ submitted.
Tuohy, IR, Dopita, MA, Mathewson, DS, Long, KS and Helfand, DJ 1982,
    ApJ, in press.
White, RL and Long, KS 1983, ApJ, in press.

DISCUSSION

CHEVALIER: The suggestion (by Blair, Kirshner and Chevalier) of
dominant magnetic pressure will not work for a high temperature X-ray
emitting gas.

FABIAN: Another way of explaining the possible inverse-correlation
of density with radius and evidence for free expansion is through

stellar winds. The remnant should expand relatively freely to the
edge of the bubble which could be 20-40pc from the explosion.

MCKEE: Ejecta expanding into a multiphase medium in which much of
the volume is at very low density can expand to fairly large diameters
(20-40pc) before thermalizing. Out to what radii are you confident
that the uniform expansion model applies?

LONG: There appears to be a break in the slope of the number
diameter relation between 30 and 40pc. However, it is very difficult
to determine to what diameter the sample is complete.

SOFT X-RAY OBSERVATION OF SUPERNOVA REMNANTS IN THE SMALL MAGELLANIC CLOUD

H.Inoue, K.Koyama, and Y.Tanaka
Institute of Space and Astronautical Science, Tokyo, Japan

The Small and Large Magellanic Clouds (SMC and LMC) are the nearest neighbouring galaxies. Their proximity enables us to investigate these galaxies in X-rays in fair detail.

Seward and Mitchell (1981) conducted a complete survey of the SMC from the *Einstein* Observatory, consisting of 40 IPC pointings with an exposure of $\sim$2000 sec each. 26 sources were detected above a threshold $L_X = 3 \times 10^{35}$ erg s^{-1}, if in the SMC. Excluding 5 foreground stars, 10 are distributed in the bar and the spiral arm regions, while the remaining 11 are widely distributed over the ourskirt of the Cloud.

We present the result of the observations of the SMC from the *Einstein* Observatory in which deep exposure of the selected regions of the SMC was performed in order to search for SNR's and SNR-related structures.

1. OBSERVATIONAL RESULTS

Our observation comprises three IPC frames with exposures in excess of 20,000 sec for each, and subsequent three HRI frames with exposures of 13,000 $\sim$ 18,000 sec. The IPC fields are chosen so as to cover the regions where HI, HII and radio structures are most prominent. Fig.1 shows a mosaic of 7 IPC maps in which 4 maps of Gull and Bruhweiler are included with their kind permission. Presence of X-ray sources besides those reported by Seward and Mitchell are evident.

In our three IPC fields, 25 significant sources are selected in total at a confidence level greater than 3σ. These sources are listed in Table I. Positional coincidence with radio sources and/or HII regions of Davies, Elliot and Meaburn (DEM) (1976) are also indicated. The tabulated X-ray luminosity values are those estimated on the assumption that 1 IPC count s^{-1} from an SMC source corresponds to $L_X \cong 2 \times 10^{37}$ erg s^{-1} in the energy range 0.15 $-$ 4.0 keV. The interstellar absorption in the SMC tends to give substantial underestimates of the luminosity and overestimates of the hardness ratio. The neutral hydrogen column density in the chosen IPC fields ranges $(2 - 4) \times 10^{21}$ cm^{-2} in the SMC

535

J. Danziger and P. Gorenstein (eds.), Supernova Remnants and their X-Ray Emission, 535–540.
© 1983 by the IAU.

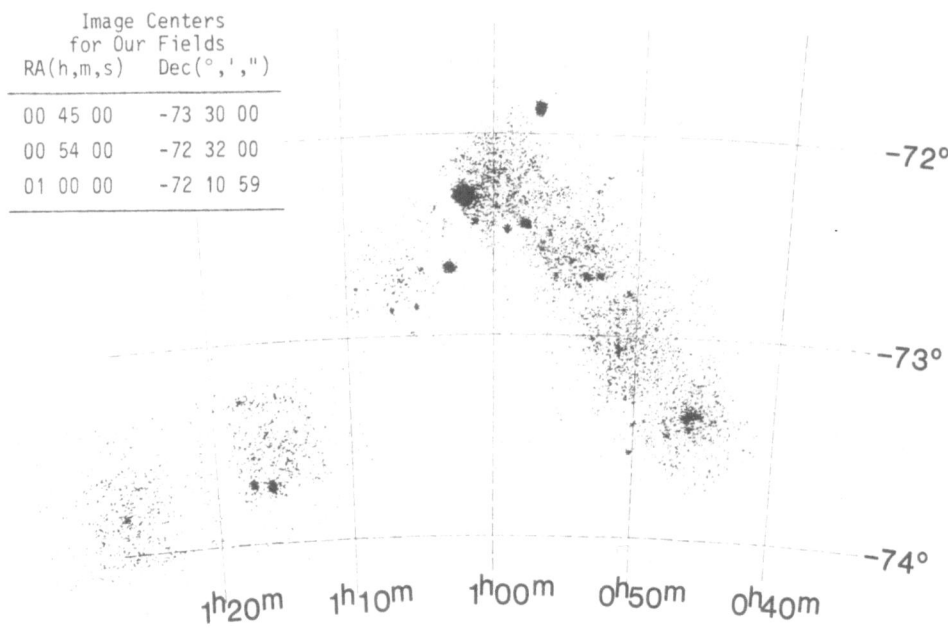

Fig.1 Mosaic of 7 IPC maps. 4 IPC maps by Gull and Bruhweiler are
included by their courtesy. (equinox: 1950)

Table I   Soft X-Ray Sources Detected

SOURCE NO.	RA (1950) (h,m,s)	Dec(1950) (°,',")	COUNT RATE (s⁻¹)	L_X(0.15−4 keV) (×10³⁵erg s⁻¹)	$\dfrac{H(1.5-4.0\text{keV})}{S(0.15-1.5\text{keV})}$	$\dfrac{H-S}{H+S}$	COMMENTS	ASSOCIATIONS
1	00 45 32	−73 28 51	0.012 ± 0.001	2.4	0.88 ± 0.20	−0.06		
2	00 45 34	−73 25 21	0.029 ± 0.002	5.8	0.31 ± 0.05	−0.56	extended(2.8') or plural	Radio SNR, DEM32
3	00 46 05	−73 21 21	0.003 ± 0.001	0.6				DEM31(N19)
4	00 46 22	−73 35 19	0.003 ± 0.001	0.6				Radio SNR, DEM42(N24)
5	00 47 15	−73 30 15	0.003 ± 0.001	0.6				DEM49?
6*	00 49 33	−73 38 08	(0.015)*	(3.0)*				Radio SNR
7	00 49 54	−73 26 48	0.023 ± 0.002	4.6	0.21 ± 0.07	−0.65		DEM60?
8	00 50 22	−73 35 09	0.015 ± 0.001	3.0	0.63 ± 0.26	−0.23		DEM70
9	00 51 13	−72 14 46	0.003 ± 0.001	0.6				
10	00 52 17	−72 42 46	0.012 ± 0.001	2.4	0.28 ± 0.19	−0.57		
11	00 53 20	−72 42 48	0.008 ± 0.001	1.3	8.0	0.78		
12	00 54 01	−72 44 37	0.003 ± 0.001	0.6				
13	00 54 27	−72 38 07	0.005 ± 0.001	1.0				DEM86
14	00 55 42	−72 42 10	0.008 ± 0.001	1.6	0.81 ± 0.49	−0.10		
15	00 55 55	−72 29 59	0.004 ± 0.001	0.8				
16	00 56 37	−72 33 55	0.005 ± 0.001	1.0				
17*	00 56 54	−71 52 03	0.085 ± 0.003	17.0	0	−1.0	HRI: point-like	Foregr.Star?
18*	00 57 46	−72 26 17	0.029 ± 0.002	5.8	0.43 ± 0.09	−0.40		Radio Em., DEM103(N66)
19	00 59 07	−72 27 46	0.008 ± 0.001	1.6				
20	01 00 05	−72 11 10	0.007 ± 0.002	1.4				extended(~4') or plural	DEM115(N74),DEM116(N75)
21*	01 01 31	−72 25 52	0.012 ± 0.002	2.4	0.93 ± 0.55	−0.04		Radio SNR
22*	01 02 25	−72 18 00	0.783 ± 0.007	157	0.38 ± 0.01		HRI: extended	Radio SNR, DEM124
23*	01 03 23	−72 39 20	0.084 ± 0.004	16.8	0.15 ± 0.03	−0.74	HRI: extended	Radio SNR, DEM125
24	01 03 45	−72 28 30	0.013 ± 0.002	2.6	0.50 ± 0.14	−0.33	extended(~3') or plural	DEM128
25	01 04 34	−72 22 10	0.017 ± 0.002	3.4	0.15 ± 0.12	−0.74		DEM131

* Sources previously detected by Seward and Mitchell (1980).

added to that in our galaxy of $(3-3.5) \times 10^{20} cm^{-2}$ (McCammon et al. 1976).
Therefore, the X-ray luminosities given in Table I may be regarded as
lower limits, apart from a spectrum-dependent uncertainty of the order
of a factor of 2. In the present observations, the lowest luminosity of
the detected sources, if in the SMC, is $L_x = 6 \times 10^{34} erg\ s^{-1}$ without taking
into account the above absorption effect. The detection of sources near
this luminosity threshold is not yet complete, however.

2. SOURCE IDENTIFICATION

1) SNR's

Of 25 sources detected, 6 (No's. 2,4,6,21,22 and 23) are identified
with the radio SNR's. Mills et al.(1982) identified 6 radio SNR's with
the Molonglo Observatory Synthesis Telescope (MOST) from the observations
of 18 targets including 16 X-ray sources of Seward and Mitchell (S.M.).
Mathewson et al.(1982) subsequently confirmed these 6 sources to be the
optical SNR's. All of these 6 radio/optical SNR's are coincident in
position with the X-ray sources, 4 of 6 being the S.M. sources. 2 of
the 6 SNR's had been previously identified as SNR's by Mathewson and
Clarke (1972, 73).

2) Suspected SNR's

Source No.18 is coincident in position with a radio source of Mills
et al. (MOST 0057-724). Although evidence for the non-thermal nature of
the radio is not firm, we tentatively identify this to be an SNR.

In addition to this, 3 more possible SNR's are selected from the
consideration of the hardness ratio. Hardness ratios of the identified
SNR's range from 0.15 to 0.9. On the other hand, compact binary X-ray
sources with kT > 3 keV would give hardness ratios no smaller than 0.7,
without taking account of the interstellar absorption in the SMC. 4
sources (No. 7, 10, 17 and 25) satisfy the criterion that the hardness
ratio be significantly smaller than 0.7. The source No.17, the second
brightest of the present catalog, is however found to be a point source
by a subsequent HRI examination. As the source also exhibits an unusually
soft spectrum (kT < 0.1 keV), we consider it to be a foreground star in
our galaxy (no identification yet).

Two more sources, No.20 and 24, are suspected as SNR's. These sources
appear extended by approximately 4' and 3', respectively, although a
possibility of closely spaced multiple sources could not be totally ruled
out. The observed soft X-ray enhancement in such extended regions as
large as 80 pc across may be analog of hot bubbles observed in our galaxy.
Thus, we have 6 identified plus 6 suspected SNR's in total.

3) Other sources

There remain 12 unidentified sources which are all of low luminosities in the range $L_x \lesssim 10^{35}$ erg s^{-1}. No cataloged galactic or extragalactic objects are coincident in position. The number of interlopers outside the SMC is estimated to be ∿5 for the lowest flux level, based on the detection of ∿0.7 serendipitous sources per square degree above 0.01 IPC counts s^{-1} with an assumed log N − log S slope of -1.5. The locations of these unidentified weak sources, when plotted on the optical image of the SMC, suggest that most of them are within the SMC and distributed along the bar and the arm. We believe that several of these sources with $L_x \lesssim 10^{35}$ erg s^{-1} are, if not all, SNR's in the SMC, since SNR's of this range of L_x are abundant in our galaxy.

3. HRI OBSERVATIONS OF SNR'S

We conducted HRI observations of three brightest sources (No.17, 22 and 23). No.22 and 23 clearly revealed shell-like structures as shown in Fig.2. No.17 was found point-like as mentioned in Section 3. Dopita, Tuohy and Mathewson (1981) reported that the source No.22 was a bright SNR in [OIII] emission with a slightly smaller diameter than that in X-rays. The radio shell of the source No.23 was also well resolved by Mills et al.(1981) and the X-ray and radio shell diameters are in good agreement with each other. The result of the HRI observation is summarized in Table II. If one assumes that these SNR's are in the adiabatic expansion phase, the X-ray luminosity L_x, shell radius r, and temperature kT would yield the initial energy E_0, age t and ambient gas density n_0 by utilizing the conventional shock wave model. n_0 can be estimated from L_x and r with a less model-dependent manner. Assuming a strong shock, $L_x \cong (\pi/3) r^3 (4n_0)^2 \Lambda_x(T)$, where $\Lambda_x(T)$ is the X-ray emissivity which is roughly constant and ∿3 x 10^{-23} erg s^{-1}cm^3 for the range kT $\gtrsim$ 0.3 keV. Table II contains so derived n_0 from the HRI result and also for other SNR's for which the shell radius can be estimated on the high-resolution radio maps of Mills et al.. Estimation of E_0 and t are model dependent.

No.22   No.23

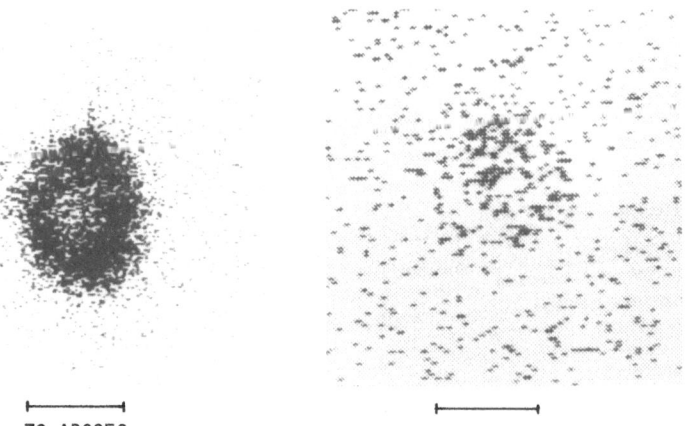

├─────────────┤   ├─────────────┤

30 ARCSEC   2 ARCMIN

Fig.2 HRI images of No.22 and 23. Both reveal shell-like structures.

Table II   X-Ray and Radio SNR's in the SMC

Source No.	RA(1950) (h,m,s)	Dec(1950) (°,',")	L_x (10^{35} erg s^{-1})	Diameter (pc)		kT (keV)	n_o (cm^{-3})	E_o (10^{51} erg)	t (10^3y)
2	00 45 34	-73 25 21	5.8	(4')(R) (2.8')(X)	(70) (50)	1.2±0.5	(0.03) (0.05)	(2.6) (1.6)	(16) (11)
4	00 46 22	-73 35 19	0.6	1.1'(R)	19		0.07		
6	00 49 33	-73 38 08	3.0	1.7'(R)	30		0.08		
21	01 01 31	-72 25 52	2.4	<40"(R)	<12		>0.27		
22	01 02 25	-72 18 00	157.	27"(X)	8	1.4±0.5	4.1	0.6	1.7
23	01 03 23	-72 39 20	17.	2.2'(X) 2.2'(R)	38	0.3±0.1	0.13	0.4	18

4. DISCUSSIONS

Long, Helfand and Grabelsky (1981) published their results of the
Einstein observations of the LMC. In the 75 sources detected in the LMC
above a threshold of 10^{35}erg s^{-1}, the identified and suspected SNR's are
25 and 11, respectively. The total number of SNR's in the LMC was
estimated to be about 55. These numbers are to be compared with 6 identified and 6 suspected SNR's in 3 IPC fields of the SMC. The luminosity
distribution of the identified and suspected SNR's in the SMC is shown
in Fig.3 in comparison with that for the LMC (Long et al. 1981). No
qualitative difference seems to exist. In the X-ray map of the SMC in
Fig.1, there appear several more low-luminosity sources in the fields of
Gull and Bruhweiler. While some of them may possibly be SNR's, the
number of SNR's in the SMC with a luminosity $L_x \gtrsim 10^{35}$erg s^{-1}may not
largely increase. We therefore conclude that the number of SNR's per
unit mass is roughly the same in both the Clouds, since the mass ratio
of the SMC to the LMC is about 1/5.

X-ray survey of SNR's in our galaxy is by far incomplete. On the
other hand, radio searches for the galactic SNR's have been fairly

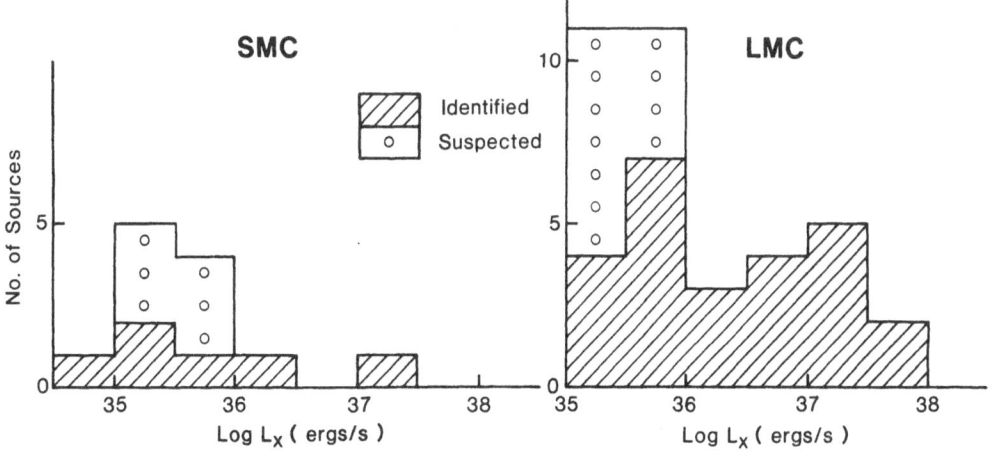

Fig.3 The luminosity distribution of the identified and suspected SNR's
in the SMC. Result for the LMC (Long et al. 1981) is also shown
for comparison.

complete. Clark and Caswell (1976) list 120 radio SNR's in our galaxy,
in which nearly 30 (and probably more by now) have so far been detected
in the X-ray band. Since every X-ray observation of a radio SNR with
the *Einstein* sensitivity yielded positive detection and few radio-silent
SNR's (except 1E1149.4-6209, Markert et al. 1981) have been found in
X-rays, we may safely rely upon the radio result and say that the number
of galactic SNR's is not much greater than 120. If the SNR's with luminosities $L_X > 10^{35}$erg s^{-1} are concerned, even a smaller number will result
considering the fact that several of the SNR's so far detected are with
luminosities less than 10^{35}erg s^{-1}.

Our galaxy is more massive by a factor of about 20 and 100 than the
LMC and the SMC, respectively. If one extrapolates the numbers of SNR's
observed in the LMC and the SMC to our galaxy in proportion to the system
mass, the expected number amounts to about 1000, which is nearly an order
of magnitude greater than the probable one. It is an important question
whether this discrepancy is due to a smaller occurrence rate of SNR's per
unit mass in our galaxy than in the Magellanic Clouds or otherwise due to
a possibility that many SNR's in our galaxy escaped detection.

We are grateful to Fred Seward for his coordination in the present
Einstein observation and to Jun Jugaku for useful information.

REFERENCES

Clark,D.H., and Caswell,J.L. 1976, M.N.R.A.S., *174*, 267
Davies,R.D., Elliott,K.H., and Meaburn,J. 1976, Mem.R.A.S., *81*, 89
Dopita,M.A., Tuohy,I.R., and Mathewson,D.S. 1981, Ap.J.(Letters), *248*,L105
Long,K.S., Helfand,K.J., and Grabelsky,D.A. 1981, Ap.J., *248*, 925
Markert,T.H., Lamb,P.C., Hartman,R.C., Thompson,D.J., and Bignami,G.F.
    1981, Ap.J.(Letters), *248*, L17
Mathewson,D.S., and Clarke,J.N. 1972, Ap.J.(Letters), *178*, L105
Mathewson,D.S., and Clarke,J.N. 1973, Ap.J., *182*, 697
Mathewson,D.S., Ford,V.L., Dopita,M.A., Tuohy,I.R., Long,K.S., and
    Helfand,D.J. 1982, Ap.J. Suppl. (in press)
McCammon,D., Meyer,S.S., Sanders,W.T., and Williamson,F.O. 1976, Ap.J.,
    209, 46
Mills,B.Y., Little,A.G., Durdin,J.M., and Kestevan,M.J. 1982, to appear
    in M.N.R.A.S.
Seward,F.D., and Mitchell,M. 1981, Ap.J., *243*, 736

OPTICAL STUDIES OF SUPERNOVA REMNANTS IN THE MAGELLANIC CLOUDS

D.S. Mathewson, V.L. Ford, M.A. Dopita and I.R. Tuohy
Mount Stromlo and Siding Spring Observatories

K.S. Long and D.J. Helfand
Columbia University

Optical identifications of 32 X-ray sources in the Magellanic Clouds confirm that they are SNRs. They are separated into four classes: the evolved, the oxygen-rich, the Balmer-dominated and the Crab-like. High velocity HI emission is observed from an extended region near 0525-66.0. It is suggested that this is produced by a possible Type III supernova which occurred out of the plane of the LMC and on the far side of the disk. The cumulative number-diameter relation for the LMC SNRs shows that they have evolved much faster than expected from the Sedov theory. It is suggested that this apparent "free-expansion" up to quite large diameters is due to the gradual conversion of the kinetic energy of the ejecta into thermal energy as they overtake the decelerating blast wave.

1. CONFIRMED SNRs IN THE MAGELLANIC CLOUDS

Optical identifications with SNRs have been made for 26 of the LMC X-ray sources in the catalog of Long et al. (1981) and for 6 of the SMC X-ray sources in the catalogs of Seward and Mitchell (1981) and Tanaka (1983). Mathewson et al. (1983) present the narrow-band images of most of the SNRs obtained using the Anglo-Australian Telescope together with the X-ray isophotes obtained with the Einstein Observatory.

Table 1 lists the SNRs in the LMC. Column 2 gives the X-ray source number from the catalog of Long et al. (1981). Column 3 gives the numbers from the catalogs of Henize (1956) and Davies, Elliott and Meaburn (1976). Column 6 gives the optical diameters except for 0525-66.0 and 0535-66.0 where the X-ray diameters are used due to the highly fragmented nature of their optical emission. Generally the optical and X-ray images are co-extensive.

Table 2 lists the SNRs in the SMC. Column 4 gives the optical diameters except for 0049-73.6 where the field is partly obscured and the radio diameter measured by Mills et al. (1982) is given. For the

541

J. Danziger and P. Gorenstein (eds.), Supernova Remnants and their X-Ray Emission, 541–549.
© 1983 by the IAU.

TABLE 1

SUPERNOVA REMNANTS IN THE LARGE MAGELLANIC CLOUD

SNR Catalog No.	X-Ray Source No.	Other Catalog No.	Position of Optical Center (1950) R.A. h m s	DEC ° ' "	Mean Diameter pc
0453-68.5	1		04 53 46	-68 34 15	36
0454-66.5		N11L	04 54 42	-66 30 22	15
0455-68.7	2	N86	04 55 53	-68 43 52	53
0500-70.2	7	N186D	05 00 20	-70 12 26	31
0505-67.9	10	DEM71	05 05 49	-67 56 38	20
0506-68.0	11	N23	05 06 03	-68 05 51	11
0509-68.7	13	N103B	05 09 13	-68 47 15	6
0509-67.5	14		05 09 36	-67 34 56	7
0519-69.7	23	N120	05 19 08	-69 42 11	28
0519-69.0	26		05 19 52	-69 05 05	8
0520-69.4	27		05 20 07	-69 28 53	32
0525-66.0	34	N49B	05 25 24	-66 01 46	37
0525-69.6	35	N132D	05 25 27	-69 41 02	29
0525-66.1	36	N49	05 25 57	-66 07 32	17
0527-65.8	39	DEM204	05 27 51	-65 52 13	56
0528-69.2	40		05 28 05	-69 14 29	32
0532-71.0	47	N206	05 32 35	-71 02 15	47
0534-69.9	53		05 34 29	-69 56 56	30
0535-70.5	54	DEM238	05 34 52	-70 35 16	46
0535-66.0	59	N63A	05 35 39	-66 03 52	19
0536-70.6	61	DEM249	05 36 43	-70 40 30	39
0538-69.1	67	N157B	05 38 10	-69 11 56	7
0540-69.3	79		05 40 34	-69 21 24	2
0543-68.9	82	DEM299	05 43 27	-69 00 05	76
0547-69.7	88	N135	05 47 37	-69 43 11	56
0548-70.4	89		05 48 24	-70 25 44	28

TABLE 2

SUPERNOVA REMNANTS IN THE SMALL MAGELLANIC CLOUD

SNR Catalog No.	Position of Optical Center (1950) R.A. h m s	DEC ° ' "	Mean Diameter pc
0045-73.4	00 45 28	-73 24 51	26
0046-73.5	00 46 38	-73 35 37	31
0049-73.6	00 49 30	-73 38 25	39
0101-72.4	01 01 40	-72 25 47	25
0102-72.3	01 02 25	-72 18 00	7
0103-72.6	01 03 38	-72 39 09	57

other SNRs the optical and radio sizes are similar.

2. SNR CLASSIFICATION

2.1 The Evolved SNRs

This class contains the majority of the SNRs. Their spectrum
which is characterised by a [SII] to Hα ratio greater than 0.7, is
recognized as due to shock waves of modest velocities (50 - 200 km s^{-1}).
There is no well-defined velocity of expansion of these SNRs and it is
thought that the optical emission arises from shocked cloudlets in the
interstellar medium.

In general the X-ray images reflect the general characteristics
of the optical images although there is not a one to one correlation.
It is believed that most of the X-ray emission also arises in shocked
cloudlets but in regions of intermediate density which are not so
dense as to have cooled too much. The displacement of the X-ray from
the optical knots reflects the density structure of the cloudlets.

Figure 1 shows the HRI X-ray isophotes superimposed on the Hα
image of 0534-69.9. A well-collimated optical jet runs out from its
western edge for about 20pc. It appears to originate near the center
of the SNR and is similar to the jet in the Crab Nebula discovered by
Gull and Fesen (1982). There is an association with the X-ray structure
which shows a central band of emission just south of the jet.

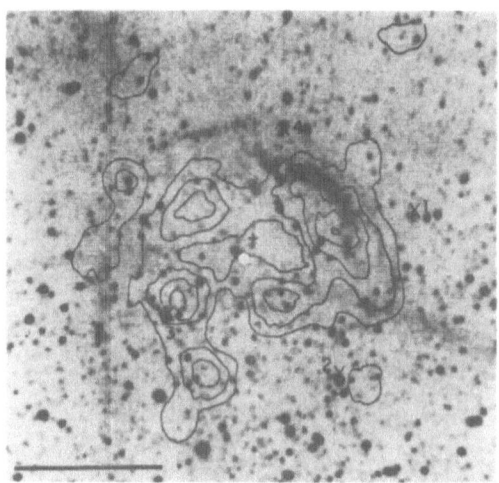

Fig. 1 SNR 0534-69.9 in the light of Hα. N to the top and E to the
left. The dot marks the optical center and the bar is one arc min.
The contour levels of the superimposed HRI X-ray isophotes are 0.006,
0.012, 0.018 and 0.024 counts s^{-1} arc min^{-2}.

2.2 The Oxygen-rich SNRs

Three examples of young oxygen-rich SNRs have been found in the
Magellanic Clouds, N132D (Danziger and Dennefeld, 1976, Lasker, 1980),
0540-69.3 (Mathewson et al. 1980) and 0102-72.3 (Dopita et al. 1981).
The oxygen-rich SN ejecta has its origin deep within the chemically
processed layers of a star greater than 25 $M_\odot$. Their spectra are
characterised by high velocity dispersions ranging from 3000 km s^{-1}
to 6000 km s^{-1}.

Surrounding 0102-72.3 is a ring of faint emission which exhibits
the high excitation line of HeII λ4686 and may be produced by the UV
flash at the time of the explosion. Between this ring and the [OIII]
filaments is a dark region, which may represent the position of the
blast wave whose high temperature would suppress the optical emission.

The three SNRs have intrinsic X-ray luminosities much greater than
CasA: indeed N132D is 16 times more luminous than CasA. This high
X-ray luminosity is a general feature of LMC SNRs as 9 of the 26 are
brighter than CasA between 0.15 and 4 keV.

2.3 The Balmer-dominated SNRs

SNRs 0505-67.9, 0509-67.5, 0519-69.0 and 0548-70.4 belong to this
class (Tuohy et al. 1983) whose main feature is that their filamentary
shells are strong in the Balmer lines of hydrogen but absent or very
weak in [OIII] and [SII]. Their Balmer spectra can be understood in
terms of a very high velocity non-radiative shock encountering gas
which is partially neutral. It is believed that the four SNRs
resulted from Type I supernovae because they have similar optical,
X-ray and radio properties to the galactic Type I supernovae, Tycho,
Kepler and SN 1006.

2.4 The Crab-like SNRs

Danziger et al. (1981) and Clark et al. (1982) suggest that N157B
is a Crab-like SNR because of its centrally condensed radio structure,
flat radio spectral index and nonthermal X-ray spectrum. The HRI
X-ray isophotes presented in Figure 2 are also centrally peaked which
supports their suggestion. The optical remnant appears to be the
shell about 7 pc in diameter visible in the [OIII] image in Figure 2;
the dot marks the center of the shell and the cross marks the [SII]
patch found by Danziger et al. (1981). The X-ray source is smaller
than this and lies on the western edge of the shell near the position
of the radio source. This region has HI surface densities of 4 x 10^{21}
atoms cm^{-2} and some of the X-ray structure may be obscured. N157B is
probably a combination of shell and filled-in center SNR similar to
Vela XYZ and MSH 15-52 in our Galaxy (Weiler 1983).

Clark et al. (1982) suggest that 0540-69.3, an oxygen-rich SNR,
may also be a Crab-like remnant as it has a smooth power law X-ray

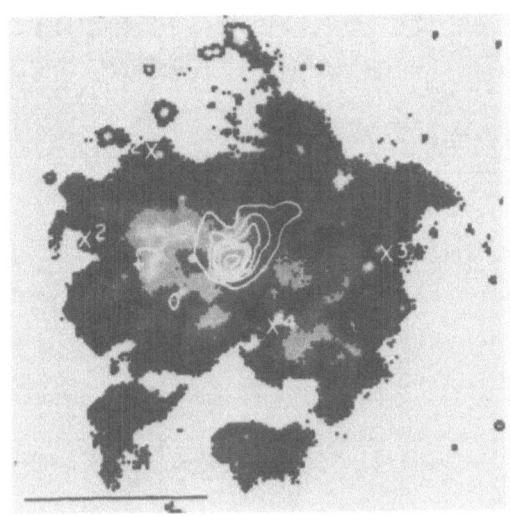

Fig. 2 SNR 0538-69.1 (N157B) in the light of [OIII]. N to the top
and E to the left. This is a black and white print of a false color
photograph to show clearly the shell of the SNR, the center of which
is marked by a dot. The bar is one arc min. The cross marks the
region strong in [SII]. The contour levels of the superimposed HRI
isophotes are 0.01, 0.03, 0.05, 0.08 and 0.15 counts s^{-1} arc min^{-2}.

spectrum consistent with a synchrotron origin. Although they point
out that a hot plasma of temperature 4 x 10^{7}K composed principally of
oxygen with absorption at low energies could mimic the observed
spectrum. The centrally peaked X-ray isophotes (Mathewson et al. 1983)
are probably due to the source being unresolved by the HRI. This and
the fact that the radio spectral index is normal argue against 0540-
69.3 being a Crab-like SNR.

3. A POSSIBLE TYPE III SNR

Mathewson and Clarke (1973) found that the nebulosity between N49
and 0525-66.0 and surrounding 0525-66.0 (see Figure 3) is collisionally
excited. During the course of an HI survey using the Parkes 64-m
radio telescope, high velocity emission was discovered from a region
containing 0525-66.0 and N49 and 270 pc by 160 pc in extent (Figure 3).
Figure 4 shows the HI profile recorded at the center of this region.
The high velocity peak is at a velocity 70 km s^{-1} greater than the
main HI peak which is representative of this area in the LMC. The
mass of the high velocity gas is calculated to be 17,000 M$_\odot$ and its
kinetic energy is 8 x 10^{50} ergs.

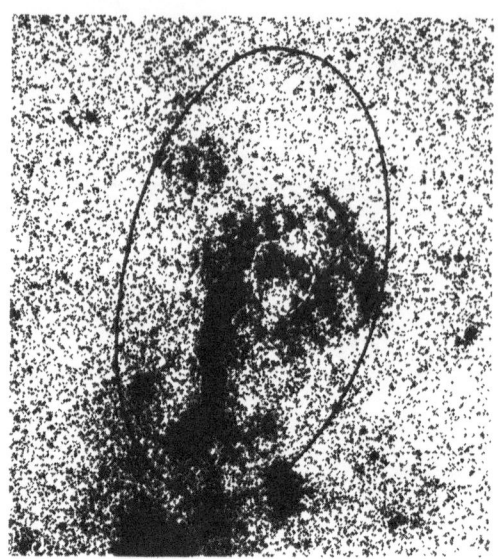

Fig. 3   A red photograph of the field of N49 and 0525-66.0.   The dashed contour is the outer HRI X-ray isophote of 0525-66.0.   The ellipse (17' x 10') encloses the region from which high velocity HI emission is observed.

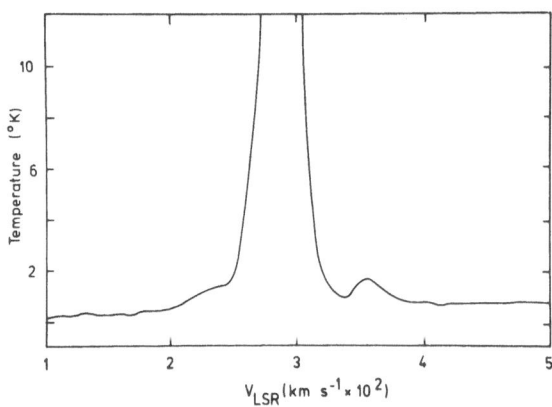

Fig. 4   The HI profile measured at the position of 0525-66.0 using the 64-m radio telescope at Parkes.   The high velocity emission peak is at V_{LSR} = 356 km s^{-1}.

It is tempting to suggest that the widespread nebulosity and high velocity HI was produced by 0525-66.0.   However the age of 0525-66.0 would certainly be less than 10,000 years and mean velocities in excess of 10^4 km s^{-1} would be necessary to produce such a phenomenon.   This seems unlikely particularly as the observed velocity of the associated HI is only 70 km s^{-1}.

A more plausible explanation is that it was an independent, much earlier SN event which occurred out of the plane of the LMC and on the far side. The bubble developed more rapidly in the tenuous outer disk region and now the swept-up material has recombined and is moving out of the plane away from the observer. The visible nebulosity was excited by the blast wave travelling transverse to the line of sight in the outer disk medium. The total energy in the explosion would have been in excess of 10^{52} ergs and therefore may have been a Type III SN. The shock waves of this fossil SNR possibly triggered off the formation of the two massive stars which produced N49 and 0525-66.0 whose blast waves may be now reheating the cavities in the inter-stellar medium formed by the old remnant.

4. THE EVOLUTION OF SNRs

The evolution of SNRs has conventionally been divided into three phases. The first is a free-expansion phase, terminated when the mass of interstellar matter that is swept up equals the mass of the ejecta. Most observed SNRs are presumed to have entered the second phase, the adiabatic blast wave phase, which is described by the Sedov similarity solution. This states that

$$ D \propto \left(\frac{E_o}{\rho_o} \right)^{1/5} t^{2/5} $$

where D is the diameter of the SNR, E_o is the energy in the initial explosion (assumed constant), ρ_o is the density of the interstellar medium and t is the age of the remnant.

The cumulative number, $N(< D)$, - diameter, D, relationship is traditionally used to investigate SNR evolution. Figure 5 shows that this relation for LMC SNRs up to a diameter of 40 pc is $N(< D) = 0.48 \, D_{pc}^{1.0 \pm 0.2}$. It is clear that SNRs do not evolve according to the Sedov solution for which the corresponding relation is $N(< D) \propto D^{5/2}$. Rather they appear to be in the free-expansion phase which is difficult to explain up to a diameter of 40 pc even assuming the three phase model of the interstellar medium of McKee and Ostriker (1977).

An explanation of this apparent free-expansion of SNRs may be as follows. When the mass of the swept-up material approaches that of the mass of the supernova material in the high velocity blast wave, the SNR will decelerate. However the ejecta travelling ballistically outward at velocities proportional to their distance from the center will overtake the decelerating blast wave and a reverse shock will be driven into the stellar ejecta. The kinetic energy of the ejecta will be transformed into thermal energy of the SNR which will prevent its

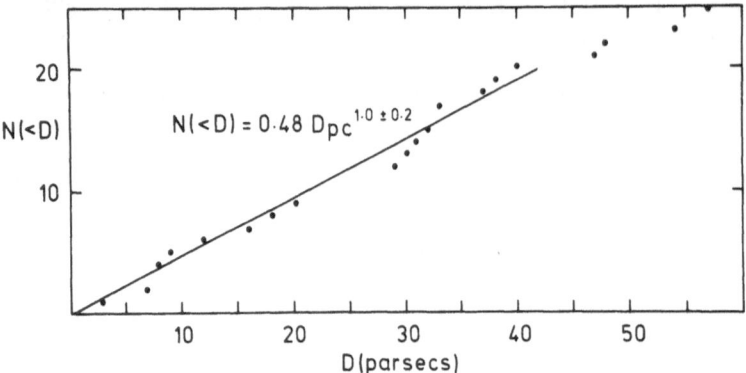

Fig. 5 The cumulative number (N < D) - diameter (D) relation for the SNRs in the LMC.

deceleration until the kinetic energy of the ejecta is exhausted. Thus the "free-expansion" phase will be maintained to quite large diameters. The clumps of reverse-shocked ejecta will be sites of high X-ray and optical emission because of their high density and chemical enrichment. Observational evidence which supports this model is the X-ray structure of Tycho and CasA which Gorenstein et al. (1983) interpret as due to an outer blast wave closely followed by the reverse shocked ejecta.

Model calculations of cooling shocks by Dopita (1979) show that the pressure just behind the shock in the interstellar clouds can be connected to the observed density using the density sensitive red [SII] line ratios. Assuming dynamic pressure equilibrium between blast wave and the shocked cloud, an estimate of E_0 can be made. This was plotted as a function of diameter for SNRs in the Galaxy, M31, M33 and the LMC (Dopita, 1979 and Blair et al. 1981). It was found that $E_0 \propto D^{2.3}$ up to diameters of 30 to 40 pc where it levelled off to a constant value of about 10^{51} ergs. In the context of the present model, this increase in E_0 is due to the gradual conversion of the kinetic energy of the SN ejecta into thermal energy of the gas in the blast wave. If this observed dependence of E_0 on diameter is substituted in the Sedov formula, it is found that $t \propto D^{1.2}$ and therefore $N(< D) \propto D^{1.2}$. This is close to that found for the LMC SNRs. The conclusion is that no longer can the Sedov solution alone be used to describe the evolution of SNRs because the value of E_0 is not constant throughout the lifetime of each remnant.

REFERENCES

Blair, W.P., Kirshner, R.P. and Chevalier, R.A.: 1981, Astrophys. J., 247, 879.

Clark, D.H., Tuohy, I.R., Long, K.S., Szymkowiak, A.E., Dopita, M.A.,
    Mathewson, D.S., and Culhane, J.H.: 1982, Astrophys. J.,
    255, 440.
Danziger, I.J. and Dennefeld, M.: 1976, Astrophys. J., 207, 394.
Danziger, I.J., Goss, W.M., Murdin, P., Clark, D.H., and Boksenberg,
    A.: 1981, Monthly Notices Roy. Astron. Soc., 195, 33P.
Davies, R.D., Elliott, K.H., and Meaburn, J.: 1976, Memoirs of the
    Roy. Astron. Soc., 81, 89.
Dopita, M.A.: 1979, Astrophys. J. Suppl., 40, 455.
Dopita, M.A., Tuohy, I.R., and Mathewson, D.S.: 1981, Astrophys. J.
    (Letters), 248, L105.
Gorenstein, P., Seward, F., and Tucker, W.: 1983, this volume, p. 1.
Gull, T.R. and Fesen, R.A.: 1982, Astrophys. J. (Letters), 260, L75.
Henize, K.G.: 1956, Astrophys. J. Suppl., 2, 315.
Lasker, B.M.: 1980, Astrophys. J., 237, 765.
Long, K.S., Helfand, D.J., and Grabelsky, D.A.: 1981, Astrophys. J.,
    248, 925.
McKee, C.F. and Ostriker, J.P.: 1977, Astrophys. J., 218, 148.
Mathewson, D.S. and Clarke, J.N.: 1973, Astrophys. J., 179, 89.
Mathewson, D.S., Dopita, M.A., Tuohy, I.R., and Ford, V.L.: 1980,
    Astrophys. J. (Letters), 242, L73.
Mathewson, D.S., Ford, V.L., Dopita, M.A., Tuohy, I.R., Long, K.S.,
    and Helfand, D.J.: 1983, Astrophys. J. Suppl., in press.
Mills, B.Y., Little, A.G., Durdin, J.M., and Kesteven, M.J.: 1982,
    Monthly Notices Roy. Astron. Soc., 200, 1007.
Seward, F.D. and Mitchell, M.: 1981, Astrophys. J., 243, 736.
Tanaka, Y.: 1983, these proceedings.
Tuohy, I.R., Dopita, M.A., Mathewson, D.S., Long, K.S., and Helfand,
    D.J.: 1983, this volume, p. 571.
Weiler, K.W.: 1983, this volume, p. 299.

DISCUSSION

FÜRST: Recently Dr. Berkhuijsen inspected the optical SNR data for
            M31 and M33. The slope of $N(< D) - D$ turned out to be also
            of the order one with a flattening at a diameter of 50 pc.

BLAIR: The $N(< D) - D$ relation in M31 is virtually useless at the
            present time because of incompleteness. The smallest
            optical SNR identified in M31 has a diameter of roughly
            20 pc.

            I would like to suggest [OI] imagery to discriminate SNRs
            from HII regions in M31 and M33. While this line is prob-
            lematical because of interference from the night sky
            emission, a test exposure of a field in M31 showed all five
            of the known SNRs in that field (unfortunately no new ones).
            This technique would not only find normal SNRs but would
            also find oxygen-rich SNRs as well.

THE RADIO PROPERTIES OF SNRs IN THE MAGELLANIC CLOUDS

B.Y. Mills
University of Sydney, NSW 2006
Australia

An understanding of the radio properties and evolution of Galactic supernova remnants has always been hampered by the difficulty of measuring
distances. A conventional wisdom has developed around a set of 'good'
calibrators but most workers involved have drawn attention to the uncertainties and the possibility of selection effects distorting results.
This major difficulty is completely overcome by studying SNRs in the
Magellanic Clouds. Although there is uncertainty in the absolute distance
scale, relative distances can be determined to better than 10% and differences of this magnitude are not significant when intercomparing SNRs.
There is, however, another set of problems associated with sensitivity
and resolution. The Clouds are an order of magnitude more distant than
the average distance of Galactic SNRs, thus many of the SNRs are close
to or below the sensitivity limits of most of the southern radiotelescopes
and, until recently, the resolution available has often been inadequate
to separate non-thermal sources from thermal HII regions, so that both
flux densities and spectra have been subject to error. Also there are
$\sim$ 1000 extragalactic background sources which can mimic the flux density
and spectra of SNRs in the Clouds, particularly when close to or behind
HII regions; as a result numerous incorrect or doubtful SNR identifications have been suggested.

The first identifications of SNRs in the Clouds were made using the
Parkes reflector at a frequency of 1410 MHz and a beamwidth of about
14 arcmin (Mathewson & Healey, 1964 ; Westerlund & Mathewson, 1966); the
three strongest radio SNRs were identified with optical nebulosities.
Subsequently there have been many radio surveys of the Clouds and specific
searches for SNRs. The principal sources of data are listed in Table 1.

Following the initial discoveries of 1964, two major advances can be
recognised. In 1972-1973, Mathewson & Clarke carried out the radio-
optical searches listed in Table 1 and identified a further ten SNRs in
the LMC and two in the SMC, bringing the total to 13 and 2 respectively.
In 1981, the first major program of the newly commissioned Molonglo
Observatory Synthesis Telescope (e.g. Mills, 1981) was to survey all
previously suggested SNR candidates. Preliminary results from this survey,

J. Danziger and P. Gorenstein (eds.), Supernova Remnants and their X-Ray Emission, 551–558.
© *1983 by the IAU.*

Table 1:   Principal sources of radio data for Magellanic Cloud SNRs

Frequency GHz	Beam size	Reference	Notes
0.408	2.8 x 3.5	Mathewson & Clarke (1972) Mathewson & Clarke (1973a) Mathewson & Clarke (1973b) Mathewson & Clarke (1973c)	Radio-optical SNR searches in both Magellanic Clouds
		Clarke et al. (1976)	A definitive catalogue of 408 MHz radio sources in both Clouds
0.843	43" x 46"	Mills et al. (1982) Mills et al. (in preparation)	Investigations of SNR candidates in both Clouds using the MOST
1.41	14'	Mathewson & Healey (1964) Westerlund & Mathewson (1966)	First identifications of SNRs in the LMC
1.415	50"	Turtle & Mills (1979)	Size data for bright SNRs in the LMC
5.0	4.1	McGee et al. (1972)	A continuum survey of the LMC
5.0 8.4	4.1 2.6	McGee et al. (1974)	A continuum survey of the SMC
8.4	2.6	McGee et al. (1978)	Measurements on selected sources in the LMC
5.0 14.7	4.1 2.2	Milne et al. (1980)	A search for SNRs in the LMC based on radio spectra

which permits clear separation of the SNRs from contaminating sources and has a high signal-to-noise ratio for the SNRs, are described in this paper.

Candidates were chosen primarily from the Einstein Observatory X-ray catalogues of Seward & Mitchell (1981) for the SMC and Long, Helford and Grebalsky (1981) for the LMC. These X-ray candidates were supplemented by other candidate SNRs previously suggested on the basis of radio and/or optical observations. Altogether 18 candidates were observed in the SMC and 67 in the LMC. Six positive identifications were made in the SMC (Mills et al., 1982). The LMC data are more extensive and complex and some further radio and optical observations will probably be needed

Table 2: <u>Radio SNRs in the Magellanic Clouds</u>

<u>SMC</u>

0045-734 (N19)	0049-736 (IE00494-7339)	0102-723 (IE0102.2-7219)
0046-735	0101-724 (IE0101.5-7226)	0103-726 (IE0103.3-7240)

<u>LMC</u>

0453-685 (X1)	0519-690 (X26)	0534-699 (X53)
0454-665 (N11L)	0520-694 (X27)	0534-705 (X54,DEM238)
0456-687 (X2, N86)	0525-660 (X34, N49B)	0535-660 (X59,N63A)
0500-702 (X7, N186D)	0525-696 (X35, N132D)	0536-706 (X61,DEM249)
0506-680 (X11, N23)	0525-661 (X36, N49)	0538-691 (X67,N157B)
0509-687 (X13, N103B)	0527-658 (X39, DEM204)	0540-693 (X79,N158A)
0509-675 (X14)	0528-692 (X40)	0547-697 (X88,N135)
0519-697 (X23, N120)	0532-710 (X47, N206)	

<u>LMC (possible SNRs)</u>

0505-679 (X10, DEM71)	0538-694	0543-679 (N70)
0536-692 (30DorC)	0543-689 (X82)	0548-704 (X89)

(Mills et al., in preparation); here I will report only on preliminary results which, however, are adequate for several valid and important statistical conclusions.

Positive identifications have been made with 23 SNRs and a further 6 are regarded as possible identifications, subject to some further investigation. At a lower level of probability several others remain to be investigated. With the exception of two SNRs in the SMC, the positive identifications could only be made with X-ray sources which also have associated visible remnants (Mathewson et al., in press). The 'possible' identifications include radio sources in which one or both of these features may be missing. Radio sources identified with SNRs are listed in Table 2; X-ray or common names are in parenthesis.

The basic results for both Clouds are presented in Figure 1 where the absolute radio luminosites of the 29 'confirmed' and the 6 'possible' SNRs are plotted against their diameters. A distance of 55 kpc has been adopted for the LMC and 63 kpc for the SMC. As the small diameter SNRs are not completely resolved by the radiotelescope, a standard procedure has been adopted to deconvolve the response. Two models comprising a thin circular ring and a uniform circular disk have been fitted to the half-power responses along the major and minor axes. The diameter quoted is based on the mean of these four results and is expected to give a good approximation to the actual size for SNRs of typical morphology. The resolution limit is estimated as about 7 pc and three of the SNRs could not be definitely resolved; 0509-687, 0509-675 and 0535-660. Their diameters have been taken from the optical values given by Mathewson et

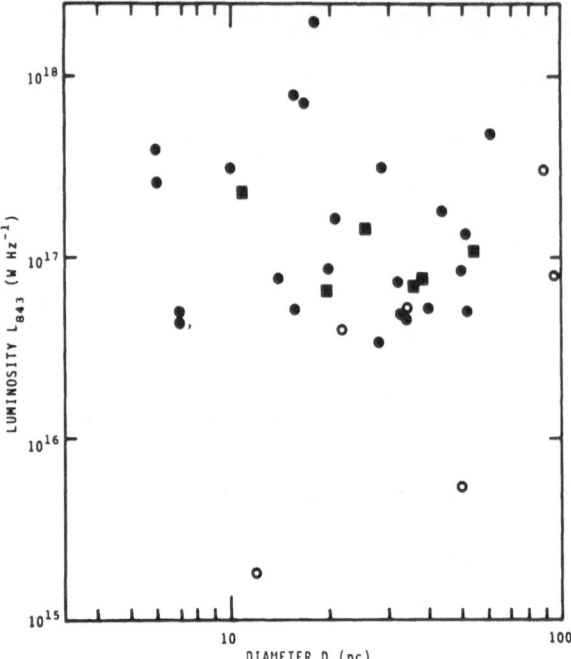

Figure 1. The distribution of radio luminosity for SNRs in both Clouds
as a function of the radio diameter. Filled circles and filled squares
represent positively identified SNRs in the LMC and SMC respectively.
Open circles are the 'possible' SNRs in the LMC.

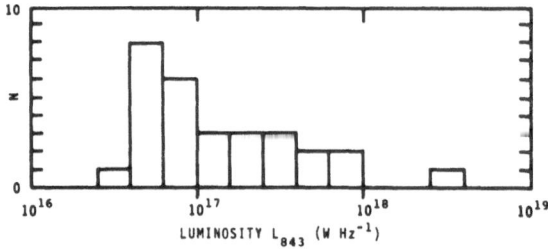

Figure 2. The histogram of radio luminosity for the positively
identified SNRs in both Clouds.

al. (in preparation), 6 pc, 7 pc and 6 pc respectively. The SNR
0540-693 which has an optical diameter quoted as 2 pc is well resolved
with a radio diameter of 10 pc.

Figure 1 is best described as a scatter diagram; there is no clear
evidence for a common evolutionary track. This is not to say that the
radio emission of SNRs does not evolve, but such evolution must be less
than the intrinsic scatter in luminosities over the range of diameters
6 - 60 pc. Over most of this range it is believed that the sample is
nearly complete; the cut-off at large diameters is almost certainly
caused by a cut-off in the X-ray candidates which, when plotted as in
Figure 2, show a steep fall-off in X-ray luminosity with increasing
diameter. The lower bound on the radio luminosity of confirmed SNRs
can hardly be instrumental as it occurs far above the radio detection
limit as shown by the detection of much weaker 'possible' SNRs.

Earlier workers found a correlation between luminosity and diameter
of the form $L \propto D^{-1}$ (corresponding to the more usual presentation
$\Sigma \propto D^{-3}$). It is now evident that such an apparent correlation arose
partly from measurement errors due to resolution effects and partly
because the low luminosity small diameter SNRs had not been detected.
As there is no valid L - D relation, luminosities at all diameters may
be combined to give the distribution for 'confirmed' SNRs shown in
Figure 3. The distribution is asymmetrical with a logarithmic mean of
$<L_{843}> = 1.1 \times 10^{17}$ W Hz^{-1}. More than 90% of luminosities lie within
a factor of 4 of the mean luminosity.

Another basic statistic is the N-D relation which describes the
expansion law of the SNRs. In Figure 3, this relation is plotted for
all the 'confirmed' SNRs and those 'possible' SNRs with luminosities
between the limits found above. There are 33 such remnants and for
diameters below 40 pc the relation may be described by the power law

$$N(<D) = 0.26D^{1.21\pm0.25},$$

where the exponent and its standard error have been obtained by a maximum

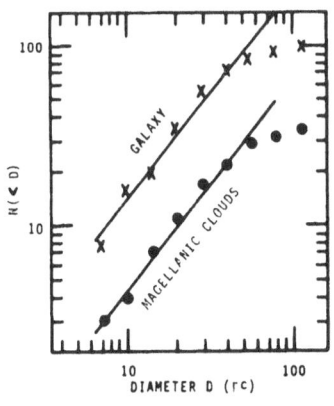

Figure 3. The N-D relation for
all SNRs in both Clouds and for
Galactic SNRs using the Cloud
distance calibration.

likelihood fit. This generally confirms the result obtained by Clarke
(1976) on a smaller sample of SNRs in the LMC and shows that the SNRs
observed are not undergoing adiabatic expansion in a uniform medium,
for which the corresponding relation is $N \propto D^{2.5}$. As suggested with
some reluctance by Clarke, interaction with the interstellar medium
must be small, but this is no longer to be seriously questioned in view
of the increasing acceptance of inhomogeneous models of the medium. It
appears that the great majority of detected SNRs in the Clouds have
expanded in the hot tenuous phase where they reach large diameters before
accreting sufficient material to slow their expansion significantly.

A further significant result from Figure 3 is the flattening of the
N-D slope for diameters greater than about 40 pc. One might well ask:
what happens to the large diameter SNRs which are deficient by an order
of magnitude at diameters around 100 pc? There is no observational
evidence for a decrease in radio luminosity below the observed cut-off.
Although we would expect their X-ray emission to have fallen below
detectable levels, they should still be detected as radio sources and
a brightening of the optical emission might be expected. There are,
in fact, many large diameter filamentary shells with strong Hα emission
visible in the Clouds (e.g. Meaburn, 1980), but these appear to have
flat radio spectra. A possible scenario involves a decay of the early
synchrotron emission combined with a buildup of free-free emission from
the developing ionized shell; some of the old SNRs may thus be masquera-
ding as shell-type HII regions. Two such apparently thermal sources
were included in our maps of the LMC and these are both designated as
possible SNRs; they are 0543-679 (N70) and 0536-692 (30DorC), with
diameters between 90 and 100 pc.

COMPARISON WITH GALACTIC SNRs

In discussing the low slope he found for the N-D relation in the
LMC, Clarke (1976) was concerned over the very different form of the
corresponding Galactic relation, which was consistent with adiabatic
expansion of the remnants. However the Galactic N-D relation was then
based on a distance scale derived from a set of calibrators of uncertain
distances. Subsequently, the Galactic result was apparently confirmed
by Clark & Caswell (1976) using a distance scale based primarily on the
Σ-D relation then current for the LMC (i.e. $\Sigma \propto D^{-3}$), together with
another set of uncertain Galactic calibrators. We have now seen that
the old Σ-D relation in the Clouds is not supported by present results.

In Figure 3, the N-D relation for the Galactic SNRs is obtained
from the catalogue of Clark & Caswell (1976) on the assumption that the
Galactic and Cloud SNRs are physically identical, i.e. the same assumption as made by Clark & Caswell. However, a distance scale defined by
the mean luminosity of the Cloud SNRs is used. At 408 MHz this scale
becomes $d_{kpc} = (1280/S_{Jy})^{\frac{1}{2}}$. The N-D relation below 40 pc diameter is
then given by

$$N(<D) = 1.05D^{1.15\pm0.14},$$

with essentially the same slope as the Cloud SNRs. The derived expansion
laws for the SNRs in the Clouds and the Galaxy are therefore consistent
and it appears that they develop in similar environments. The number
of SNRs with diameters less than 40 pc found in the Galaxy is three
times the corresponding number in the Clouds.

Clark & Caswell list a number of Galactic SNRs with measured
distances which they used as calibrators; it is instructive to compare
these with distances derived from the equation above. If we define a
ratio R between the Cloud-calibrated distance and the measured distance,
we find two results. For the optically derived distances, R = 1.5±05.;
for the kinematically derived distances, R = 0.80±0.06, with the possibility that R is smaller because many of the kinematic distances are
lower limits. There is no significant mean difference for the optically
derived distances, although the scatter is large. In deriving the
kinematic distances, 10 kpc was assumed as the distance to the Galactic
centre. The adopted Cloud distances are therefore consistent with a
distance to the Galactic centre of $\lesssim$ 8 kpc, which appears satisfactory
in the light of current estimates.

Another result is a new determination of the rate of occurrence of
SNRs in the Galaxy, which is a direct function of the expansion law.
Clark & Stephenson (1977) identify eight historical supernovae of known
age. Six of these are accepted as calibrators; SNR 1054 is rejected as
atypical and AD 360 is rejected because the derived occurrence rate is
very anomalous and the identification is queried by Clark & Stephenson.
Applying the method of Clark & Caswell to the new N-D relation, it is
found that the mean time between outbursts is 40±10 yr. If an allowance
of $\sim$ 30% is made for incompleteness, the corresponding time is $\sim$ 30 yr.
Thus the occurrence rate seems compatible with rates derived from supernovae observed in external galaxies and with estimated birthrates of
pulsars. The corresponding mean times between outbursts derived from
SNRs detected in the LMC and SMC are $\sim$ 150 yr and $\sim$ 600 yr respectively.

The mean properties of SNRs may be derived from the N-D relation
and the occurrence rate. Thus, taking for the Clouds, $N(<D) = 0.26D^{1.2}$
and a mean time between outbursts of 120 yr, we find for the 'average'
SNR with diameters between 6 and 40 pc in both Clouds and Galaxy:

$$D_{pc} = 0.057 \, t_{yr}^{0.83}$$

$$V_{km \, s^{-1}} = 3.0 \times 10^4 \, t_{yr}^{-0.17} = 1.7 \times 10^4 \, D_{pc}^{-0.2}$$

where V is the mean expansion velocity of the radio emitting shell.

These relations do not describe the evolution of some 'typical' SNR
but represent ensemble averages consistent with the present data. At
the time of observation, interaction with the interstellar medium is
proceeding and the actual expansion velocity of the radio emitting shells
will be well below the mean velocity; it might also be expected that the
optical filaments, representing either regions of high interaction or

excited knots in the interstellar medium, will have even lower velocities. It is clear that the commonly accepted 'prototype', Cas A, is exceptional in many respects.

While it is expected that further work will modify these preliminary statistics, the qualitative conclusions appear to be reasonably well established. The expansion of most recognised SNRs has been more rapid than predicted by the adiabatic law, their ages are less and their birth-rates higher.

REFERENCES

Clark, D.H. & Caswell, J.L.: 1976, M.N.R.A.S., 174, 267.
Clark, D.H. & Stephenson, F.R.: 1977, "The Historical Supernovae", (Pergamon), p. 207.
Clarke, J.N.: 1976, M.N.R.A.S., 174, 393.
Clarke, J.N., Little, A.G. and Mills, B.Y.: 1976, Aust.J.Phys.Astrophys. Suppl., No. 40, 1.
Long, K.S., Helfand, D.J. and Grebalsky, D.A.: 1981, Astrophys.J., 248, 925.
Mathewson, D.S. and Healey, J.R.: 1964, "The Galaxy and the Magellanic Clouds", (eds) F.J. Kerr and A.W. Rodgers (Canberra; Australian Academy of Science), p. 283.
Mathewson, D.S. and Clarke, J.N.: 1972, Astrophys.J.Lett., 178, L105.
Mathewson, D.S. and Clarke, J.N.: 1973a, Astrophys.J., 179, 89.
Mathewson, D.S. and Clarke, J.N.: 1973b, Astrophys.J., 180, 725.
Mathewson, D.S. and Clarke, J.N.: 1973c, Astrophys.J., 182, 697.
Mathewson, D.S., Ford, V.L., Dopita, M.A., Tuohy, I.R., Long, K.S. and Helfand, D.J.: Astrophys.J. (in press).
McGee, R.X., Brooks, J.W. and Batchelor, R.A.: 1972, Aust.J.Phys., 25, 581.
McGee, R.X., Newton, L.M. and Brooks, J.W.: 1974, Aust.J.Phys., 27, 79.
McGee, R.X., Newton, L.M. and Butler, P.W.: 1978, M.N.R.A.S., 183, 799.
Meaburn, J.: 1980, M.N.R.A.S., 192, 365.
Mills, B.Y.: 1981, Proc.astr.Soc.Aust., 4, 156.
Mills, B.Y., Little, A.G., Durdin, J.M. and Kesteven, M.J.L.: 1982, M.N.R.A.S., 200, 1007.
Mills, B.Y., Turtle, A.J., Little, A.G. and Durdin, J.M., (in preparation).
Milne, D.K., Caswell, J.L. and Haynes, R.F.: 1980, M.N.R.A.S., 191, 469.
Seward, F.D. and Mitchell, M.: 1981, Astrophys.J., 243, 736.
Turtle, A.J. and Mills, B.Y.: 1979, N.Z.J.Sci., 22, 543.
Westerlund, B.E. and Mathewson, D.S.: 1966, M.N.R.A.S., 131, 371.

DISCUSSION

MCKEE: Why does the Galactic N-D relation show a fall off at the same diameter as the Cloud relation?

MILLS: The fall off is thought to have quite different causes in Galaxy and Clouds. The similarity is curious and perhaps significant.

OPTICAL OBSERVATIONS OF FOUR BALMER-DOMINATED SUPERNOVA REMNANTS IN THE LARGE MAGELLANIC CLOUD

I.R. Tuohy, M.A. Dopita and D.S. Mathewson
Mount Stromlo and Siding Spring Observatories

K.S. Long and D.J. Helfand
Columbia University

We report the optical identification of four Balmer-dominated supernova remnants (SNRs) in the Large Magellanic Cloud. Both the Balmer-dominated spectra and the presence of a broad Hα component in one remnant can be understood in terms of a very high velocity non-radiative shock encountering gas which is partially neutral, as proposed originally by Chevalier and Raymond to account for the similar spectra of the galactic remnants, Tycho and SN1006. From a consideration of the optical and X-ray luminosities of the SNR with broad Hα emission, we infer that the fraction of neutral gas in the medium is $\lesssim$ 30%. Radio observations of the LMC remnants show that their surface brightnesses are anomalously low; this could be intrinsic to the supernova themselves, or a result of their environment. Finally, we argue that the four SNRs all resulted from Type I supernovae, in which case they are the first such remnants to be identified outside the Galaxy.

1. INTRODUCTION

A recent X-ray survey of the Large Magellanic Cloud (LMC) by Long, Helfand and Grabelsky (1981) has resulted in the detection of ∿10 new X-ray emitting supernova remnants (SNRs). As part of a program to identify the optical counterparts of these new SNRs (Mathewson et al. 1982,1983), we have isolated a group of four remnants which have optical spectra that are completely dominated by the Balmer lines of hydrogen. The four SNRs are thus very similar to the two galactic remnants, Tycho and SN1006, which have been detected only in the Balmer lines and which are believed to have resulted from Type I supernovae. In this paper we discuss the optical, X-ray and radio properties of the new LMC remnants. A more detailed account of this work can be found in Tuohy et al. (1982).

2. OPTICAL OBSERVATIONS

2.1 Results

The optical identifications were made using the Anglo-Australian

J. Danziger and P. Gorenstein (eds.), Supernova Remnants and their X-Ray Emission, 559–565.
© *1983 by the IAU.*

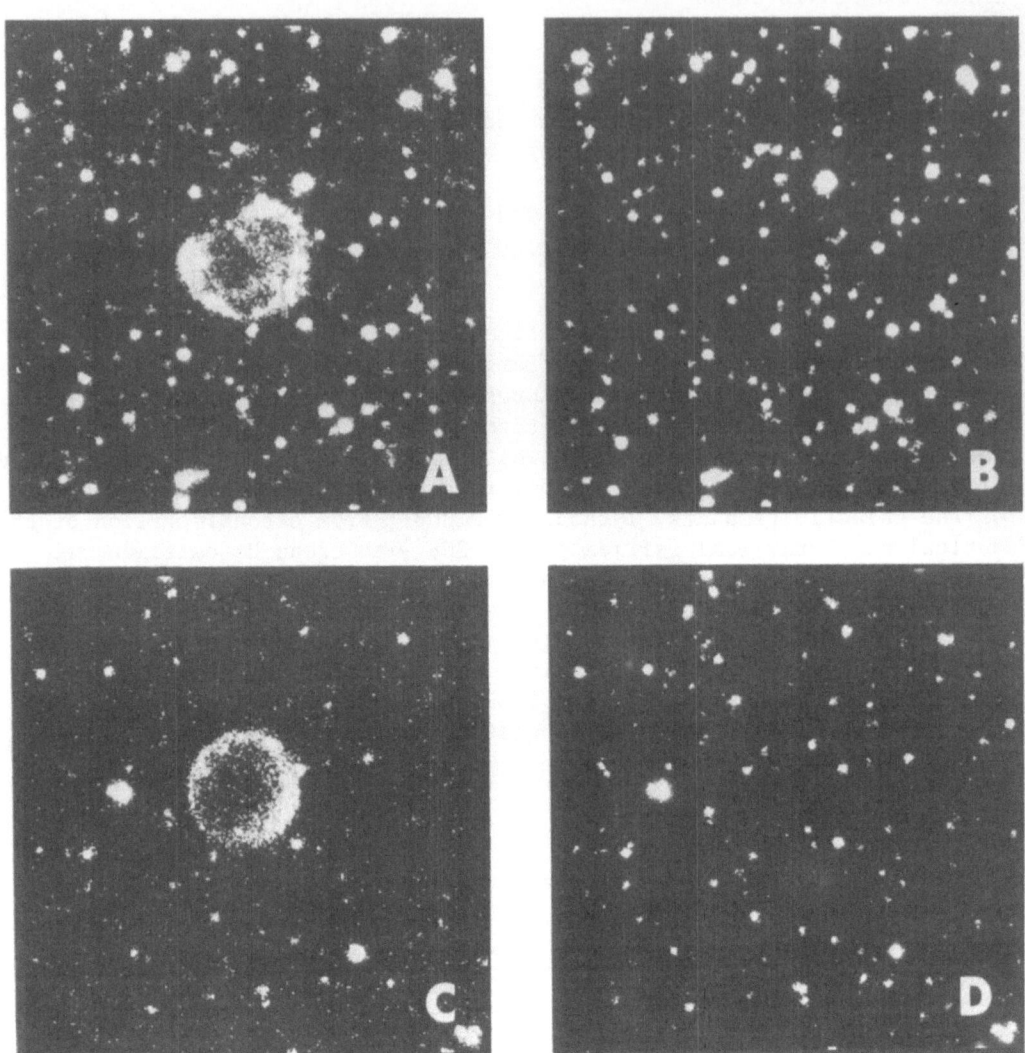

Fig. 1 Hα and [OIII] images of source 26 (A,B) and source 14 (C,D). The
scale of each image is 2.1x2.1 arcminutes. North is at the top and East
to the left.

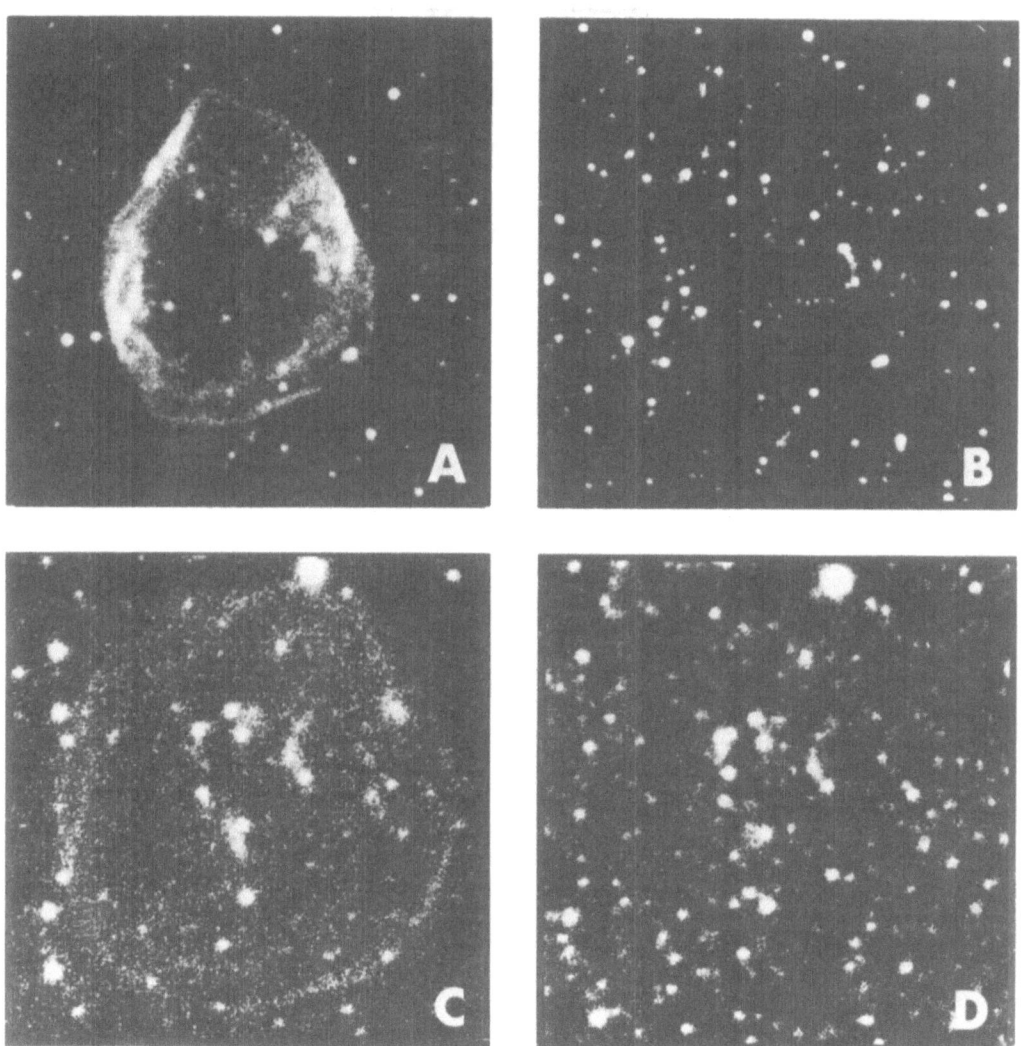

Fig. 2 Hα and [OIII] images of source 10 (A,B) and source 89 (C,D).
The scale and orientation are the same as in Fig. 1.

Telescope (AAT) with the Image Photon Counting System (IPCS).
Figures 1 and 2 show Hα and [OIII] λ5007 images of the four SNRs, namely
0505-67.9, 0509-67.5, 0519-69.0 and 0548-70.4. For convenience, we will
refer to the objects as Sources 10, 14, 26 and 89, as in the X-ray catalog of Long, Helfand and Grabelsky. Sources 14 and 26 are only visible
at Hα, and have similar angular sizes (25 and 28 arcsecs). Sources 10
and 89 are considerably larger (83 x 67 arcsec and 103 arcsec diameter),
and although predominantly visible at Hα, they show faint wisps of
[OIII] nebulosity in the inner NW and central regions respectively. The
X-ray images of the SNRs show limb-brightened shells in each case, with
diameters comparable to the optical dimensions (Mathewson et al. 1982).

Low resolution spectra of sources 14 and 26 have been obtained with
the AAT, and are shown in Figure 3. The spectra are similar, except
that a broad emission component is clearly visible beneath the Hα line
in source 26. Comparison of the Hα profile with the spectrograph
response to a night-sky line confirms that the broad component is
intrinsic to the SNR. By fitting a double Gaussian to the Hα profile,
we obtain a velocity width of 2800±300 km s^{-1}, and a ratio of the broad
to narrow emission components in the range 0.4 to 0.8.

2.2 Discussion

The four LMC remnants are the clearest examples of Balmer-
dominated SNRs yet identified, and the first of their class to be
detected outside the Galaxy. The Balmer-dominated spectra can be
understood in the context of the model developed by Chevalier and
Raymond and Chevalier, Kirshner and Raymond (1980; hereafter CKR) to

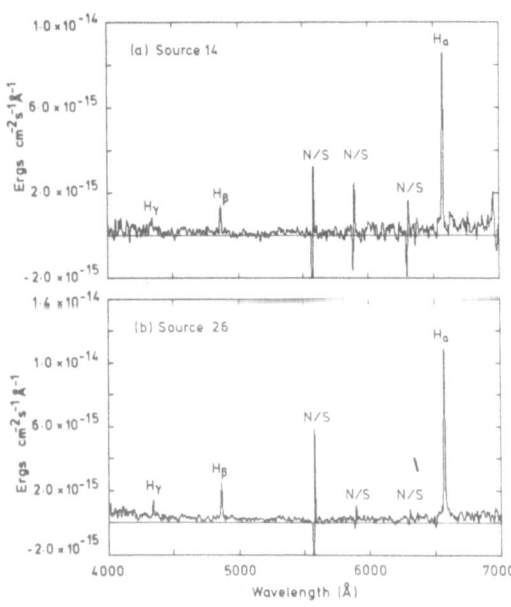

Fig. 3 Low resolution AAT
spectra obtained for Source
14(a) and Source 26(b).
Balmer lines and imperfectly
subtracted night-sky features
are indicated. The blue
continua may represent 2-
photon emission (in this case,
the slight red excess above
∿6400 Å can be attributed to
a grating leak.

explain the spectra of Tycho and SN1006. The strong Balmer emission results from the interaction of a high velocity non-radiative shock with gas that is partially neutral; the neutral hydrogen atoms have a probability of being excited several times before being ionised, and owing to the high electron temperatures behind the shock, the usual forbidden line emission typical of most SNRs is suppressed. The CKR model also accounts for the broad Balmer component which has been observed previously in Tycho (CKR), and now in source 26; this component results from charge exchange between high velocity protons and neutral hydrogen atoms. A further interesting feature of the spectra of sources 14 and 26 is the possible presence of a faint blue continuum, which may be due to hydrogen 2-photon emission.

The ratio of the broad to narrow Hα components determined above is a function of the shock velocity (v_s) in the CKR model; for source 26, we find that a value of $v_s \sim 2900\pm400$ km s^{-1} is consistent with both our ratio determination and our velocity width measurement. This shock velocity leads to a Sedov age of $\sim$500 years.

An estimate of the density of the *neutral* component of the shocked gas can be made for source 26, based on our measured shock velocity and the Hα surface brightness. We obtain a value of $n_H \sim 0.06$ cm^{-3}. This density estimate is considerably lower than the *total* density of the gas, $n_x \sim 4.7$ cm^{-3}, calculated from the X-ray luminosity ($L_x = 1.1 \times 10^{37}$ erg s^{-1}) assuming ionisation equilibrium and cosmic abundances. If both density estimates are correct, the gas must be mostly ionised (> 95%). We note that even if the gas was fully ionised by a UV flash at the time of the explosion, there has been time for the gas to partially recombine within the $\sim$500 year lifetime of the remnant.

However, the calculation of the total density from the X-ray luminosity is subject to considerable uncertainties associated with the assumptions of ionisation equilibrium and normal abundances. In particular, allowance for enrichment of the gas by metal-rich ejecta (Long, Dopita and Tuohy 1982) reduces the total density estimate to $n_x \sim 0.3$ cm^{-3}. This value is an order of magnitude less than the above estimate, and results in a considerably lower ionised fraction ($\sim$ 70%) for the medium.

The X-ray observations and the estimate of the current shock velocity allow constraints to be placed on the supernova precursor for source 26 (Tuohy et al. 1982). We can set the following limits on the energy of the explosion (E_0), the core mass of the star (M_c), the envelope mass (M_{en}), and the total mass of the star (M_*) : $1.8 \times 10^{51} \lesssim E_0 \lesssim 2.2 \times 10^{51}$ ergs, $1.0 \lesssim M_c \lesssim 1.4$ M$_\odot$, $0.0 \lesssim M_{en} \lesssim 3.0$ M$_\odot$ and $1.2 \lesssim M_* \lesssim 4.0$ M$_\odot$.

3. RADIO EMISSION

The 408 MHz fluxes of the four LMC remnants, kindly provided by

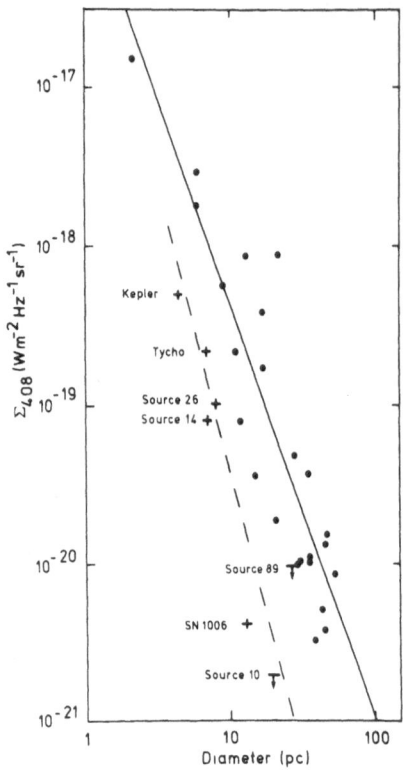

Fig. 4   The 408 MHz Σ - d
distribution for SNRs in the LMC
(solid dots) taken from Mathewson
et al. (1982,1983). Data points
for sources 10, 14, 26 and 89
(derived from Mills and Crawford
1981) and for the galactic SNRs
Kepler, Tycho and SN1006 are
superimposed.

Mills and Crawford (1981) from a more sensitive analysis of the Molonglo
survey data (Clark, Little and Mills 1976), have been used in Figure 4
to plot the radio surface brightness versus mean optical diameters.   The
plot includes the 408 MHz surface brightnesses of the remaining SNRs in
the LMC (Mathewson et al. 1982a,b), and also the corresponding points
for the three galactic SNRs, SN1006, Tycho and Kepler, each believed to
have resulted from a Type I explosion.   The four LMC remnants lie
systematically below the mean Σ-d line, with the difference for source
10 being more than a factor of 40.   Interestingly, the three galactic
SNRs also lie systematically below the line.

There are at least two possible interpretations of the anomalously
weak radio emission from the Balmer-dominated SNRs.   First, the low
surface brightnesses may result from low interstellar densities in the
environment of the SNRs (e.g., SN1006;   Caswell and Lerche 1979).   The
second possibility is that the weak radio emission is intrinsic to the
class of remnant;   e.g., due to a reduced supply of relativistic
particles, or inefficient magnetic field amplification.

4.   THE EVIDENCE FOR TYPE I SUPERNOVAE

The similarity between the optical, X-ray and radio properties of
the four LMC remnants and the galactic Type I remnants, argues that the

LMC SNRs were also produced by Type I supernovae. More quantitatively, the mass limits that we derive for source 26 (section II) rule out a Type II supernova for which substantially higher masses are expected. We conclude therefore that source 26, and by analogy, sources 10, 14 and 89 resulted from Type I explosions.

REFERENCES

Caswell, J.L., and Lerche, I.: 1979, Monthly Notices Roy. Astron. Soc.
     187, 201.
Chevalier, R.A., Kirshner, R.P., and Raymond, J.C.: 1980, Astrophys. J.
     235, 186 (CKR).
Chevalier, R.A., and Raymond, J.C.: 1978, Astrophys. J. (Letters),
     225, L27.
Clarke, J.N., Little, A.G., and Mills, B.Y.: 1976, Aust. J. Phys.,
     Astrophys. Suppl., #40.
Gronenschild, E.H.B.M., and Mewe, R.: 1981, Preprint.
Long, K.S., Dopita, M.A., and Tuohy, I.R.: 1982, Astrophys. J.
     260, 202.
Long, K.S., Helfand, D.J., and Grabelsky, D.A.: 1981, Astrophys. J.,
     248, 925.
Mathewson, D.S., Ford, V.L., Dopita, M.A., Tuohy, I.R., Long, K.S., and
     Helfand, D.J.: 1982, Astrophys. J. Suppl., in press.
Mathewson, D.S., Ford, V.L., Dopita, M.A., Tuohy, I.R., Long, K.S., and
     Helfand, D.J.: 1983, this volume, p. 553.
Mills, B.Y., and Crawford, D.F.: 1981, Private communication.
Tuohy, I.R., Dopita, M.A., Mathewson, D.S., Long, K.S., and Helfand,
     D.J.: 1982, Astrophys. J. in press.

DISCUSSION

MILLS: The results I have presented based on MOST observations show that the radio emission in two cases is normal. X89 is normal also if the identification is correct although there is some suggestion that the radio source may be extra-galactic. Only X10 is well below the normal. I also believe that the apparent differences in radio emission of galactic SNRs is a result of observational selection. All historical SNRs fall within the range of absolute luminosities we find for the Cloud SNRs, Type II as well as Type I.

TUOHY: The 408 MHz data supplied by you from the Molonglo Archival Data Bank shows that the two SNRs in question (sources 14 and 26) lie factors of 13 and 8 below the mean Σ-d line (note also that their diameters are accurately determined). We consider it significant that, without exception, the Balmer-dominated SNRs all have substantially lower surface brightnesses than the average.

OPTICAL OBSERVATIONS OF SNRs AND RING NEBULAE IN THE LMC

M. Rosado[1], Y.M. Georgelin[2], A. Laval[2] and G. Monnet[3]
[1]Instituto de Astronomia, UNAM, Mexico
[2]Observatoire de Marseille, France
[3]Observatoire de Lyon, France

The work we have performed on some of the SNRs and ring-shaped nebulae of the LMC refers mainly to: photographic plate Hα and [SII](λλ 6717 Å) imagery and Fabry-Perot (FP) interferometry.

I. THE IMAGERY

The photographic plate imagery has been performed by means of a focal reducer equiped with a 2-stage, magnetically focused, RCA image tube attached to the Cassegrain focus of the 1.52 m telescope of ESO. The photographs were calibrated, digitalized and filtered (Llebaria, 1980). From these we obtained the 2-D [SII]/Hα line-ratios of the nebula. Figure 1 is a plot showing our results on the [SII]/Hα line-ratios of some of the LMC SNRs and ring nebulae.

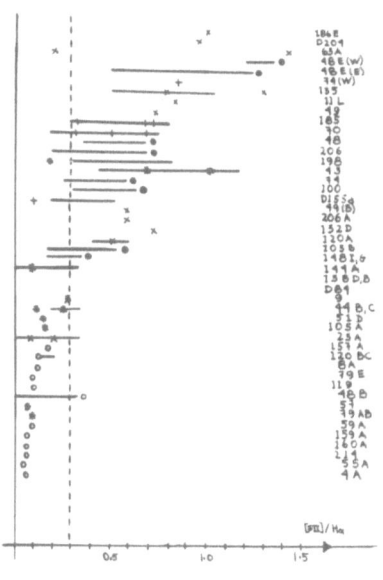

Fig. 1. [SII](λλ6717+6731)/Hα line ratios of nebulae in the LMC obtained from our imagery (straight lines) or from spectroscopic observations of different authors (points). Henize or DEM numbers are given at the right-hand side (DEM numbers are symbolized by a D preceding the number). The different symbols of the points correspond to the different nature of the nebulae (X = SNRs; + = ring nebulae of unknown origin; * = nebulae presumably formed by SSSWs; o = classical HII regions). Encircled symbols are used only to show the nature of the nebula and do not correspond to any observation.

J. Danziger and P. Gorenstein (eds.), Supernova Remnants and their X-Ray Emission, 567–571.

Figure 2 shows, as a matter of example, the [SII]/Hα line-ratio of the
nebula N 23 A which is thought to be an HII region superimposed to the
radio and X rays SNR.

Fig. 2. r = [SII](λλ6717+6731)/Hα
line-ratio distribution of the nebula
N 23. The values are as follows :
r < 0.14, weak points; 0.14 < r < 0.26,
bright points: 0.26 < r < 0.38, hori-
zontal bars and r > 0.38, vertical
bars.

        Two facts can be seen directly from the plot shown in Figure 1:
     - In general, the nebulae show large internal variations in their
[SII]/Hα line-ratios.
     - The ring nebulae of unknown origin have, in general, [SII]/Hα line-
ratios greater than those of classical HII regions ($\gtrsim$ 0.3).
        These assertions have several implications :
     - In the case of SNRs, large internal variations on the line-ratios
of some galactic SNRs (as the Cygnus Loop filaments (Fesen et al., 1982))
have been also found. These variations are interpreted as indications
of non-steady flow situations. Consequently, one must be careful with
the interpretation of relations between line-ratios and some other quantities such as diameter or galactocentric distances (the latter used in
the study of gradients in abundances) because the internal variations in
these line-ratios are, in some cases, larger than either the variations
among different SNRs or the dispersion due to the use of different shock
models. Thus one must select only the data corresponding to the brightest
filaments in order to be sure that the data correspond to radiative shocks.
     - The enhanced [SII]/Hα emission of ring nebulae has been frequently
interpreted as a consequence of shock emission. However, photoionization
models which take into account geometrical effects, such as steps in
density, may predict enhanced ratios (Stasinska, 1980). Thus, the large
internal variations of the ring nebulae may be interpreted either as
deviations to steady flows (if the emission is due to shocks) or as due
to differences in the ionization degree of the filaments (if the emission
is due to photoionization).
     - Because of the geometrical enhancement in the forbidden line-ratios
of photoionized nebulae we conclude that the only means of finding
shocks from observations at optical wavelengths are:
      i) The knowledge of the radial velocity field of the nebula (in order
         to see if there are large internal motions)
     ii) The observations of temperature sensitive line-ratios such as the
         [OIII]λλ 4363/ (5007 + 4959) line-ratio.

2. FP INTERFEROMETRY OF RING-SHAPED NEBULAE

Figure 3 shows the splitting distribution of two ring shaped nebulae of unknown origin: N 185 and N 70. These were derived from our FP interferometry performed at the Cassegrain focus of the 3.60 m telescope of ESO (Rosado et al., 1982, 1981 respectively). For N 70 we have also plotted the splittings observed by Blades et al. (1980) in their spectroscopic work. The spectroscopic observations of Dopita et al. (1981) are not shown but, in general both the spectroscopic and the interferometric data agree and complement themselves.

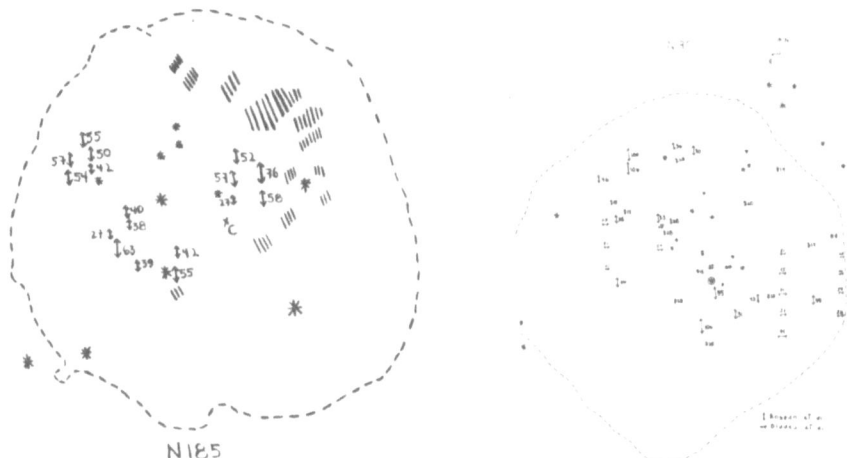

Figure 3. Distribution of splittings in N 185 (left) and N 70 (right). The numbers give the differences in radial velocities (in km s^{-1}) between the two components. The hatched zones correspond to splittings found in low surface brightness filaments but difficult to evaluate accurately.

Three models are the most favoured in explaining the origin of these ring nebulae:
- The fossil SNR model.
- The supersonic stellar wind (SSSW) driven nebula model (Weaver et al., 1977; Dyson and de Vries, 1972).
- The confined bubble (CB) model (Dopita, 1981).
The models predict differences in the kinematics which can be verified by the observations. Some of these differences are:
- The first two models may predict non-spherical shapes if the event occurs in a large scale inhomogeneous medium such as a medium with a gradient in density. Consequently, the faintest filaments must be correlated with the highest velocities (the strongest splitting). On the other hand, the CB model predicts a non-spherical shape due to the asymmetry in pressures of the collapsing cloud material when the star is not located at the center of accretion. This model may predict that the strongest splittings occur near the brightest filaments.
- The CB model predicts also, a lack of splittings in the central

band (due to the refraction of the stellar wind at the oblique shock front).

- The angular variation of the filaments located at the boundaries might be explained by the three models in their non-spherical modalities.

In the case of N 185, the splitting pattern indicates an expansion. Its density, energetics and stellar content are more compatible with the fossil SNR model (Rosado et al., 1982).

The case of N 70 is more complicated:

- The pattern show strong splittings in the central band.

- The splittings are, in the mean, of the same strength in both the faint and bright sides of the nebula (however, the strong splittings found in the weakest filaments are difficult to evaluate accurately and consequently we have an observational selection effect).

- The radial velocities of the boundaries show an angular variation.

- There are strong splittings in the bright filaments located at the boundaries.

The latter point makes these observations incompatible with any of the mentioned models. It is important to establish if there is a correlation between the brightness of a filament and the splitting strength. In any case, in a time comparable to the cluster age, a SN explosion could occur.

In conclusion, at the moment, optical observations are the only means in the discrimination of old SNRs.

REFERENCES

Blades, J.C., Elliott, K.H. and Meaburn, J.: 1980, MNRAS, 192, 101.
Dopita, M.A.: 1981, Ap. J. 246, 65.
Dopita, M.A., Ford, V.L., Mc Gregor, P.J., Mathewson, D.S. and Wilson,
    I.R.: 1981, preprint.
Dyson, J.E. and de Vries, J.: 1972, Astron. Astrophys. 20, 223.
Fesen, R.A., Blair, W.P. and Kirshner, R.P.: 1982, to appear in Ap. J.
    Vol. 262.
Llebaria, A.: 1980, Reduction Data Program. Laboratoire d'Astronomie
    Spatiale, Marseille, France.
Rosado, M., Georgelin, Y.P., Georgelin, Y.M., Laval, A. and Monnet, G.:
    1981, Astron. Astrophys. 97, 342.
Rosado, M., Georgelin, Y.P., Georgelin, Y.M., Laval, A. and Monnet, G.:
    1982, to appear in Astron. Astrophys.
Stasinska, G.: 1980, Astron. Astrophys. 84, 320.
Weaver, R., Mc Cray, R., Castor, J., Shapiro, P. and Moore, R.: 1977,
    Ap. J. 218, 377.

DISCUSSION

RAYMOND: Did you find the same small splitting in the center of N70 which was seen by Dopita et al.?

ROSADO: We find large and small splittings in the central band of N70 depending on the position. The advantage of FP interferometry is that it gives a 2-D view of the radial velocities over the entire nebula necessary in order to survey the zones of splittings. It seems that the slit position in Dopita's spectroscopy did not fall at the place of strong splittings. It is in this sense that the interferometric and spectroscopic observations are complementary.

N49: THE SITE OF A GAMMA-RAY BURST.
PRELIMINARY RESULTS FROM X-RAY OBSERVATIONS.

G. Pizzichini,
Istituto TESRE/CNR, Bologna
T.L. Cline, U.D. Desai, B.J. Teegarden
NASA/GSFC, Greenbelt
W.D. Evans, E.E. Fenimore, R.W. Klebesadel, J.G. Laros
LASL, Los Alamos
K. Hurley, M. Niel, G. Vedrenne
CESR (CNRS/UPS), Toulouse

ABSTRACT

The error box of the unusual Gamma-Ray Burst of March 5, 1979 falls completely inside the optical and radio image of the Supernova Remnant N49 in the Large Magellanic Cloud. This region was observed twice in x-rays with the High Resolution Imager of the Einstein Observatory, six weeks and nearly two years after the Gamma-Ray Burst. We show the comparison between the two observations.

The location of the unusal Gamma-Ray Burst of March 5, 1979 (Evans et al., 1979; Evans et al., 1980) coincides with N49, a well known Supernova Remnant in the Large Magellanic Cloud. The burst error box falls completely inside the remnant (Cline et al., 1982). This is, until now, the only Gamma-Ray Burst location associated with a known astrophysical object. The probability of a fortuitous alignment is very low, 4×10^{-4} (Felten, 1981).

If the event originated in N49, that is at a distance of 55 Kpc, it had a peak luminosity $\gtrsim 5 \times 10^{44}$ erg s^{-1} and an integrated energy output $\simeq 10^{45}$ erg (for a description of this event see Cline, 1980 and references therein), but at least two burst models can produce such energies (Ramaty, Lingenfelter and Bussard, 1981; Woosley and Wallace, 1982). If the burst spectrum is interpreted as a synchrotron spectrum modified by inverse comptonization of e^+-e^- pairs (Liang, 1981), the high x-ray luminosity is still acceptable.

One of the practical advantages of the coincidence between a Gamma-Ray Burst location and a Supernova Remnant was that N49 had been observed by the Imaging Proportional Counter of the Einstein Observatory one week before the totally unexpected and unforeseeable March 5, 79 event (Helfand and Long, 1979) It was observed again by the same authors, this time both with the IPC and with the High Resolution Imager, one month after the event. They give upper limits of 2.2×10^{-12} erg/(cm^{2}s) to the flux of point sources inside the

J. Danziger and P. Gorenstein (eds.), Supernova Remnants and their X-Ray Emission, 573–577.
© *1983 by the IAU.*

remnant after the event. In order to search for variations in the
x-ray flux in the error box, we observed again N49 with the HRI two
years after the burst. The results of our observation confirm those
of Helfand and Long (op. cit.). With their permission we made a
comparison of the two HRI images.

We present here our preliminary results (Table I). We give upper
limits to the flux and to the change in flux between the two HRI
observations for point sources in the remnant, which was divided into
8" x 8" squares. For each square we use as background the average of
its eight nearest neighbours. One could argue that the upper limit to
the change in flux should not exceed the upper limit to the flux of
point sources, but we must consider also the possibility that a "dip"
in the extended source has been filled, therefore we give the limits
directly as we derive them from the data.

TABLE I
X-Ray observations of the March 5, 79 Gamma-Ray Burst error box (N49)
Limits to point source fluxes and flux variations.

Observers	Date & Instrument	Duration (sec.)	CPS c/sec. $\times 10^{-3}$	L_x **	$\dot{M}$ ***	ΔL_x **	$\Delta \dot{M}$ ***
Helfand + Long (1979)	Feb. 26, 79 IPC						
"	Apr. 13, 79 IPC					<1.3	<1.5
"	" 19, 79 HRI	10200	<9.9	<1.3	<1.5		
Present work	Feb. 3, 81 HRI	7500	<7.3	<1.0	<1.1	<1.2*	<1.4
Present work + Helfand-Long	Apr. 19, 79 & Feb. 3, 81 HRI	17700	<6.7	<0.8	<1.0		

(*) See text for the difference in the upper limit to the flux of
point sources and to the change in flux.
(**) L_x and ΔL_x are in units of 10^{30} erg/s x (d/100 pc)^2.
(***) $\dot{M}$ and $\Delta \dot{M}$ are in units of 10^{10} g/sec. x(d/100 pc)^2.

By adding the two observations together we can also reduce the
upper limits to the flux of point sources inside N49. This procedure
should, however, be considered with caution, because of the two years
elapsed between the two observations, even if no variability has been
detected.

Until now, only one x-ray source has been detected in a Gamma-
Ray Burst error box (Pizzichini et al., 1981; Grindlay et al., 1982).
Upper limits to the x-ray flux of point sources have been obtained

for three more burst locations (Pizzichini et al., 1982). Both the source detected and the upper limits are one order of magnitude smaller than the upper limits we give here, but the March 5, 79 burst was a very unusual and possibly unique event. Among other things, it was the largest one even detected both in peak flux and in total energy and it has been argued that it belongs to a different class altogether (Mazets and Golenetskii, 1979), therefore we shall derive our conclusions for the location of the March 5, 79 event only from the observations of N49.

If we assume that the sources of Gamma-Ray Bursts are accreting neutron stars, as proposed by several authors (Woosley and Wallace, 1982; Bonazzola et al., 1981 and 1982; Hameury et al., 1982), our upper limits put constraints on the mass accretion rate. For a ratio of x-ray to total luminosity $\eta_x = 0.1$, using the same method and parameters of Helfand and Long (op. cit.), even at the distance of the LMC we already get a low value, $\dot{M} < 3 \times 10^{15}$ g/s or 5×10^{-11} $M_\odot$/yr. At galactic distances we have $\dot{M} < 10^{-16}$ $M_\odot$/yr x $(d/100 \text{ pc})^2$ for an interstellar hydrogen density of 1 cm^{-3}.

The accretion rates required by current models for this event vary between 3×10^{-16} and 8×10^{-13} $M_\odot$/(yr km^2). If we take a 1 km^2 polar cap, the distance to the source must be at least 170 pc for any of these models to be compatible with our upper limits, unless tne ratio of x-ray to total luminosity is much lower than usual or the accretion rate is variable.

If the burst energy is accumulated only by accretion, then the accretion rate should indeed be highly variable, as pointed out by Helfand and Long (op. cit.), because the same source produced a second, small burst on the following day and this requires an accretion rate of 1.6×10^{-13} $M_\odot$/yr x $(d/100 \text{ pc})^2$ for the time interval between the two events.

However, we have already four measurements which seem to have all been made when the source was in a low intensity state.

ACKNOWLEGEMENTS

We thank Drs. Helfand and Long for permission to use their HRI data on N49.
G. Pizzichini thanks Dr. C. Jones for her invaluable help with the data analysis at the Center for Astrophysics.

REFERENCES

Bonazzola S., Hameury J.M., Heyvaerts J. and Ventura J.: 1981, Space Sci. Rev. 30, pp. 471-474.
Bonazzola S., Hameury J.M., Heyvaerts J., Ventura J.: 1982, in Accreting Neutron Stars, Brinkmann, W. and Trümper J., eds, MPE Report No. 177, Garching bei München, pp. 241-243.
Cline T.L.: 1980, Comments on Astrophys., 9, pp. 13-22.
Cline T.L., Desai U.D., Teegarden, B.J.,, Evans W.D., Klebesadel R.W., Laros J.G., Barat C., Hurley K., Niel M., Vedrenne G., Estulin I.V., Kurt V.G., Mersov G.A. and Zenchenko V.M.: 1982,

Astrophys. J. (Letters) 255, pp. L45-L48.

Evans D., Klebesadel R., Laros J., Cline T., Desai U., Teegarden B. and Pizzichini G.: 1979, I.A.U. circ. No. 3356.

Evans W.D., Klebesadel R.W., Laros J.G., Cline T.L., Desai U.D., Pizzichini G., Teegarden B.J., Hurley K., Niel M., Vedrenne G., Estoolin I.V., Kouznetsov A.V., Zenchenko V.M. and Kurt V.G.: 1980, Astrophys. J. (Letters) 237, pp. L7-L9.

Felten J.E.: 1981, Proceedings of the 17th International Cosmic Ray Conference, Paris, Late papers volume No. 52-55.

Grindlay J.E., Cline T., Desai U.D., Teegarden B.J., Pizzichini G., Evans W.D., Klebesadel R.W., Laros J.G., Hurley K., Niel M. and Vedrenne G.: 1982, CFA preprint series No. 1734, to appear in Nature.

Hameury J.M., Bonazzola S., Heyvaerts J. and Ventura J.: 1982, Astron. and Astrophys. 111, 242-251.

Helfand D.J. and Long K.S.: 1979, Nature 282, pp. 589-591.

Liang E.P.T.: 1981, Nature 292, pp. 319-321.

Mazets E.P. and Golenetskii S.V.: 1979, preprint Academy of Sciences of the USSR No. 632.

Pizzichini G., Danziger J., Grosbøl P., Tarenghi M., Cline T.L., Desai U.D., Mushotzky R., Teegarden B.J., Evans W.D., Klebesadel R.W., Laros J.G., Barat C., Hurley K., Niel M., Vedrenne G., Estulin I.V., Mersov G., Zenchenko V. and Kurt V.: 1981, Space Sci. Rev. 30, pp. 467-470.

Pizzichini G., Cline T.L., Desai U.D., Teegarden B.J., Hurley K., Niel M., Vedrenne G., Evans W.D., Fenimore E.E., Klebesadel R.W. and Laros J.G.: 1982, in Accreting Neutron Stars, Brinkmann, W. and Trümper J., eds, MPE report No. 177, Garching bei München, pp. 237-240.

Ramaty R., Lingenfelter R.E. and Bussard R.W.: 1981, Astrophys. and Space Sci., 75, pp. 193-203.

Ventura J.: 1982, preprint 82-II-AV1, COSPAR meeting, Ottawa, invited paper.

Woosley S.E. and Wallace R.K.: 1982, Astrophys. J. 258, pp. 716-732.

DISCUSSION

BISNOVATY-KOGAN:
I want to mention an additional difficulty in identification of 5 March 1979 Gamma-Ray Burst with SNR in LMC. The luminosity of this source in the quite hard x-ray pulsar phase is much greater than the Eddington optical luminosity, so the outer layers of the neutron star will be thrown away with a velocity almost equal to c. In this case it is very improbable to obtain the regular hard x-ray pulsations which have been observed.

KOCH-MIRAMOND:
Are there existing or planned observations of the Gamma-Ray Burst error box at higher energy x-rays?

PIZZICHINI:
Not to my knowledge.

DUROUCHROUX:
Your argument for correlation of the Gamma-Ray Burst with N49 is mainly based on variations of the x-ray flux before and after the burst. Is this phenomenon very unusual in the Einstein observations of the same region in the sky separated by a few months? (My question is not related to a sky region where a burst took place but to any region observed at least two times).

PIZZICHINI:
No; in fact no flux variations hav been detected. My argument in favour of the association of the March 5, 1979 Burst with N49 is based on the coincidence between the remnant and the burst error box (see Felten, op. cit.) and on the fact that it is possible to account for an origin of the burst in the LMC (Ramaty et al., 1981, Liang, 1981). Helfand and Long (op. cit.) who had IPC observations of N49 before and after the event find only an upper limit to the change in flux. We also find only an upper limit to the change between 1 month and 2 years after the event.

There are of course other objects which have been observed several times by the Einstein Observatory and found to be variable, but their locations do not coincide with Gamma-Ray Burst locations.

X-RAY, OPTICAL AND UV OBSERVATIONS OF THE YOUNG SUPERNOVA REMNANT IN THE IRREGULAR GALAXY NGC 4449

W. P. Blair[1], R. P. Kirshner[2], P. F. Winkler, Jr.[3],
J. C. Raymond[1], R. A. Fesen[4] and T. R. Gull[4]

1. Harvard-Smithsonian Center for Astrophysics
2. Department of Astronomy, The University of Michigan
3. Department of Physics, Middlebury College
4. Laboratory for Solar and Stellar Physics, NASA Goddard
   Spaceflight Center

Abstract: A powerful young supernova remnant (SNR) similar to Cas A has recently been discovered in the irregular galaxy NGC 4449. We have obtained X-ray, optical and ultraviolet data which allow us to investigate possible models for this object and estimate its age. Several lines of argument indicate a massive star of order 25 $M_\odot$ as the precursor to this remnant. If the x-ray emission is attributed to a reverse shock in the ejecta, the remnant should be $\sim$ 120 years old.

In the past few years, several new SNRs with characteristics similar to Cas A have been detected. The study of such objects is important because it provides an opportunity to investigate not only the evolution of young SNRs, but the relation of these objects to nucleosynthesis and the distribution of heavy elements in the interstellar medium (ISM). Here we present data on one such object and discuss a self-consistent model that accounts for all of the observations presently available.

The SNR in the irregular galaxy NGC 4449 was first detected at radio wavelengths by Seaquist and Bignell (1978), who found a strong, unresolved non-thermal source about 1' north of the nucleus of the galaxy. At the presumed distance of NGC 4449 (5 Mpc; Sandage and Tammann 1975), the source is 25 times more luminous at 2.7 GHz than Cas A. Subsequent radio observations with the VLA (Seaquist, private communication) have only been able to place an upper limit on the diameter of the object of < 0".2, which corresponds to a linear diameter < 5 pc. The radio source is coincident with an emission region whose optical spectrum shows a composite of narrow emission lines belonging to a normal H II region and broad lines (full width $\sim$ 7000 km s^{-1}) which belong to the SNR (Balick and Heckman 1978, Kirshner and Blair 1980). However, broad components were initially detected only at the positions of [O I] $\lambda\lambda$6300,6364, [O II] $\lambda\lambda$7320,7330 and [O III] $\lambda\lambda$4363,4959,5007.

579

J. Danziger and P. Gorenstein (eds.), Supernova Remnants and their X-Ray Emission, 579–582.

We have extended this early work by obtaining an additional high
quality optical spectrum as well as observations with the HEAO-B
(Einstein X-ray Observatory) and International Ultraviolet Explorer
(IUE) satellites. The optical spectrum was obtained with the 2.1m
telescope at Kitt Peak National Observatory and shows additional broad
lines which belong to the SNR, including [O II] λ3727, [Ne III]
$\lambda\lambda$3869,3968 and [S II] λ4070.

Although only a handful of lines belonging to the SNR are evident, we
can still estimate the physical conditions in the optically emitting
gas. The ratio of [O II] λ3727/λ7325 can be used to infer a density
of roughly log $N_e \approx 5.5$, assuming the lines are formed in a region
with $T \approx 15,000$ K. The ratio of [O III] λ4363/λ4959+λ5007 indicates T
$\approx 50,000$ K in the O^{++} region; however, the density is high enough that
the nebular lines should be partially de-excited and $T \approx 40,000$ K is
probably more realistic. The assumption of pressure equilibrium
between the O°, O^+ and O^{++} zones allows the mass of oxygen in the
currently cooling gas to be estimated using the method of Peimbert
(1971); we find $M(O) \approx 0.01\ M_\odot$, which is about a factor of 50 higher
than for Cas A.

Comparison of these observations to shock model calculations is
difficult because of the distinctly non-solar abundances in this SNR.
Itoh (1981) has calculated some shock models for a gas of pure oxygen
composition and finds that a shock velocity of at least 140 km s^{-1} is
needed to produce strong [O III] emission; however, none of these
models incorporates a density as high as is indicated for this SNR.

Crude estimates of the abundances of neon and sulfur relative to
oxygen can be made using the observed ratios of [Ne III]/[O III] and
[S II]/[O II]. These estimates can be compared to the final
abundances from the stellar evolution models of Weaver and Woosley
(1980) for 15 $M_\odot$ and 25 $M_\odot$ stars, which include the effects of
explosive nucleosynthesis. The relative Ne and S abundances are both
consistent with a precursor star of $\approx 25\ M_\odot$, a conclusion which is
supported (at least indirectly) by the presence of massive stars in
the H II region itself.

The IUE spectrum represents 17.7 hours of integration on the SNR. A
very weak feature with $\Delta v \approx 7000$ km s^{-1} appears at λ1660. If this is
the O III] λ1664 line from the SNR, it is within a factor of two of the
strength one would expect using Itoh's (1981) high density model "C"
(assuming $A_v = 0.7$, which is determined from the narrow Hα and Hβ
lines belonging to the H II region). We have been unable to entirely
exclude the possibility that this is a weak camera feature, although
two other "blank" long exposures have been checked and no comparable
feature seems to be present. Even if this feature represents an upper
limit to the O III] λ1664 strength, the absence of the (normally much
stronger) C IV λ1550 and C III] λ1909 lines argues for a depleted
abundance of carbon relative to oxygen. This is at least
qualitatively in agreement with the 25 $M_\odot$ model discussed earlier.

TABLE 1
NGC 4449 SNR Models

Model Parameter	ISM-Dominated Reverse Shock (Model A)	Ejecta-Dominated Reverse Shock (Model B)	Blast Wave
T_x (Assumed)	6×10^6 K	6×10^6 K	$\gtrsim 10^8$ K
L_x (0.2-4 keV)	0.8×10^{39} erg s^{-1}	0.8×10^{39} erg s^{-1}	1.7×10^{39} erg s^{-1}
Radius	1.2 pc	0.4 pc	14 pc
X-ray Emitting Mass	36 $M_\odot$	2 $M_\odot$	3000 $M_\odot$
Swept-up Mass	30 $M_\odot$	0.2 $M_\odot$	3000 $M_\odot$
Ambient ISM Density	150 cm^{-3}	25 cm^{-3}	11 cm^{-3}
Age	140 yr	120 yr	1600 yr

The X-ray data consist of a 32,000 sec HRI exposure that encompasses all of NGC 4449. Although low level emission from the galaxy is apparent in this image, three point sources are clearly visible, one of which corresponds precisely with the radio and optical positions of the SNR. Since no spectral information is available from the HRI detector, the interpretation of the X-ray data rest on the choice of a model for the X-ray emission.

There are two types of shocks which are likely to produce X-rays in young SNRs: a blast wave propagating outward through the ISM, or a reverse shock (cf. Gull 1975). Both phenomena may be present in the NGC 4449 SNR, but it is likely that the reverse shock emission is the dominant source of X-rays in the 0.2-4 keV HRI band, as is also the case in Cas A (Fabian et al. 1980). With this assumption, we estimate a temperature in the X-ray gas of $T_x = 6 \times 10^6$ K and derive L_x (0.2-4 keV) = 8×10^{38} ergs s^{-1}.

If the emission is dominated by a reverse shock, we can investigate two models which should bracket the real situation. In model A, we assume the material heated by the reverse shock is swept-up ISM material with cosmic abundances and in model B we assume a plasma made of undiluted ejecta from the supernova (for which we have assumed the abundances from the 25 $M_\odot$ model of Weaver and Woosley 1980). The emissivity is much higher ($\sim$ 20 x) in the ejecta-dominated model B, so much less material is needed to create the X-ray emission.

Table 1 summarizes the results from these two models as well as showing the predictions of the blast wave model mentioned earlier. We have assumed a simplified geometry whereby the X-rays occur in a uniform density plasma in a thin spherical shell of radius R and thickness R/12 (filling factor f = 0.25). The X-ray luminosity can then be expressed,

$$L_x = 4/3 \pi R^3 f n_x^2 P'(\Delta E, T) \tag{1}$$

where n_x is the electron density in the X-ray emitting gas and
$P'(\Delta E, T)$ is the emission function, which depends on the electron
temperature and plasma composition. The electron density n_x is
obtained by assuming pressure equilibrium between the X-ray and
optically emitting gas, i.e. $n_x T_x \sim n_o T_o$. From the optical data, n_o
$T_o \sim 4.5 \times 10^9$, so using T_x from above, $n_x^o \sim 700$ cm^{-3}.

Equation (1) can be solved for the radius, R, with the results shown
in Table 1. Both reverse shock models lead to predicted radii below
the upper limit given by the VLA observations, but the radius from the
blast wave model is too large. These predicted radii allow estimates
of the mass of X-ray emitting gas, the swept-up mass (both $\propto R^3$) and
the age of the remnant ($\propto R$) all shown in Table 1.

All things considered, something close to model B appears to be most
plausible, as summarized below. The Hβ flux from the H II region
implies that many massive stars must be present in the H II region to
keep it photoionized. About 100-150 years ago, one of these stars
exploded sending forth 10-30 M$_\odot$ of ejecta enriched by a factor of 5-50
in heavy elements. The surrounding H II region is dense enough that a
reverse shock has now developed in the fast moving ejecta. Rapid
cooling occurs behind this shock and knots condense to become optical
filaments. The cooling time,

$$t_c = \frac{n_e kT_e}{n_e^2 P'(\Delta E, T)} \tag{2}$$

is only about 60 years for model B, so substantial changes may be
evident over a period of a few years.

This project gratefully acknowledges support from the following
grants: NSF AST 81-05050, and NASA NAG 8341, NAG 8389 and NAG 5-87.

REFERENCES

Balick, B. and Heckman, T. 1978, Ap. J. (Letters), 226, L7.
Fabian, A. C., Willingale, R., Pye, J. P., Murray, S. S., and
     Fabbiano, G. 1980, M.N.R.A.S., 193, 175.
Gull, S. F. 1975, M.N.R.A.S., 171, 263.
Itoh, H. 1981, Pub. A.S.J., 33, 1.
Kirshner, R. P. and Blair, W. P. 1980, Ap. J., 236, 135.
Peimbert, M. 1971, Ap. J., 170, 261.
Sandage, A. and Tammann, G. A. 1975, Ap. J., 196, 313.
Seaquist, E. R. and Bignell, R. C. 1978, Ap. J. (Letters), 226, L5.
Weaver, T. A. and Woosley, S. E. 1980, Ann. N. Y. Acad. Sci., 336,
     335.

LUMINOUS SUPERNOVA REMNANT CANDIDATES IN M82 AND THE PHYSICAL
PROPERTIES OF THE VARIABLE RADIO SOURCE 41.9+58

P.P. Kronberg
David Dunlap Observatory and Scarborough College,
University of Toronto

P. Biermann
Max-Planck-Institut für Radioastronomie, Bonn

ABSTRACT

Large numbers ($\sim$30) of supernova remnant candidates have been
revealed in new VLA maps of the inner 600 pc of M82 with subarcsecond
resolution. The flux density decrease of the bright source 41.9+58 in
the M82 nuclear region is found to be occurring with an approximately
constant spectral index of -0.9 at $\nu \gtrsim 3$ GHz, and the low frequency
turnover has moved from $\sim$700 MHz in 1974-1975 to <400 MHz in 1979-81.

A model of 41.9+58 is proposed in which it is, at least initially
confined by its own hot ($\sim 10^8$K) pre-SN ejecta, and its relativistic
particles are accelerated by a pulsar at its centre.

1.   FULL-RESOLUTION VLA MAPS OF THE M82 NUCLEUS

The M82 nucleus has been observed with the full resolution of the
VLA (0".4) at λ6 cm (5 GHz). At this resolution, as Figure 1 shows, a
large number of compact ($\lesssim$0".2) sources are visible from 100 mJy down to
0.5 mJy - the current limit of dynamic range at λ6 cm. Within this
flux density range, at least 30 discrete sources can be identified, 16
of which are stronger than 2 mJy. If the majority of these are supernova remnants, they are all more radio luminous than Cassiopeia A which,
at the assumed 3.2 Mpc distance of M82, would have a 5 GHz flux density
of 0.8 mJy and an angular size of $\sim$0.2 arc seconds.

A preliminary spectral index comparison between 5 and 15 GHz for
the stronger sources in Figure 1 shows that the vast majority have steep
spectra, which is consistent with their being powerful supernova
remnants. Their luminosities range from 1 $\rightarrow$ 200 times that of Cas A.
In an earlier astrometric comparison between the radio hotspots and
optical knots, Kronberg, Pritchet and van den Bergh (1971) and O'Connell
and Mangano (1978) determined that there was little, if any, correspondence between radio knots at $\sim$2" resolution and compact features in the
optical and near-infrared. A preliminary comparison at the present,
higher radio resolution indicates that this is still the case, which is

583

J. Danziger and P. Gorenstein (eds.), Supernova Remnants and their X-Ray Emission, 583–590.
© *1983 by the IAU.*

not too surprising in view of the high optical obscuration within most
of the inner 600 pc. A more detailed analysis of the new radio maps
is in progress, and comparison of the radio features with several optical
bands (Kronberg, Biermann and Schwab, 1983) will be published elsewhere.

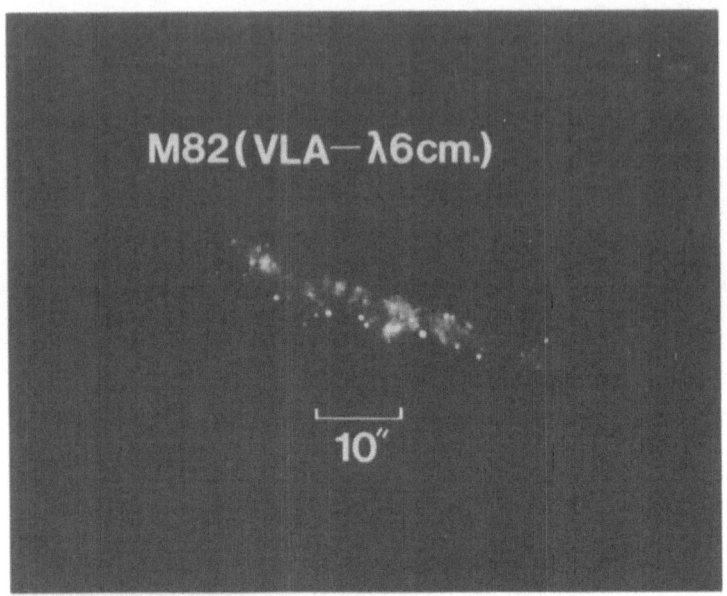

Figure 1.   The radio emission from the inner 600 pc of the M82 nucleus,
            mapped at 0''.4 resolution with the NRAO Very Large Array in
            1981. The brightest point source is 41.9+58. The faintest
            point sources visible in this picture have radio luminosities
            comparable with that of Cassiopeia A.

O'Connell and Mangano (1978) concluded from the spectral type and
colours observed in parts of the M82 disk that a primary encounter with
M81 - the likely trigger for M82's current phase of violent activity -
occurred somewhat over $\sim 2 \times 10^8$ years ago. When we compare this "maximum
lifetime" of M82's active phase with that of $\lesssim 5 \times 10^7$ years for the
massive stars which are ionizing the central HII regions in M82 (Recillas-
Cruz and Peimbert (1970), O'Connell and Mangano (1978)) and also the
free-fall time of some of the dense clouds, $\sim 10^6$ yrs for $n \sim 1000$ cm^{-3}, it
is clear that several generations of massive star formation have occurred
recently. This is the likely explanation for the compact radio sources -
many are probably supernova - and the ambient synchrotron-emitting
cosmic ray gas which is evident in the radio map. The VLA radio observations enable us to deduce a further timescale, namely the time it will
take the current cosmic ray clouds to disperse from the central 200 pc
of M82. Our VLA maps show that the minimum cosmic ray energy density
(assuming equipartition with the magnetic field), ε_{CR}, $\simeq 10^{-9}$ erg cm^{-3}.
This gas, being bounded by i/s clouds having $10 \lesssim n \lesssim 2000$ cm^{-3}, will
advance with a ram-pressure determined velocity of ~ 100 km/s (we ignore

adiabatic losses). We can thus calculate that the present configuration of radio emitting clouds will have dispersed in $\sim$$10^6$ years, a time which agrees with the other two independently determined timescales mentioned above (Kronberg, Biermann and Schwab, 1981). Thus the observations support a scenario in which chaotic and expanding cosmic ray clouds envelop and perhaps statically compress dense i/s gas clouds which are induced to collapse into massive stars. The end phases of stellar evolution generate directly, through supernovae, pulsars, etc., and/or indirectly through interstellar turbulence, the next generation of cosmic ray clouds. A self-perpetuating (for a time) scenario of this sort is the most likely explanation of the active condition of the M82 nuclear region. The large implied numbers of massive stars make the M82 nucleus an interesting laboratory for supernovae and their remnants - which we appear to be seeing in large numbers and with very high luminosities, by galactic standards. The next section discusses the properties of the most luminous - presumably stellar - object, namely the variable source labelled 41.9+58 by Kronberg and Wilkinson (1975).

2. THE COMPACT, VARIABLE SOURCE, 41.9+58

This source is by far the most luminous source within the 600 pc long M82 nuclear region. Its spectrum was first defined by Kronberg and Wilkinson (1975) between 408 MHz, where it was optically thick, and 8085 MHz where its spectral slope was found to be steep and close to -0.9 ($S\sim\nu^{\alpha}$). Successive NRAO interferometer measurements made between 1971 and 1975 showed further that the 8085 MHz flux was decreasing with time (Kronberg and Clarke 1978), which fact was later confirmed by VLA measurements in 1978 and 1981. VLBI measurements in 1974 by Geldzahler et al. (1977) revealed the remarkable result that, at least at that time, most of 41.9+58's 8 GHz flux came from a region only 1.5 milliarcseconds in diameter ($\sim$1 light month). Figure 2 shows the spectrum of 41.9+58 at 2 epochs; that closest to Geldzahler et al's VLBI measurement, and then in 1981 from simultaneous 20, 6 and 2 cm measurements with the VLA. A repeat by Conway in 1979.1 of Kronberg and Wilkinson's 408 MHz flux measurement indicates that the optically thick flux is increasing, whereas at the higher frequencies it is decreasing with nearly constant slope with an e-folding time of $\sim$15 years. Furthermore, the low frequency turnover (τ=1) frequency, 700 MHz in ca. 1975 has decreased to $\sim$400 MHz in 1981. The spectral index in 1981 is -0.9, so that the index of the relativistic electron energy spectrum is 2.8. The earliest measurement of 41.9+58's flux in 1966 at 11 cm (Bash, 1968) is consistent with a backwards extrapolation of the spectrum in Fig. 2, and sets a conservative lower limit to the age of 9 years in 1975, and also an upper limit to the average velocity of expansion which is 1270 km/s. Unfortunately no definitive VLBI map has been made since, although a more recent low frequency (18 cm) observations by Jones, Sramek and Terzian (1981) indicate elongated structure up to $\sim$15 mas in p.a. 56° (epoch 1979.4). It is unfortunate that since 1974 no detailed VLBI map has been obtained at ν>3 GHz (where interstellar scattering will not likely obscure the true radio size).

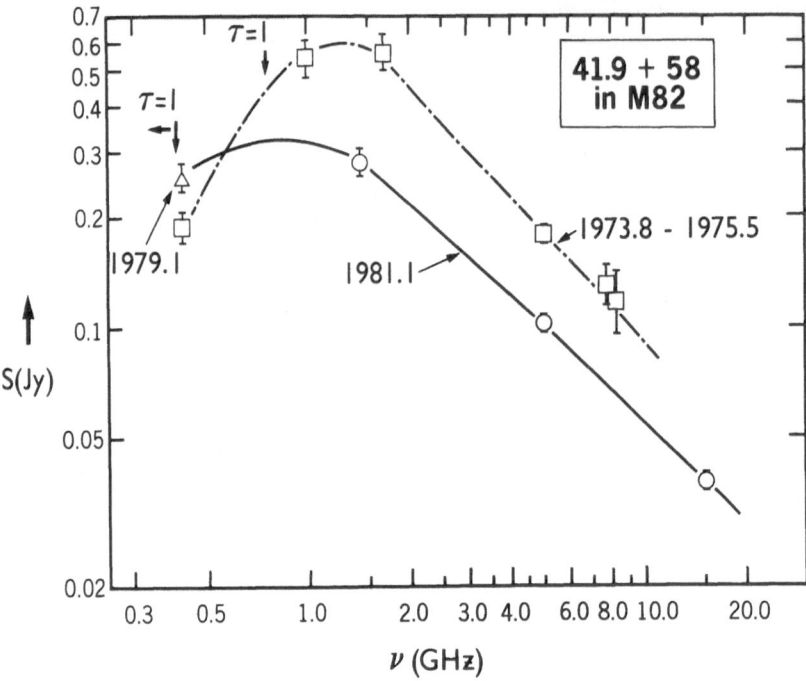

Figure 2.   The radio spectrum of 41.9+58 for 2 epochs. The earlier
            epoch was chosen to include flux density measurements which
            are relatively close in time to that of the VLBI measurement
            of the angular size (1974.6) by Geldzahler et al.

     If we assume incoherent synchrotron radiation and use the measured
ca. 1975 spectrum (Fig. 2) and 7.8 GHz size, some interesting requirements must be imposed on this source's physical parameters (cf. Kronberg
and Clarke (1978) and Brown and Neff (1980)): The very well defined low
frequency spectral turnover (Fig. 2) establishes a firm upper limit on
the synchrotron self absorption (ssa) frequency (ν_{ssa}), as well as the
combination of thermal absorption (ν_{th}) both within and in front of the
radiating volume. If the value of $\tau=1$ at 700 MHz is due to ssa then
the 1.5 mas size and observed luminosity of 6×10^{37} erg/s (integrating
between 10^7 and 2.10^{10} Hz) require that the magnetic field strength,
$B \sin \theta \approx 2 \times 10^{-4}$ gauss. This, being far below the equipartition field
(0.13 gauss for equal proton and electron energies (k=1), or 0.4 gauss
if k=100), requires a high energy density of relativistic particles,
$E_p = 3.2 \times 10^{47}$ ergs (k=1), or 1.6×10^{-3} ergs cm^{-3}.

     The fact that any non-relativistic electrons within the source must
also cause $\tau_{th}=1$ at $\nu \lesssim 700$ MHz requires that at most $2.5\ T_8^{3/4} M_\odot$ of non-
relativistic gas exists within the source, i.e. $n_{th} < 1.16 \times 10^7\ T_8^{3/4}\ cm^{-3}$.
This puts an upper limit of 6.3×10^{49} erg at $T = 10^8 K$ to its thermal
energy content. Comparing n_{th} with the likely density of relativistic
electrons, we conclude that the thermal and relativistic particles are
present in roughly comparable numbers.

It must be noted here that both the total particle energy and the ratio of n_r/n_{th} are highly sensitive to the dimensions of the radiating volume, and these may have changed significantly from 1975 to 1981. The fact that the frequency at which $\tau=1$ moved significantly downward between 1975 and 1981 (Fig. 2) suggests that the sharp turnover in the 1975 spectrum was indeed due to synchrotron self-absorption, and that 41.9+58 has expanded and decreased in surface brightness. We can then conclude that the $\tau=1$ frequency due to thermal absorption in front of the source (and probably within the source) lies below 400 MHz (the $\tau=1$ frequency in 1981), on the grounds that the optical depth of foreground material is unlikely to change so quickly with time, being less sensitive to the source's radius than the ssa frequency.

Assuming that all the foreground ionized gas exists in a shell of hot thermal gas having thickness ΔR, $\tau_{th}=1$ at $\nu<400$ MHz now requires that the density in the shell has an upper limit given by

$$n_{shell} \lesssim \frac{6.7 \times 10^6 T_8^{3/4}}{\Delta R^{1/2} (3 \cdot 10^{16} \text{cm})} \quad \text{cm}^{-3}$$

The shell must be sufficiently hot, and dense to restrain the expansion of the energetic relativistic electron cloud and at the same time have less than unity optical depth at 400 MHz. We find that, for $T=10^8$K and $n_{shell} \sim 10^7$, these conditions are just satisfied for tne epoch 1975 source parameters. This also limits the mass of the shell at $M_{shell} \lesssim 20 T_8^{3/4} M_\odot$. A mass in the range of $1 \rightarrow 10\ M_\odot$ is probably more realistic, in which case $\Delta R \sim R$ in 1975.

This leads us to propose a model for 41.9+58 in which, in 1975, most of the synchrotron radio emission was within a diameter of $\sim 7 \times 10^{16}$cm (~ 25 light days) and confined by a hot shell of density 10^7 cm^{-3} and temperature $\sim 10^8$K. (Figure 3a). The radio source is expanding slowly (at $v \lesssim 1000$ km/s) and contains a magnetized rotator, probably a pulsar whose rapidly rotating magnetosphere efficiently accelerates the radiating relativistic electrons. The mass of the surrounding thermal shell is too great to be swept up interstellar matter. We postulate that it is the ejecta from the massive star which is undergoing a supernova-like explosion.

The configuration in Fig. 3a is not stable, and the hot, light relativistic gas will expand into instabilities in the surrounding shell as Fig. 3b illustrates schematically. An asymmetrical source and/or jets will develop quickly, and such a phenomenon might explain the asymmetrical source shape observed using 18 cm VLBI by Jones, Sramek and Terzian (1981).

The confining shell would be a source of bremsstrahlung X-rays. We note that the extrapolation of the synchrotron spectrum in Fig. 2 to X-rays would give only 1/20 of the value of $\sim 10^{39}$ erg sec^{-1} reported by Griffiths (1980), and likely associated with 41.9+58. A shell near $2M_\odot$ having a temperature between 10^7 and 10^8 K in our model would provide

A MODEL FOR 41.9+58 IN M82

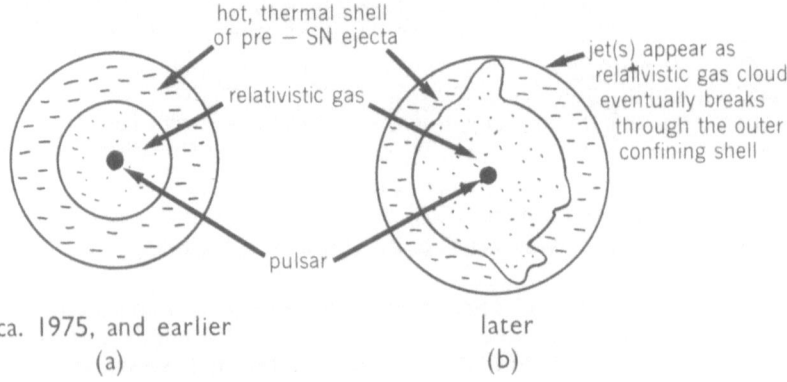

Figure 3.   A suggested physical model for 41.9+58 (left) which is
            consistent with the physical parameters as measured in ca.
            1975. Instabilities will develop quickly in the surrounding
            hot ejecta which will be "punctured" by the relativistic gas,
            and an asymmetric growth of overall radio dimensions will
            develop within a short period of time (right side).

about this much X-ray emission, however it is also possible that inverse
compton emission, at least at epoch 1975, also contributed to the X-ray
flux. Our calculations serve to show that the X-ray flux from 41.9+58
will be an important guide in a more detailed model than we have
described here. A more refined model will require definitive VLBI maps,
which are now needed to provide the crucial observational clues to the
interesting nature of 41.9+58. A further monitoring of its radio spectrum
and X-ray flux is also being undertaken.

     It is premature to classify 41.9+58 with other galactic supernovae,
since its luminosity exceeds that of Cas A by ∿200x. Combined with the
large number of luminous (relative to Cas A) objects in Fig. 1 it
appears that a much more luminous supernova-like object may commonly
exist, at least in regions of very active star formation. It is likewise premature to compare 41.9+58 with the comparably luminous supernova-like radio source recently discovered in M100 (Weiler et al. 1981).
Whatever the type of star associated with 41.9+58, it provides us with
one of the first opportunities to study the early phase of an explosive
stellar event.

ACKNOWLEDGEMENTS

     We thank the Director and Staff of the U.S. National Radio Astronomy
Observatory for the granting of VLA time, and their hospitality and
assistance. This research was also supported by the Natural Sciences
and Engineering Council of Canada (NSERC) (P.P.K.), and the Deutsche
Forschungsgemeinschaft, under project SFB131 (P.B.).

REFERENCES

Bash, F.N., 1968, Astrophys. J. (Suppl.) 16, pp. 373-404.
Brown, R.L., and Neff, S.G., Astrophys. J., 241, pp. 561-566.
Geldzahler, B.J., Kellermann, K.I., Shaffer, D.B., and Clark, B.G.,
    1977, Astrophys. J. (Letters) 215, pp. L5-L6.
Griffiths, R.E. Highlights of Astronomy, 1980, 5, pp. 641-651.
Jones, D.L., Sramek, R.A., and Terzian, Y., 1981, Astrophys. J. 246,
    pp. 28-37.
Kronberg, P.P., Pritchet, C.J. and van den Bergh, S. 1972, Astrophys.
    J. (Letters), 173, pp. L47-L50.
Kronberg, P.P., and Wilkinson, P.N., 1975, Astrophys. J. 200, pp. 430-438.
Kronberg, P.P., and Clarke, J.N., 1978, Astrophys. J. (Letters), 224,
    pp. L51-L54.
Kronberg, P.P., Biermann, P., Schwab, F., 1981, Astrophys. J. 246,
    pp. 751-760.
Kronberg, P.P., Biermann, P., and Schwab, F., 1983 in preparation.
O'Connell, R.W., and Mangano, J.J., 1978, Astrophys. J. 221, pp. 62-79.
Weiler, K.W., van der Hulst, J.M., Sramek, R.A., and Panagia, N.,
    1981, Astrophys. J. (Letters) 243, pp. L151-L156.

DISCUSSION

DICKEL: These sources are clearly a different class of object than SNR
in our galaxy or the other local group galaxies.   They are on the
average over 100x the brightest one in the Milky Way which says
different types of galaxies produce qualitatively different supernovae
and remnants.   Except for position what is the evidence that 41.9+58
is not just a standard variable quasar?

KRONBERG: Apart from the a priori unlikelihood of finding a background
source with S_6>100mJy 3" from the centroid of the M82 radio nucleus,
to our knowledge no quasar which varies so quickly with a constant,
steep spectrum has been found thus far.

GOSS: (1)   What is known about the continuum spectral indices of the
SNR candidates in M82?   (2)   What is the radio structure of the dynamical centre of M82?   Is there a nuclear source?

KRONBERG: (1)   The majority we have measured thus far, at least among
the stronger ones, have steep spectra.   Our values are very preliminary
but range from $\sim$-0.4 to $\sim$-0.9.   (2)   The position of the dynamical
centre is ill-defined, though approximately at the centre of the radio
complex.   There is an amorphous complex of radio emission in this area
(3"-5" east of 41.9+58) but no obvious radio source at the likely
dynamical nucleus.

PACINI: Obviously this source in M82 is probably of the same nature
as the other radio supernovae discovered recently.   Its long lifetime
combined with the known frequency of SN suggests the possibility that
similar sources may be detected in a large fraction of galaxies.   The
signature would be a very compact radio source combined with a transient
X-ray source.   For the rest, the models for radio supernovae by
Chevalier or by Salvati and myself can be tested against the source in
M82 and similar systems.   Data on combined radio and X-ray evolution
are essential in this respect.

THE X-RAY EMISSION OF A CLUMPY IRREGULAR GALAXY FROM THOUSANDS OF SUPERNOVA REMNANTS?

D.S. Heeschen[1] and J. Heidmann[2]
1. National Radio Astronomy Observatory
   Charlottesville, VA, USA
2. Observatoire de Paris Meudon, France

1. OBSERVATIONS

Clumpy irregular galaxies contain 5-10 "clumps" which are hyperactive HII complexes each equivalent to 100 giant HII regions of the 30 Doradus type (Heidmann 1982). We observed one of them, Mkn 325 (= NGC 7673), with the Einstein IPC in Dec.1980 (seq.no 10201) for 3,200 s. The reduction was made kindly by D.E.Harris. The source was localized at
23h 25m 12.2s, + 23° 18' 25" (1950)
in agreement with the optical position, at a quite weak level (14 counts in the 1.4-2.9 kev range). Thus we could not get valuable spectral information, only that the spectrum is rather not soft. Correction for galactic absorption N_H = $5x10^{20}$ at.cm^{-2} (Heyles 1975) is applied. Fits of power law spectra happen to all go through the point with flux density $4.5x10^{-5}$ mJy at $3.0x10^{17}$ Hz (1.24 kev) and they yield a flux (1.1 ± 0.3) $x10^{-13}$ erg cm^{-2} s^{-1} inside 1-3 kev and an X-luminosity $(2.2 \pm 0.3)x10^{41}$ erg s^{-1} inside 0.5-4.5 kev, for a distance 49 Mpc.

2. OVERALL SPECTRUM

Combining this X-ray result to published or unpublished data in other spectral domains we obtain the overall global spectrum given in the Table and Figure. Tokunaga's IR measurements refer to a 10" aperture and, being partial, give only a spectral index: α = 0.53. Huchra's UBV are also partial and give α = -1.56 (a galactic absorption A_B = 0.31 correction was applied). The UV fluxes falling inside the IUE large slit were converted to global values using the photometry of the clumps by Coupinot et al. (1982); galactic absorption was also corrected for.

3. DISCUSSION

We first recall that these same UV results show that Mkn 325 may contain $4.2x10^4$ O8V + $1.10x10^5$ B8I stars (even more if there is internal absorption) and that at 155 nm it radiates $1.2x10^{28}$ erg s^{-1} Hz^{-1}, i.e.

J. Danziger and P. Gorenstein (eds.), Supernova Remnants and their X-Ray Emission, 591–595.
© 1983 by the IAU.

frequency (Hz)	flux density (mJy)	reference
1.415×10^{9}	39 ± 4	Bieging et al.1977
2.700×10^{9}	26 ± 4	Bieging et al.1977
5.000×10^{9}	$\leqslant 18$	Biermann et al.1980
1.070×10^{10}	10.3 ± 0.7	Heidmann et al.1982
8.333×10^{13}	7.2 ± 0.65	Heidmann and Tokunaga private com.
1.364×10^{14}	9.6 ± 0.15	Heidmann and Tokunaga private com.
1.364×10^{14}	30	Heidmann and Shaya private com.
2.400×10^{14}	12.5 ± 0.25	Heidmann and Tokunaga private com.
5.425×10^{14}	20.7	Huchra 1977
6.818×10^{14}	16.2	Huchra 1977
6.818×10^{14}	32	de Vaucouleurs et al.1976
8.197×10^{14}	10.8	Huchra 1977
1.176×10^{15}	5.8	Benvenuti et al.1982b
1.935×10^{15}	4.2	Benvenuti et al.1982a
3.000×10^{17}	4.5×10^{-5}	this paper

Table. Global emission of Mkn 325 versus frequency

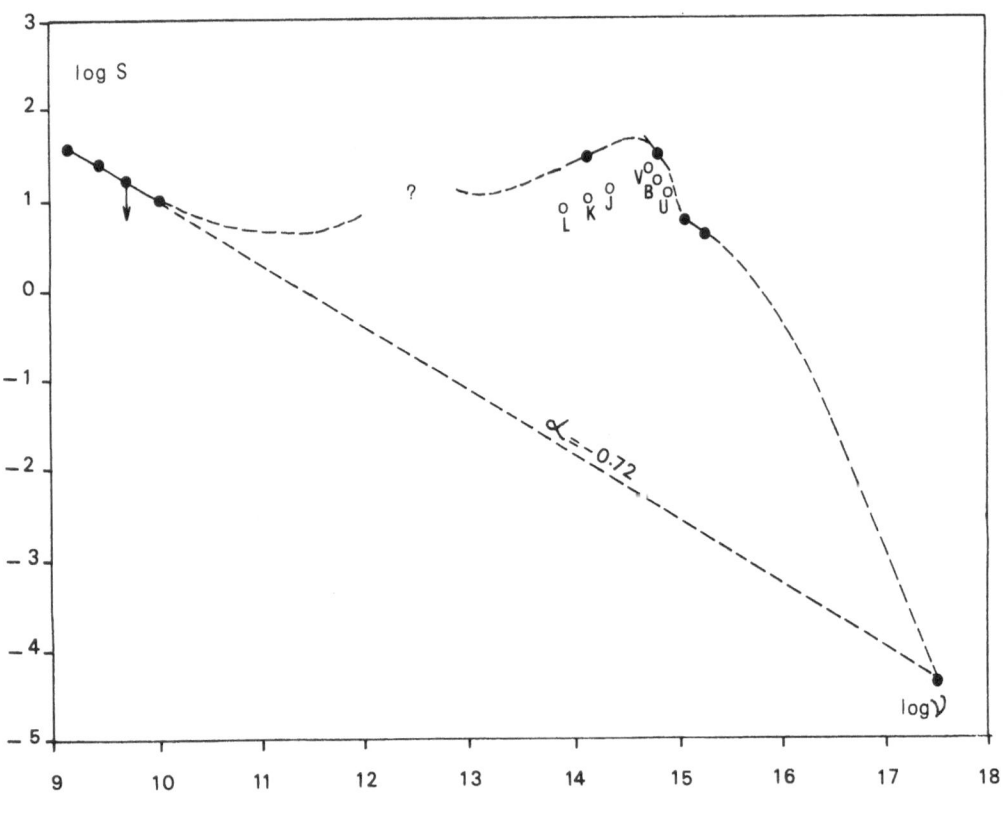

Figure. Overall spectrum of Mkn 325 in mJy vs Hz

670 times more than the giant HII region 30 Dor. It is with these yard-sticks of extreme star formation activity that we have to appraise the X-luminosity of Mkn 325.

O and B stars, class I-V, have X-luminosities (always in the same kev range) 10^{30}-10^{34} erg s^{-1} (Vaiana et al.1981); then they fall short. Supernova remnants reach easily 10^{37} erg s^{-1} (Van Speybroeck et al.1979, Agraval and Riegler 1980, Dopita et al.1981, Long et al.1981). Then about 10^4 such SNRs could account for the Mkn 325 X-emission. This is a tremendous number and corresponds to one SNR for 10 early OB-stars. X-ray binary stars are more powerful, reaching above 10^{38} erg s^{-1} (Long et al.1981, Seward and Mitchell 1981) so that 1,000 X-binaries could also account for the emission, i.e. one for 100 OB stars.

We note an interesting fact: the strongest X-source in the LMC, the binary LMC X-1, may be physically associated with the giant HII region 30 Dor. If we scale up its X-luminosity $2x10^{38}$ erg s^{-1} with the factor 670 derived above from UV data we get $1.4x10^{41}$ erg s^{-1}, which is practically the Mkn 325 observed value. Thus, as well in X-rays as in the UV, this galaxy can be considered as a collection of $\sim$700 giant HII regions.

The overall spectrum shows that the X-flux density is in line with the extrapolated radio spectrum: spectral index from radio to X α = -0.72, radio index α = -0.66 $\pm$ 0.01 (Heidmann et al.1982). The same fact was also observed by Fabbiano et al.(1982) for peculiar galaxies.

Physically these relationships between radio, UV and X emissions may be understood through the initial mass functions of the gigantic bursts which link SNRs and/or young supernovae of the SN1979c type (radio), OB stars (UV) and X-binaries (X).

4. TRIGGERING OF THE BURSTS OF STAR FORMATION

We did not detect X-emission from the nearby paired galaxy Mkn 326 (= NGC 7677). This sharpens most interestingly the problem of the triggering of the huge bursts of star formation in Mkn 325. First we note that with the radial velocity difference 157 km s^{-1} and the separation 95 kpc found by Bottinelli et al.(1975), an eventual tidal interaction in the pair occured $\sim$ $6x10^8$ years ago, which is an order of magnitude larger than the presumed age of the bursts. Second, Mkn 326 is comparable to Mkn 325 in luminosity (1.8 times less from Huchra, 1977), in neutral hydrogen mass (3 times more), in total mass (4 times more from Bottinelli et al.,1975) and in diameter (equal from de Vaucouleurs et al.,1976). Third, in spite of this similarity, CFH 3.6m telescope plates (Coupinot et al.1983) show that Mkn 326 is a spiral with no morphological distur-bance; further it has normal B-V (0.71) and U-B (0.06) colors while Mkn 325 is much bluer (0.35 and -0.43, Huchra, 1977). All this, combined with the fact that we did not detect X-emission from Mkn 326, is against a ti-dal triggering of the bursts of Mkn 325 as proposed by Duflot and Alloin (1981).

The fact that, on the other hand, the velocity field in Mkn 325 does not show signs of a merging process (Duflot and Alloin 1981) leaves quite open the interesting question of the origin of the extreme star bursting activity in clumpy irregular galaxies which was recently discussed by Boesgaard et al.(1982).

The incompleteness of this Einstein observation, made at the end of its career (3,500 s out of 40,000 allocated),reduced its potentiality to get X-spectral slope and maping, informations which would be first-hand for the investigation of the high-mass part of the initial mass function of the bursts and of the exact nature of the end-products of short-lived stars in the clumps, especially through comparison with high resolution Space Telescope optical and VLA radio imageries.

Aknowledgments: J.H. acknowledges the renewal of his NATO grant no 1962.

REFERENCES

Agraval,P.C.and Riegler,G.R.1980, Ap.J.237, L33
Benvenuti,P.,Casini,C.and Heidmann,J.1982a, M.N.R.A.S.198, 825
Benvenuti,P.,Casini,C.and Heidmann,J.1982b, Advances in Ultraviolet Astronomy: four years of IUE research, Goddard Space Flight Center,p.156
Bieging,J.H.et al.1977,Astron.Astrophys.60, 353
Biermann,P.et al.1980, Astron.Astrophys.81, 235
Boesgaard,A.M.,Edwards,S.and Heidmann,J.1982, Ap.J.252, 487
Bottinelli,L.et al.1975, Astron.Astrophys.41, 61
Coupinot,G.,Hecquet,J.and Heidmann,J.1982, M.N.R.A.S.199, 451
Coupinot,G.et al.1983, in preparation
Dopita,M.A.,Tuohy,I.R. and Mathewson,D.S.1981, Ap.J.248, L105
Duflot,M.and Alloin,D.1981, Astron.Astrophys.112, 257
Fabbiano,G.,Feigelson,E.and Zamorani,G.1982, Ap.J.256, 397
Heidmann,J.1982, Highligts of Astronomy, 6, p. 611.
Heidmann,J.,Klein,U.and Wielebinski,R.1982, Astron.Astrophys.105, 188
Heyles,C.1975, Astron.Astrophys.Supp.20,37
Huchra,J.P.1977, Ap.J.Supp.35,171
Long,K.S.,Helfand,D.J.and Grabelsky,D.A.1981, Ap.J.248, 925
Seward,F.D.and Mitchell,M.1981, Ap.J.243, 736
Van Speybroeck,L.,et al.1979, Ap.J.234, L49
Vaiana,G.S.et al.1981, Ap.J.245,163
Vaucouleurs,G.de,Vaucouleurs,A.de,and Corwin,H.G.Jr,1976, Second Reference Catalogue of Bright Galaxies, Austin

DISCUSSION

FABIAN: We have recently published X-ray observations of a similar
burst of star formation, in a small galaxy 2 magnitudes intrinsically
fainter than the SMC (Stewart et al.M.N.R.A.S.Aug.82). This galaxy, NGC
5408, radiates $\sim 10^{40}$ erg s^{-1} in X-rays. We conclude that massive binaries produce the X-rays and not SNRs.

HEIDMANN: Yes. But I should stress again that clumpy galaxies are 100
times more luminous intrinsically and this could lead to different
physical conditions; see e.g. our failure to detect the CO mm line
(Gordon et al. 1982, P.A.S.P. 94, 415).

DENNEFELD: What does the visible spectrum of Mkn 325 look like? From
what you said one should expect either spectral features typical of SNRs
(like [SII] for example) or Wolf-Rayet characteristics of hot stars as
seen in giant HII regions, or both.

HEIDMANN: We obtained optical spectra of clumpy galaxies at Mauna Kea
(see Ap.J. 252, 487): in addition to typical HII features there is
indeed enhancement of the [SII] lines which indicates presence of SNRs.
With A. Pitault we are looking for WR stars features, which at first
sight are not prominent.

BLAIR: With regards to whether X-ray binaries or SNRs are mainly responsible for the X-ray emission you see: with a burst of star formation,
one would expect many massive stars to form, and the remnants of these
massive stars seem to be prodigious X-ray emitters (cf.the NGC 4489 SNR
with $L_x \sim 10^{39}$ erg s^{-1} and 41.9 + 58 in M82 with the same luminosity, as
discussed in the previous talk by Kronberg). Even though their lifetimes
are quite short, there may be enough of these objects after a burst of
star formation to dominate the X-ray emission.

HEIDMANN: Yes. In fact we hope to get information from a combination
of X-ray and of radio VLA observations. We may also think about new
types of X-ray emitters in our hyperactive bursts. Indeed Heeschen,
Yin and I, in a VLA study of clumpy galaxies, just discovered the first
known case of a strong, compact, variable radiosource which is not
nuclear. Such a feature should be investigated for its X-ray properties
with future instruments.

A CATALOGUE OF GALACTIC SUPERNOVA REMNANTS

Sidney van den Bergh
Dominion Astrophysical Observatory
Herzberg Institute of Astrophysics
Victoria, B.C. V8X 4M6 CANADA

At this conference results have been presented on a number of individual galactic supernova remnants, but many others remain unstudied. It therefore seemed worthwhile to present a catalogue of all presently known SNR's in the Galaxy. Objects of which the true nature is not yet well established have, as far as possible, been omitted. Remnants which have been detected at optical wavelengths are marked by an asterisk in Table 1. Data on the optical identifications are from van den Bergh (1978), supplemented by recent results of Zealey, Elliot and Malin (1979), Reich, Kallas and Steube (1979), Downes, Pauls and Salter (1980), van den Bergh (1981) and Reich and Braunsfurth (1981). Also marked in the table are supernova remnants that have been detected in x-rays. These x-ray identifications are from miscellaneous sources.

Of the 135 supernova remnants in the catalogue 40 have been seen optically and 33 have been observed in x-rays. The distribution of supernova remnants in galactic longitude is shown in Fig. 1. Neither the optical nor the x-ray remnants exhibit the sharp peak towards the galactic centre that is shown by the radio supernova remnants. The reason for this is, of course, that optical remnants can only be seen if they are relatively nearby and suffer low absorption. Furthermore many distant x-ray remnants are too faint to be observed with currently available instrumentation.

J. Danziger and P. Gorenstein (eds.), Supernova Remnants and their X-Ray Emission, 597–604.
© *1983 by the IAU.*

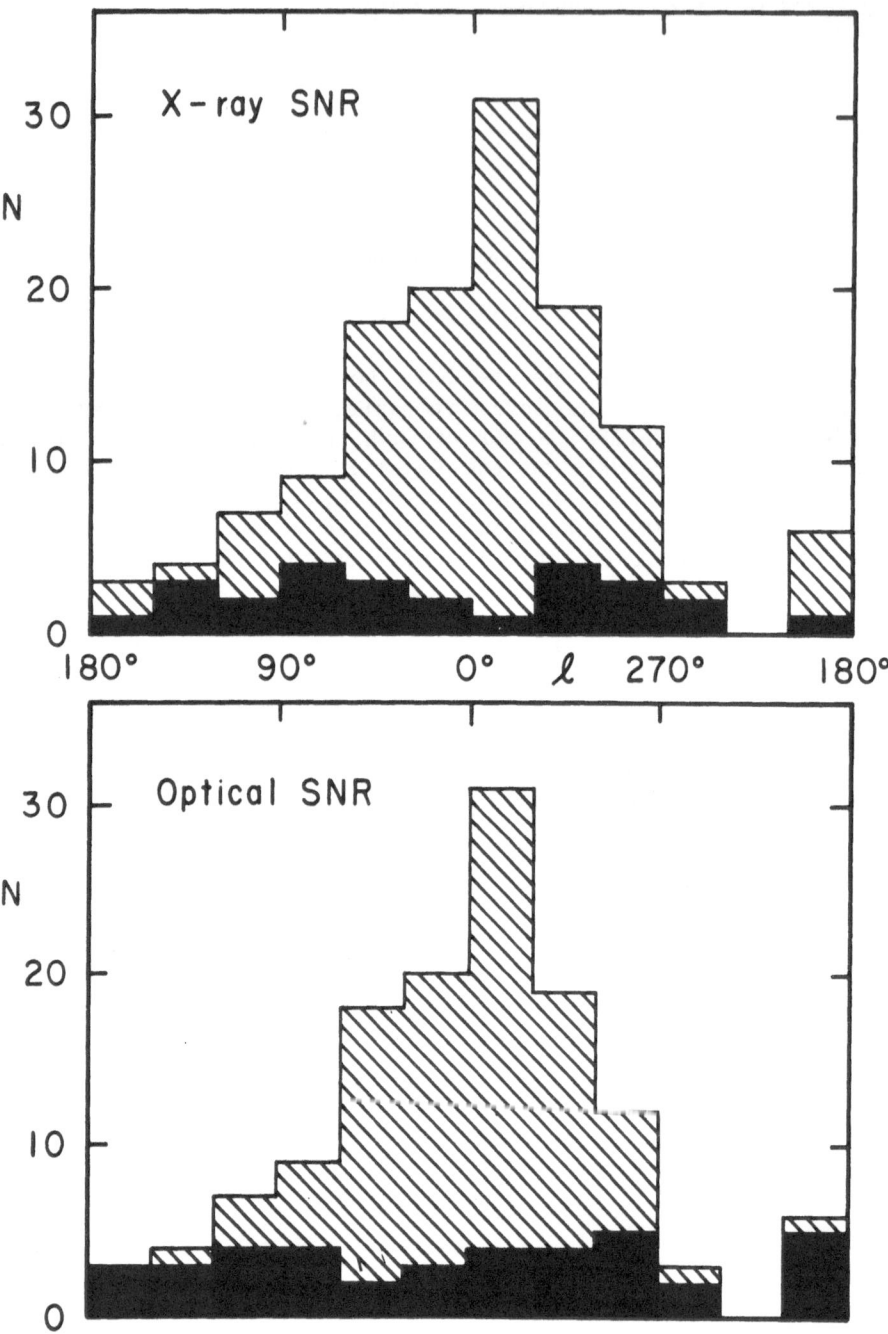

Fig. 1.   Distribution in galactic longitude of radio supernova
remnants compared with that of x-ray remnants (upper panel)
and optically visible remnants (lower panel).

TABLE 1

CATALOGUE OF GALACTIC SUPERNOVA REMNANTS

Designation	Name	$\alpha(1950)\delta$				Optical Remnant	X-ray Source
G 0.0 - 0.0	Sgr A (East)	$17^h42^m37^s$	$-28°$	$59'$		–	–
G 4.5 + 6.8	Kepler SNR	17 27 41	-21	27		*	*
G 5.3 - 1.1	Milne 56	17 58 30	-24	50		*	–
G 6.4 - 0.1	W 28	17 57 36	-23	25		*	*
G 7.7 - 3.7		18 14 15	-24	05		–	–
G 10.0 - 0.3		18 05 42	-20	25		–	–
G 11.2 - 0.3		18 08 31	-19	26		–	–
G 11.4 - 0.0		18 07 37	-19	04		–	–
G 12.0 - 0.1		18 09 16	-18	38		–	–
G 15.9 + 0.2		18 15 50	-15	02		–	–
G 18.8 + 0.3	Kes 67	18 21 00	-12	25		–	–
G 21.5 + 0.9		18 30 47	-10	36		–	*
G 21.8 - 0.6	Kes 69	18 30 05	-10	09		–	–
G 22.7 - 0.2		18 30 35	-09	13		–	–
G 23.3 - 0.3	W41	18 31 40	-08	51		–	–
G 23.6 + 0.3		18 30 25	-08	14		–	–
G 24.7 + 0.6		18 31 30	-07	08		–	–
G 24.7 - 0.6		18 35 30	-07	36		–	–
G 27.4 + 0.0	Kes 73	18 38 36	-04	59		–	–
G 29.7 - 0.2	Kes 75	18 43 48	-03	02		–	*
G 31.9 + 0.0	3C391 = Kes 77	18 46 47	-00	59		–	
G 32.0 - 4.9	3C396.1	19 04 30	-03	06		–	–
G 33.2 - 0.6		18 54 34	-00	06		–	–
G 33.7 + 0.0	Kes 79	18 50 04	+00	36		–	–
G 34.6 - 0.5	W44	18 53 35	+01	17		–	*
G 35.6 - 0.4		18 55 18	+02	05		–	–
G 39.2 - 0.3	3C396	19 01 35	+05	22		–	–
G 39.7 - 2.0	W50	19 09 15	+05	12		*	*
G 40.5 - 0.5		19 04 38	+06	25		–	–
G 41.1 - 0.3	3C397	19 05 08	+07	04		–	–
G 41.9 - 4.1	PKS 1920 + 06	19 20 00	+06	00		–	–
G 43.3 - 0.2	W49B	19 08 43	+09	01		–	*
G 46.8 - 0.3		19 15 45	+12	04		–	–
G 47.6 + 6.1	CTB63	18 54 00	+15	45		–	–
G 49.2 - 0.5	W51	19 21 30	+13	57		–	–
G 53.6 - 2.2	3C400.2	19 36 30	+17	08		*	–
G 54.4 - 0.3		19 31 00	+18	55		–	–
G 55.7 + 3.4		19 19 40	+21	37		–	–
G 65.3 + 5.7	S91 + S94	19 31 00	+31	10		*	*
G 65.7 + 1.2		19 50 05	+29	17		–	–
G 68.8 + 2.6	CTB 80	19 51 03	+32	45		*	*

TABLE 1 (continued)

Designation	Name	α(1950)δ				Optical Remnant	X-ray Source
G 74.3 − 8.5	Cygnus Loop	$20^h49^m30^s$	+30°	45'		*	*
G 74.9 + 1.2	CTB 87	20 14 05	+37	04		−	*
G 78.2 + 2.1		20 19 25	+40	18		*	*
G 82.2 + 5.3	W63	20 17 23	+45	24		−	−
G 84.2 − 0.8		20 51 35	+43	15		−	−
G 89.0 + 4.7	HB21	20 43 20	+50	29		−	−
G 93.3 + 6.9		20 50 55	+55	10		−	−
G 93.6 − 0.2	CTB 104A	21 26 50	+50	33		−	−
G 94.0 + 1.0	3C434.1	21 23 30	+51	40		−	−
G109.2 − 1.0	CTB 109	22 59 51	+58	39		*	*
G111.7 − 2.1	Cas A	23 21 10	+58	32		*	*
G114.3 + 0.3		23 24 45	+61	38		−	−
G116.5 + 1.1		23 51 17	+62	58		−	−
G116.9 + 0.2	CTB 1	23 56 45	+62	10		*	−
G119.5 + 9.8	CTA 1	00 04 18	+72	04		*	*
G120.1 + 1.4	Tycho SNR	00 22 33	+63	52		*	*
G126.2 + 1.6		01 18 16	+64	01		*	−
G127.1 + 0.5		01 27 00	+62	58		−	−
G130.7 + 3.1	3C58	02 01 53	+64	35		*	*
G132.7 + 1.3	HB 3	02 14 00	+62	18		*	*
G160.4 + 2.8	HB 9	04 57 00	+46	36		*	*
G166.1 + 4.4	OA 184	05 15 38	+41	46		*	−
G166.3 + 2.5	VRO 42.05.01	05 23 21	+43	00		*	−
G180.3 − 1.7	S147	05 36 45	+27	44		*	−
G184.6 − 5.8	Crab	05 31 31	+21	59		*	*
G189.0 + 3.0	IC 443	06 14 06	+22	37		*	*
G193.3 − 1.5	PKS 0607 + 17	06 05 50	+16	40		−	−
G205.6 − 0.1	Monoceros	06 35 00	+06	30		*	−
G206.9 + 2.3	PKS 0646 + 06	06 46 00	+06	30		*	−
G260.4 − 3.4	Pup A	08 20 30	−42	50		*	*
G261.9 + 5.5	PKS 0902 − 38	09 02 22	−38	29		−	−
G263.4 − 3.0	Vela XYZ	08 32 00	−45	00		*	*
G284.2 − 1.7	MSH 10 5̲3	10 16 06	−58	37		*	−
G287.8 − 0.5		10 45 00	−59	23		−	−
G290.1 − 0.8	MSH 11 − 61A	11 00 52	−60	37		*	−
G291.0 − 0.1	MSH 11 − 6̲2̲	11 09 49	−60	22		−	−
G292.0 + 1.8	MSH 11 − 5̲4̲	11 22 22	−58	59		*	*
G293.8 + 0.6		11 33 05	−60	36		−	−
G296.1 − 0.7		11 48 15	−62	27		*	*
G296.5 + 10.0	PKS 1209 − 5̲2̲	12 07 00	−52	07		*	*
G296.8 − 0.3		11 55 48	−62	18		−	−
G298.5 − 0.3		12 09 58	−62	36		−	−
G298.6 + 0.0		12 11 18	−62	18		−	−

TABLE 1 (continued)

Designation	Name	α(1950)δ			Optical Remnant	X-ray Source
G299.0 + 0.2		$12^h15^m05^s$	$-62°$	08'	–	–
G302.3 + 0.7		12 42 54	-61	51	–	–
G304.6 + 0.1	Kes 17	13 02 35	-62	26	–	–
G308.7 + 0.0		13 38 05	-62	01	–	–
G309.2 − 0.6		13 43 00	-62	36	–	–
G309.8 + 0.0		13 47 03	-61	50	–	–
G311.5 + 0.0		14 01 58	-61	43	–	–
G315.4 − 0.3		14 32 00	-60	22	–	*
G315.4 − 2.3	RCW 86	14 39 08	-62	15	*	–
G316.3 − 0.0	MSH 14 − 57	14 37 43	-59	47	–	–
G320.3 − 1.2	MSH 15 − 52	15 10 00	-58	57	*	*
G321.9 − 0.3		15 16 45	-57	23	–	–
G322.3 − 1.2	Kes 24	15 23 05	-57	56	–	–
G323.5 + 0.1		15 25 05	-56	12	–	–
G326.3 − 1.8	MSH 15 − 56	15 48 50	-56	00	*	–
G327.1 − 1.1		15 50 35	-54	58	–	*
G327.4 − 0.4	Kes 27	15 44 54	-53	39	–	*
G327.6 + 14.5	SN 1006	14 59 30	-41	45	*	*
G328.0 + 0.3		15 49 33	-53	19	–	–
G328.4 + 0.2	MSH 15 − 57	15 51 45	-53	08	–	–
G330.0 + 15.0	Lupus Loop	15 09 00	-39	00	–	*
G330.2 + 1.0		15 57 20	-51	25	–	–
G332.0 + 0.2		16 09 23	-50	49	–	–
G332.4 + 0.1	MSH 16 − 51	16 11 38	-50	32	–	–
G332.4 − 0.4	RCW 103	16 13 54	-50	56	*	*
G335.2 + 0.1		16 23 50	-48	36	–	–
G336.7 + 0.5		16 28 30	-47	13	–	–
G337.0 − 0.1	CTB 33	16 32 08	-47	30	–	–
G337.2 − 0.7		16 35 45	-47	44	–	–
G337.3 + 1.0	Kes 40	16 29 05	-46	29	–	–
G337.8 − 0.1	Kes 41	16 35 15	-46	53	–	–
G338.2 + 0.4		16 34 40	-46	16	*	–
G338.3 − 0.1		16 37 25	-46	27	–	–
G338.5 + 0.1		16 37 20	-46	12	–	–
G339.2 − 0.4		16 43 00	-44	34	*	–
G340.4 + 0.4		16 43 00	-44	34	–	–
G340.6 + 0.3		16 44 05	-44	30	–	–
G341.9 − 0.3	MSH 16 − 48	16 51 20	-43	54	–	–
G342.1 + 0.1	Kes 45	16 50 11	-43	30	*	–
G343.2 − 5.6		17 20 00	-46	00	–	–
G344.7 − 0.1		17 00 15	-41	37	–	–
G346.6 − 0.2		17 06 45	-40	06	–	–
G348.5 + 0.1	CTB 37A	17 11 12	-38	26	–	–

TABLE 1 (continued)

Designation	Name	$\alpha(1950)\delta$			Optical Remnant	X-ray Source
G348.7 + 0.3	CTB 37B	$17^h10^m45^s$	$-38°$	06'	–	–
G349.7 + 0.2		17 14 37	-37	23	–	–
G350.0 − 1.8		17 23 45	-38	20	–	–
G350.1 − 0.3		17 17 40	-37	24	–	–
G351.2 + 0.1		17 19 00	-36	09	–	–
G352.7 − 0.1		17 24 20	-35	05	–	–
G355.9 − 2.5		17 42 30	-33	43	–	–
G357.7 − 0.1	MSH 17 − 39	17 37 04	-30	56	–	–

REFERENCES

Downes, A.J.B., Pauls, T., and Salter, C.J.: 1980
       Astron. Astrophys. **92**, 47.
Reich, W., Kallas, E., and Steube, R.: 1979 Astron. Astrophys. **78**, L13.
Reich, W., and Braunsfurth, E.: 1981 Astron. Astrophys. **99**, 17.
van den Bergh, S.: 1978 Astrophys. J. Suppl. **38**, 119.
van den Bergh, S.: 1981 Vistas Astron. **25**, 109.
Zealey, W.J., Elliott, K.H., and Malin, D.F.: 1979
       Astron. Astrophys. Suppl. **38**, 39.

DISCUSSION

DENNEFELD: Has the most recent analysis of proper motions provided a probable date of explosion different from the one you derived in 1976?

VAN DEN BERGH: In Kamper and van den Bergh (1976) we obtained an explosion date $T_0 = 1657 \pm 3$, compared to $T_0 = 1658 \pm 3$ derived in the present paper.

KIRSHNER: Do you have any comment on the possible observation by Flamsteed?

VAN DEN BERGH: Our present observations show a deceleration coefficient $k = 0.0010 \pm 0.0019$ (m.e.) yr^{-1}, in which it was assumed that $a = -kv$. With this value we cannot yet exclude the possibility that Flamsteed observed Cas A.

JONES: Is it not reasonable to assume that the apparent lack of deceleration of the fast knots is due to their just having enountered the reverse shock? The knots would then be encountering previously shocked ejecta which would be expected to be hydrogen-poor; and the knots would not have had time to decelerate appreciably. In response to a previous question about deceleration you said there wasn't any which "wasn't surprising because one sees no hydrogen emission from the gas that would decelerate the knots".

VAN DEN BERGH: The best data on deceleration of knots comes from the fast-moving knots in the "jet". These lie well outside the main SNR shell so that they would be exposed to hydrogen-rich interstellar matter, yet no hydrogen contamination is seen in their spectra.

BEGELMAN: What is your best guess for the total mass of material that is radiating at optical wavelengths at any one time in the fast-moving knots?

VAN DEN BERGH: The total mass of the optical knots is only a fraction of a solar mass. An estimate of this mass is given in Peimbert and van den Bergh (1971).

TENORIO-TAGLE: Do you have any further information on the jets?

VAN DEN BERGH: I have the impression that the knots in the jet are on average fainter now than they were 25 years ago. Photometric observations are, however, required to confirm this.

TUFFS: Two points: (1) I should like to point out that the small cluster of QSF's is outside the plateau edge of the radio remnant which is very well defined in the S.S.W. Maybe we have to reconsider our conventional interpretation that the plateau edge represents a shock front. (2) I have a new, accurate proper motion for the radio knot Bell 38 which is coincident with the QSF "Feature A". This proper motion is $\mu_x =$

+0.01 ± 0.01 arc sec per year, μ_y = -0.11 ± 0.01 arc sec per year. This
is significantly different from the optical proper motion of this
feature, and I would like to suggest that this discrepancy is due to the
fact that although the radio knot is < 1" in size, the optical feature
is clearly resolved. Thus we are measuring different emission regions,
and there seems no reason why the proper motions should agree, bearing
in mind the interpretation of the QSF's as pre-existing circumstellar
material. The random component of the motion of this radio knot cannot
be due to morphological changes within the radio knot.

VAN DEN BERGH: (1) Yes, (2) Maybe.

Page references are to the first pages of the relevant papers.

Page references are to the first pages of the relevant papers.